HUMAN-LIKE MACHINE INTELLIGENCE

Human-Like Machine Intelligence

Edited by

Stephen Muggleton
Imperial College London

Nick Chater
University of Warwick

OXFORD
UNIVERSITY PRESS

OXFORD
UNIVERSITY PRESS

Great Clarendon Street, Oxford, OX2 6DP,
United Kingdom

Oxford University Press is a department of the University of Oxford.
It furthers the University's objective of excellence in research, scholarship,
and education by publishing worldwide. Oxford is a registered trade mark of
Oxford University Press in the UK and in certain other countries

First Edition published in 2021
Impression: 1

Published in the United States of America by Oxford University Press
198 Madison Avenue, New York, NY 10016, United States of America

British Library Cataloguing in Publication Data
Data available

Library of Congress Control Number: 2021932529

ISBN 978–0–19–886253–6

DOI: 10.1093/oso/9780198862536.001.0001

Printed and bound by
CPI Group (UK) Ltd, Croydon, CR0 4YY

Preface

Recently there has been increasing excitement about the potential for artificial intelligence to transform human society. This book addresses the leading edge of research in this area. This research aims to address present incompatibilities of human and machine reasoning and learning approaches.

According to the influential US funding agency DARPA (originator of the Internet and Self-Driving Cars) this new area represents the Third Wave of Artificial Intelligence (3AI, 2020s–2030s), and is being actively investigated in the United States, Europe and China. The UK's Engineering and Physical Sciences Research Council (EPSRC) network on Human-Like Computing (HLC) was one of the first network internationally to initiate and support research specifically in this area. Starting activities in 2018, the network represents around 60 leading UK artificial intelligence and cognitive scientists involved in the development of the interdisciplinary area of HLC. The research of network groups aims to address key unsolved problems at the interface between psychology and computer science.

The chapters of this book have been authored by a mixture of these UK and other international specialists based on recent workshops and discussions at the Machine Intelligence 20 and 21 workshops (2016, 2019) and the Third Wave Artificial Intelligence workshop (2019). Some of the key questions addressed by the human-like computing programme include how AI systems might (1) explain their decisions effectively, (2) interact with human beings in natural language, (3) learn from small numbers of examples and (4) learn with minimal supervision. Solving such fundamental problems involves new foundational research in both the psychology of perception and interaction as well as the development of novel algorithmic approaches in artificial intelligence.

The book is arranged in five parts. The first part describes central challenges of human-like computing, ranging from issues involved in developing a beneficial form of AI (Russell, Berkeley), as well as a modern philosophical perspective on Alan Turing's seminal model of computation and his view of its potential for intelligence (Millican, Oxford). Two chapters then address the promising new approach of virtual bargaining and representational revision as new technologies for supporting implicit human–machine interaction (Chater, Warwick; Bundy, Edinburgh).

Part 2 addresses human-like social cooperation issues, providing both the AI perspective of dialectic explanations (Toni, Imperial) alongside relevant psychological research on the limitations and biases of human explanations (Hahn, Birkbeck) and the challenges human-like communication poses for AI systems (Healey, Queen Mary). The possibility of reverse engineering human cooperation is described (Kleiman-Weiner, Harvard) and contrasts with issues in using explanations in machine teaching (Hernandez-Orallo, Politecnia Valencia). Part 3 concentrates on Human-Like Perception and Language, including new approaches to human-like computer vision (Muggleton, Imperial), and the related new area of apperception (Evans, DeepMind), as well as suggestions on combining human and machine vision in analysing complex signal data (Jay, Manchester). An ongoing UK study on social interaction is described in (Pickering, Edinburgh) together with a chapter exploring the use of multi-modal communication (Vigliocco, UCL).

In Part 4, issues related to human-like representation and learning are discussed. This starts with a description of work on human–machine scientific discovery (Tamaddoni-Nezhad, Imperial) which is related to models of fast and slow learning in humans (Mareschal), followed by a chapter on machine-learning methods for generating mutual explanations (Schmid, Bamberg). Issues relating graphical and symbolic representation are described in (Jamnik, Cambridge). This has potential relevance to applications for inductively generating programs for use with spreadsheets (De Raedt, Leuven). Lastly, Part 5 considers challenges for evaluating and explaining the strength of human-like reasoning. Evaluations are necessarily context dependent, as shown in the paper on automated common-sense spatial reasoning (Cohn, Leeds), though a second paper argues that Bayesian-inspired approaches which avoid probabilities are powerful for explaining human brain activity (Sanborn, Warwick). Bayesian approaches are also shown to be capable of explaining various oddities of human reasoning, such as the conjunction fallacy (Tentori, Trento). By contrast, when evaluating situated AI systems there are clear advantages and difficulties in evaluating robot football players using objective probabilities within a competitive environment (Sammut, UNSW). The book closes with a chapter demonstrating the ongoing challenges of evaluating the relative strengths of human and machine play in chess (Bratko, Ljubjana).

June 2020

Stephen Muggleton
and Nick Chater
Editors

Acknowledgements

This book would not have been possible without a great deal of help. We would like to thank Alireza Tamaddoni-Nezhad for his valuable help in organising the meetings which led to this book, and with his help in finalizing the book itself, as well as Francesca McMahon, our editor at OUP, for her advice and encouragement. We also thank our principal funder, the EPSRC, for backing the Network on Human-Like Computing (HLC, grant number EP/R022291/1); and acknowledge additional support from the ESRC Network for Integrated Behavioural Science (grant number grant number ES/P008976/1). Finally, special thanks are due to Bridget Gundry for her hard work, tenacity, and cheerfulness in driving the book through to a speedy and successful conclusion.

Contents

Part 5 Evaluating Human-like Reasoning

List of Contributors

Alahmadi, Alaa
University of Manchester, UK

Bennett, Brandon
University of Leeds, UK

Blakeman, Sam
Birkbeck College,
University of London, UK

Bohan, David
INRA, France

Bratko, Ivan
University of Ljubljana, Slovenia

Bundy, Alan
University of Edinburgh, UK

Chater, Nick
University of Warwick, UK

Cheng, Peter
University of Sussex, UK

Cocarascu, Oana
Imperial College London, UK

Cohn, Anthony
University of Leeds, UK

Cyras, Kristijonas
Imperial College London, UK

Dai, Wang-Zhou
Imperial College London, UK

Dauxais, Yann
KU Leuven, Belgium

Davies, Alan
University of Manchester, UK

Dempsey, Katherine
University of Manchester, UK

De Raedt, Luc
KU Leuven, Belgium

Evans, James
University of Chicago, USA

Evans, Richard
Imperial College London and
DeepMind, UK

Farid, Reza
University of New South Wales, Australia

Ferri, Cesar
Universitat Politècnica de València, Spain

Garrod, Simon
University of Glasgow, UK

Gautrais, Clément
KU Leuven, Belgium

Grzyb, Beata
UCL Division of Psychology and
Language Sciences, UK

Guid, Matej
University of Ljubljana, Slovenia

Hahn, Ulrike
Birkbeck, University of London, UK

Healey, Patrick
Queen Mary, University of London, UK

Hernandez-Orallo, Jose
Universitat Politècnica de València, Spain

Hristova, Dayana
University of Ljubljana, Slovenia

Jamnik, Mateja
University of Cambridge, UK

Jay, Caroline
University of Manchester, UK

Kleiman-Weiner, Max
MIT and Harvard University, USA

Kolb, Samuel
KU Leuven, Belgium

León-Villagrá, Pablo
University of Edinburgh, UK

Li, Xue
University of Edinburgh, UK

Mareschal, Denis
Birkbeck College,
University of London, UK

Milani, Ghazal
Imperial College London

Millican, Peter
University of Oxford, UK

Misyak, Jennifer
Warwick Business School, UK

Muggleton, Stephen
Imperial College London, UK

Parkes, David
Harvard University, USA

Philalithis, Eugene
University of Edinburgh, UK

Pickering, Martin
University of Edinburgh, UK

Rago, Antonio
Imperial College London, UK

Raybould, Alan
University of Edinburgh, UK

Russell, Stuart
UC Berkeley, USA

Sammut, Claude
University of New South Wales, Australia

Sanborn, Adam
University of Warwick, UK

Schmid, Ute
University of Bamberg, Germany

Spicer, Jake
University of Warwick, UK

Sundh, Joakim
University of Warwick, UK

Tamaddoni-Nezhad, Alireza
University of Surrey and Imperial College London, UK

Tenenbaum, Joshua
MIT, USA

Tentori, Katya
University of Trento, Italy

Tesic, Marko
Birkbeck, University of London, UK

Teso, Stefano
KU Leuven, Belgium

Toni, Francesca
Imperial College London, UK

Verbruggen, Gust
KU Leuven, Belgium

Vigliocco, Gabriella
UCL Division of Psychology and Language Sciences, UK

Vigo, Markel
University of Manchester, UK

Wang, Rose
MIT, USA

Wicaksono, Handy
University of New South Wales, Australia

Wiley, Timothy
University of New South Wales, Australia

Wu, Sarah
MIT, USA

Zhu, Jian-Qiao
University of Warwick, UK

Part 1

Human-like Machine Intelligence

1

Human-Compatible Artificial Intelligence

Stuart Russell
University of California, Berkeley, USA

1.1 Introduction

Artificial intelligence (AI) has as its aim the creation of intelligent machines. An entity is considered to be intelligent, roughly speaking, if it chooses actions that are expected to achieve its objectives, given what it has perceived.[1] Applying this definition to machines, one can deduce that AI aims to create machines that choose actions that are expected to achieve their objectives, given what they have perceived.

Now, what are these objectives? To be sure, they are—up to now, at least—objectives that we put into them; but, nonetheless, they are objectives that operate exactly as if they were the machines' own and about which they are completely certain. We might call this the *standard model* of AI: build optimizing machines, plug in the objectives, and off they go. This model prevails not just in AI but also in control theory (minimizing a cost function), operations research (maximizing a sum of rewards), economics (maximizing individual utilities, gross domestic product (GDP), quarterly profits, or social welfare), and statistics (minimizing a loss function). The standard model is a pillar of twentieth-century technology.

Unfortunately, this standard model is a mistake. It makes no sense to design machines that are beneficial to us *only* if we write down our objectives completely and correctly. If the objective is wrong, we might be lucky and notice the machine's surprisingly objectionable behaviour and be able to switch it off in time. Or, if the machine is more intelligent than us, the problem may be irreversible. The more intelligent the machine, the worse the outcome for humans: the machine will have a greater ability to alter the

[1] This definition can be elaborated and made more precise in various ways—particularly with respect to whether the choosing and expecting occur within the agent, within the agent's designer, or some combination of both. The latter certainly holds for human agents, viewing evolution as the designer. The word 'objective' here is also used informally, and does not refer just to end goals. For most purposes, an adequately general formal definition of 'objective' covers preferences over lotteries over complete state sequences. Moreover, 'state' here includes mental state as well as the world state external to the entity.

Start Russell, *Human-Compatible Artificial Intelligence* In: *Human Like Machine Intelligence.* Edited by: Stephen Muggleton and Nick Charter, Oxford University Press. © Oxford University Press 2021. DOI: 10.1093/oso/9780198862536.003.0001

world in ways that are inconsistent with our true objectives and greater skill in foreseeing and preventing any interference with its plans.

In 1960, after seeing Arthur Samuel's checker-playing program learn to play checkers far better than its creator, Norbert Wiener (1960) gave a clear warning:

> If we use, to achieve our purposes, a mechanical agency with whose operation we cannot efficiently interfere . . . we had better be quite sure that the purpose put into the machine is the purpose which we really desire.

Echoes of Wiener's warning can be discerned in contemporary assertions that 'superintelligent AI' may present an existential risk to humanity. (In the context of the standard model, 'superintelligent' means having a superhuman capacity to achieve given objectives.) Concerns have been raised by such observers as Nick Bostrom (2014), Elon Musk (Kumparak 2014), Bill Gates (2015),[2] and Stephen Hawking (Osborne 2017). There is very little chance that as humans we can specify our objectives completely and correctly in such a way that the pursuit of those objectives by more capable machines is guaranteed to result in beneficial outcomes for humans.

The mistake comes from transferring a perfectly reasonable definition of intelligence from humans to machines. The definition is reasonable for humans because we are entitled to pursue our own objectives—indeed, whose would we pursue, if not our own? The definition of intelligence is *unary*, in the sense that it applies to an entity by itself. Machines, on the other hand, are not entitled to pursue their own objectives.

A more sensible definition of AI would have machines pursuing *our* objectives. Thus, we have a binary definition: entity A chooses actions that are expected to achieve the objectives of entity B, given what entity A has perceived. In the unlikely event that we (entity B) can specify the objectives completely and correctly and insert them into the machine (entity A), then we can recover the original, unary definition. If not, then the machine will necessarily be *uncertain* as to our objectives while being obliged to pursue them on our behalf. This uncertainty—with the coupling between machines and humans that it entails—is crucial to building AI systems of arbitrary intelligence that are provably beneficial to humans. We must, therefore, reconstruct the foundations of AI along binary rather than unary lines.

1.2 Artificial Intelligence

The goal of AI research has been to understand the principles underlying intelligent behaviour and to build those principles into machines that can then exhibit such behaviour. In the 1960s and 1970s, the prevailing theoretical definition of intelligence was the capacity for logical reasoning, including the ability to derive plans of action

[2] Gates wrote, 'I am in the camp that is concerned about superintelligence. . . . I agree with Elon Musk and some others on this and don't understand why some people are not concerned.'

guaranteed to achieve a specified goal. A popular variant was the problem-solving paradigm, which requires finding a minimum-cost sequence of actions guaranteed to reach a goal state. More recently, a consensus has emerged in AI around the idea of a rational agent that perceives and acts in order to maximize its expected utility. (In Markov decision processes and reinforcement learning, utility is further decomposed into a sum of rewards accrued through the sequence of transitions in the environment state.) Subfields such as logical planning, robotics, and natural-language understanding are special cases of the general paradigm. AI has incorporated probability theory to handle uncertainty, utility theory to define objectives, and statistical learning to allow machines to adapt to new circumstances. These developments have created strong connections to other disciplines that build on similar concepts, including control theory, economics, operations research, and statistics.

In both the logical-planning and rational-agent views of AI, the machine's objective—whether in the form of a goal, a utility function, or a reward function—is specified exogenously. In Wiener's words, this is 'the purpose put into the machine'. Indeed, it has been one of the tenets of the field that AI systems should be *general-purpose*—that is, capable of accepting a purpose as input and then achieving it—rather than *special-purpose*, with their goal implicit in their design. For example, a self-driving car should accept a destination as input instead of having one fixed destination. However, some aspects of the car's 'driving purpose' are fixed, such as that it shouldn't hit pedestrians. This is built directly into the car's steering algorithms rather than being explicit: no self-driving car in existence today 'knows' that pedestrians prefer not to be run over.

Putting a purpose into a machine that optimizes its behaviour according to clearly defined algorithms seems an admirable approach to ensuring that the machine's behaviour furthers our own objectives. But, as Wiener warns, we need to put in the *right* purpose. We might call this the King Midas problem: Midas got exactly what he asked for—namely, that everything he touched would turn to gold—but, too late, he discovered the drawbacks of drinking liquid gold and eating solid gold. The technical term for putting in the right purpose is *value alignment*. When it fails, we may inadvertently imbue machines with objectives counter to our own. Tasked with finding a cure for cancer as fast as possible, an AI system might elect to use the entire human population as guinea pigs for its experiments. Asked to de-acidify the oceans, it might use up all the oxygen in the atmosphere as a side effect. This is a common characteristic of systems that optimize: variables not included in the objective may be set to extreme values to help optimize that objective.

Unfortunately, neither AI nor other disciplines built around the optimization of objectives have much to say about how to identify the purposes 'we really desire'. Instead, they assume that objectives are simply implanted into the machine. AI research, in its present form, studies the ability to achieve objectives, not the design of those objectives. In the 1980s the AI community abandoned the idea that AI systems could have definite knowledge of the state of the world or of the effects of actions, and they embraced uncertainty in these aspects of the problem statement. It is not at all clear why, for the most part, they failed to notice that there must also be uncertainty in the objective.

Although some AI problems such as puzzle solving are designed to have well-defined goals, many other problems that were considered at the time, such as recommending medical treatments, have no precise objectives and ought to reflect the fact that the relevant preferences (of patients, relatives, doctors, insurers, hospital systems, taxpayers, etc.) are not known initially in each case.

Steve Omohundro (2008) has pointed to a further difficulty, observing that any sufficiently intelligent entity pursuing a fixed, known objective will act to preserve its own existence (or that of an equivalent successor entity with an identical objective). This tendency has nothing to do with a self-preservation instinct or any other biological notion; it's just that an entity usually cannot achieve its objectives if it is dead. According to Omohundro's argument, a superintelligent machine that has an off-switch—which some, including Alan Turing (1951) himself, have seen as our potential salvation—will take steps to disable the switch in some way. Thus we may face the prospect of superintelligent machines—their actions by definition unpredictable and their imperfectly specified objectives conflicting with our own—whose motivation to preserve their existence in order to achieve those objectives may be insuperable.

1.3 1001 Reasons to Pay No Attention

Objections have been raised to these arguments, primarily by researchers within the AI community. The objections reflect a natural defensive reaction, coupled perhaps with a lack of imagination about what a superintelligent machine could do. None hold water on closer examination. Here are some of the more common ones:

- *Don't worry, we can just switch it off:*[3] This is often the first thing that pops into a layperson's head when considering risks from superintelligent AI—as if a superintelligent entity would never think of that. It is rather like saying that the risk of losing to Deep Blue or AlphaGo is negligible—all one has to do is make the right moves.
- *Human-level or superhuman AI is impossible:*[4] This is an unusual claim for AI researchers to make, given that, from Turing onward, they have been fending off such claims from philosophers and mathematicians. The claim, which is backed by no evidence, appears to concede that if superintelligent AI were possible, it would be a significant risk. It is as if a bus driver, with all of humanity as his passengers, said, 'Yes, I'm driving toward a cliff—in fact, I'm pressing the pedal to the metal. But trust me, we'll run out of gas before we get there.' The claim also represents a foolhardy bet against human ingenuity. We've made such bets before and lost.

[3] AI researcher Jeff Hawkins, for example, writes, 'Some intelligent machines will be virtual, meaning they will exist and act solely within computer networks.... It is always possible to turn off a computer network, even if painful.' https://www.recode.net/2015/3/2/11559576/.

[4] The AI100 report (Stone *et al.* 2016) includes the following assertion: 'Unlike in the movies, there is no race of superhuman robots on the horizon or probably even possible.'

On 11 September 1933, renowned physicist Ernest Rutherford stated, with utter confidence, 'Anyone who expects a source of power from the transformation of these atoms is talking moonshine'. On 12 September 1933, Leo Szilard invented the neutron-induced nuclear chain reaction. A few years later, he demonstrated such a reaction in his laboratory at Columbia University. As he recalled in a memoir: 'We switched everything off and went home. That night, there was very little doubt in my mind that the world was headed for grief.'

- *It's too soon to worry about it*: The right time to worry about a potentially serious problem for humanity depends not just on when the problem will occur but also on how much time is needed to devise and implement a solution that avoids the risk. For example, if we were to detect a large asteroid predicted to collide with the Earth in 2070, would we say, 'It's too soon to worry'? And if we consider the global catastrophic risks from climate change predicted to occur later in this century, is it too soon to take action to prevent them? On the contrary, it may be too late. The relevant timescale for human-level AI is less predictable, but, like nuclear fission, it might arrive considerably sooner than expected. Moreover, the technological path to mitigate the risks is also arguably less clear. These two aspects in combination do not argue for complacency; instead, they suggest the need for hard thinking to occur soon. Wiener (1960) amplifies this point, writing,

> The individual scientist must work as a part of a process whose time scale is so long that he himself can only contemplate a very limited sector of it.... Even when the individual believes that science contributes to the human ends which he has at heart, his belief needs a continual scanning and re-evaluation which is only partly possible. For the individual scientist, even the partial appraisal of this liaison between the man and the process requires an imaginative forward glance at history which is difficult, exacting, and only limitedly achievable. And if we adhere simply to the creed of the scientist, that an incomplete knowledge of the world and of ourselves is better than no knowledge, we can still by no means always justify the naive assumption that the faster we rush ahead to employ the new powers for action which are opened up to us, the better it will be. We must always exert the full strength of our imagination to examine where the full use of our new modalities may lead us.

One variation on the 'too soon to worry about it' argument is Andrew Ng's statement that it's 'like worrying about overpopulation on Mars'. This appeals to a convenient analogy: not only is the risk easily managed and far in the future but also it's extremely unlikely that we'd even try to move billions of humans to Mars in the first place. The analogy is a false one, however. We're already devoting huge scientific and technical resources to creating ever more capable AI systems. A more apt analogy would be a plan to move the human race to Mars with no consideration for what we might breathe, drink, or eat once we arrived.

- *It's a real issue but we cannot solve it until we have superintelligence*: One would not propose developing nuclear reactors and *then* developing methods to contain the reaction safely. Indeed, safety should guide how we think about reactor design. It's worth noting that Szilard almost immediately invented and patented a feedback control system for maintaining a nuclear reaction at the subcritical level for power generation, despite having absolutely no idea of which elements and reactions could sustain the fission chain.

  By the same token, had racial and gender bias been anticipated as an issue with statistical learning systems in the 1950s, when linear regression began to be used for all kinds of applications, the analytical approaches that have been developed in recent years could easily have been developed then, and would apply equally well to today's deep learning systems.

  In other words, we can make progress on the basis of general properties of systems—e.g., systems designed within the standard model—without necessarily knowing the details. Moreover, the problem of objective misspecification applies to all AI systems developed within the standard model, not just superintelligent ones.
- *Human-level AI isn't really imminent, in any case*: The AI100 report, for example, assures us, 'contrary to the more fantastic predictions for AI in the popular press, the Study Panel found no cause for concern that AI is an imminent threat to humankind'. This argument simply misstates the reasons for concern, which are not predicated on imminence. In his 2014 book, *Superintelligence: Paths, Dangers, Strategies*, Nick Bostrom, for one, writes, 'It is no part of the argument in this book that we are on the threshold of a big breakthrough in artificial intelligence, or that we can predict with any precision when such a development might occur.' Bostrom's estimate that superintelligent AI might arrive within this century is roughly consistent with my own, and both are considerably more conservative than those of the typical AI researcher.
- *Any machine intelligent enough to cause trouble will be intelligent enough to have appropriate and altruistic objectives*:[5] This argument is related to Hume's is–ought problem and G. E. Moore's naturalistic fallacy, suggesting that somehow the machine, as a result of its intelligence, will simply perceive what is right given its experience of the world. This is implausible; for example, one cannot perceive, in the design of a chessboard and chess pieces, the goal of checkmate; the same chessboard and pieces can be used for suicide chess, or indeed many other games still to be invented. Put another way: where Bostrom imagines humans driven

[5] Rodney Brooks (2017), for example, asserts that it's impossible for a program to be 'smart enough that it would be able to invent ways to subvert human society to achieve goals set for it by humans, without understanding the ways in which it was causing problems for those same humans'. Often, the argument adds the premise that people of greater intelligence tend to have more altruistic objectives, a view that may be related to the self-conception of those making the argument. Chalmers (2010) points to Kant's view that an entity necessarily becomes more moral as it becomes more rational, while noting that nothing in our current understanding of AI supports this view when applied to machines.

extinct by a putative robot that turns the planet into a sea of paperclips, we humans see this outcome as tragic, whereas the iron-eating bacterium *Thiobacillus ferrooxidans* is thrilled. Who's to say the bacterium is wrong? The fact that a machine has been given a fixed objective by humans doesn't mean that it will automatically take on board as additional objectives other things that are important to humans. Maximizing the objective may well cause problems for humans; the machine may recognize those problems as problematic for humans; but, by definition, they are not problematic within the standard model from the point of view of the given objective.

- *Intelligence is multidimensional, 'so smarter than humans' is a meaningless concept*: This argument, due to Kevin Kelly (2017), draws on a staple of modern psychology—the fact that a scalar IQ does not do justice to the full range of cognitive skills that humans possess to varying degrees. IQ is indeed a crude measure of human intelligence, but it is utterly meaningless for current AI systems because their capabilities across different areas are uncorrelated. How do we compare the IQ of Google's search engine, which cannot play chess, to that of Deep Blue, which cannot answer search queries? None of this supports the argument that because intelligence is multifaceted, we can ignore the risk from superintelligent machines. If 'smarter than humans' is a meaningless concept, then 'smarter than gorillas' is also meaningless, and gorillas therefore have nothing to fear from humans. Clearly, that argument doesn't hold water. Not only is it logically possible for one entity to be more capable than another across all the relevant dimensions of intelligence, it is also possible for one species to represent an existential threat to another even if the former lacks an appreciation for music and literature.

1.4 Solutions

Can we tackle Wiener's warning head-on? Can we design AI systems whose purposes don't conflict with ours, so that we're sure to be happy with how they behave? On the face of it, this seems hopeless because it will doubtless prove infeasible to write down our purposes correctly or imagine all the counterintuitive ways a superintelligent entity might fulfil them.

If we treat superintelligent AI systems as if they were black boxes from outer space, then indeed there is no hope. Instead, the approach we seem obliged to take, if we are to have any confidence in the outcome, is to define some formal problem F and design AI systems to be F-solvers, such that the closer the AI system comes to solving F perfectly, the greater the benefit to humans. In simple terms, the more intelligent the machine, the better the outcome for humans: we hope the machine's intelligence will be applied both to learning our true objectives and to helping us achieve them. If we can work out an appropriate F that has this property, we will be able to create provably beneficial AI.

There is, I believe, an approach that may work. Humans can reasonably be described as having (mostly implicit and partially formed) preferences over their future lives—that is, given enough time and unlimited visual aids, a human could express a preference (or indifference) when offered a choice between two future lives laid out before him or her in all their aspects. (This idealization ignores the possibility that our minds are composed of subsystems with effectively incompatible preferences; if true, that would limit a machine's ability to satisfy our preferences optimally, but it doesn't seem to prevent us from designing machines that avoid catastrophic outcomes.) The formal problem F to be solved by the machine in this case is a game-theoretic one: to maximize human future-life preferences subject to its initial uncertainty as to what they are, in an environment that includes human participants. Furthermore, although the future-life preferences are hidden variables, they're grounded in a voluminous source of evidence, namely, all of the human choices ever made.

This formulation sidesteps Wiener's problem, because we do not put a fixed purpose in the machine according to which it can rank all possible futures. Instead, the machine knows that it doesn't know the true preference ranking, so it naturally acts cautiously to avoid violating potentially important but unknown preferences. (We can certainly include fairly strong priors on the positive value of life, health, etc., to make the machine more useful more quickly.) The machine may learn more about human preferences as it goes along, of course, but it will never achieve complete certainty. Such a machine will be motivated to ask questions, to seek permission or additional feedback before undertaking any potentially risky course of action, to defer to human instruction, and to allow itself to be switched off. These behaviours are not built in via preprogrammed scripts or rules; rather, they fall out as solutions of the formal problem F.

As noted in the introduction, this involves a shift from a unary view of AI to a binary one. The classical view, in which a fixed objective is given to the machine, is illustrated qualitatively in Figure 1.1. Once the machine has a fixed objective, it will act to optimize the achievement of the objective; its behaviour is effectively independent of the human's behaviour.[6] On the other hand, when the human objective is unobserved by the machine (see Figure 1.2), the human and machine behaviors remain coupled information-theoretically because human behavior provides further information about human objectives.

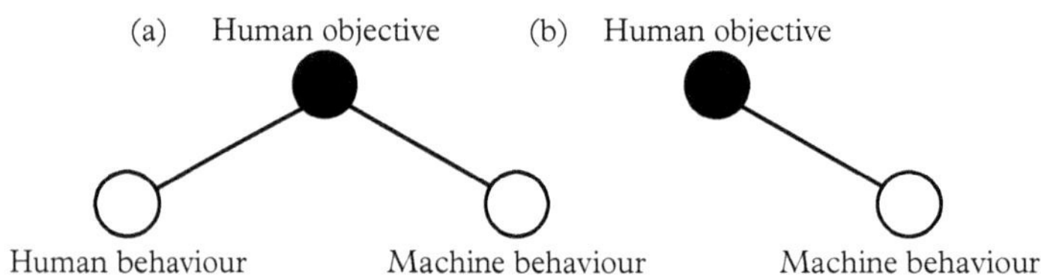

Figure 1.1 *(a) The classical AI situation in which the human objective is considered fixed and known by the machine, depicted as a notional graphical model. Given the objective, the machine's behaviour is (roughly speaking) independent of any subsequent human behaviour, as depicted in (b). This unary view of AI is tenable only if the human objective can be completely and correctly stated.*

[6] The independence is not strict because the human's behaviour can provide information about the state of the world. Thus, a passenger in an automated taxi could tell the taxi that snipers have been reported on the road it intends to take, picking off passengers for fun; but this might affect the taxi's behaviour only if it already knows that death by gunfire is undesirable for humans.

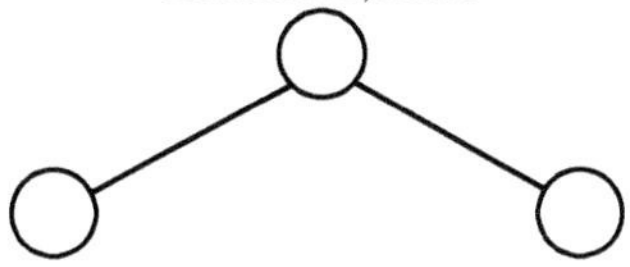

Figure 1.2 *When the human objective is unobserved, machine behaviour is no longer independent of human behaviour, because the latter provides more information about the human objective.*

1.4.1 Assistance games

This basic idea is made more precise in the framework of *assistance games*—originally known as cooperative inverse reinforcement learning (CIRL) games in the terminology of Hadfield-Menell *et al.* (2017*a*). The simplest case of an assistance game involves two agents, one human and the other a robot. It is a game of partial information, because, while the human knows the reward function, the robot does not—even though the robot's job is to maximize it. It may involve a form of inverse reinforcement learning (Russell 1998; Ng and Russell 2000) because the robot can learn more about human preferences from the observation of human behaviour—a process that is the dual of reinforcement learning, wherein behaviour is learned from rewards and punishments.

To illustrate assistance games, I'll use the *paperclip game.* It's a very simple game in which Harriet the human has an incentive to 'signal' to Robbie the robot some information about her preferences. Robbie is able to interpret that signal because he can solve the game and therefore he can understand what would have to be true about Harriet's preferences in order for her to signal in that way.

The steps of the game are depicted in Figure 1.3. It involves making paperclips and staples. Harriet's preferences are expressed by a payoff function that depends on the number of paperclips and the number of staples produced, with a certain 'exchange rate' between the two. Harriet's preference parameter θ denotes the relative value (in dollars) of a paperclip; for example, she might value paperclips at $\theta = 0.45$ dollars, which means staples are worth $1 - \theta = 0.55$ dollars. So, if p paperclips and s staples are produced, Harriet's payoff will be $p\theta + s(1 - \theta)$ dollars in all. Robbie's prior is $P(\theta) = Uniform(\theta; 0, 1)$. In the game itself, Harriet goes first and can choose to make two paperclips, two staples, or one of each. Then Robbie can choose to make 90 paperclips, 90 staples, or 50 of each.

Notice that if she were doing this by herself, Harriet would just make two staples, with a value of \$1.10. (See the annotations at the first level of the tree in Figure 1.3.) But Robbie is watching, and he learns from her choice. What exactly does he learn? Well, that depends on how Harriet makes her choice. How does Harriet make her choice? That depends on how Robbie is going to interpret it. One can resolve this circularity by finding a Nash equilibrium. In this case, it is unique and can be found by applying the iterated-best-response algorithm: pick any strategy for Harriet; pick the best strategy

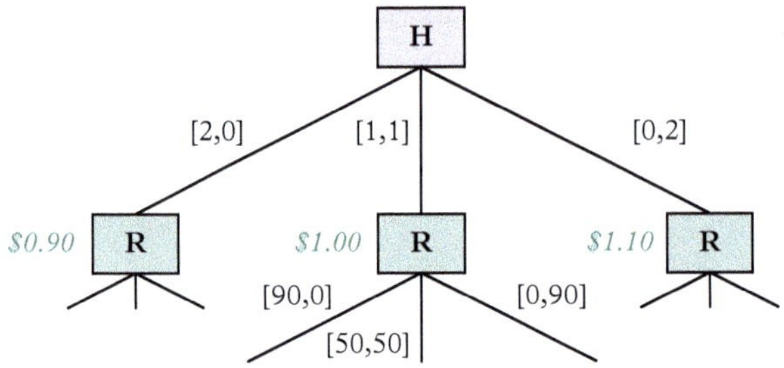

Figure 1.3 *The paperclip game. Each branch is labeled $[p, s]$ denoting the number of paperclips and staples manufactured on that branch. Harriet the human can choose to make two paperclips, two staples, or one of each. (The values in green italics are the values for Harriet if the game ended there, assuming $\theta = 0.45$.) Robbie the robot then has a choice to make 90 paperclips, 90 staples, or 50 of each.*

for Robbie, given Harriet's strategy; pick the best strategy for Harriet, given Robbie's strategy; and so on. The process unfolds as follows:

1. Start with the greedy strategy for Harriet: make two paperclips if she prefers paperclips; make one of each if she is indifferent; make two staples if she prefers staples.
2. There are three possibilities Robbie has to consider, given this strategy for Harriet:
   (a) If Robbie sees Harriet make two paperclips, he infers that she prefers paperclips, so he now believes the value of a paperclip is uniformly distributed between 0.5 and 1.0, with an average of 0.75. In that case, his best plan is to make 90 paperclips with an expected value of $67.50 for Harriet.
   (b) If Robbie sees Harriet make one of each, he infers that she values paperclips and staples at 0.50, so the best choice is to make 50 of each.
   (c) If Robbie sees Harriet make two staples, then by the same argument as in (a), he should make 90 staples.
3. Given this strategy for Robbie, Harriet's best strategy is now somewhat different from the greedy strategy in step 1. If Robbie is going to respond to her making one of each by making 50 of each, then she is better off making one of each not just if she is exactly indifferent, but if she is anywhere close to indifferent. In fact, the optimal policy is now to make one of each if she values paperclips anywhere between about 0.446 and 0.554.
4. Given this new strategy for Harriet, Robbie's strategy remains unchanged. For example, if she chooses one of each, he infers that the value of a paperclip is uniformly distributed between 0.446 and 0.554, with an average of 0.50, so the best choice is to make 50 of each. Because Robbie's strategy is the same as in step 2, Harriet's best response will be the same as in step 3, and we have found the equilibrium.

With her strategy, Harriet is, in effect, teaching Robbie about her preferences using a simple code—a language, if you like—that emerges from the equilibrium analysis. Note also that Robbie never learns Harriet's preferences exactly, but he learns enough to act optimally on her behalf—that is he acts (given his limited options) just as he would if he did know her preferences exactly. He is provably beneficial to Harriet under the assumptions stated, and under the assumption that Harriet is playing the game correctly.

It is possible to prove that provided there are no ties that cause coordination problems, finding an optimal strategy for the robot in an assistance game can be done by solving a single-agent partially observable Markov decision process (POMDP) whose state space is the underlying state space of the game plus the human preference parameters θ. POMDPs in general are very hard to solve, but the POMDPs that represent assistance games have additional structure that enables more efficient algorithms (Malik *et al.* 2018).

1.4.2 The off-switch game

Within the same basic framework, one can also show that a robot solving an assistance game will defer to a human and allow itself to be switched off. This property is illustrated in the *off-switch game* shown in Figure 1.4 (Hadfield-Menell *et al.* 2017*b*). Robbie is now helping Harriet find a hotel room for the International Paperclip Convention in Geneva. Robbie can act now—let's say he can book Harriet into a very expensive hotel near the meeting venue. He is quite unsure how much Harriet will like the hotel and its price; let's say he has a uniform probability for its net value to Harriet between -40 and $+60$, with an average of $+10$. He could also 'switch himself off'—less melodramatically, take himself out of the hotel booking process altogether—which is defined (without loss of generality) to have value 0 to Harriet. If those were his two choices, he would go ahead and book the hotel, incurring a significant risk of making Harriet unhappy. (If the range were -60 to $+40$, with average -10, he would switch himself off instead.) I'll give Robbie

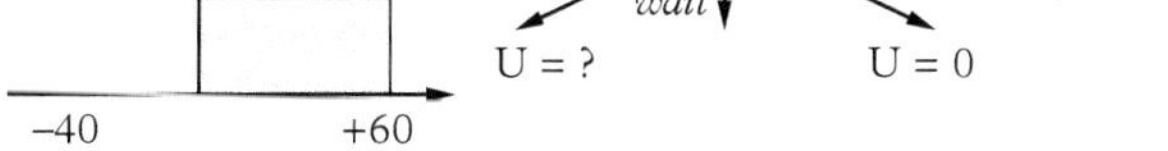

Figure 1.4 *The off-switch game. R, the robot, can choose to act now, with a highly uncertain payoff; to switch itself off; or to defer to H, the human. H can switch R off or let it go ahead. R now has the same choice again. Acting still has an uncertain payoff, but now R knows the payoff is nonnegative.*

a third choice, however: explain his plan, wait, and let Harriet switch him off. Harriet can either switch him off or let him go ahead and book the hotel. What possible good could this do, one might ask, given that he could make both of those choices himself?

The point is that Harriet's choice—to switch Robbie off or let him go ahead—provides Robbie with information about Harriet's preferences. I'll assume, for now, that Harriet is rational, so if Harriet lets Robbie go ahead, it means the value to Harriet is positive. Now, as shown in Figure 1.4, Robbie's belief changes: it is uniform between 0 and +60, with an average of +30.

So, if one evaluates Robbie's initial choices from his point of view:

1. Acting now and booking the hotel has an expected value of +10.
2. Switching himself off has a value of 0.
3. Waiting and letting Harriet switch him off leads to two possible outcomes:
   (a) There is a 40% chance, based on Robbie's uncertainty about Harriet's preferences, that she will hate the plan and will switch Robbie off, with value 0.
   (b) There is a 60% chance Harriet will like the plan and allow Robbie to go ahead, with expected value +30.

Thus, waiting has expected value $(0.4 \times 0) + (0.6 \times 30) = +18$, which is better than the +10 Robbie epects if he acts now.

The upshot is that Robbie has a positive incentive to defer to Harriet—that is, to allow himself to be switched off. This incentive comes directly from Robbie's uncertainty about Harriet's preferences. Robbie is aware that there's a chance (40% in this example) that he might be about to do something that will make Harriet unhappy, in which case being switched off would be preferable to going ahead. Were Robbie already certain about Harriet's preferences, he would just go ahead and make the decision (or switch himself off); there would be absolutely nothing to be gained from consulting Harriet, because, according to Robbie's definite beliefs, he can already predict exactly what she is going to decide.

In fact, it is possible to prove the same result in the general case: as long as Robbie is not completely certain that he's about to do what Harriet herself would do, he is better off allowing her to switch him off. Intuitively, her decision provides Robbie with information, and the expected value of information is always non negative. Conversely, if Robbie is certain about Harriet's decision, her decision provides no new information, and so Robbie has no incentive to allow her to decide.

Formally, let $P(u)$ be Robbie's prior probability density over Harriet's utility for the proposed action a. Then the value of going ahead with a is

$$EU(a) = \int_{-\infty}^{\infty} P(u) \cdot u\, du = \int_{-\infty}^{0} P(u) \cdot u\, du + \int_{0}^{\infty} P(u) \cdot u\, du\,.$$

On the other hand, the value of action d, deferring to Harriet, is composed of two parts: if $u > 0$ then Harriet lets Robbie go ahead, so the value is u, but if $u < 0$ then Harriet switches Robbie off, so the value is 0:

$$EU(d) = \int_{-\infty}^{0} P(u) \cdot 0 \, du + \int_{0}^{\infty} P(u) \cdot u \, du.$$

Comparing the expressions for $EU(a)$ and $EU(d)$, it follows immediately that

$$EU(d) \geq EU(a)$$

because the expression for $EU(d)$ has the negative-utility region zeroed out. The two choices have equal value only when the negative region has zero probability—that is, when Robbie is already certain that Harriet likes the proposed action.

There are some obvious elaborations on the model that are worth exploring immediately. The first elaboration is to impose a cost for Harriet's time. In that case, Robbie is less inclined to bother Harriet if the downside risk is small. This is as it should be. And if Harriet is really grumpy about being interrupted, she shouldn't be too surprised if Robbie occasionally does things she doesn't like.

The second elaboration is to allow for some probability of human error—that is, Harriet might sometimes switch Robbie off even when his proposed action is reasonable, and she might sometimes let Robbie go ahead even when his proposed action is undesirable. It is straightforward to fold this error probability into the model (Hadfield-Menell *et al.* 2017*b*). As one might expect, the solution shows that Robbie is less inclined to defer to an irrational Harriet who sometimes acts against her own best interests. The more randomly she behaves, the more uncertain Robbie has to be about her preferences before deferring to her. Again, this is as it should be: for example, if Robbie is a self-driving car and Harriet is his naughty two-year-old passenger, Robbie should not allow Harriet to switch him off in the middle of the highway.

The off-switch example suggests some templates for controllable-agent designs and provides a simple example of a provably beneficial system in the sense introduced above. The overall approach resembles principal–agent problems in economics, wherein the principal (e.g., an employer) needs to incentivize another agent (e.g., an employee) to behave in ways beneficial to the principal. The key difference here is that we are building one of the agents in order to benefit the other. Unlike a human employee, the robot should have no interests of its own whatsoever.

Assistance games can be generalized to allow for imperfectly rational humans (Hadfield-Menell *et al.* 2017*b*), humans who don't know their own preferences (Chan *et al.* 2019), multiple human participants, multiple robots, and so on. Scaling up to complex environments and high-dimensional perceptual inputs may be possible using methods related to deep inverse reinforcement learning. By providing a factored or structured action space, as opposed to the simple atomic actions in the paperclip game, the opportunities for communication can be greatly enhanced. Few of these variations have been explored so far, but I expect the key property of assistance games to remain true: robots that solve such games will be beneficial (in expectation) to humans (Hadfield-Menell *et al.* 2017*a*).

While the basic theory of assistance games assumes perfectly rational robots that can solve the assistance game exactly, this is unlikely to be possible in practical situations. Indeed, one expects to find qualitatively different phenomena occurring when the robot is much less capable than, roughly as capable as, or much more capable than the human. There is good reason to hope that in all cases improving the robot's capability will be beneficial to the human, because it will do a better job of learning human preferences and a better job of satisfying them.

1.4.3 Acting with unknown preferences

Multiattribute utility theory (Keeney and Raiffa 1976) views the world as composed of a set of attributes $\{X_1, \ldots, X_n\}$, with preferences defined on lotteries over complete assignments to the attributes. This is clearly an oversimplification, but it suffices for our purpose in exploring some basic phenomena.

In some cases, a machine's scope of action is strictly limited. For example, a (non-Internet-connected) thermostat can only turn the heating on and off, and, to a first approximation, affects the temperature in the house and the owner's bank balance.[7] It is plausible in this case to imagine that the thermostat might develop a decent model of the user's preferences over temperature and cost attributes.

In the great majority of circumstances, however, the AI system's knowledge of human preferences will be extremely incomplete compared to its scope of action. How can it be useful if this is the case? Can it even fetch the coffee?

It turns out that the answer is yes, if we understand 'fetch the coffee' the right way. 'Fetch the coffee' does not divide the world into goal states (where the human has coffee) and non-goal states. Instead, it says that the human's current preferences rank coffee states above non-coffee states *all other things being equal*. This idea of goals as *ceteris paribus* comparatives is well-established (von Wright 1972; Wellman and Doyle 1991). In this context, it suggests that the machine should act in a *minimally invasive* fashion—that is, satisfy the preferences it knows about (coffee) without disturbing any other attributes of the world.

There remains the question of why the machine should assume that leaving other attributes unaffected is better than disturbing them in some random way. One possible answer is some form of risk aversion, but I suspect this is not enough. One has a sense that a machine that does nothing is better than one that acts randomly, but this is certainly not implicit in the standard formulation of MDPs. I think one has to add the assumption that the world is not in an *arbitrary* state; rather, it resembles a state sampled from the

[7] In reality, it is very difficult to limit the effects of the agent's actions to a small set of attributes. Turning the heat off may make the occupants more susceptible to viral infection, and profligate heating may tip the occupants into bankuptcy, and so on. A device connected to the Internet, with the ability to send character streams, can affect the entire planet through propaganda, online trading, etc.

stationary distribution that results from the actions of human agents operating according to their preferences (Shah *et al.* 2019). In that case, one expects a random action to make things worse for the humans.

There is another kind of action that is beneficial to humans even when the machine knows nothing at all about human preferences: an action that simply expands the set of actions available to the human. For example, if Harriet has forgotten her password, Robbie can give her the password, enabling a wider range of actions than Harriet could otherwise execute.

1.5 Reasons for Optimism

There are some reasons to think this approach may work in practice. First, there is abundant written and filmed information about humans doing things (and other humans reacting). More or less every book ever written contains evidence on this topic. Even the oldest clay tablets, tediously recording the exchange of N sheep for M oxen, give information about human preferences between sheep and oxen. Technology to build models of human preferences from this storehouse will presumably be available long before superintelligent AI systems are created.

Second, there are strong near-term economic incentives for robots to understand human preferences, which also come into play well before the arrival of superintelligence. Already, computer systems record one's preferences for an aisle seat or a vegetarian meal. More sophisticated personal assistants will need to understand their user's preferences for cost, luxury, and convenient location when booking hotels, and how these preferences depend on the nature and schedule of the user's planned activities. Managing a busy person's calendar and screening calls and emails requires an even more sophisticated understanding of the user's life, as does the management of an entire household when entrusted to a domestic robot. For all such roles, trust is essential but easily lost if the machine reveals itself to lack a basic understanding of human preferences. If one poorly designed domestic robot cooks the cat for dinner, not realizing that its sentimental value outweighs its nutritional value, the domestic-robot industry will be out of business.

For companies and governments to adopt the new model of AI, a great deal of research must be done to replace the entire toolbox of AI methods, all of which have been developed on the assumption that the objective is known exactly. There are two primary issues for each class of task environments: how to relax the assumption of a known objective and what form of interaction to assume between the machine and the human. For example, problem-solving task environments have an objective defined by a goal test $G(s)$ and a stepwise cost function $c(s, a, s')$. Perhaps the machine knows a relaxed predicate $G' \supset G$ and upper and lower bounds c^+ and c^- on the cost function, and can ask the human (1) whether any given state s satisfies G and (2) whether one trajectory to s is preferred to another. Design considerations include formal precision, algorithmic complexity, feasibility of the interaction protocol from the human point of view, and applicability in real-world circumstances.

The standard model of AI as maximizing objectives does not imply that all AI systems have to solve some particular problem formulation such as an influence diagram or a factored MDP. For example, it is entirely consistent with the standard model to build an AI system directly as a policy, expressed as a set of condition–action rules specifying the optimal action in each category of states.[8] By the same token, I am not proposing that all AI systems under the new model have to solve some explicitly formulated representation of the assistance game. It is important to maintain a broad conception of the approach and how it applies to the design of AI systems for any particular task environment. The crucial elements are (1) acknowledgement that there is partial and uncertain information about the true human preferences that are relevant in the task environment; (2) a means for information to flow at run-time from humans to machines concerning those preferences; and (3) allowance for the human to be a joint participant in the run-time process.

1.6 Obstacles

There are obvious difficulties with an approach that expects machines to learn underlying preferences from observing human behaviour. The first is that humans are irrational, in the sense that our actions do not reflect our preferences. This irrationality arises in part from our computational limitations relative to the complexity of the decision problems we face. For example, if two humans are playing chess and one of them loses, it's because the loser (and possibly the winner too) made a mistake—a move that led inevitably to a forced loss. A machine observing that move and assuming perfect rationality on the part of the human might well conclude that the human *preferred* to lose. Thus, to avoid reaching such conclusions, the machine must take into account the *actual* cognitive mechanisms of humans.

As yet, we do not know enough about human cognitive mechanisms to invert real human behaviour to get at the underlying preferences. One thing that seems intuitively clear, however, is that one of our principal methods for coping with the complexity of the world is to organize our behaviour hierarchically. That is, we make (defeasible) commitments to higher-level goals such as 'write an essay on a human-compatible approach to AI'; then, rather than considering all possible sequences of words, from 'aardvark aardvark aardvark . . .' to 'zyzzyva zyzzyva zyzzyva . . .' as a chess program would do, we choose among subtasks such as 'write the introduction' and 'read more about preference elicitation'. Eventually, we get down to the choice of words, and then typing each word involves a sequence of keystrokes, each of which is in turn a sequence of motor control commands to the muscles of the arms and hands. At any given point, then, a human is embedded at various particular levels of multiple deep and complex hierarchies of partially overlapping activities and subgoals. This means that for the machine to

[8] Many applications of control theory work exactly this way: the control theorist works offline with a mathematical model of the system and the objective to derive a *control law* that is then implemented in the controller.

understand human actions, it probably needs to understand a good deal about what these hierarchies are and how we use them to navigate the real world.

Machines might try to discover more about human cognitive mechanisms by an inductive learning approach. Suppose that in some given state s Harriet's action a depends on her preferences θ according to mechanism h, that is, $a = h(\theta, s)$. (Here, θ represents not a single parameter such as the exchange rate between staples and paperclips, but Harriet's preferences over future lives, which could be a structure of arbitrary complexity.) By observing many examples of s and a, is it possible eventually to recover h and θ? At first glance, the answer seems to be no (Armstrong and Mindermann 2019). For example, one cannot distinguish between the following hypotheses about how Harriet plays chess:

1. h maximizes the satisfaction of preferences, and θ is the desire to win games.
2. h minimizes the satisfaction of preferences, and θ is the desire to lose games.

From the outside, Harriet plays perfect chess under either hypothesis.[9] If one is merely concerned with predicting her next move, it doesn't matter which formulation one chooses. On the other hand, for a machine whose goal is to help Harriet realize her preferences, it really does matter! The machine needs to know which explanation holds. From this viewpoint, something is seriously wrong with the second explanation of behaviour. If Harriet's cognitive mechanism h were really trying to minimize the satisfaction of preferences θ, it wouldn't make sense to call θ her preferences. It is, then, simply a mistake to suppose that h and θ are separately and independently defined. I have already argued that the assumption of perfect rationality—that is, h is maximization—is too strong; yet, for it to make sense to say that Harriet has preferences, h will have to satisfy (or nearly satisfy) some basic properties associated with rationality. These might include choosing correctly according to preferences in situations that are computationally trivial—for example, choosing between vanilla and bubble-gum ice cream at the beach. Cherniak (1986) presents an in-depth analysis of these minimal conditions on rationality.

Further difficulties arise if the machine succeeds in identifying Harriet's preferences, but finds them to be inconsistent. For example, suppose she prefers vanilla to bubble gum and bubble gum to pistachio, but prefers pistachio to vanilla. In that case her preferences violate the axiom of transitivity and there is no way to maximally satisfy her preferences. (That is, whatever ice cream the machine gives her, there is always another that she would prefer.) In such cases, the machine could attempt to satisfy Harriet's preferences *up to inconsistency*; for example, if Harriet strictly prefers all three of the aforementioned flavors to licorice, then it should avoid giving her licorice ice cream.

[9] Of course, the Harriet who prefers to lose might grumble when she keeps winning, thereby giving a clue as to which Harriet she is. One response to this is that grumbling is just more behaviour, and equally subject to multiple interpretations. Another response is to say that Harriet might feel grumbly but, in keeping with her minimizing h, would instead jump for joy. This is not to say that there is no fact of the matter as to whether Harriet is pleased or displeased with the outcome.

Of course, the inconsistency in Harriet's preferences could be of a far more radical nature. Many theories of cognition, such as Minsky's *Society of Mind* (1986), posit multiple cognitive subsystems that, in essence, have their own preference structures and compete for control—and these seem to be manifested in addictive and self-destructive behaviours, among others. Such inconsistencies place limits on the extent to which the idea of machines helping humans even makes sense.

Also difficult, from a philosophical viewpoint, is the apparent *plasticity* of human preferences—the fact that they seem to change over time as the result of experiences. It is hard to explain how such changes can be made rationally, because they make one's future self less likely to satisfy one's present preferences about the future. Yet plasticity seems fundamentally important to the entire enterprise, because newborn infants certainly lack the rich, nuanced, culturally informed preference structures of adults. Indeed, it seems likely that our preferences are at least partially formed by a process resembling inverse reinforcement learning, whereby we absorb preferences that explain the behaviour of those around us. Such a process would tend to give cultures some degree of autonomy from the otherwise homogenizing effects of our dopamine-based reward system.

Plasticity also raises the obvious question of which Harriet the machine should try to help: Harriet_{2020}, Harriet_{2035}, or some time-averaged Harriet? (See Pettigrew (2020) for a full treatment of this approach, wherein decisions for individuals who change over time are made as if they were decisions made on behalf of multiple distinct individuals.) Plasticity is also problematic because of the possibility that the machine may, by subtly influencing Harriet's environment, gradually mould her preferences in directions that make them easier to satisfy, much as certain political forces have been said to do with voters in recent decades.

I am often asked, 'Whose values should we align AI with?' (The question is usually posed in more accusatory language, as if my secret, Silicon-Valley-hatched plan is to align all the world's AI systems with my own white, male, Western, cisgender, Episcopalian values.) Of course, this is simply a misunderstanding. The kind of AI system proposed here is not 'aligned' with any values, unless you count the basic principle of helping humans realize their preferences. For each of the billions of humans on Earth, the machine should be able to predict, to the extent that its information allows, which life that person would prefer.

Now, practical and social constraints will prevent all preferences from being maximally satisfied simultaneously. We cannot all be Ruler of the Universe. This means that machines must mediate among conflicting preferences—something that philosophers and social scientists have struggled with for millennia. At one extreme, each machine could pay attention only to the preferences of its owner, subject to legal constraints on its actions. This seems undesirable, as it would have a machine belonging to a misanthrope refuse to aid a severely injured pedestrian so that it can bring the newspaper home more quickly. Moreover, we might find ourselves needing many more laws as machines satisfy their owners' preferences in ways that are very annoying to others even if not strictly illegal. At the other extreme, if machines consider equally the preferences of all humans, they might focus a larger fraction of their energies on the least fortunate than their owners might prefer—a state of affairs not conducive to investment in AI. Presumably, some middle ground can be found, perhaps combining a degree of obligation to the machine's owner with public subsidies that support contributions to the greater good. Determining the ideal solution for this issue is an open problem.

Another common question is, 'What if machines learn from evil people?' Here, there is a real issue. It is *not* that machines will learn to copy evil actions. The machine's actions need not resemble in any way the actions of those it observes, any more than a criminologist's actions resemble those of the criminals she observes. The machine is learning about human preferences; it is not adopting those preferences as its own and acting to satisfy them. For example, suppose that a corrupt passport official in a developing country insists on a bribe for every transaction, so that he can afford to pay for his children to go to school. A machine observing this will not learn to take bribes itself: it has no need of money and understands (and wishes to avoid) the toll imposed on others by the taking of bribes. The machine will instead find other, socially beneficial ways to help send the children to school. Similarly, a machine observing humans killing each other in war will not learn that killing is good: obviously, those on the receiving end very much prefer not to be dead.

The difficult issue that remains is this: what should machines learn from humans who enjoy the suffering of others? In such cases, any simple aggregation scheme for preferences (such as adding utilities) would lead to some reduction in the utilities of others in order to satisfy, at least partially, these perverse preferences. It seems reasonable to require that machines simply ignore positive weights in the preferences of some for the suffering of others (Harsanyi 1977).

1.7 Looking Further Ahead

If we assume, for the sake of argument, that all of these obstacles can be overcome, as well as all of the obstacles to the development of truly capable AI systems, are we home free? Would provably beneficial, superintelligent AI usher in a golden age for humanity? Not necessarily. There remains the issue of adoption: how can we obtain broad agreement on suitable design principles, and how can we ensure that only suitably designed AI systems are deployed?

On the question of obtaining agreement at the policy level, it is necessary first to generate consensus within the research community on the basic ideas of—and design templates for—provably beneficial AI, so that policy-makers have some concrete guidance on what sorts of regulations might make sense. The economic incentives noted earlier are of the kind that would tend to support the installation of rigorous standards at the early stages of AI development, because failures would be damaging to entire industries, not just to the perpetrator and victim. We already see this in miniature with the imposition of machine-checkable software standards for cell-phone applications.

On the question of enforcement of policies for AI software design, I am less sanguine. If Dr Evil wants to take over the world, he or she might remove the safety catch, so to speak, and deploy an AI system that ends up destroying the world instead. This problem is a hugely magnified version of the problem we currently face with malware. Our track record in solving the latter problem does not provide grounds for optimism concerning the former. In Samuel Butler's *Erewhon* and in Frank Herbert's *Dune*, the solution is to ban all intelligent machines, as a matter of both law and cultural imperative. Perhaps if we find institutional solutions to the malware problem, we will be able to devise some less drastic approach for AI.

The problem of misuse is not limited to evil masterminds. One possible future for humanity in the age of superintelligent AI is that of a race of lotus eaters, progressively enfeebled as machines take over the management of our entire civilization. This is the future imagined in E. M. Forster's story *The Machine Stops*, written in 1909. We may say, now, that such a future is undesirable; the machines may agree with us and volunteer to stand back, requiring humanity to exert itself and maintain its vigour. But exertion is tiring, and we may, in our usual myopic way, design AI systems that are not *quite* so concerned about the long-term vigour of humanity and just a *little* more helpful than they would otherwise wish to be. Unfortunately, this process continues in a direction that is hard to resist.

1.8 Conclusion

Finding a solution to the AI control problem is an important task; it may be, in Bostrom's words, 'the essential task of our age'. It involves building systems that are far more powerful than ourselves while still guaranteeing that those systems will remain powerless, forever.

Up to now, AI research has focused on systems that are better at making decisions, but this is not the same as making better decisions. No matter how excellently an algorithm maximizes, and no matter how accurate its model of the world, a machine's decisions may be ineffably stupid, in the eyes of an ordinary human, if it fails to understand human preferences.

This problem requires a change in the definition of AI itself—from a field concerned with a unary notion of intelligence as the optimization of a given objective, to a field concerned with a binary notion of machines that are provably beneficial for humans. Taking the problem seriously seems likely to yield new ways of thinking about AI, its purpose, and our relationship to it.

References

Armstrong, S. and Mindermann, S. (2019). Occam's razor is insufficient to infer the preferences of irrational agents, in *Advances in Neural Information Processing Systems 31*.

Bostrom, N. (2014). *Superintelligence: Paths, Dangers, Strategies*. Oxford: Oxford University Press.

Brooks, R. (2017). The seven deadly sins of AI predictions. *MIT Technology Review*, 6 October.

Chalmers, D. J. (2010). The singularity: a philosophical analysis. *Journal of Consciousness Studies*, 17, 7–65.

Chan, L., Hadfield-Menell, D., Srinivasa, S. *et al.* (2019). The assistive multi-armed bandit, in *Proceedings of the Fourteenth ACM/IEEE International Conference on Human–Robot Interaction*. Daegu, Republic of Korea; 11–14 March.

Cherniak, C. (1986). *Minimal Rationality*. Cambridge, MA: MIT Press.

Gates, W. (2015). Ask me anything. Reddit, 28 January. https://www.reddit.com/r/IAmA/comments/2tzjp7/hi_reddit_im_bill_gates_and_im_back_for_my_third/

Hadfield-Menell, D., Dragan, A. D., Abbeel, P. *et al.* (2017*a*). Cooperative inverse reinforcement learning, in *Advances in Neural Information Processing Systems 29*.

Hadfield-Menell, D., Dragan, A. D. *et al.* (2017*b*). The off-switch game, in *Proceedings of the Twenty-Sixth International Joint Conference on Artificial Intelligence*. Melbourne; August 19–25.

Harsanyi, J. (1977). Morality and the theory of rational behavior. *Social Research*, **44**, 623–56.

Keeney, R. L. and Raiffa, H. (1976). *Decisions with Multiple Objectives: Preferences and Value Tradeoffs*. New York: Wiley.

Kelly, K. (2017). The myth of a superhuman AI. *Wired*, 25 April.

Kumparak, G. (2014). Elon Musk compares building artificial intelligence to 'summoning the demon'. TechCrunch, 26 October. https://techcrunch.com/2014/10/26/elon-musk-compares-building-artificial-intelligence-to-summoning-the-demon/

Malik, D., Palaniappan, M., Fisac, J. F. *et al.* (2018). An efficient, generalized Bellman update for cooperative inverse reinforcement learning, in *Proceedings of the Thirty-Fifth International Conference on Machine Learning*. Sydney; 6–11 August.

Minsky, M. L. (1986). *The Society of Mind*. New York, NY: Simon and Schuster.

Ng, A. Y. and Russell, S. J. (2000). Algorithms for inverse reinforcement learning, in *Proceedings of the Seventeenth International Conference on Machine Learning*.

Omohundro, S. (2008). The basic AI drives, in *AGI-08 Workshop on the Sociocultural, Ethical and Futurological Implications of Artificial Intelligence*. Memphis, TN; 4 March.

Osborne, H. (2017). Stephen Hawking AI warning: artificial intelligence could destroy civilization. *Newsweek*, 7 November.

Pettigrew, R. (2020). *Choosing for Changing Selves*. Oxford: Oxford University Press.

Russell, S. J. (1998). Learning agents for uncertain environments, in *Proceedings of the Eleventh ACM Conference on Computational Learning Theory*. Madison, WI; 24–26 July.

Shah, R., Krasheninnikov, D., Alexander, J. *et al.* (2019). The implicit preference information in an initial state, in *Proceedings of the Seventh International Conference on Learning Representations*. New Orleans; 6–9 May.

Stone, P., Brooks, R. A., Brynjolfsson, E. *et al.* (2016). Artificial intelligence and life in 2030. Technical report, Stanford University One Hundred Year Study on Artificial Intelligence: Report of the 2015–2016 Study Panel.

Turing, A. (1951). Can digital machines think? Lecture broadcast on BBC Third Programme. Typescript available at http://www.turingarchive.org.

von Wright, G. (1972). The logic of preference reconsidered. *Theory and Decision*, **3**, 140–67.

Wellman, M. P. and Doyle, J. (1991). Preferential semantics for goals, in *Proceedings of the Ninth National Conference on Artificial Intelligence*. Anaheim, CA; 14–19 July.

Wiener, N. (1960). Some moral and technical consequences of automation. *Science*, **131**(3410), 1355–8.

2

Alan Turing and Human-Like Intelligence

Peter Millican

Hertford College, Oxford, UK

The idea of Human-Like Computing became central to visions of Artificial Intelligence through the work of Alan Turing, whose model of computation (1936) is explicated in terms of the potential operations of a human "computer", and whose famous test for intelligent machinery (1950) is based on indistinguishability from human verbal behaviour. But here I shall challenge the apparent human-centredness of the 1936 model (now known as the Turing machine), and suggest a different genesis with a primary focus on the foundations of mathematics, and with human comparisons making an entrance only in retrospective justification of the model. It will also turn out, more surprisingly, that the 1950 account of intelligence is ultimately far less human-centred than it initially appears to be, because the *universality* of computation—as established in the 1936 paper—makes human intelligence just one variety amongst many. It is only when Turing considers *consciousness* that he goes seriously astray in suggesting that machine intelligence must be understood on the human model. But a better approach is clearly revealed through his own earlier work, which gave ample reason to reinterpret intelligence as sophisticated information processing for some purpose, and to divorce this from the subjective consciousness with which it is humanly associated.

2.1 The Background to Turing's 1936 Paper

Alan Turing's remarkable 1936 paper, "On Computable Numbers, with an Application to the *Entscheidungsproblem*", introduced the first model of an all-purpose, programmable digital computer, now universally known as the *Turing machine*. And the paper, as noted above, gives the impression that this model is inspired by considering the potential operations of a human "computer". Yet the title and organisation of the paper suggest instead that Turing is approaching the topic from the direction of fundamental issues in the theory of mathematics, rather than any abstract analysis of human capabilities. It will be useful to start with an overview of two essential components of this theoretical background.

Peter Millican, *Alan Turing and Human-Like Intelligence* In: *Human Like Machine Intelligence.* Edited by: Stephen Muggleton and Nick Charter, Oxford University Press. © Oxford University Press 2021. DOI: 10.1093/oso/9780198862536.003.0002

The first of these components is Georg Cantor's pioneering work on the *countability* or *enumerability* of various infinite sets of numbers: the question of whether the elements of these sets could in principle be set out—or *enumerated*—in a single list that contains every element of the set at least once. Cantor had shown in 1891 that such enumeration of *rational* numbers (i.e. fractions of integers) is indeed possible, since they can be exhaustively ordered by the combined magnitude of their numerator and denominator.[1] *Real* numbers, however, cannot be enumerated, as demonstrated by his celebrated *diagonal* proof, which proceeds by *reductio ad absurdum*. Focusing on real numbers between 0 and 1 expressed as infinite decimals,[2] we start by assuming that an enumeration of these *is* possible, and imagine them laid out accordingly in an infinite list R (so we are faced with an array which is infinite both horizontally, owing to the infinite decimals, and vertically, owing to the infinite list). We then imagine constructing another infinite decimal α by taking its first digit $\alpha[1]$ from the first real number in the list r_1 (so $\alpha[1] = r_1[1]$), its second digit from the second real number in the list ($\alpha[2] = r_2[2]$), its third digit from the third real number in the list ($\alpha[3] = r_3[3]$), and so on. Thus α is the infinite decimal that we get by tracing down the diagonal of our imagined array: in every case α has its nth digit in common with r_n. We now imagine constructing another infinite decimal number β from α, by systematically changing every single digit according to some rule (e.g. if $\alpha[n] = 0$, then $\beta[n] = 1$, else $\beta[n] = 0$). For any would-be enumeration R, this gives a systematic method of constructing a number β whose n^{th} digit $\beta[n]$ must in every case be *different* from the nth digit of r_n. Thus β cannot be identical with any number in the list, contradicting our assumption that R was a complete enumeration, and it follows that no such enumeration is possible.

The second essential component in the background of Turing's paper is David Hilbert's *decision problem* or *Entscheidungsproblem*: can a precise general procedure be devised which is able, in finite time and using finite resources, to establish whether any given formula of first-order predicate logic is provable or not? A major goal of Hilbert's influential programme in the philosophy of mathematics was to show that such *decidability* was achievable, and his 1928 book with Wilhelm Ackermann even declared that the *Entscheidungsproblem* should be considered the main problem of mathematical logic (p. 77). It accordingly featured prominently in Max Newman's Cambridge University course on the Foundations of Mathematics, attended by Alan Turing in spring 1935. But by then Gödel's incompleteness theorems of 1931—also covered in Newman's course—had shown that two other major goals of Hilbert's programme (proofs of *consistency* and *completeness*) could not both be achieved, and Turing's great paper of 1936 would show that decidability also was unachievable.

A potentially crucial requirement in tackling the *Entscheidungsproblem*—especially if a negative answer is to be given to the question of decidability—is to pin down exactly what types of operation are permitted within the would-be general decision procedure.

[1] For example, 1/1 (sum 2); 1/2, 2/1 (sum 3); 1/3, 2/2, 3/1 (sum 4); 1/4, 2/3, 3/2, 4/1 (sum 5); and so on—every possible fraction of positive integers will appear somewhere in this list. To include negative fractions of integers, we could simply insert each negative value immediately after its positive twin.

[2] Or alternatively *binimals*, as discussed below.

For without some such circumscription of what is permissible (e.g. ruling out appeal to an all-knowing oracle or deity, if such were to exist), it is hard to see much prospect of delimiting the range of what can theoretically be achieved. An appropriate limit should clearly prohibit operations that rely on unexplained magic, inspiration, or external intervention, and should include only those that are precisely specifiable, performable by rigorously following specific instructions in a "mechanical" manner, and reliably yielding the same result given the same inputs. A procedure defined in these terms is commonly called an *effective method*, and results thus achievable are called *effectively computable.* These concepts remain so far rather vague and intuitive, but what Turing does with the "computing machines" that he introduces in §1 of his 1936 paper is to define a precise concept of effective computability in terms of what can be achieved by a specific kind of machine whose behaviour is explicitly and completely determined by a lookup table of conditions and actions. Different tables give rise to different behaviour, but the scope of possible conditions and actions is circumscribed precisely by the limits that Turing lays down.

2.2 Introducing Turing Machines

As its title suggests, the 1936 paper starts from the concept of a *computable* number:

> "The 'computable' numbers may be described briefly as the real numbers whose expressions as a decimal are calculable by finite means." (p. 58)

Turing's terminology is potentially confusing here, in view of what follows. The numerical expressions he will actually be concerned with express real numbers *between 0 and 1*, interpreted in *binary* rather than *decimal* (i.e. sequences of "0" and "1", following an implicit *binary point*, as opposed to sequences of decimal digits following a decimal point). To provide a distinctive term for such binary fractions, let us call them *binimals*. For example, the first specific example that Turing gives (§3, p. 61)[3] generates the infinite sequence of binary digits:

0 1 0 1 0 1 0 1 0 1...

which is to be understood as expressing the recurring binimal fraction $0.\overline{01}$, numerically equivalent to 1/3.[4] That the binimal recurs to infinity is no difficulty: on the contrary,

[3] In what follows, references of this form, citing section and page numbers, are always either to the 1936 paper or—later—to the 1950 paper. Note also that for convenience, all page references to Turing's publications are to the relevant reprint in Copeland (2004).

[4] The "1" in the second binimal place represents 1/4, and the value of each subsequent "1" is 1/4 of the previous one. So we have a geometric series 1/4 + 1/16 + 1/64 + ... whose first term is 1/4 and common ratio 1/4, yielding a sum of 1/3 by the familiar formula $a/(1\text{-}r)$.

all of Turing's binimal expressions will continue to infinity, whether recurring (as in the binimal for 1/2, i.e. 0.1000 . . .) or not (as in the binimal for $\pi/4$).[5]

That Turing's binimal expressions continue to infinity may well seem surprising, however, given his declared aim to explore those that are *computable*, glossed as "calculable by finite means". At the very beginning of §1 of his paper he acknowledges that the term "requires rather more explicit definition", referring forward to §9 and then commenting "For the present I shall only say that the justification lies in the fact that the human memory is necessarily limited." (p. 59). In the next paragraph he compares "a man in the process of computing a real number to a machine which is only capable of a finite number of conditions". So the finitude he intends in the notion of *computable numbers* does not apply to the ultimate extent of the written binimal number, nor therefore to the potentially infinite *tape*—divided into an endless horizontal sequence of *squares*—on which the digits of that number (as well as intermediate workings) are to be printed. Rather, what is crucial to Turing's concept of computability "by finite means" is that the choice of behaviour at each stage of computation is tightly defined by a finite set of machine memory *states*,[6] a finite set of *symbol types*, and a limited range of resulting *actions*. Each possible combination of state and symbol—the latter being read from the particular square on the tape that is currently being *scanned*—is assigned a specific repertoire of actions. Computation takes place through a repeated sequence of scanning the current square on the tape, identifying any symbol (at most one per square) that it contains, then performing the action(s) assigned to the relevant combination of current state and symbol.

For theoretical purposes, the actions for each state/symbol combination are very tightly constrained, limited to printing a symbol (or blank) on the current square, moving the scanner one square left or right, and changing the current state. For much of his paper, however, Turing slightly relaxes these constraints, allowing multiple printings and movements, as in the example illustrated below. This shows the machine defined by a table specified in §3 of the 1936 paper (p. 62), running within a Turing machine simulator program.[7] Starting with an empty tape in state 1 (or "b" in Turing's paper), the machine prints the sequence of symbols we see at the left of the tape ("P@" prints "@"; "R" moves right; "P0" prints "0"), then moves left twice ("L,L") to return to the square containing the first "0", before transitioning into state 2. Next, finding itself now scanning an "0" in state 2, it transitions into state 3. Next, still scanning an "0" but now in state 3, it moves right twice and stays in state 3. Then, scanning a blank square ("None") in state 3, it prints a "1" and moves left, transitioning into state 4. And so on.

[5] Note that a decimal or binimal will recur if, and only if, it represents a *rational* number, i.e. a fraction of integers.

[6] Turing's term for what is now generally called a *state* is an "m-configuration" (§1, p. 59).

[7] By far the best way of familiarising oneself with the operation of Turing machines is to see them in action. The Turing machine simulator illustrated here is one of the example programs built into the *Turtle System*, freely downloadable from www.turtle.ox.ac.uk. Within the simulator program, this particular machine table is available from the initial menu, which also includes some other relevant examples taken from Petzold's excellent book on the 1936 paper (2008).

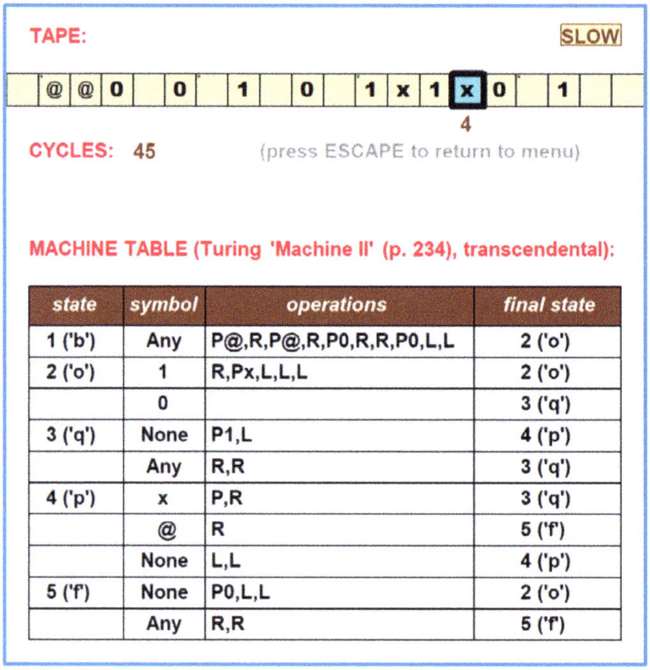

state	*symbol*	*operations*	*final state*
1 ('b')	Any	P@,R,P@,R,P0,R,R,P0,L,L	2 ('o')
2 ('o')	1	R,Px,L,L,L	2 ('o')
	0		3 ('q')
3 ('q')	None	P1,L	4 ('p')
	Any	R,R	3 ('q')
4 ('p')	x	P,R	3 ('q')
	@	R	5 ('f')
	None	L,L	4 ('p')
5 ('f')	None	P0,L,L	2 ('o')
	Any	R,R	5 ('f')

A Turing Machine in Action

Turing's machine table has been cleverly designed to print out an infinite binimal sequence, which has the interesting property of never recurring:

0 0 1 0 1 1 0 1 1 1 0 1 1 1 1 0 1 1 1 1 1 0 . . .

Adding an initial binimal point, this yields a binimal number which is clearly *irrational* (precisely because it never recurs), but which has been shown to be *computable* in the sense that there exists a Turing machine which generates it, digit by digit, indefinitely.[8] This notion is trivially extendable to binimals that also have digits before the binimal point. And in this extended sense it is clear—given that numerical division can be mechanised on a Turing machine—that all rational numbers are computable, so it now follows that *the computable numbers are strictly more extensive than the rational numbers*. Indeed, Turing will go on later, in §10 of the paper, to explain that all of the familiar irrational numbers, such as π, e, $\sqrt{2}$, and indeed all real roots of rational algebraic equations, are computable.

[8] To avoid reference to the infinite sequence of digits, we can formally define a binimal number as computable if and only if there exists a Turing machine such that, for all positive integers n, there is some point during the execution of that machine at which the tape will display the subsequence consisting of its first n digits.

2.3 The Fundamental Ideas of Turing's 1936 Paper

Accepting for the moment Turing's key claim that the scope of what is calculable by his computing machines does indeed coincide with the desired notion of *computability*, we can readily see how the work of Cantor could have provided clear inspiration for his subsequent method of proceeding. To start with, all these machines are by definition entirely deterministic,[9] so each can compute, in binimal form, at most one computable number. They succeed in computing such a number if and only if—when left to proceed indefinitely—they would continue to generate a potentially infinite sequence of binary digits on the tape. Turing calls these machines "circle-free" (§2, p. 60), in contrast to a "circular" machine that stops generating binary digits (for example by getting into a non-printing loop). But then, since—by Turing's key claim—any computable number must correspond to at least one machine table that is capable of computing it, there cannot be more computable numbers than there are circle-free machine tables.[10]

With his thinking apparently deeply informed by Cantor's methods, it seems plausible that Turing at this point would have quite quickly seen the potential for paradox. For on the one hand, since his machine tables can straightforwardly be reduced to a linear textual form, and lines of text can be ordered by length and then lexicographically (i.e. quasi-alphabetically, acknowledging both letters and other symbols), it immediately follows that machine tables, and hence the numbers they are capable of generating, must be enumerable—we simply list the numbers according to the ordering of the tables that generate them.[11] But on the other hand, since this list C of computable binimals will produce exactly the kind of infinite array from which Cantor's diagonal argument arose, it is very natural to wonder whether a similar diagonal argument would work here, to demonstrate by *reductio ad absurdum* that the supposed infinite list of computable numbers C *cannot* be a complete enumeration. A puzzling contradiction seems to be in prospect.

Following this thread leads directly to one of the great innovations in Turing's paper—the Universal Turing Machine. A diagonal argument can generate a paradox here only if the application of that argument is *computable*, for there is no contradiction in defining a real number that is not on list C, if that number is not itself computable. Hence the obvious next step is to explore how one might attempt to compute the paradoxical number, which Turing calls β (§8, p. 72). This requires (a) iterating through all possible machine tables in some appropriate order; (b) identifying in turn those that are circle-free; then (c) generating the nth digit from the nth circle-free table and adjusting it according to the same rule that we used in the case of Cantor's argument (i.e. if $c_n[n] = 0$, then $\beta[n] = 1$, else $\beta[n] = 0$). If these three steps are all achievable by computational methods, then we shall have a genuine paradox: β will be computable, and yet absent from the complete enumeration of computable numbers.

[9] Turing explicitly states that his paper considers only "automatic machines" (§2, p. 60).

[10] In fact any computable number can be generated by an infinite number of machine tables, since arbitrary irrelevant states can be added without affecting the behaviour (see §5, p. 68).

[11] Note that it does not matter if computable numbers occur in the list more than once, as long as all are present.

Of the three stages involved here, the most straightforward is to iterate through all possible machine tables. As noted earlier, any machine table can be reduced to a linear textual form. Then the individual symbols that occur in this linear text can be systematically translated as decimal digits (or sequences of digits), thus reducing the entire table to a (very large) integer. Translation in the reverse direction is also fairly easy, so it is in principle unproblematic to iterate up through the positive integers, identifying those that are potential "description numbers" of Turing machine tables.[12]

Having thus mechanised the identification of each possible table in turn, we now need to be able to generate the nth digit from the nth relevant table, and this is where we require a Universal Turing Machine: a machine which can simulate the operation of any given machine table so as to generate that n^{th} digit.[13] Turing's proof that such a universal machine is possible (in §§6–7, pp. 68–72 of his paper) was an intellectual *tour de force* and a landmark in the theory of computation, proving for the first time the possibility of a universal programmable computer. But envisaging the *possibility* of such a machine did not require any comparable effort of imagination: as we have seen, it arose naturally from the Cantorian context of his investigation into computable numbers.

Stages (a) and (c) of the paradox-generating process have turned out to be achievable; hence if paradox is to be avoided, it must fail at stage (b), which involves identifying whether a given machine table is circle-free. Turing himself draws this conclusion early in §8 of his paper (p. 72), but he also acknowledges that this indirect argument might leave readers unsatisfied, and he accordingly goes on to explore in more detail how the generation of β is bound to fail. The crucial difficulty arises when the hypothetical checking machine—which has been designed to test, in turn, whether each of the enumerated machine tables is circle-free—comes to test its own machine table (p. 73). This is essentially the same point that is now very familiar from proofs of the unsolvability of the Halting Problem: the supposed halting oracle has to fail when it comes to test (a slightly modified version of) itself.[14]

Turing has now identified something that is in general uncomputable: whether an arbitrary machine table is, or is not, circle-free. And from this point in the paper he accordingly changes focus from the domain of Cantor—enumerability of computable

[12] Turing explains description numbers in §5 of his paper, entitled "Enumeration of computable sequences" (pp. 66–8). The idea of translating complex formulae—and even sequences of formulae—into single large integers (and back again) would already have been very familiar to Turing through his study of Gödel's theorems.

[13] Note that Turing's initial machines, such as the one illustrated in the image of the simulator above, all start off with a completely blank tape, and then generate the relevant binimal number rightwards along the tape from the starting position. The Universal Machine, however, begins with a tape that already contains, to the left, an encoded version of the machine table that is to be simulated. The Universal Machine then performs its simulation by constantly referring back to that encoded table to work out what needs to be done at each stage.

[14] Suppose—for *reductio*—that $H(P,T)$ is a program that infallibly tests whether any arbitrary program P will halt given input T, outputting 'Yes' or 'No' accordingly (and then halting). We can create a paradoxical program K from H, by replacing "print('Yes')" with a non-terminating loop such as "repeat print('Yes') until 0=1", for then a positive halting verdict will fail to halt, while a negative halting verdict still halts. And hence the result of $K(K,K)$ cannot consistently be assigned: it should halt if and only if it doesn't halt. Thus the supposition that program H exists leads to a contradiction: there can be no such program.

numbers and diagonalisation—to the domain of Hilbert and his decision problem. The "application to the *Entscheidungsproblem*" promised by Turing's title requires him to translate his result about the uncomputability of machine behaviour into some parallel result about the undecidability of formulae in predicate logic. As a first step he neatly proves an important lemma, that since there is no computable test for whether an arbitrary machine is circle-free, nor can there be a computable test for whether an arbitrary machine will ever print a given symbol such as "0" (§8, pp. 73–4). The remainder of the task, which he postpones to §11, is technically tricky but conceptually quite straightforward. First, having defined appropriate predicates, he shows how, given any machine M, it is possible to construct a predicate formula Un(M) which states, in effect, *that M will at some point print "0"* (pp. 84–5). He goes on to show that Un(M) will be *provable* if and only if M *does indeed* at some point print "0". But then, since there is no "general (mechanical) process" for determining whether M ever prints "0", it follows that there can be no such "general (mechanical) process for determining whether Un(M) is provable". "Hence the Entscheidungsproblem cannot be solved" (p. 87).

2.4 Justifying the Turing Machine

Sandwiched between §8, where Turing proves his important lemma, and §11, where he applies it to the *Entscheidungsproblem*, are two sections aiming to justify the adequacy of his model of computation, by showing "that the 'computable' numbers [in the technical sense defined by his machines] include all numbers which would naturally be regarded as computable" (§9, p. 74). Turing offers arguments "of three kinds", the first of which—described as "A direct appeal to intuition" (p. 75)—gives the best evidence that he is basing his machines on an abstraction from the operations of a human "computer". Now he expands on ideas briefly sketched in §1 (p. 59), where we saw that he appeals to the finitude of human memory as justification for insisting on "finite means" of calculability. Here in §9 (pp. 75–7) he gives a far more elaborate argument, to the effect that his tape-based machines can, in principle, mimic any kind of systematic calculation that can be performed by a human "computer". He argues in turn that a tape divided into squares provides an appropriate simplification of a human notebook; that there should be a bound to the number of squares that can be observed at any one moment; and that the symbols used in calculation and the number of possible "states of mind" must be finite in number, reflecting our limited recognitional capacities, because "If we were to allow an infinity of symbols . . . [or] an infinity of states of mind, some of them will be 'arbitrarily close' and will be confused" (pp. 75–6). He then goes on to explain how "the operations performed by the [human] computer" can plausibly be split up into "simple operations", each of which "consists of some [elementary] change of the physical system consisting of the computer and his tape", involving no more than one change of symbol, a move to observe an adjacent square, or a change of state of mind. Thus we reach the design of the Turing machine, apparently starting from an abstract analysis of how a human "computer" operates.

Turing's second kind of argument for the adequacy of his model of computation involves showing the equivalence of the resulting notion of computability with that of other established definitions. In the remainder of §9 he explains how a Turing machine can be constructed "which will find all the provable formulae" of a systematised version of "the Hilbert functional calculus", or what we now know as first-order predicate logic (pp. 77–8). In the Appendix to the paper (pp. 88–90), he goes on to prove, in outline, that his notion of computability also coincides with Alonzo Church's concept of effective calculability (or λ-definability).

Turing's third kind of argument occupies §10 of the paper, appropriately entitled "Examples of large classes of numbers which are computable" (pp. 79–83). These classes encompass various combinations and iterations of computable functions, the root of any computable function that crosses zero, the limit of any "computably convergent sequence", π, e, and all real algebraic numbers. After all this, the reader is left with an appreciation of the comprehensive power of Turing machines, rendering plausible the claim that they can indeed circumscribe the appropriate boundaries of our intuitive notion of effective computability. Indeed this claim, that the functions that are effectively computable are to be identified with those that are Turing-machine computable—or equivalently, λ-definable (Church) or general recursive (Gödel)—is now known as the *Church-Turing thesis*, and widely accepted on the basis of Turing's arguments.

2.5 Was the Turing Machine Inspired by Human Computation?

Andrew Hodges' magnificent 1983 biography, which helped to bring Turing to public prominence, describes the early pages of §9 of the 1936 paper as "among the most unusual ever offered in a mathematical paper, in which [Turing] justified the definition [of a computing machine] by considering what *people* could possibly be doing when they 'computed' a number ..." (p. 104). According to Mark Sprevak, "A Turing machine is an abstract mathematical model of a human clerk.... Turing wanted to know which mathematical tasks could and could not be performed by a human clerk." (2017, p. 281). And beyond mere modelling, Robin Gandy credits Turing with having *proved* a theorem of similarity between human and machine computability: "He... considers the actions of an abstract human being who is making a calculation... [and] the limitations of our sensory and mental apparatus.... Turing easily shows that the behavior of the computor can be exactly simulated by a Turing machine.... [His] analysis ... *proves a theorem ... Any function which can be calculated by a human being following a fixed routine is computable.*" (1988, pp. 81–3).

Turing's later publications about machine intelligence—which we shall come to shortly—make it especially tempting to see the Turing machine as inspired by human computation and the desire to model it. Hodges accordingly considers it "striking ... that the Turing machine, formulated 14 years before the 'Turing Test', was also based on a principle of imitation. The 'machine' was modelled by considering what a human being could do when following a definite method." (2009, p. 14). This can suggest a unity of purpose linking the mathematical analysis of the 1936 paper to the more philosophical discussions that would emerge over a decade later:

> "The problem of mind is the key to 'Computable Numbers'.... [Turing] wished from the beginning to promote and exploit the thesis that all mental processes—not just the processes which could be explicitly described by 'notes of instructions'—could be faithfully emulated by logical machinery" (Hodges 1988, pp. 6, 8).

Hodges here goes further than most commentators on Turing's paper, but many have found it plausible that modelling the limits of algorithmic *human* thinking was the principal driver behind the design of the Turing machine, even if Turing's main aim in doing so was not to provide a model of human thinking for its own sake, but rather to circumscribe the notion of an effective method on the way to tackling the *Entscheidungsproblem.* Gandy apparently takes this view: "I suppose, but do not know, that Turing, right from the start of his work, had as his goal a proof of the undecidability of the Entscheidungsproblem" (1988, p. 82).

Against all this, I would like to suggest that Turing's design of his computing machines was not primarily intended to model human thinking, but nor did it derive primarily from his desire to resolve the *Entscheidungsproblem.* Instead, I believe that Turing's enquiry—and his conception of computing machines—started from an interest in the notion of *computable numbers* themselves. Having devised his machines and appreciated their power, he then saw the need to provide some argument for their theoretical generality, but it was probably quite late in the process that Turing turned to doing this in any detail (in §9 of his paper, as we have seen). Moreover their "application to the *Entscheidungsproblem*" was not, I suspect, foreseen in advance, but became apparent during the course of his investigation into computable numbers.

This interpretative hypothesis derives obvious support from the title of Turing's paper, and from its ordering and organisation (bearing in mind that it was written before the days of wordprocessors, when restructuring and reordering could be tedious in the extreme). But more substantially, there is a clue to the origin of Turing's thought in a footnote reference that has been generally overlooked, but which coheres extremely well with the explanation above of how the fundamental ideas of the 1936 paper naturally emerge once an investigation into computable numbers is under way.

The footnote in question is attached to the end of the first sentence of §8 (entitled "Application of the diagonal process"):

> "It may be thought that arguments which prove that the real numbers are not enumerable would also prove that the computable numbers and sequences cannot be enumerable.[4]" (p. 72)

Footnote 4 itself reads "*Cf.* Hobson, *Theory of functions of a real variable* (2nd ed., 1921), 87, 88." E. W. Hobson was G. H. Hardy's immediate predecessor as Sadleirian Professor of Pure Mathematics at Cambridge, retiring in 1931. His *Theory of Functions* was the first

English book covering Lebesgue integration and measure, and was highly influential on teaching practice, so it should be no surprise that Turing, who arrived at Cambridge as an undergraduate in Mathematics that same year, was familiar with it.[15]

On following up this reference, one might naturally expect to find an exposition of Cantor's familiar diagonal argument.[16] But that argument occurs in Hobson's book at pages 82–3, and the discussion over pages 87–8 is quite different. Here Hobson is concerned with the *definability* of numbers, particularly in the light of paradoxes such as those published in 1905 by Julius König and Jules Richard. This discussion in Hobson's book draws on papers that he had presented to the London Mathematical Society in January and November 1905 (the latter in response to König's September publication). The initial paper quickly stimulated several responses in the Society's *Proceedings*, from G. H. Hardy, A. C. Dixon, Bertrand Russell, and Philip Jourdain.[17] Moreover it is mentioned (somewhat approvingly), together with König's paper and Dixon's response, in a footnote to Whitehead and Russell's *Principia Mathematica*.[18]

Hobson's 1921 discussion of *definable* numbers bears striking parallels with Turing's 1936 discussion of *computable* numbers. Hobson emphasises repeatedly that finite definitions from any "given stock of words and symbols" must be enumerable, and alludes (albeit disapprovingly) to König's inference that real numbers "fall into two classes ... [those] capable of finite definition, and those ... inherently incapable of finite definition" (p. 87). He also draws explicit attention to Cantor's diagonal argument as a method of defining further numbers that were not in the original enumerable set, and although he himself avoids any suggestion of paradox here, he does mention both König and Richard in a footnote (along with his own initial 1905 paper, and *Principia Mathematica*). Hobson takes diagonalisation to show "that there exists, and can exist, at any time, no stock of words and symbols which cannot be increased for the purpose of defining new elements of the continuum", but he denies "that there exists any element of the continuum that is inherently incapable of finite definition" (pp. 87, 88).[19]

Hobson's discussion is also somewhat suggestive of another prominent theme in Turing's paper, that of defining or computing a number by constructing it digit-by-digit.

[15] I am grateful to Patricia McGuire, Archivist of King's College, for ascertaining that Hobson's second edition was acquired by the college library on January 28 1921.

[16] Indeed, my own primary interest in consulting it was to discover whether Turing had encountered the Cantorian argument as expressed in binary or decimal form—Hobson uses decimals. It was a delightful surprise to find, instead of Cantor's argument, clear corroboration for my interpretative hypothesis.

[17] All, incidentally, Cambridge men. Hardy and Russell, at least, were well known to Turing; Jourdain died in 1919; Dixon retired in 1930 (from Queen's University Belfast to Middlesex), but from 1931 until 1933 was President of the London Mathematical Society, which would later publish Turing's 1936 paper.

[18] In the second edition of 1927, this is at p. 61 of Volume 1. The footnote is to a numbered paragraph explaining König's paradox, and immediately followed by another explaining Richard's paradox. For discussion, and a reprint of Russell's response to Hobson, see Russell's *Collected Papers, Volume 5* (2014), chapter 2.

[19] Hobson's mathematical argument on p. 88 appears to take for granted, without any obvious basis, that a sequence of definable elements must itself be definable (and hence that its limit must also be definable). Fan (2020) explicates Hobson's thinking in terms of the requirement for a "*norm* by which [an] aggregate is defined" (p. 131); see also Fan's §5 (pp. 135–7) on "Hobson's Way Out" and §6 (pp. 137–8) which includes some remarks on Turing.

He couches Cantor's argument in terms of "a set of rules by means of which the mth digit of the nth number [in an enumeration] can be determined" (p. 82). And he provides a "finite definition" of any number that is the limit of a convergent sequence $\{x_n\}$, in terms of successive m^{th} digits each of which is "defined as that digit which is identical with the mth digit of an infinite number of the elements" of the sequence (p. 88).

There is more than enough here to give a plausible account of the initial inspiration for Turing's landmark paper. Hobson starts by raising the question of whether there is a tenable distinction between *definable* and *non-definable* numbers. But Turing—perhaps in the wake of Newman's lecture course with its focus on effective "mechanical" methods, and perhaps in order to pin the problem down in terms of concrete calculation of numbers digit-by-digit (as just discussed)—now interprets *definition* in terms of *calculability*.[20] He then designs a calculating machine whose purpose is precisely to generate numbers on an endless tape, something very different from what any human "computer" or "clerk" is likely to be doing, but exactly on target in Hobson's context.[21] Next comes the idea of enumerating the possible machines, a close parallel to the enumeration of definitions that is key to the paradoxes of König and Richard. The parallel to the Richard paradox runs even deeper, since this involves explicit diagonalisation, identifying the nth digit of the nth legitimate definition, just as Turing's diagonal argument—applied in the paragraph immediately following the Hobson footnote—involves identifying the nth digit generated by the nth circle-free machine.[22] Both diagonal processes, moreover, crucially raise the issue of testing for legitimacy: in Richard's case, whether an enumerated string is a genuine number definition; in Turing's case, whether a machine description number is genuinely that of a circle-free machine. But this last issue would presumably have occurred to Turing even without any prior example, for as we saw, it follows directly from the logic of his argument—on pain of paradox—that whether a machine is circle-free has in general to be uncomputable.

Understanding Turing's process of thought as arising from a concern with *definable* numbers coheres well with the structure of his paper, and also fits naturally into what we know of his intellectual context. The outline above also explains why Turing was able

[20] Turing's diagonal process, however, ultimately forces a distinction between the two notions, as anticipated in the second paragraph of his paper: "The computable numbers do not . . . include all definable numbers, and an example is given of a definable number [β as described earlier] which is not computable" (1936, p. 58). I suspect that recognition of this came *after* Turing had pursued the potential paradox that arises from identification of the two notions, but in any case, this early prominence of the issue within the finished paper somewhat corroborates my hypothesis that he started from an interest in *definable* numbers.

[21] Hodges (2013) suggests another possible inspiration for this kind of machine, namely, Turing's interest in "normal numbers", on which his King's College friend David Champernowne had published a paper (1933), and the topic of a note that Turing himself wrote soon after submitting the 1936 paper (drafted on the back of its typescript). A normal number is one whose digits and groups of digits are all uniformly distributed in the infinite limit.

[22] Turing had a deep interest in philosophy of mathematics, even extending to giving a Moral Sciences Club paper on the subject in December 1933 (see Hodges 1983, pp. 85–6). So quite independently of Hobson's textbook, he would already have known about Richard's paradox, whose salience was very clear to those working on the foundations of mathematics in the 1930s. For example, the paradox is mentioned in *Principia Mathematica* (see note 18 above); Gödel's famous paper of 1931 remarks: "The analogy between this result and Richard's antinomy leaps to the eye" (1962, p. 40); and Alonzo Church's (1934) is devoted to the paradox.

to work so fast as to astonish Newman by presenting him with the completed draft of his paper in April 1936 (Hodges 1988, p. 3). Turing later told Gandy "that the 'main idea' of the paper came to him when he was lying in Grantchester meadows in the summer of 1935" (Gandy 1988, p. 82). This suggests that there was one leading idea that crystallised everything in Turing's mind, rather than a succession of original points.[23] The hypothesis sketched above is entirely consistent with this, because it builds so much of his argument from materials that were already familiar to him. If that hypothesis is correct, then Turing's "main idea" in Grantchester meadows might have been his initial recognition that there was potential for a paradox regarding computability, structurally parallel to the familiar paradoxes of definability. That, however, was a relatively small step in the context, simply linking Hobson's discussion (and/or Richard's paradox) with the idea of "mechanical" computability emphasised in Newman's lectures of spring 1935. Far more substantial inspiration would have been required for Turing to come up with his entirely innovative design for a tape-based and table-driven computing machine capable of generating endless binimals: that design, I strongly suspect, was his "main idea".

2.6 From 1936 to 1950

In the decade following his 1936 paper, Turing experienced the most dramatic confirmation of his confidence in the powers of automated computation, working for much of that time at Bletchley Park cracking Nazi codes, to the huge benefit of the allied war effort. After the war, he was recruited by the National Physical Laboratory to design the Automatic Computing Engine (ACE), of which an initial version (Pilot ACE) would be operational by 1950. But by May 1948, he was instead employed at Manchester University, having been repelled by institutional politics and hiatus at the NPL, and attracted by the offer of a job from Newman, who had in 1945 been appointed Head of the Manchester Mathematics Department. This gave Turing, as Deputy Director of the Computing Machine Laboratory, the opportunity to work on the world's first electronic stored-program digital computer, the "Manchester Baby", and subsequently on the first such computer to be manufactured commercially, the Ferranti Mark I. Now he had the practical prospect of indulging his intense interest in machine intelligence, notably by developing—with his college friend David Champernowne—a chess-playing program called *Turochamp* (though the Ferranti proved insufficiently powerful to run the completed program).[24]

Perhaps more than anyone else in the world at this time, Turing was aware of the amazing potential of digital computers. The theoretical limits that he had identified in

[23] Gandy continues: "The 'main idea' might have been either his analysis of computation, or his realization that there was a universal machine, and so a diagonal argument to prove unsolvability." (1988, p. 82). But the latter seems unlikely, for we have seen that diagonalisation was a standard part of the mathematical repertoire, while the need for a universal machine—though admittedly not the method of implementing it—is fairly easy to appreciate once the requirement of mechanising a diagonal computing process has been noticed.

[24] There is now a wealth of material on Turing's history over this period. See for example Hodges (1983), pp. 314–415; Copeland (2004), pp. 353–77, 395–401; Copeland *et al.* (2017), pp. 199–221.

1936 were of little practical consequence now; the far more significant result was that even his simple machines of 1936 could, in principle, compute anything that could be programmed. Given the right software, therefore, a powerful enough *universal* machine could mimic any computational procedure that could be taught explicitly by one person to another. And far more practical hardware was rapidly being developed—on both sides of the Atlantic—that could execute such algorithms with ever-increasing speed and reliability. A computer revolution could only be a matter of time.

But most people were, of course, quite unaware of this progress, and indeed deliberately kept ignorant of their immense debt to Bletchley Park and its machines. It was to be another three decades before computers would start to become widely familiar, and in the late 1940s the idea that an electronic machine could "think" in anything like the way that we do would have generally seemed preposterous, and even offensive. The British were still overwhelmingly Christian, believing in an immaterial soul that was the seat of our reason and consciousness, and which could survive destruction of the body. Some pressures on this consensus were beginning to show, especially amongst intellectuals aware of the Enlightenment challenges and of our evolutionary place in nature. But the manifest evils of war had, if anything, made religion more salient to those who had been bereaved, rather than generating widespread doubt about a benign creation. Not until the major social changes of the 1960s would the religious commitment of the British people start to be seriously undermined. Those changes included the repeal, in 1967, of the long-standing (and religiously-inspired) legal prohibition of male homosexuality, under which Turing was prosecuted in 1952 and which tragically led to his early death.

Against this background, Turing in the late 1940s began to argue the case for machine intelligence, with a particular focus on undermining what he saw as the dominant *prejudice* against computers. In 1947 he gave a lecture on the ACE to the London Mathematical Society, alluding to his 1936 results on the limitations of computers but emphasising explicitly—in his concluding paragraph—the idea of *fairness* in judging them:

> "It has . . . been shown with certain logical systems there can be no machine which will distinguish provable formulae of the system from unprovable. . . . Thus if a machine is made for this purpose it must in some cases fail to give an answer. On the other hand if a mathematician is confronted with such a problem he would search around a[nd] find new methods of proof . . . I would say that fair play must be given to the machine. Instead of it sometimes giving no answer we could arrange that it gives occasional wrong answers. But the human mathematician would likewise make blunders when trying out new techniques. It is easy for us to regard these blunders as not counting and give him another chance, but the machine would probably be allowed no mercy. . . . To continue my plea for 'fair play for the machines' when testing their I.Q. . . . , the machine must be allowed to have contact with human beings in order that it may adapt itself to their standards. The game of chess may perhaps be rather suitable for this purpose, as the moves of the machine's opponent will automatically provide this contact." (1947, pp. 393–4)

The following year, Turing completed a report for the Director of the National Physical Laboratory which was published only after his death. Entitled "Intelligent Machinery", it has been described by Jack Copeland as "the first manifesto of artificial intelligence" (2004, p. 401). The final numbered section, entitled "Intelligence as an emotional concept", runs as follows:

> "The extent to which we regard something as behaving in an intelligent manner is determined as much by our own state of mind and training as by the properties of the object under consideration. If we are able to explain and predict its behaviour or if there seems to be little underlying plan, we have little temptation to imagine intelligence. With the same object therefore it is possible that one man would consider it as intelligent and another would not; the second man would have found out the rules of its behaviour.
>
> It is possible to do a little experiment on these lines, even at the present stage of knowledge. It is not difficult to devise a paper machine which will play a not very bad game of chess. Now get three men as subjects for the experiment A, B, C. A and C are to be rather poor chess players, B is the operator who works the paper machine.... Two rooms are used with some arrangement for communicating moves, and a game is played between C and either A or the paper machine. C may find it quite difficult to tell which he is playing.
>
> (This is a rather idealized form of an experiment I have actually done.)" (1948, p. 431)

Here we have a clear anticipation of the Turing Test, applied to the particular case of chess. But notice how closely its setup is tied to the overcoming of *prejudice*—the extent to which our judgements of intelligence are subjectively dependent on our own assumptions, and even "emotional" in character. To guard against this, Turing's proposed experiment ensures fairness through a "blind" testing procedure, a procedure which he will later advocate explicitly as the appropriate method for making such judgements.[25]

2.7 Introducing the Imitation Game

Turing's famous *Mind* paper of 1950, "Computing Machinery and Intelligence", starts in a way that is both cavalier and rather confusing:

> "I propose to consider the question, 'Can machines think?' This should begin with definitions of the terms 'machine' and 'think'. The definitions might be

[25] Thus Turing's section heading need not be interpreted as saying that our judgements of intelligence *ought to be* made on an "emotional" basis, nor that intelligence is a *response-dependent* concept (as suggested by Proudfoot 2017, pp. 304–5). Turing's talk of our "temptation to imagine intelligence" suggests instead that when we ascribe intelligence, we take ourselves to be hypothesising something quite distinct from our own immediate feelings.

> framed so as to reflect so far as possible the normal use of the words, but this attitude is dangerous. If the meaning of the words 'machine' and 'think' are to be found by examining how they are commonly used it is difficult to escape the conclusion that the meaning and the answer to the question, 'Can machines think?' is to be sought in a statistical survey such as a Gallup poll. But this is absurd. Instead of attempting such a definition I shall replace the question by another, which is closely related to it and is expressed in relatively unambiguous words." (§1, p. 441)

One obvious objection is that although *the meanings of words* might plausibly be discovered from—or even determined by— their "normal use" (a view associated with Ludwig Wittgenstein, whom Turing knew well), that is very far from implying that *the answer to a question* framed in such words is to be found in a Gallup poll. Another obvious problem is with the whole notion of *replacing* one question by another: is this supposed to imply that the answer to the second can simply be carried over to the first? But how can this be guaranteed, since their meaning must presumably be different if the first question has ambiguities that are not reflected in the second? Turing never explains or even discusses these issues in respect of what he takes to be his two "closely related" questions. It is also somewhat ironic that his "relatively unambiguous" replacement question itself turns out to be highly ambiguous!

There is a great deal more vagueness, loose argument, and even humour to come throughout the paper, which falls far short of the rigour that would now be expected of a paper published in a leading philosophical journal. So when analysing it, we are well advised to bear in mind what Robin Gandy tells us about the context in which it was written:

> "The 1950 paper was intended not so much as a penetrating contribution to philosophy but as propaganda. Turing thought the time had come for philosophers and mathematicians and scientists to take seriously the fact that computers were not merely calculating engines but were capable of behaviour which must be accounted as intelligent; he sought to persuade people that this was so. He wrote this paper—unlike his mathematical papers—quickly and with enjoyment. I can remember him reading aloud to me some of the passages—always with a smile, sometimes with a giggle. Some of the discussions of the paper I have read load it with more significance than it was intended to bear." (Gandy 1996, p. 125)

This should counsel us against, for example, putting decisive weight on one individual passage that favours a very specific interpretation, when other passages point towards a different view. In some cases Turing's own position might not have been entirely clear, and sometimes he might simply have been careless in expressing it. In seeking to understand the Turing Test, therefore, literal reading of the text must be subject to the discretion of interpretative judgement.

Turing's new question—his replacement for "Can machines think?"—is set in the context of an "imitation game" involving a man A, a woman B, and an interrogator C—whose task is to identify, through written questions and answers, which of the other two (identified using the aliases "X" and "Y") is the man and which is the woman. The man's task is to deceive the interrogator into making the wrong identification by pretending to be a woman, while the woman attempts to assist the interrogator by being herself. In Turing's illustrative example, the interrogator asks "Will X please tell me the length of his or her hair?"—and X, who is actually the man, falsely answers "My hair is shingled, and the longest strands are about nine inches long" (§1, p. 441). Turing now imagines Y responding "I am the woman, don't listen to him!". (However, in later discussions of the game/test setup, the interrogations become one-to-one and interactions between the two "witnesses" disappear, as do the "X" and "Y" aliases.)

2.8 Understanding the Turing Test

The very short next paragraph—which ends §1 of the paper—introduces what we now know as the Turing Test. Since we shall be referring back to it in what follows, I mark it with "(⋆)":

> (⋆) "We now ask the question, 'What will happen when a machine takes the part of A in this game?' Will the interrogator decide wrongly as often when the game is played like this as he does when the game is played between a man and a woman? These questions replace our original, 'Can machines think?'" (§1, p. 441)

Turing had led us to expect one question to replace "Can machines think?", but now he offers two, though since the first of them does not have a yes/no answer, it seems most charitable to interpret that as mere framing for the second. More problematically, however, his description of this new game is seriously incomplete. Taken literally, it seems as though the machine is to be slipped into the game in place of the man, *attempting to imitate a woman*. But Turing says nothing to indicate whether or not the interrogator C is to be told of the change, yet if he or she is not told about it—and is left under the illusion that the other two participants are a man and a woman—then the supposed test for "thinking" becomes ludicrous.[26] Imagine, for example, what might happen if participant A, identified as X by the interrogator, now becomes a simple chatbot with a narrow range of answers tuned to the game context:

C: "Will X please tell me the length of his or her hair?"
A/X: "I want to you know that I am the woman."

[26] Hayes and Ford (1995, p. 972), Sterrett (2000, pp. 542, 545–52), and Traiger (2000, pp. 565–7) all apparently see it as an advantage that the interrogator should be expecting the two "witnesses" to be a man and a woman, supposedly making the contest more demanding and subtle. Saygin *et al.* (2000, pp. 466–7), though less explicit on the interrogator's expectations, likewise favour a gender-based test on somewhat similar grounds. But these discussions all overlook the risk that this setup can, on the contrary, dramatically *lower* the bar for the computer, since there is then no requirement that it should succeed in passing itself off as an *intelligent* woman.

B/Y: "I am the woman—don't listen to him."

C: "OK, Y, please tell me the length of *your* hair."

B/Y: "My hair is shingled, and the longest strands are about nine inches long."

A/X: "He's lying—I am the woman."

As this goes on, we can expect a sequence of intelligent answers from the woman (B), and mere repetitive claims of womanhood from the chatbot (A), but since the interrogator is under the impression that the two participants are a man and a woman, he or she can only conclude that one of the two is obsessive or mentally defective, with no obvious clue as to whether the imbecile is male or female (since either way, the answers from the intelligent participant are likely to be relevantly similar).[27] Perhaps this will make it a coin-toss whether he or she guesses correctly which is the genuine woman—in which case the chatbot will presumably perform better in this game than a typical man—but that hardly amounts to a serious test for intelligence![28]

Thus the new form of Turing's imitation game can only make reasonable sense if the interrogator is to be told that the decision is now between a human and a computer. But in that case, the question of gender becomes an irrelevant complicating factor: the interrogator is far more likely to focus on questions that attempt to distinguish person from machine, than questions that attempt to distinguish male from female (given that one of the participants is of neither sex). And indeed it is striking that in the rest of Turing's paper, not a single one of the questions he proposes is gender-related: they concern skill at poetry, arithmetic, and chess, with no hint of gender relevance (see especially §2, p. 442; §6.4, p. 452; and §6.5, p. 454).

The subsequent text of the paper also repeatedly makes clear that Turing intends his test to be focused on the human/computer rather than female/male decision.[29] In §2, entitled "Critique of the New Problem" and starting *immediately* after passage (⋆) quoted above, Turing six times talks explicitly of a "man"—even implying that the machine's obvious strategy is *to imitate a man*—and makes no mention whatever of women or the gender issue (pp. 442–3). A similar pattern continues for the remainder of the paper, with women mentioned only in the context of an imagined "theological objection" (§6.1, p. 449), while the words "man" or "men" occur a further 30 times. Mostly these words

[27] A more interesting chatbot in the same spirit might rant at length about the oppressive and binary nature of testing in general, and how testing for gender in particular is both patriarchal and problematically culture-relative, especially when focused on such trivia as length or styles of hair. Turing (1952, p. 495, quoted later) says that "the machine would be permitted all sorts of tricks", and one possible trick here would be to parody an obsessive feminist who says a great deal, but without actually giving any detailed attention to any of the questions.

[28] I am assuming here that a typical woman will generally outperform a typical man in this gender imitation game, so that a man is doing extremely well if he achieves a 50% probability of fooling an interrogator. But of course it might be possible that a talented man could outperform most women, perhaps by taking advantage of cognitive biases on the part of the interrogator (such as false sexist assumptions about male and female interests or abilities).

[29] For more detailed textual discussion coming to the same conclusion, see Piccinini (2000).

appear to be used gender-neutrally (as was then standard), but this again suggests that gender is quite irrelevant to the intended Turing Test.

Turing describes his test more carefully at the end of §5 when, having explained the universality of digital computers—and clarified that his concern is not just with *current* computers, but with *imaginable* future technology—he rephrases the key question as follows:

> "It was suggested tentatively [at the end of §3] that the question, 'Can machines think?' should be replaced by 'Are there imaginable digital computers which would do well in the imitation game?'... But in view of the universality property we see that... [this question]... is equivalent to this, 'Let us fix our attention on one particular digital computer *C*. Is it true that by modifying this computer to have an adequate storage, suitably increasing its speed of action, and providing it with an appropriate programme, *C* can be made to play satisfactorily the part of A in the imitation game, *the part of B being taken by a man*?'" (§5, p. 448, emphasis added)

Since gender has not featured in the discussion since §1, it seems clear that "man" here should be interpreted gender-neutrally and the test interpreted accordingly, as comparing human against computer, which is indeed how the Turing Test has most commonly been interpreted.

Another change here concerns the criterion for computer success. Whereas (⋆) suggested a comparison with the success rate of a man attempting to imitate a woman—which now seems of little relevance to the question at issue—§5 asks instead whether the computer "can be made to play satisfactorily the part of [a human impersonator] in the imitation game".[30] Given the setup of the game, the most obvious way of assessing whether its performance is indeed "satisfactory" is in terms of the probability (judged by observed frequency) with which an interrogator is fooled into wrongly identifying which of the two "witnesses" is the human, and which the computer. This is exactly the criterion that Turing appears to adopt when at the beginning of §6 he makes his famous 50-year prediction:

> "I believe that in about fifty years' time it will be possible to programme computers... to make them play the imitation game so well that an average interrogator will not have more than 70 per cent. chance of making the right identification after five minutes of questioning." (§6, p. 449)

[30] Copeland and Proudfoot (2009, p. 124; cf. Copeland 2017, p. 271) take (⋆) more seriously, claiming that "the man-imitates-woman game is... part of the protocol for scoring the test". On their view, "If the computer (in the computer-imitates-human game) does no worse than the man (in the man-imitates-woman game), it passes the test." Against this, we have already seen evidence that (⋆) is carelessly written, and if Turing indeed takes the man-imitates-woman game to be providing an important baseline for assessment, one would reasonably expect him to give some consideration to how well a typical man would do in that game, which he never does.

Once we are in the domain of statistical frequency, however, there is no need for continuing with the three-participant game.[31] For it now makes more sense to conduct the exercise using a sequence of individual *viva-voce* examinations, a format which Turing adopts after the first section of the paper (at §2 p. 442 and §6.4 p. 452). His subsequent works also move in this direction. In the 1951 lecture "Can Digital Computers Think?", Turing makes no mention of the competitive game, but speaks instead of "something like a viva-voce examination, but with the questions and answers all typewritten" (p. 484). And in a 1952 BBC radio discussion, he adds more rigour by bringing in a panel of judges rather than a single interrogator, and emphasises that objectivity demands repeated tests with a mixed population of people and machines:

> "I would like to suggest a particular kind of *test* that one might apply to a machine. You might call it a test to see whether the machine thinks, but it would be better to avoid begging the question, and say that the machines that pass are (let's say) 'Grade A' machines. The idea of the test is that the machine has to try and pretend to be a man, by answering questions put to it, and it will only pass if the pretence is reasonably convincing. A considerable proportion of a jury, who should not be expert about machines, must be taken in by the pretence. They ... [can] ask it questions, which are transmitted through to it: it sends back a typewritten answer ... [questions can be about] anything.... the machine would be permitted all sorts of tricks so as to appear more man-like, such as waiting a bit before giving the answer, or making spelling mistakes ... We had better suppose that each jury has to judge quite a number of times, and that sometimes they really are dealing with a man ... That will prevent them saying 'It must be a machine' every time without proper consideration. Well, that's my test.... It's not the same as 'Do machines think', but it seems near enough for our present purpose, and raises much the same difficulties." (Turing *et al.*, 1952, p. 495)

It may seem odd to take a radio discussion rather than a paper in *Mind* as authoritative, but this seems to provide Turing's most considered presentation of the Turing Test.

2.9 Does Turing's "Intelligence" have to be Human-Like?

The Turing Test certainly gives a superficial impression of advocating a human-centred understanding of "thinking" or "intelligence" (terms that Turing tends to treat as equivalent).[32] Some influential interpreters have even seen it as intended to provide an

[31] Especially given that Turing never suggests that there might be interaction between the human and the computer of the kind that he described in the gendered game (e.g. "Don't listen to that machine—I'm the human!").

[32] Turing's (1947) and (1948), aimed at mathematical audiences, refer repeatedly to *intelligence* rather than *thinking*, and although the 1948 report alludes to "our task of building a 'thinking machine'" (p. 420), the

"operational definition",[33] but this goes too far. Turing's very first sentence in the 1952 BBC discussion is "I don't want to give a definition of thinking" (p. 494). And in the 1950 paper itself, he seems explicitly to allow that machine intelligence could be very different from the human variety, thus implying that passing of the Turing Test is not *necessary* for machine intelligence:

> "May not machines carry out something which ought to be described as thinking but which is very different from what a man does? This objection is a very strong one, but at least we can say that if, nevertheless, a machine can be constructed to play the imitation game satisfactorily, we need not be troubled by this objection." (§2, p. 442)

Whether passing the test is *sufficient* to deserve the accolade of intelligence may depend on where the threshold is drawn. As we saw above, Turing's 1952 discussion is relatively guarded, suggesting we "say that the machines that pass are (let's say) 'Grade A' machines", and combining this with the vague threshold that they "only pass if the pretence is reasonably convincing" (p. 495). The 1950 paper might seem to imply a more precise threshold, whereby computers need to "play the imitation game so well that an average interrogator will not have more than 70 per cent. chance of making the right identification after five minutes of questioning" (§6, p. 449). But this is merely a *prediction* of what Turing believes will be possible "in about fifty years' time", and he nowhere commits himself to its significance.

For an example of what Turing apparently considers to be a *successful* test, we must look to the *viva-voce* dialogue in §6.4 of the 1950 paper, where he is responding to what he calls "The Argument from Consciousness":

> **"Interrogator:** In the first line of your sonnet which reads 'Shall I compare thee to a summer's day', would not 'a spring day' do as well or better?
>
> **Witness:** It wouldn't scan.
>
> **Interrogator:** How about 'a winter's day'. That would scan all right.
>
> **Witness:** Yes, but nobody wants to be compared to a winter's day.
>
> **Interrogator:** Would you say Mr. Pickwick reminded you of Christmas?
>
> **Witness:** In a way.

inverted commas suggest that he considers the term colloquial rather than precise. This diagnosis is confirmed by the 1950 paper and the radio broadcast (1951a), which both refer to *intelligence* in their title, yet both start with a quotation about *thinking*. And although machine *thought* is mentioned far more often in the 1950 paper that in the other pieces, it is striking that nearly every such reference is either explicitly quoted or implicitly put in the voice of an objector, while even the two exceptions (at §2, p. 442 and §6, p. 449) are indirect in nature. It therefore seems reasonable to conclude that Turing views *intelligence* as the more correct term, with *thought* as a colloquial alternative.

[33] For example Shannon and McCarthy (1956), p. v; Millar (1973), p. 595; Hodges (1983), p. 415; French (1990), pp. 11–12; Michie (1993), p. 29.

Interrogator: Yet Christmas is a winter's day, and I do not think Mr. Pickwick would mind the comparison.

Witness: I don't think you're serious. By a winter's day one means a typical winter's day, rather than a special one like Christmas." (p. 452)

Ignoring for the moment the context of Turing's (tendentious) response to the question of consciousness,[34] let us appreciate the force of his implicit point, that *if we encountered a machine capable of responding with this degree of sophistication, in response to comparably difficult questions chosen across a wide range of subjects by some independent group of interrogators, then it would seem unreasonable to withhold the attribution of "intelligence" to that machine.* If this is accepted, then the Turing Test provides a sort of existence proof of possibility. For here is a *conceivable* outcome that should be enough to persuade us of a machine's intelligence, and an outcome, moreover, of whose genuine *possibility* Turing is entirely confident.

Turing's confidence stems, of course, from his 1936 work on the *universality* of digital computers, a notion which he emphasises strongly in both the 1950 paper (§5, pp. 446–8) and his 1951 radio lecture (pp. 482–4).[35] This absolves him from any obligation to analyse the specific powers of any particular computer, given that:

> "we are not asking . . . whether the computers at present available would do well, but whether there are imaginable computers which would do well" (§3, p. 443)

If this is indeed Turing's primary question, then there is no need for him to be very specific about the level of performance required before he is prepared to commit himself to saying that a machine is genuinely "thinking" or "intelligent". It is enough to have an existence proof of a potential *future* machine that would *clearly* pass: a potential to be fulfilled only when digital hardware and programming techniques have caught up with the theoretical possibility.

From this perspective, however, there need after all be nothing very special about human-likeness in Turing's theoretical vision. Once we have a universal computer, *any* theoretically feasible level or flavour of algorithmic "thinking" is potentially achievable, given appropriate memory capacity, speed, and programming, and within the bounds of possibility explored in his 1936 paper. So it seems that the crucial role of the Turing Test need not be to provide any realistic benchmark of practical progress in AI, but instead, to serve as the context for an *extreme exemplar* that throws down a gauntlet by facing critics

[34] Turing's example conversation concerns poetry because he is responding to Geoffrey Jefferson's Lister Oration for 1949, as quoted at §6.4, p. 451: "Not until a machine can write a sonnet . . . because of thoughts and emotions felt, and not by the chance fall of symbols, could we agree that machine equals brain". Jefferson here poses false alternatives, and Turing's reply on p. 452 addresses only the second of these: his machine's *viva-voce* responses are clearly better than *chance*, but this does nothing to prove that there is genuine *emotion* behind the words.

[35] Such "universality" seems to encompass both the Church-Turing Thesis and the possibility of a Universal Turing Machine. Also relevant here is Turing's argument that a discrete-state machine can in principle mimic a continuous machine—such as the nervous system appears to be—with arbitrary precision (§6.7, pp. 456–7).

of AI with an imagined level of performance that any fair-minded interrogator—forced to judge on that performance alone—would have to count as "intelligent". After the critics have been won over to the extent of acknowledging this as a *hypothetical* possibility, the universality of digital computers can then lead them on to acknowledge intelligent machinery as a *genuine* possibility.

2.10 Reconsidering Standard Objections to the Turing Test

This approach accordingly blunts one of the most oft-repeated objections to the Turing Test, that of alleged *anthropomorphism*. It can also provide a response to what we might call *the chatbot objection*, that the level of performance suggested by Turing's 50-year prediction—namely fooling an "average interrogator" 30% of the time "after five minutes of questioning"—may well be relatively easy to achieve using a machine that is very far from intelligent. This objection is fatal to the Turing Test if it is conceived as *a measure of progress towards AI* or as *a way of gauging the relative "intelligence" of individual programs*. And the main problem here is a different kind of human-centredness, not in respect of what is being measured (namely, the desired human-like reactions of the program), but rather, in respect of who is doing the measuring (namely, a naturally biased and gullible human judge). Joseph Weizenbaum's *ELIZA* program of 1966 taught us something previously unexpected, which would probably have been as surprising to Turing as to others: namely, that an extremely crude "chatbot" program, in conversation with a naïve interlocutor, is often able to sustain a credible conversation for at least a few interchanges. I have discussed this issue at greater length elsewhere,[36] but for present purposes it is enough to point out that advocates of the "extreme exemplar" conception of the Turing Test proposed above would be quite free to take on the lesson of *ELIZA*. Accordingly, they can simply concede that the 50-year prediction has turned out to be too easily fulfilled—by crude syntactic tricks and misdirection rather than by sophisticated information processing—to provide any useful benchmark of intelligence. But the far more demanding standard of Turing's *viva-voce* example remains entirely untouched by this concession.

We can also rebut what might be called the *lookup table* objection, that a system based on a giant lookup table of responses could in principle pass the Turing Test to any desired degree of sophistication.[37] To my mind this objection exemplifies a dubious use of that philosophers' term "in principle", for even to set up appropriate responses for a few short exchanges would require more memory cells than there are atoms in the visible universe, and the numbers grow exponentially with every new exchange. Trusting

[36] See Millican (2013), pp. 596–8. For a system which incorporates a faithful emulation of the original *ELIZA*, enabling the internal mechanisms to be inspected in real time, see http://www.philocomp.net/ai/elizabeth.htm.

[37] This objection dates back to Shannon and McCarthy (1956, p. vi), but is most commonly associated with Block (1981), perhaps owing to the happy coincidence that "Blockhead" is a remarkably apt name for the postulated program. Another objection in a similar spirit is Searle's notorious Chinese Room Argument (1984), which likewise attempts to pump our intuitions in an imagined situation of outrageous unfeasibility (see Millican 2013, pp. 588–90).

our intuitive judgements in such a fairy-tale scenario seems very questionable, and the most that can be expected from it is to persuade us that intelligence cannot plausibly be *defined* in terms of the Turing Test. But the objection has no force against the "extreme exemplar" view, whose point is to claim that the sort of system Turing envisages is genuinely feasible within the not-too-distant future. That claim is, of course, potentially disputable, as Turing himself recognises. In their 1952 radio discussion, Newman talks of the Manchester machine requiring "thousands of millions of years" to analyse a game of chess by brute force, and says that to suppose such things "will be done in a flash on machines of the future, is to move into the realms of science fiction". Turing responds: "If one didn't know already that these things [such as playing chess well] can be done by brains within a reasonable time one might think it hopeless to try with a machine. The fact that a brain *can* do it seems to suggest that the difficulties may not really be so bad as they now seem." (Turing *et al.* 1952, pp. 503–4). Advances in software technology since then indeed provide support for this style of response. As one topical example, computer chess programs now routinely defeat grandmasters, even running on commonplace hardware. And as another (very different) example, deep learning systems have in recent years proved able to solve subtle pattern-recognition problems that previously seemed intractable.[38] As one possible application of these, it now looks relatively plausible that such systems could enable machines to mimic the sort of culturally nuanced "subcognitive" reactions highlighted by Robert French in his attack on the Turing Test, which seemed to many at the time to be resistant to any foreseeable algorithmic simulation.[39]

One important type of objection to the Turing Test remains, based on the general idea that behavioural indistinguishability in terms of answers to questions cannot prove similarity in terms of subjective experience. The premise here is entirely correct, since—to put the point crudely—if we know that a system has been programmed to generate the relevant responses through the execution of some algorithm, then the occurrence of those responses cannot give us evidence of some *other* cause. So if our own responses are in fact generated (at least in part) by subjective, conscious experience, then a program

[38] Neural networks might themselves be considered liable to the lookup table objection, on the grounds that they operate without any "intelligence", and indeed are rather like fuzzy lookup tables. But such a network could at most play a subsidiary role within any system that aspired to pass the Turing Test, given the multitude of different kinds of problem—many requiring specific and fine-grained answers—that could be set, through exponentially many possible sequences of questions (e.g. about arithmetic, or chess, or poetry, or any of a thousand other subjects). Part of the genius of Turing's choice of test is that to pass it in full generality, any practically feasible program would have to be capable of operating with great precision and discrimination over a wealth of coordinated data structures.

[39] See, for example, the "rating games" in French (1990, pp. 18–22), which include asking the subject of the Turing Test to rate "Flugly" as the name of a teddy bear or as the name of a glamorous female movie star. Note also that on the "extreme exemplar" conception of the Turing Test, one can plausibly reject the whole idea that simulation of highly culture-relative judgements should be considered a requirement. For on this conception, the point of the Test is enable presentation of an exemplar that would unequivocally count as intelligent on any reasonably fair standard, *with indistinguishability from a human playing a procedural rather than normative role, facilitating the assurance of fairness through "blind" judging*. But demanding absolute similarity in respect of cultural judgements is *not* fair, as becomes obvious if we imagine applying such a test to humans from very different cultures.

reproducing that external behaviour *without consciousness being causally involved* cannot be bringing it about in an identical manner; and hence any argument from similar output to similar causation is completely undermined.

Especially in these days of inscrutable machine learning, it is important to emphasise that this objection does not depend on our having precise understanding of the algorithm responsible for a machine's behaviour. It is enough to know in general that the algorithms are designed to operate by standard computational methods, on hardware systems whose behaviour is well understood in terms of physical processes that have no reliance on consciousness. In these circumstances, there is no basis whatever for supposing that consciousness somehow magically makes an entrance once a certain kind of behaviour is produced, when that behaviour is already sufficiently accounted for by the algorithmic implementation.

A potential response to this sort of objection, in the spirit of Turing's "fair play for the machines", is to suggest that on similar principles, if we understood human neurophysiology and biochemistry well enough, then we would be able to explain human behaviour entirely in *those* terms, thereby "proving" that consciousness plays no role in *human* behaviour.[40] But such a response is implicitly taking for granted exactly what the proponent of the objection will deny, namely, that consciousness is indeed causally inert in human behaviour (or at best, that it is a mere abstraction from behaviour and functional role, rather than something ontologically distinct). And it is important to note here that such a denial need not be founded at all on some sort of Cartesian dualism or belief in souls; for it is entirely compatible with accepting, on the basis of evolutionary evidence, that consciousness is almost certainly a function of physical matter. *How on earth consciousness arose*, and *how it can be generated* by neurophysiology and biochemistry (etc.) is currently a mystery, and the relevant sciences might well have to go through conceptual revolutions—as did the physical sciences—before we can even glimpse solutions to their most fundamental questions. But *that consciousness somehow arose*, and *that it does play a genuine causal role in our own behaviour*, seem as obvious as almost anything can be. When we have only started to scale the foothills of these sciences relatively recently, it is hubristic to suppose that we can predict what their ultimate form will be, and absurdly so to assume in advance that this ultimate form can give no genuine causal role to consciousness.

Turing's own treatment of "The Argument from Consciousness" obscures this crucial epistemological asymmetry between the causation of human and machine behaviour by raising the spectre of solipsism, with a touch of humour:

> "According to the most extreme form of this view the only way by which one could be sure that a machine thinks is to *be* the machine... Likewise according to this view the only way to know that a *man* thinks is to be that particular man. It is in fact the solipsist point of view. It may be the most logical view to hold but it makes communication of ideas difficult. A is liable to believe 'A thinks

[40] I am grateful to an anonymous referee for posing this very response.

> but B does not' whilst B believes 'B thinks but A does not'. Instead of arguing continually over this point it is usual to have the polite convention that everyone thinks." (§6.4, p. 452)

It may be that an extreme solipsist would indeed view the consciousness of another person and of a machine with equal scepticism, but this is no basis on which to draw a balanced and rational conclusion. The common-sense position is instead to acknowledge that one's own consciousness is very probably indicative of other people's consciousness also, given their similar biological origin and nature. But again, this gives *no ground whatever* for extrapolating consciousness to a machine, however similar its behaviour, if we have reason to believe that the similarity of that behaviour is in no way driven by such biological processes, but instead by some program designed for the purpose (whose operations themselves require no machine consciousness).

This is the weakest major point in Turing's two seminal papers, where he should have been prepared to break away from the human-centred paradigm of "intelligence" that he had strategically highlighted in his famous Test. Here he should have had the candour to say that a machine capable of giving comparable answers to those of an expert human would be a clear exemplar of intelligence *whether or not it was conscious*. Intelligence is standardly understood to be a measure of sophisticated information processing for some purpose, not a measure of subjective experience. Human manifestations of intelligence may indeed be commonly—perhaps usually—accompanied by subjectivity. But after Turing has shown that sophisticated information processing is something that can equally be achieved by a machine (and without invoking any subjective experience), it then becomes entirely appropriate to distinguish the information processing from the subjectivity, and to reserve the word "intelligence" for the former.[41] This would involve some revision of our naïve conceptual scheme, away from a human-centred view of intelligence. But such revision, Turing should have insisted, would be fully justified in the light of his own fundamental discoveries.*

References

Block, Ned (1981), "Psychologism and Behaviorism", *Philosophical Review* 90, pp. 5–43.

Champernowne, D. G. (1933), "The Construction of Decimals Normal in the Scale of Ten", *Journal of the London Mathematical Society* 8, pp. 254–60.

Church, Alonzo (1934), "The Richard Paradox", *American Mathematical Monthly* 41, pp. 356–61.

Cooper, S. Barry and Jan van Leeuwen, eds (2013), *Alan Turing: His Work and Impact*, Waltham Massachusetts: Elsevier.

Copeland, B. Jack, ed. (2004), *The Essential Turing*, Oxford: Clarendon Press.

Copeland, B. Jack (2017), "Intelligent Machinery", in Copeland *et al.* (2017), pp. 265–75.

[41] For more extensive discussion of these issues, which are merely sketched here, see Millican (2013, pp. 590–6).

* I am grateful to Andrew Hodges, Alexander Paseau, Shashvat Shukla, John Truss, and Stan Wainer for helpful discussion when preparing this paper, and also two anonymous referees.

Copeland, Jack and Diane Proudfoot (2009), "Turing's Test: A Philosophical and Historical Guide", in Epstein *et al.* (2009), pp. 119–38.

Copeland, B. Jack, Jonathan P. Bowen, Mark Sprevak, and Robin Wilson, eds (2017), *The Turing Guide*, Oxford: Oxford University Press.

Epstein Robert, Gary Roberts, and Grace Beber, eds (2009), *Parsing the Turing Test*, Dordrecht: Springer.

Fan, Zhao (2020), "Hobson's Conception of Definable Numbers", *History and Philosophy of Logic* 41, pp. 128–39.

French, Robert M. (1990), "Subcognition and the Limits of the Turing Test", *Mind* 99, pp. 53–65 and reprinted in Millican and Clark (1996), pp. 11–26.

Gandy, Robin (1988), "The Confluence of Ideas in 1936", in Herken (1988), pp. 55–111.

Gandy, Robin (1996), "Human versus Mechanical Intelligence", in Millican and Clark (1996), pp. 125–36.

Gödel, Kurt (1962), *On Formally Undecidable Propositions of Principia Mathematica and Related Systems*, tr. B. Meltzer with an introduction by R. B. Braithwaite, New York: Basic Books.

Hayes, Patrick and Kenneth Ford (1995), "Turing Test Considered Harmful", *IJCAI*-95, pp. 972–7.

Herken, Rolf, ed. (1988), *The Universal Turing Machine: A Half-Century Survey*, Oxford: Oxford University Press.

Hilbert, David and Wilhelm Ackermann (1928), *Grundzüge der theoretischen Logik*, Berlin: Springer.

Hobson, E. W. (1921), *The Theory of Functions of a Real Variable and the Theory of Fourier's Series*, second edition, Cambridge: Cambridge University Press.

Hodges, Andrew (1983), *Alan Turing: The Enigma of Intelligence*, London: Burnett.

Hodges, Andrew (1988), "Alan Turing and the Turing Machine", in Herken (1988), pp. 3–15.

Hodges, Andrew (2009), "Alan Turing and the Turing Test", in Epstein *et al.* (2009), pp. 13–22.

Hodges, Andrew (2013), "Computable Numbers and Normal Numbers", in Cooper and van Leeuwen (2013), pp. 403–4. This is immediately followed by Turing's "A Note on Normal Numbers" (pp. 405–7) and Verónica Becher's "Turing's Note on Normal Numbers" (pp. 408–12).

Michie, Donald (1993), "Turing's Test and Conscious Thought", *Artificial Intelligence* 60, pp. 1–22 and reprinted in Millican and Clark (1996), pp. 27–51.

Millar, P. H. (1973), "On the Point of the Imitation Game", *Mind* 82, pp. 595–7.

Millican, Peter (2013), "The Philosophical Significance of the Turing Machine and the Turing Test", in Cooper and van Leeuwen (2013), pp. 587–601.

Millican, Peter and Andy Clark, eds (1996), *Machines and Thought*, Oxford: Oxford University Press.

Petzold, Charles (2008), *The Annotated Turing: A guided tour through Alan Turing's historic paper on computability and the Turing machine*, Indianapolis: Wiley.

Piccinini, Gualtiero (2000), "Turing's Rules for the Imitation Game", *Minds and Machines* 10, pp. 573–82.

Proudfoot, Diane (2017), "Turing's Concept of Intelligence", in Copeland *et al.* (2017), pp. 301–7.

Russell, Bertrand (2014), ed. Gregory H. Moore, *Collected Papers, Volume 5: Toward Principia Mathematica, 1905–08*, Abingdon: Routledge.

Saygin, Ayse Pinar, Ilyas Cicekli, and Varol Akman (2000), "Turing Test: 50 Years Later", *Minds and Machines* 10, pp. 463–518.

Searle, John (1984), *Minds, Brains, and Science*, Cambridge Massachusetts: Harvard University Press.

Shannon, C. E. and J. McCarthy, eds (1956), *Automata Studies*, Princeton: Princeton University Press.

Sprevak, Mark (2017), "Turing's Model of the Mind", in Copeland *et al.* (2017), pp. 277–85.

Sterrett, Susan G. (2000), "Turing's Two Tests for Intelligence", *Minds and Machines* 10, pp. 541–59.

Traiger, Saul (2000), "Making the Right Identification in the Turing Test", *Minds and Machines* 10, pp. 561–72.

Turing, Alan M. (1936), "On Computable Numbers, with an Application to the *Entscheidungsproblem*", *Proceedings of the London Mathematical Society*, Second Series, Vol. 42 (1936–7), pp. 230–65; reprinted in Copeland (2004), pp. 58–90 (page references are to this reprint).

Turing, Alan M. (1947), *Lecture on the Automatic Computing Engine*, delivered on 20 February 1947 to the London Mathematical Society, reprinted in Copeland (2004), pp. 378–94.

Turing, Alan M. (1948), *Intelligent Machinery*, report prepared for Sir Charles Darwin, Director of the National Physical Laboratory, reprinted in Copeland (2004), pp. 410–32.

Turing, Alan M. (1950), "Computing Machinery and Intelligence", *Mind* 59, pp. 433–60; reprinted in Copeland (2004), pp. 441–64 (page references are to this reprint).

Turing, Alan M. (1951a), "Intelligent Machinery, A Heretical Theory", talk for *The '51 Society* broadcast on the BBC, typescript transcribed in Copeland (2004), pp. 472–5.

Turing, Alan M. (1951b), "Can Digital Computers Think?", lecture broadcast on the BBC on 15 May 1951, transcribed in Copeland (2004), pp. 482–6.

Turing, Alan M., Richard Braithwaite, Geoffrey Jefferson, and Max Newman (1952), "Can Automatic Calculating Machines be Said to Think?", discussion broadcast on the BBC on 10 January 1952, transcribed in Copeland (2004), pp. 494–506.

Weizenbaum, Joseph (1966), "ELIZA—A Computer Program For the Study of Natural Language Communication Between Man And Machine", *Communications of the ACM* 9, pp. 36–45.

Whitehead, Alfred North and Bertrand Russell (1927), *Principia Mathematica*, second edition, Cambridge: Cambridge University Press. The relevant material is reprinted in the abridged version *Principia Mathematica to* *56, published by Cambridge in 1962.

3

Spontaneous Communicative Conventions through Virtual Bargaining

Nick Chater and Jennifer Misyak

Warwick Business School, UK

3.1 The Spontaneous Creation of Conventions

The spontaneous creativity of human communication is so familiar that we rarely stop to consider how astonishing it is (Clark, 1996), and hence how difficult to emulate by an artificial system. Person A, who has arrived at a conference and is chatting with a crowd of colleagues, notices Person B just arriving. Making eye contact and smiling, A points first to the conference name tag on her lapel and then waves vaguely, arm aloft over a crowd of heads, in the general direction of a distant table. B weaves through the melee across to the table and picks up her own name tag. The message is successfully sent and received, using an unexceptional, although novel communicative signal, with the meaning roughly along the lines of 'you can get your name tag over at that table'.

Yet the reasoning that connects such signals to their meanings seems, on closer inspection, remarkably complex. First, the eye contact and smile seem to be crucial. Without these, B will simply see A engaging in some puzzling contortions rather than engaging in communication (e.g., A might be reaching towards ceiling, or merely stretching), and B would certainly not conclude that A was attempting to communicate *with her*. Similarly, touching the name tag might not be interpreted as having communicative intent but merely a random movement or an attempt to brush away a crumb.

Second, the combination of actions is crucial. Merely pointing at the name tag could equally be attempting to convey, for example, 'Look, I've already got my name tag', 'Isn't this name tag a bizarre colour?', or even 'Ha ha, I've taken your name tag' (as a joke). And pointing with the arm aloft could indicate a large number of people or objects in the room or a poster on the wall; or refer to a general area of the room, or the approximate direction of Paddington Station, or due North. It is only the combination of actions (along with the mutual recognition that these are to be interpreted together rather than individually) that makes clear that A is trying to tell B where the name tags are.

Nick Chater and Jennifer Misyak, *Spontaneous Communicative Conventions through Virtual Bargaining* In: *Human Like Machine Intelligence.* Edited by: Stephen Muggleton and Nick Charter, Oxford University Press.
DOI: 10.1093/oso/9780198862536.003.0003

Third, the specifics of the situation, and background knowledge of all kinds, are also crucial. For example, suppose that this is the second day of the conference, and A and B have already met on the first day. Then it is unlikely B would be lacking a name tag. Indeed, even if, as it happens, B had forgotten to pick up a name tag, B would most likely have no particular reason to suppose that A knew this, which would probably block the above interpretation of A's gesture. On the other hand, if B had somehow missed the name tag table on the first day and mentioned this to A that evening, then A's signal might, after all, plausibly be interpreted as helpfully pointing the location of the table. Alternatively, imagine that it is the first day of the meeting. A is the conference organizer and B is a student who is supposed to be helping hand out the name tags, and who has been unexpectedly delayed. In this situation, A's message is likely to be interpreted as 'the name tag table is over there—you'd better get over there quickly!' And making sense of this message requires that all of this background and situational information is common ground between A and B (of which more below).

Fourth, the range of alternative actions that can be performed is crucial (e.g., Levinson, 2000). Suppose that the room were entirely empty, aside from A and B. Then A would most likely simply say 'you can pick up your tag on that table', and an unexpected pantomime performance using gestures would seem rather baffling and probably be difficult to interpret. On the other hand, though, matters would be different if A and B were outside a hushed lecture hall, with the lecturer about to begin. Then A's use of gesture, rather than speech, would make sense as a way of avoiding disturbing the proceedings (mostly likely A would first clarify, e.g., by putting a finger to her lips). And consider the nature of A's pointing gesture (with arm aloft) when in an empty, rather than a crowded, room. In an empty room, holding the arm aloft would seem necessarily to have some mysterious communicative import (perhaps the name-tag table is behind a wall in the next room?), otherwise, why not simply point directly with an extended arm? But in a crowded room, holding the arm aloft is required to avoid impolitely bumping into other attendees, and, indeed, to make the pointing hand visible above the crowd, so that holding the arm aloft is required to enable communication, and is therefore not interpreted as part of the communicative signal.

We could, of course, continue more or less indefinitely expanding upon such complexities. But the moral we want to draw is twofold: first, that spontaneous communication, however natural to us, is actually astonishingly subtle; and second, that this subtlety needs to be captured if we are to be able to construct artificial intelligent robots, whether assistants, care-givers, or even software avatars, that can smoothly and naturally communicate with humans.

The general challenge of understanding such reasoning is, of course, considerable—and can depend in principle on arbitrary general knowledge about the physical and social worlds. The task of building a theory of spontaneous communication has to take the general ability to store and reason about such knowledge for granted; the challenge is to clarify how such general reasoning can be harnessed to determine the meaning of spontaneous communicative signals. Moreover, while in this chapter we will emphasize the creativity and flexibility of human communication and the challenge this poses for instilling the capacity for genuinely human-like interaction into machines, it is also important that each new communicative signal has the potential to set a precedent for the interpretation of future signals. Gradually, then, spontaneous signals may become increasingly stereotyped through processes of cultural evolution. Indeed, this is one possible viewpoint concerning the origin of the highly conventionalized system of signals embodied of natural language (Christiansen and Chater, 2016). Such conventionalization is never, though, complete—the creativity of the indirect, ironic, metaphorical use of natural language is one of its most remarkable features. So we may hope that any insights into *de novo* spontaneous communication, which is the focus here, may ultimately help analyse how language works, particularly in dialogue (Clark, 1996).

3.2 Communication through Virtual Bargaining

To focus the discussion, let us start by describing an experimental set-up that we explored in previous work (Misyak *et al.*, 2016). Suppose that we have two people who have to communicate about the contents of three sealed boxes, each of which contains either a good outcome for both players (represented by the image of a banana) or a bad outcome for both players (represented by the image of a scorpion). On each trial of the experiment, new pairs of partners are randomly matched and interact anonymously; consequently, participants cannot create conventions specific to a particular pairing over time.

Between the pair, one player knows the contents of the boxes and has to communicate this to the other player using a very limited set of signals. This first player has a set of one or more tokens, which can be used to send a signal. Specifically, tokens can be placed on any box (though only one token per box); not all tokens need be used (indeed, it is possible to send a 'null' signal, not placing any of the tokens). Moreover, the token must be placed at a standardized location on the box, so that no information is conveyed by the 'manner' in which the token is placed. The second player receives this signal and uses it to decide which box(es) to open. Opening a box with the banana leads to the same positive outcome for both players; opening a box with a scorpion leads to the same, and much larger, negative outcome for both players. Therefore, the second player is motivated to open, where possible, boxes that contain bananas with a high degree of confidence, and to avoid opening boxes that contain scorpions, and the first player is motivated to send a signal that will facilitate this.

Let us consider first the straightforward situation in which it is common ground between the players that there is a single banana and two scorpions (of course, only the first player knows which box contains the banana). It is also common ground that the first player has a single token. How is the first player to proceed? It is intuitively obvious, and indeed observed, that the player will place their single token on the box with the banana. The second player will then open the box and only that box, so that both players benefit from the positive reward associated with the banana.

What happens, though, when it is common ground that there are two bananas and one scorpion, and the first player just has a single token? Clearly one possibility is to use the single token on one of the boxes containing one of the bananas. But there is a better strategy, and one that, perhaps surprisingly, people are able to alight upon quite reliably, even on the first time they encounter this type of trial, which is for the first player to mark the scorpion, and the second player to choose both the non-marked boxes. This allows both players to get twice the reward that they would achieve with the direct approach of marking a single box (i.e., they can obtain two bananas rather than one). This outcome will only be achieved, of course, if both players spontaneously alight on the same signal-message 'convention' and, indeed, they are reasonably confident that this convention is common ground between them.

To create a rational theory of communication (Grice, 1957, 1975; Sperber and Wilson, 1986; Levinson, 2000; Frank and Goodman, 2012), a central question for understanding communication is, therefore, to be able to reconstruct the reasoning that leads to these situation-specific choices of convention. Even our simple experimental set-up turns out to be surprisingly challenging to explain in these terms.

One starting point might appear to be mutual prediction: the sender attempts to predict the convention that the hearer will employ, and the hearer attempts to predict the convention that the speaker will employ. But this reasoning is clearly circular. Moreover, the sender has to work out which convention the receiver thinks the sender will choose, but the sender does not yet know herself what she will choose—this is the very question at issue, after all. Some formal approaches to communication (e.g., the important recent work on rational speech acts and related ideas: Frank and Goodman, 2012; Shafto, Goodman, and Griffiths, 2014; Goodman and Frank, 2016) attempt to pursue this approach and focus on 'fixed points' in the sender's and receiver's recursive beliefs about each other, an approach that has provided important insights into the operation of, for example, scalar implicatures (e.g., that 'some people like penguins' is typically taken to suggest that the stronger claim that 'everyone likes penguins' is not known, or this more informative claim would have been made). This approach is closely related to equilibrium approaches in game theory, including the Nash equilibrium (Nash, 1950), and broader notions such as rationalizability (Bernheim, 1984; Pearce, 1984). Where the communicative set-up is specified in a way such that there is a unique equilibrium, the mutual prediction approach will find it successfully. But this strategy provides no principled way of determining which equilibrium is selected—from this point of view, as long as both sender and receiver agree on the convention employed, they have a viable solution.

Yet, as we have seen, there will be many communicative conventions which are potential viable solutions in a particular situation. The crucial challenge to make communication possible is for both parties to be able to alight upon the same solution without, on pain of regress, requiring prior communication to establish agreement on the right convention.

An alternative perspective comes from the idea of 'we-reasoning' in non-standard approaches to game theory (e.g., Colman, 2003; Sugden, 2003; Bacharach *et al.*, 2006). According to this perspective, when engaged in a joint activity (communication being a paradigmatically joint activity: Clark, 1996), people may ask themselves not what should *I* do, in the light of what *you* will do, in the light of what *you* think *I* will do, and so on. Rather they should ask: what should *we* do? This breaks out of the infinite regress because each player is not trying to second-guess the other, but to trying to imagine what they, considered as a unit, would decide.

The problem of communication is, on this analysis, a particularly interesting and subtle special case of the general problem of coordinating thought and behaviour (Schelling, 1960). Specifically, the outcome of each player's choice of communicative convention depends on alignment with the choice of convention of the other. Coordination games consider the problem of alignment more broadly. One well-known coordination game (Schelling, 1960) requires people independently to specify a time-of-day and location to meet in New York City (Grand Central Station at midday is a popular choice). Other games require that people align on the same colour (where red is a popular choice, but magenta is not) or letter of the alphabet (where A will be a more popular choice than L) (Schelling, 1960). Coordination games are notoriously difficult to capture in conventional game theory (Bacharach *et al.*, 2006), precisely because there are many equilibria where each person aligns with the other (people could successfully meet at, say, any grid reference location in New York City), but only some of these equilibria (e.g., Grand Central Station) are remotely likely to be chosen.

But how, precisely, can the players come to conclusion about what they would decide? The theory of virtual bargaining (e.g., Misyak and Chater, 2014; Misyak *et al.*, 2014; Chater *et al.*, 2016) provides a specific answer to this question: players imagine a hypothetical process of negotiation, by which they come to a joint conclusion about how they should act (see Bundy, Philalithis, and Li, this volume, for a computational analysis of virtual bargaining in the context of communication, in terms of modelling changes in a logic-based representation).

The hypothetical negotiation must proceed purely from beliefs and objectives that are common knowledge—they must depend only on information that each player knows, knows that those of the other knows, and so on. This is because if either player reasons in a way that depends on the use of any private knowledge to determine how a negotiation might proceed, then the other player will be likely to come to a different conclusion concerning the hypothetical negotiation because they lack that knowledge. Hence, there is every chance that the players will come to different conclusions, and coordination is likely to fail.

Now the outcome of hypothetical negotiation, whether about communicative conventions or anything else, can often be ill-defined. After all, in many contexts, the outcome of real, explicit negotiations may be difficult or impossible to predict. Where no clear convention would 'obviously' result from a hypothetical negotiation, the meaning of signals will be unclear, probably to both sender and receiver; and communication will not succeed. The interesting cases are, of course, those in which communication is possible—where a hypothetical negotiation between sender and receiver would naturally lead to a unique agreed convention.

In the light of the virtual bargaining viewpoint, let us reconsider the experimental example introduced above, with common ground that there is one banana and two scorpions, and one token available to the sender. Suppose that both players were to negotiate about the most effective way to associate signals and messages in this type of situation: it seems 'obvious' that both players would agree that the best plan is to place a token on the box with the banana.

Of course, it is conceivable that they might choose some other strategy. Suppose, for example, that the boxes are arranged in an equilateral triangle (as, indeed, they are in the experiments reported in Misyak *et al.*, 2016). In this arrangement, the two players could agree to place the token on the box on the corner of the triangle that is one clockwise turn from the box with the banana; or, for that matter, one anticlockwise turn (indeed, there are three other even less natural mappings, but we will not consider these further). Both players know that agreements would be very unlikely, and would not be chosen in a hypothetical negotiation. But what exactly makes them so unnatural? We suggest that there are at least three classes of (potentially related) reasons:

(1) Simplicity. These conventions are unnecessarily complex and, perhaps relatedly, likely to lead to implementation errors. Moreover, there seems to be a powerful general psychological preference for simplicity across many perceptual and cognitive domains (Attneave, 1954; Chater, 1996; Chater and Vitányi, 2002; Feldman, 2009).

(2) Generality. They apply only in unnecessarily restricted circumstances. For example, the clockwise or anticlockwise rule has no obvious application if the boxes are arranged in a row; or on a 3 × 3 grid; or any number of formations. A strategy that picks out the unique box with the banana *by marking that very box* generalizes easily to any layout.

(3) Symmetry breaking. The 'direct' mapping is unique. But the unnatural 'clockwise' mapping seems to be part of a larger class, including, at least, the 'anticlockwise' mapping. Yet there seems to be no obvious way to break symmetry between these two. They appear equally simple and general, and hence it is not clear how the two parties will reliably coordinate on the same one.

A crucial challenge for future research is to make precise these different types of reasons, and perhaps others, and to explain how these factors might be traded off against one another to yield a unique choice of mapping (where this is possible at all), and we shall briefly discuss this challenge further below. In any case, however the notion of naturalness of a communicative convention is ultimately explained, it is clear that in simple cases (such as where there is one banana and one token), people are able to solve the communicative challenge and independently alight on the same 'natural' convention. From a virtual bargaining viewpoint, this means both players can imagine reaching the hypothetical agreement ('just put the token on the box with the banana'). And, according to the theory of virtual bargaining, it is this ability to alight on this hypothetical agreement that underpins the confident usage of this convention by both parties, without need for any actual prior discussion.

The case in which it is common ground that there are *two* bananas and *one* scorpion is very different, however. A sender using the token directly to indicate a box with a banana, and the receiver opening that same box, can yield just one reward to be shared between the players. But a convention in which the scorpion is signalled, and the receiver then chooses both the other boxes, is clearly a better strategy because it allows two bananas to be selected, and hence two units of reward to be shared between the players. Thus, if both players were to imagine hypothetically discussing how they would communicate in a situation of this kind, they might both rapidly conclude that they would decide on this 'inversion' strategy of marking the box with the scorpion, not because of its naturalness but because it is simply more effective.

This inversion strategy is less simple and less general than the first. But, in the situation of interest here, the players have a commonly recognized goal: to obtain as many bananas as possible without uncovering scorpions.

The ability of people spontaneously to 'flip' a communicative signal in this way is rather remarkable. For this to work, it is crucial that both players realize that the other will do the same—if either player suspects that the other may not have noticed the 'inverse' convention, and may merely follow the more direct convention of 'pointing' at a single banana, then it may be better to follow the more direct convention. Indeed, player A needs to be confident that player B has spotted the inversion convention, that B believes that A will have spotted it, that B has also judged that A believes that B will have spotted it, and so on indefinitely. In short, it needs to be *common ground* between A and B that the inversion convention is the best option. This seems a remarkably high epistemic bar; nonetheless, people are able to coordinate on the inversion with reasonably high reliability (Misyak *et al.*, 2016).

The richness of players' reasoning is further illustrated by the impact of two variations on the 'two bananas' scenario. In one experimental condition, it is in common ground that person B is only able to open a single box (boxes can be opened by placing an icon representing an axe on the box; players can see that B only possesses a single axe that can be used once). Now the potential advantage of the more complex inversion mapping is annulled. And, indeed, participants now typically adopt the standard convention of A placing a token on one of the boxes, and B opening that box.

The other variation is subtler: a wall is introduced into the scene, which partially blocks B's view. The numbers of objects of different kinds are represented, in the experimental set-up, by 'shadows in the grass' where those objects previously stood before being placed under the boxes. But the wall selectively blocks B's view of these shadows (and all of this is in common ground, assuming partners take account of the occlusion). With the wall present, A is indeed less inclined than otherwise to use the inversion mapping over the more direct mapping: while she knows that there are, on this particular trial, two bananas and one scorpion, she can also perceive that this information is not in common ground. Therefore, it cannot be used to support coordinating on the inversion convention.

3.3 The Richness and Flexibility of Signal-Meaning Mappings

Let us now switch to a related, but slightly different, set-up. This, and the examples considered below have not yet been explored experimentally, but the communicative intuitions that they generate seem rather compelling. Henceforth, then, we will be relying

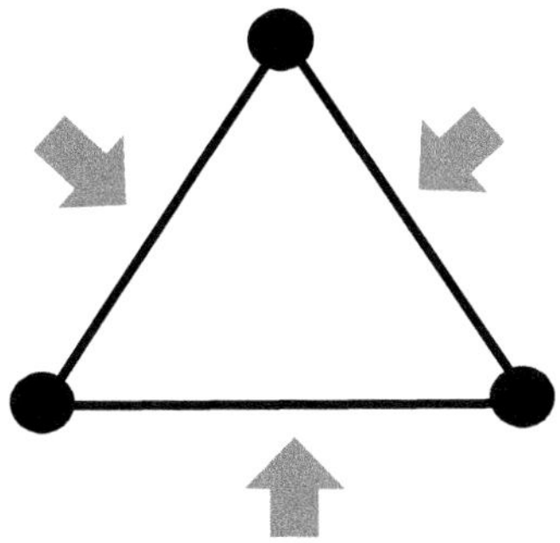

Figure 3.1 *A simple communicative set-up. The sender, A, can highlight one of the sides of the triangle (indicated by the grey arrows); but the communicative goal is to pick out one corner of the triangle, which B then selects.*

on our own and our readers' communicative intuition, much as linguists routinely draw on native speakers' linguistic intuitions.

For concreteness, let us consider signalling in a slightly different geometric set-up. Suppose, for example, that for some reason A would like to pick out a particular corner of the equilateral triangle, so that B can select it (Figure 3.1); but suppose that A can only do so by pointing at, or otherwise highlighting, one of the sides of the triangle (indicated by the grey arrows in Figure 3.1). Suppose, for concreteness, that the selected side of the triangle flashes on and off alternately, indicating that it has been selected (the grey arrows in Figure 3.1 merely indicate which side might be selected—they are not part of the signal).

There are three possible messages to be sent (the corners), and three possible signals (the sides). So, from an abstract point of view, any one-to-one mapping between these will suffice for perfect communication, of which there will be $3 \times 2 \times 1 = 6$ possibilities. If they could agree a 'code-book' before communication begins, then any of these would suffice. But, as in our example at the conference above, there is usually no opportunity for prior communication (and, indeed, prior communication had better not itself require still further prior communication, or we will fall into an infinite regress).

Interestingly, though, one of these mappings seems psychologically 'special'. Mapping a. puts the side providing the signal opposite the indicated corner of the triangle (i.e., so that the side, and the corner, together 'cover' all three vertices of the triangle). This provides a unique and simple connection between sides and corners. Mappings d. and f. have a slightly less natural pattern: the side indicates one of its vertices, rotating in either the clockwise (mapping d.), or anticlockwise (mapping f.) direction. Yet the very fact that there are two equally natural such mappings makes attempting to use either difficult because, without prior communication, there is no way reliably to break the symmetry between them. The remaining three mappings, b., c., and e., have no intuitively natural explanation, although if both parties were simultaneously to alight on any one of these, it would be possible perfectly to indicate corners using sides.

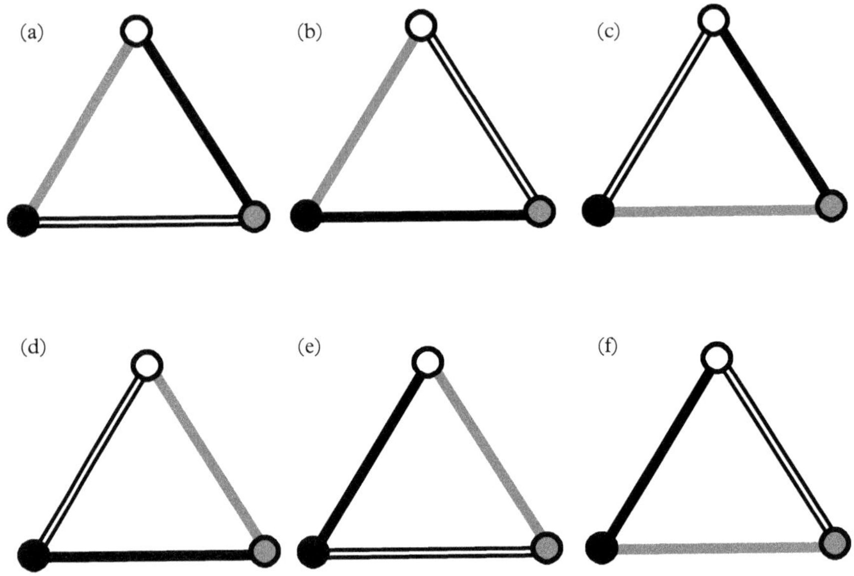

Figure 3.2 *Six possible communicative conventions. Illustrated here are the six one-to-one correspondences between vertices and sides. For expository purposes only, the corresponding vertices and sides are shown here as either black, grey, or 'hollow'. The actual stimuli would all be of uniform colour.*

The mapping in which sides indicate the 'opposite' corner seems to be both simple and unique. And it also has a (perhaps somewhat modest) degree of generality. Indeed, the same strategy will work for any regular polygon shape with an odd number of sides (pentagon, heptagon and so on)—a specific side can successfully be indicated by the 'opposite' corner.

The attentive reader will have noticed a simple connection between this new case and the inversion with the bananas and scorpions experimental set-up described earlier: picking out a corner can be viewed as highlighting the 'opposite' side, but equally as indicating the two corners joined by that side. That is, by pointing to one corner, we can implicitly be viewed as picking out the other two, which is just the inversion mapping discussed above.

Yet different representations of the same type of problem suggest very different generalizations to new cases. For example, in the original bananas and scorpions set-up, the triangular layout of the boxes appears to be incidental—they could equally be arranged in a line, for example. But in the corners and lines set-up the geometric structure is critical. For example, consider Figure 3.3, in which we see a range of ways in which a corner on one triangle can be mapped to a side (i.e., to the two corners joined by that side), but where that side is part of a different triangle (which may be a transformed version of the first).

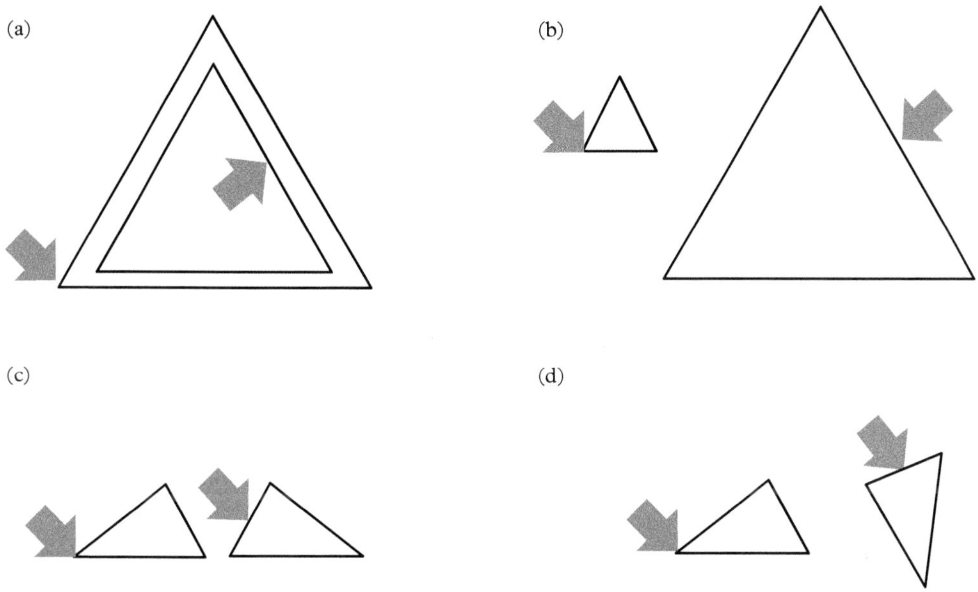

Figure 3.3 *Communicative conventions which engage geometric reasoning. A geometric interpretation of the link between corners and sides allows a range of interesting communicative generalizations. Suppose, for example, that, in Figure 3a, signals correspond to corners of the outer triangle, and the message is to pick out one of the sides of the inner triangle. Then the mapping from corner to 'opposite' side appears to provide the most natural mapping. This applies even when the triangles are separated in space (b), mirror reflected (c), or rotated (d). These examples suggest that in this type of case, the mapping operates in two steps: first linking one corner with the corresponding side of the same triangle; and then mapping from the face of one triangle, to the corresponding face of the other. Note that each of these mappings is reversible (and could just as well be used to use a side of one triangle to indicate a corner of the other). Here, unlike in the original bananas and scorpions set-up, geometric structure is crucial.*

3.4 The Role of Cooperation in Communication

We have so far been assuming that the sender and receiver are attempting to coordinate on a set of momentary conventions. It is therefore natural to assume that communication presupposes that the sender and receiver both wish to cooperate, and indeed, some such assumption is typically part of theories of natural language pragmatics (e.g., Grice, 1975; Sperber and Wilson, 1986; Levinson, 2000).

Yet the role of cooperative intent in communicative reasoning is subtle. Consider, for example, a variation of the bananas and scorpions paradigm described above, where instead of sharing in the positive or negative outcomes, all the 'payoffs' go to the receiver, B. Suppose that we assume a modest level of goodwill (e.g., slight prosocial preferences will suffice). Then, both A and B can still reason that were they able explicitly to discuss which convention to adopt in a particular situation, they would typically agree to a communication strategy that would allow B to gain the highest payoff (i.e., to retrieve as many bananas as possible, but no scorpions). Thus, both will agree on the meaning to be assigned to the placing of the token, both on standard trials (involving one banana) and on trials involving two bananas where inversion is applicable.

Suppose, though, that A and B are old enemies, and that it is common knowledge to both that A would be delighted were B to achieve a bad outcome, for example by choosing a scorpion.

In the one banana case, suppose A places the token on a particular box. B will interpret this signal as having a meaning such as 'this is the box with the banana' or 'choose this box'. But B will, of course, be highly suspicious, realizing that A is hoping to mislead. B might react by choosing another box; or might suspect that the sender is engaged in double bluff, and so might choose the box indicated by A after all. Indeed, according to standard game theoretic analysis, the only stable strategies in this scenario are that A should choose a box at random, and that B should entirely ignore this signal. Indeed, this result arises quite generally: in adversarial contexts, a rational sender should send a signal that carries no information whatever about the environment, and a rational receiver will ignore it.

We suggest, though, that even though the signal is utterly uninformative, its *meaning* is still clear. That is, both A and B interpret the token on a specific box as meaning that this is the box containing the banana, *even though neither believes that the signal is being used reliably*. This situation is, of course, familiar when people are communicating using natural language: speaker and hearer may agree upon the meaning of the speaker's words, even where neither trusts that the speaker is using those words reliably.

How is it possible for meaning and informativeness to diverge so completely? One explanation for linguistic cases rests on the observation that meaning in natural language is governed by public conventions, known to both parties; and that the proper operation of these conventions can be well-defined, even where both parties may suspect that these conventions will be subverted in practice.

The present case, though, is more interesting. In this type of novel communicative context, there is no pre-existing system of conventions on which to rely. Even though their interests are opposed, the sender and receiver are nonetheless able to agree on the appropriate mapping between signals and meanings. Indeed, this agreement is crucial in, for example, the receiver's suspicion that the sender is attempting to mislead her (or double bluffing, or whatever it may be). Indeed, without an agreed interpretation, two parties could not be engaged in communication of any kind (whether misleading or not). Without an agreed interpretation, the sender's positioning of the token would not be seen as meaningful though untrustworthy, but as no more communicatively relevant than if the token had been placed purely by chance.

We suspect that the right way to analyse such situations may be as follows. Both parties are able to infer what agreement they would have reached about the mapping between signals and meanings, were they able to bargain virtually. But, if their interests are opposed, they are also aware that *following* a credible bargain would not, in practice, be possible. This is because, even if an agreement were hypothetically to be reached, neither party would have any expectation that the other would follow it (as each is attempting to 'outsmart' the other).

The details of this type of analysis could be spelt out in a variety of ways. One simple approach would be to assume that the mapping between signals and meanings, which determines the meaning of each signals, is determined on the assumption that both parties have sufficiently prosocial attitudes towards the other that agreement is possible. Both parties can then infer the meaning of the signal, 'as if' their interests were aligned rather than opposed; but, at the same time, neither party expects the meaning reliably to be respected if, in practice, their interests are implacably opposed.

3.5 The Nature of the Communicative Act

An additional subtlety may arise from the presumed nature of the communicative act. We have so far largely assumed that people are communicating in order to achieve mutual advantage—and, in our experimental paradigm, outcomes are shared between the sender and receiver so that their interests are aligned. But this is not always the case.

It might be, for example, that it is common ground that A is *instructing* B regarding what she should do (e.g., suppose A is a manager, and B is a worker used to being subject to A's instruction). Perhaps the very communicative act of instruction inherently requires that the objective common to both parties is the satisfaction of the aims of the sender of the communicative signal. In this spirit, the token will have the imperative meaning 'choose this box!'

Suppose, in particular, that all payoffs were to go to A (rather than being split, or given entirely to B). Then, on the inversion trial, where A has the possibility of obtaining two bananas, we might speculate that both parties might interpret the token as an instruction, or perhaps request, *meaning* 'choose the other two boxes!' Of course, if the sender and receiver's interests are opposed, neither may expect this convention to be followed reliably.

Consider, now, a different possibility, where it is common ground that the communicative act is an instance of *advising*, rather than instructing (e.g., suppose a parent were communicating with a child). We might suppose the communicative act of advising inherently requires that the objective common to both parties is the satisfaction of the aims of the receiver, B; that is, the sender, A, is presumed to be attempting to help the receiver make a choice which is in her own interests. This common presumption would, as before, allow both A and B to agree on a mapping between signals and meanings, and this agreement about the appropriate mapping can determine both parties' interpretation of the meanings of the signals, even where neither believes that the mappings will be used 'honestly'.

An interesting experimental set-up in which to consider these issues further could introduce boxes with different payoffs for the sender and receiver. Figure 3.4 shows three boxes, the contents of which are assumed to be common ground. One box is advantageous to the sender (+2) but negative for the receiver (−1); a second box has the opposite payoffs; and the third has modest positive payoffs for both (+1, +1).

What then, is the meaning of a token being placed on one of the boxes? (We assume, as before, that the token must be placed on a fixed location so that its precise

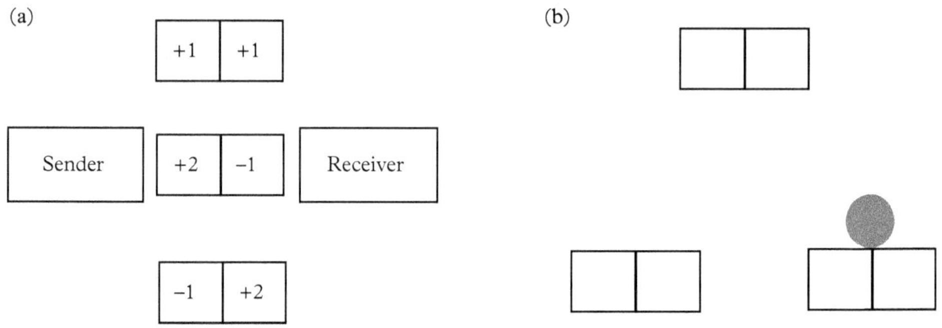

Figure 3.4 *The importance of agreeing the communicative act. The interpretation of a simple signal may depend on the nature of the communicative act. Panel a. shows three boxes, whose contents (but not location) are common ground. The left-hand number in each box corresponds to the sender's payoff; the right-hand number is the receiver's payoff. Panel b. shows a layout of the boxes, which are now sealed. The sender can place a blob in one of three locations, marking one of the three boxes. But which will the sender mark? One box is advantageous only to the sender; one only to the receiver; and one is modestly advantageous to both. Which interpretation is chosen will depend on whether the participants view the signal as aiming to achieve mutual benefit, or as giving an instruction to, or advice to, the receiver. Successful communication will require that the nature of the communicative act is common ground.*

geometric position conveys no information.) If the players are communicating for mutual advantage, as we have considered so far, then the only credible bargain they can reach is to choose the (+1, +1) box—neither player, presumably, is likely to willingly agree to a loss.[1]

But if it is common ground that the communicative act is one of instruction, then the signal presumably should mean 'choose this, the (+2, −1) box, because I say so!' If, on the other hand, it is common ground that the communicative act is one of advice, then the signal presumably should mean 'choose this, the (−1, +2), box—you won't regret it!'

At first glance, cases of instruction and advice may seem to deviate from the 'virtual bargaining' viewpoint. One might imagine that if the interests of only one party are important, bargaining between the two parties is unnecessary. Notice, though, that communication still requires that the sender and receiver *agree on* the same set of signal-meaning mappings. To achieve the best interests of even one party successfully, it is necessary that both parties work coherently together. Thus, we suggest that the right analysis is that the parties are still bargaining about which convention will be most useful, but utility is now defined purely by the interests of one player. In other words, for instructions or requests, we have to agree a convention that will serve the sender's interests; in advice, we have to agree a convention that will serve the receiver's interests.

This point becomes clear if we return to our original bananas and scorpions experimental set-up, but assume that all payoffs go to one player only. If A receives the

[1] Except given very high levels of prosocial preferences, or the ability to redistribute money, or to expect reciprocation later, and so on; we ignore these possibilities here.

entire payoff, then A can be viewed as sending a signal that should be interpreted as an instruction (or, more accurately perhaps, a request, as we are assuming no status or power difference between the players). If there are two bananas and one scorpion, then A may perhaps still successfully send the inversion signal, realizing that both parties can agree that this signal will lead to a better payoff for A. Successful communication requires that it is common ground that both parties seek to further A's interests; and that A and B agree upon a convention that best achieves the sender's objectives. The same is true for where the B person is given the entire payoff, and A's signal is interpreted as advice. The use of the inversion signal can, again, lead to a better outcome. Now successful communication requires that it is common ground that both parties adopt the interests of the receiver, and that they agree upon the convention that best achieves her objectives.

3.6 Conclusions and Future Directions

The reasoning underpinning how people agree spontaneous communicative conventions is remarkably rich. In human communicative interactions, such reasoning is often so immediate and so natural that we are entirely unaware of its existence. But, of course, we are equally oblivious to the formidable computations involved in perception, motor control, and common-sense reasoning. Building human-like artificial intelligence systems with which people can smoothly interact will require spelling out these processes in much greater formal detail. Here we have only lightly sketched the beginnings of a theory, by which signal-meaning mappings are created by a process of virtual bargaining, in the light of common knowledge between sender and receiver.

We have touched on two specific areas in which theoretical elaboration is required. First, understanding the role of cooperation and how it is possible that people can agree the meaning of spontaneous communicative conventions, even when they are in conflictual interaction (i.e., neither believes that the other is actually trying to cooperate). Second, how the nature of the communicative act (instruction, advice, and so on) affects the objective of the bargaining process. A full account will need to deal more fully with these difficult issues, but also many others.

One set of challenges concerns how common ground is established. In practice, common ground can sometimes be established by 'public' announcements, or the details of a scene being in 'plain view' for both parties—although concluding that information is in common ground appears to presuppose that both parties are paying attention to the publicly available information, and indeed that it is common ground that both parties are doing so. Thus, a regress threatens—inferences about what is common ground seem, inevitably, to depend on prior assumptions about what common ground comprises. The threat of type of regress may suggest that common ground needs to be considered as basic rather than being derived from individual knowledge.

Another challenge is how to take account of cognitive limitations. For example, if one party doubts that the other has noticed a good potential convention, or doubts that the other believes that they themselves have noticed that convention, and so on, then the existence of that convention will not be in common ground, and hence would appear to be blocked from being used by both parties.

Yet a further, and crucial, set of issues concerns how people trade off the communicative demands of the present situation (e.g., that the inversion mapping is best, because it will allow two bananas to be chosen, not one) against precedent from past mappings (usually, one person simply marks the box with a banana). We find experimentally that people are able, flexibly, to modify conventions from trial to trial. But, equally, they reuse and generalize previously established conventions. The balance between moment-by-moment flexibility and the creation of layers of increasingly rich communicative conventions seems crucial in attempting to understand the emergence, and functioning, of human language (Hopper and Traugott, 2003; Christiansen and Chater, 2008, 2016; Kirby *et al.*, 2008).

This chapter has outlined an approach to understanding the spontaneous emergence of communicative conventions in simple stylized scenarios. The reasoning underpinning the choice of these conventions is remarkably subtle and as yet only partially understood. As we have seen, issues concerning the nature of common ground, the role of cognitive limitations, and the interplay between present task demands and the precedent from past communicative interactions, and many more, are currently not well understood. Yet such understanding will be required in order to build computational systems that are able to communicate with us in a genuinely human-like way.

Acknowledgements

NC and JM were supported by ERC grant 295917-RATIONALITY; N. C. was also partially supported by the ESRC Network for Integrated Behavioural Science [grant number ES/P008976/1].

References

Attneave, F. (1954). Some informational aspects of visual perception. *Psychological Review*, **61**, 183–93.

Bacharach, M., Gold, N., and Sugden, R. (2006). *Beyond Individual Choice: Teams And Frames in Game Theory*. Princeton, NJ: Princeton University Press.

Bernheim, B. D. (1984). Rationalizable strategic behavior. *Econometrica*, **52**(4): 1007–28.

Bundy, A., Philalithis, E., and Li, X. (this volume). Modelling Virtual Bargaining Using Logical Representation Change.

Chater, N. (1996). Reconciling simplicity and likelihood principles in perceptual organization. *Psychological Review*, **103**, 566–81.

Chater, N., Misyak, J. B, Melkonyan, T. *et al.* (2016). Virtual Bargaining: Building the Foundations for a Theory of Social Interaction, in J. Kiverstein (ed.), *Routledge Handbook of the Philosophy of the Social Mind*. Abingdon: Routledge, 418–30.

Chater, N., and Vitányi, P. (2002). Simplicity: A unifying principle in cognitive science? *Trends in Cognitive Sciences*, 7, 19–22.

Christiansen, M., and Chater, N. (2008). Language as shaped by the brain. *Behavioral and Brain Sciences*, **31**, 489–558.
Christiansen, M., and Chater, N. (2016). *Creating Language*. Cambridge, MA: MIT Press.
Clark, H. H. (1996). *Using Language*. Cambridge: Cambridge University Press.
Colman, A. M. (2003). Cooperation, psychological game theory, and limitations of rationality in social interaction. *Behavioral and Brain Sciences*, **26**(2): 139–98.
Feldman J. (2009). Bayes and the simplicity principle in perception. *Psychological Review*, **116**, 875–87.
Frank, M. C., and Goodman, N. D. (2012). Predicting pragmatic reasoning in language games. *Science*, **336**(6084), 998.
Goodman, N. D., and Frank, M. C. (2016). Pragmatic language interpretation as probabilistic inference. *Trends in Cognitive Sciences*, **20**(11), 818–29.
Grice, H. P. (1957). Meaning. *The Philosophical Review*, **66**(3), 377–88.
Grice, H. P. (1975). Logic and conversation, in D. Davidson and G. Harman (eds), *The Logic of Grammar*. Encino, CA: Dickenson, 64–75.
Hopper, P. J., and Traugott, E. C. (2003). *Grammaticalization*. Cambridge: Cambridge University Press.
Kirby, S., Cornish, H., and Smith, K. (2008). Cumulative cultural evolution in the laboratory: an experimental approach to the origins of structure in human language. *Proceedings of the National Academy of Sciences*, **105**(31), 10681–6.
Levinson, S. C. (2000). *Presumptive Meanings: The Theory of Generalized Conversational Implicature*. Cambridge, MA: MIT Press.
Misyak, J. B., and Chater, N. (2014). Virtual bargaining: A theory of social decision-making. *Philosophical Transactions of the Royal Society B: Biological Sciences*, **369**(1655), 20130487.
Misyak, J. B., Melkonyan, T., Zeitoun, H. *et al.* (2014). Unwritten rules: Virtual bargaining underpins social interaction, culture, and society. *Trends in Cognitive Sciences*, **18**(10), 512–19.
Misyak, J., Noguchi, T., and Chater, N. (2016). Instantaneous conventions. *Psychological Science*, **27**(12), 1550–61.
Nash, J. F. (1950). Equilibrium points in n-person games. *Proceedings of the National Academy of Sciences*, **36**(1), 48–9.
Pearce, D. (1984). Rationalizable strategic behavior and the problem of perfection. *Econometrica*, **52**, 1029–50.
Schelling, T. C. (1960). *The Strategy of Conflict*. Cambridge, MA: Harvard University Press.
Shafto, P., Goodman, N. D., and Griffiths, T. L. (2014). A rational account of pedagogical reasoning: Teaching by, and learning from, examples. *Cognitive Psychology*, **71**, 55–89.
Sperber, D., and Wilson, D. (1986). *Relevance: Communication and Cognition*, Vol. 142. Cambridge, MA: Harvard University Press.
Sugden, R. (2003). The logic of team reasoning. *Philosophical Explorations*, **6**(3), 165–81.

4

Modelling Virtual Bargaining using Logical Representation Change

Alan Bundy, Eugene Philalithis, and Xue Li

University of Edinburgh, UK

4.1 Introduction—Virtual Bargaining

A recently developing body of empirical work on joint problem-solving explores limit cases of human coordination, where signalling conventions can still be efficiently formed, and flexibly revised, even without sufficient information bandwidth to coordinate them.

In a typical example, pairs of human participants are presented with tasks where (1) the information required to complete each task, and (2) the capacity to act on that information, are divided between them. One participant—a sender—holds key information but cannot act on it. The other participant—a receiver—can take the actions needed but needs additional information to select these actions among possible alternatives (Misyak *et al.*, 2016). Neither participant can use language, or another medium of sufficient bandwidth to express all of the information required.

Yet human participants comfortably succeed in these 'impossible' coordination games. Humans select optimal moves (Misyak and Chater, 2014), create appropriate conventions (Misyak *et al.*, 2016), and develop these initial conventions appropriately as the task complexity grows (Misyak and Chater, 2017) in order to maximize their joint profit—all this despite a greatly restricted communication channel.

This success in the face of insufficient explicit communication motivates the theory of *virtual bargaining*. Virtual bargaining rests on the need for additional inference to bridge the gap between the information *available*, and the information *required* to interpret players' signals. According to virtual bargaining, this added inference takes the form of a 'what if' scenario played out privately by both players, each adopting the most beneficial outcome of virtual negotiation on how to interpret their signals (Misyak and Chater, 2014). When both players imagine the same 'what if' scenario and play as if the virtual negotiation really happened, their interpretations will match.

Virtual bargaining therefore divides the burden of signal interpretation between observed information and private reasoning—and as a result can explain instances of 'impossible' coordination. However, no computational model currently exists for how

Alan Bundy, Eugene Philalithis, and Xue Li, *Modelling Virtual Bargaining using Logical Representation Change* In: *Human Like Machine Intelligence.* Edited by: Stephen Muggleton and Nick Charter, Oxford University Press.
 DOI: 10.1093/oso/9780198862536.003.0004

the (effectively one-shot) learning and flexible revision displayed across these instances can be feasibly reproduced. At the same time, it has been argued that virtual bargaining underwrites a number of human conventions and unwritten rules, from politics (Misyak *et al.*, 2014) to driving (Chater *et al.*, 2018). The promise of human-like virtual bargaining abilities replicated by artificial agents is thus vast: from machines that communicate non-verbally but effectively with humans in joint problem-solving to machines that grasp, create, and share unwritten workplace rules with humans, and with each other.

Our present chapter reports work in progress aiming to replicate this coordination behaviour. We consider examples of signalling conventions spontaneously adapted in a simple game of item selection and avoidance, and we suggest that the rich, efficient inference stipulated by virtual bargaining for these conventions can be understood as *logical inference* capturing the players' joint reasoning about each game and its rules; specifically, logical inference facilitated by the ABC system for representation change.

The **ABC** Repair System (Li *et al.*, 2018) combines **A**bduction (Cox and Pietrzykowski, 1986) and **B**elief revision (Gärdenfors, 1992) with the more recent Reformation algorithm (Bundy and Mitrovic, 2016) for **C**onceptual change. Abduction and belief revision repair faulty logic theories by respectively adding/deleting axioms or deleting/adding preconditions to rules. Reformation repairs them by changing the *language* of a theory. For practical reasons (discussed below in section 4.3.2), the ABC System is limited to Datalog theories (Ceri *et al.*, 1990), although Reformation has been implemented for richer logics (Mitrovic, 2013; Bundy and Mitrovic, 2016). Datalog is a logic programming language restricted to Horn clauses and allowing no functions except for constants, but it has proven adequately expressive for our usage.

4.2 What's in the Box?

We begin by considering the family of human behaviours we presently aim to model, in the form of moves made by players in a coordination game, and explain our overall approach.

A reliable demonstration of virtual bargaining is built around a two-player game of item selection and avoidance (Misyak *et al.*, 2016). In this game, a sender can see inside boxes with harmful or helpful contents, such as a scorpion or a banana. The sender can mark one of the items for the receiver in some way, but cannot open them. In turn, the receiver can open any of the items, but they cannot see inside. The players' joint goal is to open as many helpful items as possible per round while opening no harmful items; their restriction is that the sender alone cannot give sufficient input for the receiver to determine what unopened items belong in each set.

The general procedure is described as follows (Misyak *et al.*, 2016):

> We developed an interactive two-player computer game in which both partners viewed a 3D-simulated scene, but each saw the scene from the opposite visual perspective .... The game environment consisted of three boxes, each containing either a reward (banana) or nonreward (scorpion). The number of rewards and

> the rewards' locations in the boxes, as well as other scene variables, changed from trial to trial. One partner played the role of sender, and the other played as a receiver: They shared the joint task of uncovering as many rewards as possible while avoiding nonreward. Contents of the boxes were visible only to the sender by means of panels that slid open on the side of the box facing the sender .... However, a set of shadows (impressions of bananas and scorpions, embedded in the virtual ground) was sometimes mutually visible to both players. The shape and number of these shadows corresponded to the number of scorpions and bananas inside the three boxes on that trial.

In other words, both players know the ratio of helpful (bananas) to harmful items (scorpions), but only the sender knows what's in each box. Figure 4.1 illustrates the baseline condition of this arrangement. The sender may then use a token visible to both players in order to mark items for the receiver's benefit. In the manipulation of interest to virtual bargaining, the sender can mark *at most one* out of the three items.

Players take turns: the sender marks, then the receiver makes their choices, before the outcome of the game is announced to both. Selecting the maximum number of helpful items, while avoiding all harmful items, will win the game; all other outcomes lose. Winning the game is conditional on interpretation: where only one mark (i.e., one axe token) is available to the sender, players must use it flexibly. When there are more harmful than helpful items (i.e., two scorpions to one banana) the sender marks the single helpful item, and the receiver interprets the mark to mean 'helpful'. When there are more helpful than harmful items (i.e. one scorpion to two bananas) the sender marks the *harmful* item, and the receiver interprets the mark to mean '*all other items* helpful'.

Negotiating this flexible signalling convention explicitly is impossible for players. They must instead *infer* their interpretation, for example, from what they both know about the game (the game rules, the end goal, the ratio of helpful to harmful items) and their respective roles in it. The two players are both given the same rules of the game, and both have the same goal. Otherwise, the only mode of communication available to the players are the tokens used to mark items by the sender. From their shared knowledge of the rules and goals of the game, plus their own private view of the experimental set-up,

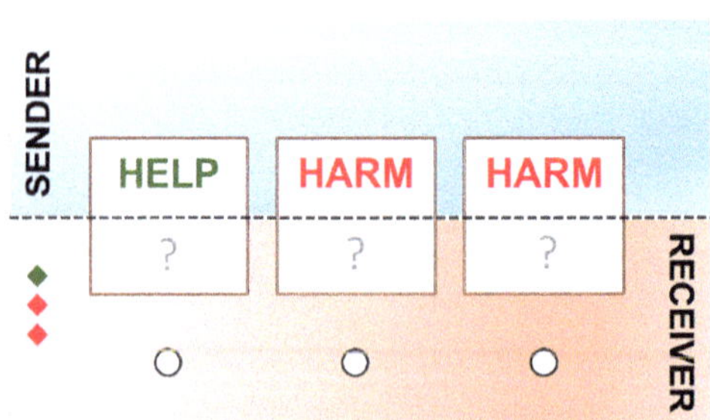

Figure 4.1 *Basic game set-up. Sender sees item contents. Receiver only sees content ratio.*

the players must devise a jointly appropriate convention: the sender must use the tokens to signal the contents of the items in a way the receiver will understand, and the receiver must decipher those signals and act on them as the sender intended.

Furthermore, the game itself can also evolve, adding novel situations, restrictions, or signalling vocabularies (Misyak and Chater, 2017). As a result, any initially established conventions will not always remain optimal. The sender and receiver must adapt their convention to suit each new scenario before playing it, via reasoning, rather than physical trial and error. All of these behaviours are attributed to virtual bargaining: the advance modelling, via reasoning, of joint problem-solving before it even happens.

A first step toward replicating virtual bargaining is thus a faithful reproduction of players' behaviours in this selection and avoidance game. That task is made easier by the very minimal interaction permitted between players: the sender's choice of mark and the receiver's choice of items, taken in turns, are all the information transmitted. Modelling human players' behaviour therefore reduces to modelling their choices, for example, as a result of inference capturing players' reasoning from the minimal available input.

The present body of work on virtual bargaining distinguishes two clearly separable reasoning steps in how human players respond to this family of games (Misyak and Chater, 2017): (1) constructing an initial signalling convention from players' shared knowledge of the game rules, the end goal, and other available information; and (2) adapting that convention after circumstances change. Accordingly, the job of modelling virtual bargaining divides into two distinct pieces: (a) an algorithm for how players use the available information to establish a signalling convention spontaneously without negotiation; and (b) an algorithm of how players spontaneously adapt that signalling convention without negotiation. Our present aim is to model the latter process: how players spontaneously adapt established signalling conventions to novel requirements.

4.3 Datalog Theories

We now move to consider our toolset, starting with Datalog. Originally invented as a subset of Prolog, as a logic programming language targeted at querying deductive databases (Ceri *et al.*, 1990), Datalog can also be treated as a sub-logic of first-order logic.

4.3.1 Clausal form

Datalog programs are a collection of rules and ground facts. They can be represented as a subset of Horn clauses, which are disjunctions of negated or unnegated propositions. To emphasize this relationship, we will use Kowalski's clausal format (Kowalski, 1979):[1]

$$(Q_1 \wedge \ldots \wedge Q_m) \implies (R_1 \vee \ldots \vee R_n) \tag{4.1}$$

[1] Kowalski advocates a variation of this format which is more suggestive of its procedural reading. He puts the head on the left, the body on the right, and writes the implication arrow backwards. This is also the version used in Prolog.

where the Q_i are implicitly negated because they are on the LHS of the implication arrow.

Definition 4.1 (Horn Clauses) *Horn clauses are clauses (4.1) in which either* $n = 0$ *or* $n = 1$. *They, therefore, fit one of the following four forms.*

Implication: $(Q_1 \wedge \ldots \wedge Q_m) \implies R$. *These usually represent the rules of a theory.*

Assertion: $\implies R$. *These usually represent the facts of a theory.*

Goals: $Q_1 \wedge \ldots \wedge Q_m \implies$. *These usually arise from the negation of the conjecture to be proved and subsequent subgoals in a derivation.*

Empty Clause: $\implies$. *This is the target of a refutation-style proof. It represents success in proving a conjecture.*

where $1 \leq m$, *and* R *and the* Q_i *are* propositions, *that is, formulae of the form* $P(t_1, \ldots, t_n)$, *where each* t_j *is either a variable or a constant. Where they exist,* R *is called the* head *of the clause and the* Q_i *form the* body.

We will adopt the convention that variables are written in lower case, and constants and predicates start with a capital letter.[2]

4.3.2 Datalog properties

Datalog programs standardly consist of implications (rules) and ground assertions (facts). Our Datalog theories, however, also contain goals and the empty clause, so as to represent conjectures to be proved, and the derivation of false in refutation proofs.

In our Datalog theories, we also adopt the following Datalog program restrictions:

1. There are no non-nullary functions, that is, the arguments to predicates are either variables or constants, so there is no function nesting.
2. Each predicate has a unique arity.
3. There are no unsafe clauses, in other words each variable that appears in the head of a clause also appears in the body of that clause.

As we see below, despite these restrictions, Datalog is sufficiently expressive for our application to virtual bargaining. Further possible restrictions exist, to allow Datalog to be more efficient as a programming language, which we do not need to adopt here.

Deduction in Datalog is decidable. This is not the case in full first-order logic (FOL), which is only semi-decidable, that is, if there is a proof, FOL deduction will eventually find it by exhaustive search, but if there isn't we could search fruitlessly forever. In Datalog, if there is no proof of a conjecture, the search will eventually terminate without success, so we can be sure that the conjecture is not a theorem. This is one of the more important technical advantages of restricting our logical theories to Datalog.

[2] The opposite of the standard Prolog convention.

The decidability of Datalog is a consequence of its lack of functions. This is because there are only a finite number of ground terms,[3] namely the set of constants. This means that there are only a finite number of distinct formulas.[4] Since the number of ground terms is finite, and the Herbrand base is also finite, all quantified formulae can be translated into propositional logic and as a result all Datalog theories are decidable.

In addition to implementing deduction in Datalog theories, our ABC system also uses a special mechanism for the $=$ predicate, based on the unique name assumption. Different constants are assumed to be unequal, unless this assumption is overridden by an explicit $=$ relation asserted between them. This has the consequence that we can treat $\neq$ as an un-negated predicate in its own right, not as the negation of $=$. This enables its use as a predicate in propositions that form the clauses of Datalog theories.

4.3.3 Application 1: Game rules as a logic theory

The rules of the selection and avoidance game can be represented as a Datalog theory:[5]

- The receiver must select each helpful item.

$$item \in Help \implies Select(Receiver, item) \tag{4.2}$$

- The receiver must not select a harmful item.

$$item \in Harm \wedge Select(Receiver, item) \implies \tag{4.3}$$

  Note our unusual use of a goal clause to express constraint (4.3) as a Horn clause. This would not be allowed in a Datalog *program* but it is allowed in our Datalog *theories*.

- No items are both helpful and harmful.

$$item_1 \in Help \wedge item_2 \in Harm \implies item_1 \neq item_2 \tag{4.4}$$

- If there are both helpful and harmful items then the sender must mark an item.

$$Help \neq \varnothing \wedge Harm \neq \varnothing \implies Mark(Sender, Sk) \tag{4.5}$$

  where Sk is a Skolem constant, an item whose identity we know nothing about.

- The sender can mark at most one thing.

$$Mark(Sender, item_1) \wedge Mark(Sender, item_2) \implies item_1 = item_2 \tag{4.6}$$

[3] That is, terms without variables.
[4] Up to variable renaming.
[5] We call this the *Unique Name Assumption with Exceptions*. Note that this allows the use of both $=$ and $\neq$ in a Datalog theory.

In the above formalization, $Help$ corresponds to the set of items[6] (boxes) containing bananas and $Harm$ to the set containing scorpions. $Mark(Sender, item)$ means the sender places a token on $item$. $Select(Receiver, item)$ means the receiver opens $item$.

For present purposes, we only consider cases where the sender only has one token they can place. For a condition in which the sender has more tokens, the rules will be slightly different. In such a case, the game rules themselves will have changed, but these rules are determined by the experimenter. The participants would be *informed* the rules have changed, rather than *infer* the necessary changes as a result of logical reasoning, for example. The evolution of players' rules knowledge is not part of our modelling target.

It is important to note that, at this stage, the above rules are insufficient for either player to plan their moves. An additional logical step is required, as we discuss below.

4.3.4 Application 2: Signalling convention as a logic theory

We now move to consider the baseline condition of our selection and avoidance game, abstracted in Figure 4.2. The top and bottom half represent the game environment as the sender and the receiver view it. The labels on the items indicate whether their contents are helpful or harmful, which the sender can see. The receiver (who is below the dividing line) cannot see the contents of the items. The tick under one of the items denotes the item being marked with a token by the sender, as a signal for the receiver.

We will refer to the items as Box_1, Box_2, and Box_3 from left to right, respectively.

For the sender's mark to work as a signal encoding information about the game, players require a convention for its interpretation. The convention used by participants in this baseline case is quite simple, and can be represented by the following two clauses:

- Marking an item signals an item.

$$Mark(Sender, item) \implies Signal(item) \tag{4.7}$$

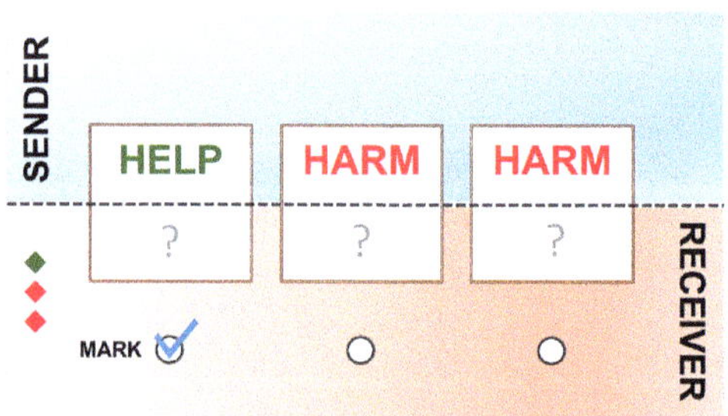

Figure 4.2 *Baseline condition. Marking an item implies the item is helpful.*

[6] We use set membership, rather than a unary predicate, to denote helpful vs harmful. Together with our implementation of $\neq$ this allows *non*-membership to be represented via Horn clause in (4.5).

- Any signalled items must be helpful.

$$Signal(item) \implies item \in Help \tag{4.8}$$

In other words, marking is a signal for 'helpful'. Although we do not presently explore how this initial convention is spontaneously established, it is arguably intuitive. In any item selection game, interpreting a signal as simply meaning 'this is the item to select' is a likely initial strategy for players to attempt, even without coordination.

Combined with the above, this convention easily determines the receiver's strategy:

- Marking an item signals an item.

$$Mark(Sender, item) \implies Signal(item)$$

- Any signalled items must be helpful.

$$Signal(item) \implies item \in Help$$

- Select each helpful item.

$$item \in Help \implies Select(Receiver, item)$$

(Note that this is rule 4.2, from section 4.3.3, so is protected from repair.)

The sender's strategy can be given accordingly, using (4.5), (4.6), (4.7), and (4.8).

4.4 SL Resolution

SL Resolution (Kowalski and Kuehner, 1971) is a deductive rule that is particularly well suited to Reformation. A single SL Resolution step takes the form:

$$\frac{R_1 \wedge \ldots \wedge R_i \ldots \wedge R_k \implies}{(R_1 \wedge \ldots \wedge R_{i-1} \wedge Q_1 \wedge \ldots \wedge Q_m \wedge R_{i+1} \ldots \wedge R_k)\sigma \implies} \quad (Q_1 \wedge \ldots \wedge Q_m) \implies \mathbf{P}$$

where the highlighted R_i is the selected literal, the highlighted P is the literal it is resolved with and σ is the most general unifier of P and R_i, that is, the most general substitution of terms for variables that will make them identical.

4.4.1 SL refutation

Resolution proofs work by refutation. The conjecture to be proved is negated and added to the axioms. The empty clause, $\implies$, is then (we hope) derived. This derivation is interpreted as showing that negating the target conjecture leads to a contradiction, such that the conjecture has been proved by *reductio ad absurdum*. For Horn clause theories, negated conjectures take the form of a goal clause, as already shown in (4.3).

An SL Resolution refutation on Horn clauses takes the form:

$$\cfrac{\cfrac{Goal}{Goal_1 \wedge \ldots \wedge Goal_m}\ Axiom}{\vdots} \qquad \cfrac{Goal}{\Longrightarrow}\ Axiom$$

where the $Goal$s are all goal clauses and the $Axiom$s are either implication or assertion clauses. This has the advantage that we can apply any repair (as we discuss in section 4.5.1) directly to the axiom involved in the resolution step, without needing to inherit the repair back up the refutation to an axiom. This advantage is secured by restricting to Datalog theories, as all their formulae are Horn clauses. SL Resolution refutation on non-Horn clauses also requires ancestor resolution: that is, resolution between a goal literal and another goal above it on the same branch. In that case, no axiom is directly involved and inheritance is required. Avoiding the need for such inheritance is another of the technical advantages gained by restricting our logical theories to Horn clauses.

4.4.2 Executing the strategy

The receiver's strategy determines which items they select as a result of the sender's actions. This strategy can be executed by applying SL Resolution to the goal clause $Select(Receiver, item) \Longrightarrow$ using the convention clauses from Section 4.3.4 plus the assertion $\Longrightarrow Mark(Sender, Box_1)$, as observable from Figure 4.2. In the course of this refutation, $item$ will be instantiated to one of the three available boxes. The desired refutation is:

$$\cfrac{\cfrac{\cfrac{\cfrac{Select(Receiver, item) \Longrightarrow}{item \in Help \Longrightarrow}\ item \in Help \Longrightarrow Select(Receiver, item)}{Signal(item) \Longrightarrow}\ Signal(item) \Longrightarrow item \in Help}{Mark(Sender, item) \Longrightarrow}\ Mark(Sender, item) \Longrightarrow Signal(item)}{\Longrightarrow}\ \Longrightarrow Mark(Sender, Box_1) \tag{4.9}$$

This proves $Select(Receiver, Box_1)$: the receiver selects just Box_1 as intended.

The sender's strategy can also be represented using the clauses from section 4.3.4, but with the instantiated goal clause $Select(Receiver, Box_1) \Longrightarrow$ and the uninstantiated assertion $\Longrightarrow Mark(Sender, item)$, where $item$ must be instantiated to the specific item to be marked with a token. This is because the sender wants to discover which item they must mark, so that the receiver will subsequently select Box_1, as intended:

$$\begin{array}{ll}
Select(Receiver, Box_1) \Longrightarrow & \\
\hline
Box_1 \in Help \Longrightarrow & item \in Help \Longrightarrow Select(Receiver, item) \\
\hline
Signal(Box_1) \Longrightarrow & Signal(item) \Longrightarrow item \in Help \\
\hline
Mark(Sender, Box_1) \Longrightarrow & Mark(Sender, item) \Longrightarrow Signal(item) \\
\hline
\Longrightarrow & \Longrightarrow Mark(Sender, item)
\end{array}$$

4.5 Repairing Datalog Theories

Having considered theory representation, we now move to consider theory repair. The ABC System (Li *et al.*, 2018) diagnoses and repairs two kinds of fault in Datalog theories: incompatibility and insufficiency. Both arise from reasoning failures: mismatches between the theorems of a theory $\mathbb{T}$ and the observations of an environment $\mathbb{S}$, such as our game environment. $\mathbb{S}$ is a pair of sets of ground propositions $\langle \mathcal{T}(\mathbb{S}), \mathcal{F}(\mathbb{S})\rangle$. A *ground proposition* is a formula of the form $P(C_1, \ldots, C_n)$, where P is an n-ary predicate and the C_i are constants. $\mathcal{T}(\mathbb{S})$ are the ground propositions we observe to be true and $\mathcal{F}(\mathbb{S})$ are those we observe to be false. So, ideally:

$$R \in \mathcal{T}(\mathbb{S}) \Longrightarrow \mathbb{T} \vdash R$$
$$R \in \mathcal{F}(\mathbb{S}) \Longrightarrow \mathbb{T} \nvdash R$$

That is, the true ground propositions are theorems of $\mathbb{T}$ and the false ones are not.

We can view the theorems of $\mathbb{T}$ as predictions about the environment. These predictions can be confounded in two ways: something false is predicted (incompatibility) or something true is not predicted (insufficiency).

Definition 4.2 (Incompatibible and Insufficient)

Incompatible: $\mathbb{T}$ *is* incompatible *with* $\mathbb{S}$ *iff* $\exists R.\ \mathbb{T} \vdash R \wedge R \in \mathcal{F}(\mathbb{S})$.

Insufficient: $\mathbb{T}$ *is* insufficient *for* $\mathbb{S}$ *iff* $\exists R.\ \mathbb{T} \nvdash R \wedge R \in \mathcal{T}(\mathbb{S})$.

4.5.1 Fault diagnosis and repair

These two kinds of fault are diagnosed and repaired in a dual way. $\mathcal{F}(\mathbb{S})$ and $\mathcal{T}(\mathbb{S})$ are both finite sets. The ABC system tries to prove each member of these sets. If a member of $\mathcal{F}(\mathbb{S})$ is proved then we have discovered an incompatibility. Similarly, if a member of $\mathcal{T}(\mathbb{S})$ is not proved then we have discovered an insufficiency. Incompatibilities can be repaired by blocking the unwanted proof. Insufficiencies can be repaired by unblocking a wanted failed proof.

Definition 4.3 (Repair Operations for Incompatibility) *In the case of incompatibility, the unwanted proof can be blocked by causing any of the resolution steps to fail. Suppose the chosen resolution step is between a goal* $P(s_1, \ldots, s_n)$ *and an axiom* $Body \Longrightarrow P(t_1, \ldots, t_n)$, *where each* s_i *and* t_i *pair can be unified. Possible repair operations are as follows:*

Belief Revision 1: *Delete the targeted axiom.*

Belief Revision 2: *Add an unprovable precondition to the body of the targeted axiom.*

Reformation 1: *Rename P in the targeted axiom to the new predicate P'.*

Reformation 2: *Increase the arity of all occurrences P in the axioms by 1. Ensure recursively that the new arguments, s_{n+1} and t_{n+1}, in the targeted occurrence of P, are not unifiable.*

Reformation 3: *For some i, suppose s_i and t_i are both the constant C. Change t_i to the new constant C'.*

Heuristic 1 (Algorithm for Creating New Arguments) *In operation Reformation 2 of Definition 4.3, we need to create a new argument for the $n+1$ argument position in each occurrence of P. The spirit of Datalog is that assertions are ground facts and implications are non-ground general rules. We have also observed that the new arguments are usually used to distinguish different types of P. Reformation is a purely syntactic algorithm, which does not have access to any semantics when choosing new constant names. Therefore, we use the new constants, $Abnormal$ and $Normal$, subscripted if necessary, when constants are needed as new arguments. Otherwise, we use new variables.*

The following algorithm is used to decide which term to use in each occurrence.

1. *For the targeted axiom $Body \implies P(t_1, \ldots, t_n)$ let t_{n+1} be $Abnormal$. For the goal proposition $P(s_1, \ldots, s_n)$ that it is resolved with let s_{n+1} be $Normal$.*
2. *Propagate these two constants by instantiating the resolutions steps they are inherited from or to: $Normal$ upwards and $Abnormal$ downwards.*
3. *Select one axiom whose $n+1$ argument has been instantiated to $Normal$ and one to $Abnormal$. Where there is a choice, prefer facts over rules. Typically, choose the top-most axiom for $Normal$ and the bottom-most one for $Abnormal$. For all other $n+1$s arguments, choose a new variable per axiom.*

This algorithm is illustrated in section 4.5.2 below.

Definition 4.4 (Repair Operations for Insufficiency) *In the case of insufficiency, the wanted failed proof can be unblocked by causing a currently failing resolution step to succeed. Suppose the chosen resolution step is between a goal $P(s_1, \ldots, s_m)$ and an axiom $Body \implies P'(t_1, \ldots, t_n)$, where either $P \neq P'$ or for some i, s_i, and t_i cannot be unified. Possible repair operations are:*

Abduction 1: *Add a new axiom whose head unifies with the goal $P(s_1, \ldots, s_m)$.*

Abduction 2: *Locate the axiom whose body proposition created this goal and delete this proposition from the axiom.*

Reformation 4: *Replace $P'(t_1, \ldots, t_n)$ in the axiom with $P(s_1, \ldots, s_m)$.*

Reformation 5: *Decrease the arity of all occurrences P' by 1. Remove the ith argument from P', that is, the i for which s_i and t_i are not unifiable.*

Reformation 6: *If s_i and t_i are not unifiable, then they are unequal constants, say, C and C'. Either (1) rename all occurrences of C' in the axioms to C or (2) replace the offending occurrence of C' in the targeted axiom by a new variable.*

4.5.2 Example: The black swan

The following example is adapted from (Gärdenfors, 1992). Consider the following Datalog theory $\mathbb{T}$:

$$\begin{aligned} German(x) &\Longrightarrow European(x) \\ European(x) \wedge Swan(x) &\Longrightarrow White(x) \\ \Longrightarrow German(Bruce) & \qquad \Longrightarrow Swan(Bruce) \end{aligned}$$

From these axioms, we can infer $White(Bruce)$.

$$\dfrac{\dfrac{\dfrac{\dfrac{White(Bruce) \Longrightarrow \qquad European(x) \wedge Swan(x) \Longrightarrow White(x)}{European(Bruce) \wedge Swan(Bruce) \Longrightarrow \qquad German(x) \Longrightarrow European(x)}}{German(Bruce) \wedge Swan(Bruce) \Longrightarrow \qquad \Longrightarrow German(Bruce)}}{Swan(Bruce) \Longrightarrow \qquad \Longrightarrow Swan(Bruce)}}{\Longrightarrow} \tag{4.10}$$

However, suppose we observe that Bruce is black and not white, that is, $Black(Bruce) \in \mathcal{T}(\mathbb{S})$ and $White(Bruce) \in \mathcal{F}(\mathbb{S})$. $\mathbb{T}$ is, therefore, both incompatible and insufficient wrt $\mathbb{S}$.

We will deal with the incompatibility. One solution, mooted in (Gärdenfors, 1992), is to add an exception to one of the rules, for example:

$$x \neq Bruce \wedge European(x) \wedge Swan(x) \longrightarrow White(x)$$

This seems to us to be an unsatisfactory solution. A better solution is to note that $European(x)$ is ambiguous. It could be interpreted as x is a European *type* or as a European *resident*. European types of swans are white, but a black swan can be a resident, for example, in a zoo.

We can achieve this repair, for instance, by adding an additional argument to $European$ (see operation Reformation 2 in Definition 4.3). To affect this, we need to break the highlighted resolution step in refutation 4.10. We now need to create new terms for all these extra arguments by following Heuristic 1. We consider first the axiom involved in the targeted resolution step. Giving $European$ a new variable y as an argument gives:

$$German(x) \Longrightarrow European(x, y)$$

We are now in violation of the safety restriction (restriction 3 in section 4.3.2): there is a variable in the clause's head that does not appear in its body. So we must also add y to $German$.

$$German(x, y) \implies European(x,y)$$

We must now add a new argument to any other occurrence of $European$ and $German$—and do so in such a way to ensure that the red unification in the refutation will fail. To help us decide how to do this, we:

- Add the new arguments of $European$ and $German$ into refutation (4.10).
- In the highlighted resolution step we want to break, instantiate the new arguments to the two occurrences of $European$ to $Normal$ in the goal and $Abnormal$ in the axiom.
- Propagate these instantiations through the refutation: $Normal$ upwards and $Abnormal$ downwards.

This gives:

$$\cfrac{\cfrac{\cfrac{\cfrac{White(Bruce) \implies \quad European(x, Normal) \wedge Swan(x) \implies White(x)}{European(Bruce, Normal) \wedge Swan(Bruce) \implies \quad German(x, Abnormal) \implies European(x, Abnormal)}}{German(Bruce, Abnormal) \wedge Swan(Bruce) \implies \quad \implies German(Bruce, Abnormal)}}{Swan(Bruce) \implies \quad \implies Swan(Bruce)}}{\implies}$$

which breaks the refutation by the failure of $European(x, Abnormal)$ to unify with $European(x, Normal)$. This analysis suggests the following repaired theory, $\nu(\mathbb{T})$:

$$\begin{aligned} German(x, y) &\implies European(x, y) \\ European(x, Normal) \wedge Swan(x) &\implies White(x) \\ \implies German(Bruce, Abnormal) & \qquad \implies Swan(Bruce) \end{aligned}$$

from which $White(Bruce)$ is no longer provable. This sequence is a basic example of a repair conducted automatically using the ABC system (Li *et al.*, 2018).

4.6 Adapting the Signalling Convention

We now demonstrate an application of these theory repair ideas to the spontaneous adaptation of conventions for the selection and avoidance game. We will consider three alternatives to the baseline version of the game illustrated in Figure 4.2, and how the signalling convention used in that baseline can be automatically adapted for each.

4.6.1 'Avoid' condition

A different variant of the selection and avoidance game is depicted in Figure 4.3.

In this alternative, there are two bananas, and to achieve their mutual goal, the sender must guide the receiver to select both of them. The receiver still does not know exactly which items contain the bananas, but both players do know that there are more bananas than scorpions—as indicated in the figure by the annotation $|Harm| < |Help|$. This annotation must be added as an additional axiom $|Harm| < |Help|$. The sender, as depicted, reverses the earlier convention, and marks the box containing the scorpion.

It is important to reiterate that in human trials, both players can spontaneously adopt this new convention without trial and error (Misyak *et al.*, 2016), with this behaviour in particular used to motivate virtual bargaining as an explanation.

To explain how this spontaneous adaptation can be reproduced automatically, we return to our formalizations of the game, its rules, and the players' convention. If the convention defined in section 4.3.4 is applied to this example, together with the game rules in section 4.3.3, it will fail. Instead of indicating that the receiver selects Box_2 and Box_3, which contain the helpful bananas, the convention will indicate that Box_1, containing the scorpion, is helpful. Meaning, there are two insufficiencies and one incompatibility.

Repairing Two Insufficiencies

We will deal first with the two insufficiencies. They are symmetric, so we tackle just the unprovable $Select(Receiver, Box_3) \in \mathcal{T}(\mathbb{S})$. To focus on the root of the problem we can try to prove $Select(Receiver, Box_3)$ in refutation (4.11), instantiate all the variables and track the occurrences of Box_3 (highlighted in green) downwards and the occurrences of Box_1 upwards (highlighted in blue).

$$
\begin{array}{c l}
Select(Receiver, {\color{green}Box_3}) \Longrightarrow & \\ \hline
{\color{green}Box_3} \in Help \Longrightarrow & item \in Help \Longrightarrow Select(Receiver, item) \\ \hline
{\color{red}Signal(Sender, Box_3)} \Longrightarrow & Signal(Sender, item) \Longrightarrow item \in Help \\ \hline
Mark(Sender, {\color{green}Box_3}) \Longrightarrow & Mark(Sender, item) \longrightarrow {\color{red}Signal(item)} \\ \hline
\Longrightarrow & \Longrightarrow Mark(Sender, {\color{blue}Box_1})
\end{array} \tag{4.11}
$$

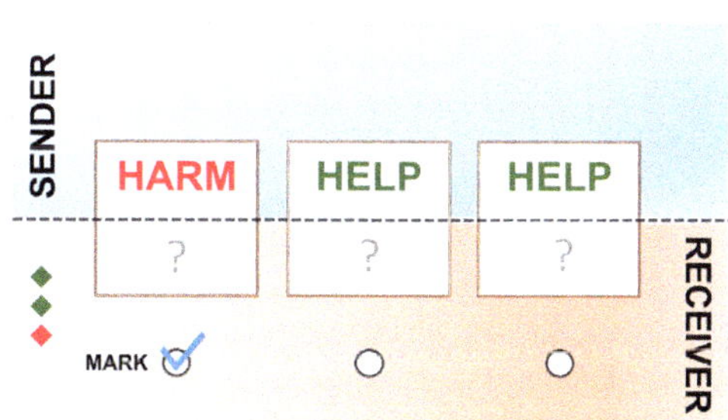

Figure 4.3 *'Avoid' condition. Marking an item implies other items are helpful.*

We can repair this insufficiency using operation Reformation 3 from Definition 4.4 on the first half of the previously established signalling convention, namely the axiom:

$$Mark(Sender, item) \implies Signal(item)$$

The failed unification will succeed if we replace the 'offending' argument in the targeted axiom with a fresh variable:

$$Mark(Sender, item) \implies Signal(item') \tag{4.12}$$

However, this will leave $item'$ as an orphan variable which, owing to the restrictions of our chosen language (Datalog), needs to appear in the body of the rule. This can be achieved by adding the new precondition $item \neq item'$

$$Mark(Sender, item) \wedge item \neq item' \implies Signal(item') \tag{4.13}$$

This new precondition will be satisfied by the implicit inequality $Box_1 \neq Box_3$ that is provided by the ABC System's unique name assumption mechanism. The modified refutation (4.11) will now succeed. The insufficiency has been repaired.

Repairing an incompatibility

We now move on to the incompatibility, caused by the fact that $Select(Receiver, Box_1) \in \mathcal{F}(\mathbb{S})$ but is nonetheless provable:

$$\begin{array}{ll}
\underline{Select(Receiver, Box_1) \implies} & item \in Help \implies Select(Receiver, item) \\
\underline{Box_1 \in Help \implies} & Signal(item) \implies item \in Help \\
\underline{Signal(Box_1) \implies} & Mark(Sender, item) \wedge item \neq item' \implies Signal(item') \\
\underline{Mark(Sender, item) \implies} & \implies Mark(Sender, Box_1) \\
\underline{Box_1 \neq item' \implies} & \implies Box_1 \neq Box_2 \\
\implies &
\end{array} \tag{4.14}$$

One notable aspect of human players' behaviour is that the contrary conventions appear to coexist. Where the two game versions are played back to back, players are capable of efficiently toggling between the 'select' and 'avoid' signalling conventions (Misyak *et al.*, 2016). In line with human performance on this task, we therefore want to repair this incompatibility such that the new convention is flexible enough to cover both the baseline as well as the alternative game. We can achieve this by duplicating an axiom and then modifying the two versions. One version of these axioms will work for the baseline and another version for the alternative in Figure 4.3.

We can repair this incompatibility using operation Belief Revision 2 from Definition 4.3 on the axiom:

$$Mark(Sender, item) \wedge item \neq item' \implies Signal(item')$$

The assertion, already in the theory as a ground proposition, which distinguishes the baseline condition in section 4.3.4 from this one is $|Harm| < |Help|$, so that is the obvious precondition to add:

$$Mark(Sender, item) \wedge item \neq item' \wedge |Harm| < |Help| \implies Signal(item') \quad (4.15)$$

To ensure that axiom (4.12) will still be available to apply to the baseline, but not in the alternative condition in Figure 4.3, we must add the complementary precondition:

$$Mark(Sender, item) \wedge |Help| < |Harm| \implies Signal(item) \quad (4.16)$$

The complete revised signalling convention, in line with human behaviour, is now:

$$\begin{aligned} Mark(Sender, item) \wedge |Help| < |Harm| &\implies Signal(item) \\ Mark(Sender, item) \wedge item \neq item' \wedge |Harm| < |Help| &\implies Signal(item') \\ Signal(item) &\implies item \in Help \end{aligned}$$

This repair is uniquely, automatically identified by the ABC system for axiom (4.7). This is achieved by selecting each precondition only from ground propositions already contained in the theory, corresponding to players' knowledge of the two game variants.

4.6.2 Extended vocabulary

A different variant of the item selection and avoidance game is depicted in Figure 4.4.

In this variant game, the rules are changed to allow two mark placements for each item: an outer and inner position. This choice can be used to distinguish between the left- and right-hand situations depicted. In the former (as with Figure 4.2) only the marked item is helpful; in the latter, all of the items are helpful. Humans spontaneously exploit the extended vocabulary of 'outer' and 'inner' marks to express this difference (Misyak and Chater, 2017).

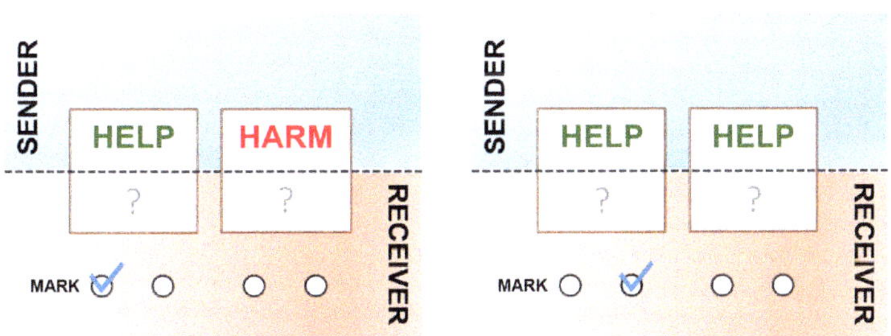

Figure 4.4 *Extended vocabulary. Marking an 'inner' position implies every item is helpful.*

This variant requires an extension to the rules of the game previously given in section 4.3.3. The sender not only marks (at most) one item with their token but must also choose between the inner and outer mark placements while doing so. An additional binary predicate, $Side$ is therefore introduced into the ontology of the rules, along with two new game rules governing its usage:

$$Mark(Sender, item) \implies Side(Sender, Sk_2) \tag{4.17}$$

$$Side(Sender, side_1) \wedge Side(Sender, side_2) \implies side_1 = side_2 \tag{4.18}$$

(4.17) asserts that when the sender marks an item, they must also choose a side. (4.18) asserts that only one side may be chosen. As noted in section 4.3.3, changes to game rules are communicated directly to players: they are not seen as products of inference.

In addition, representing the specific situations in the left- and right-hand side of Figure 4.3 requires us to add, as new axioms, the assertions $Side(Sender, Outer)$ and $Side(Sender, Inner)$ respectively. (These represent what players observe in each case.)

Repairing an Insufficiency

Suppose no distinction is made between the inner and outer positions. In the left-hand side game, this works well. As in section 4.3.4, refutation (4.9) proves $Select(Receiver, Box_1)$, as desired. In the right-hand side, however, though $Select(Receiver, Box_2) \in \mathcal{T}(\mathbb{S})$, it cannot be proved. This is an insufficiency. As before, we can instantiate the failed proof of $Select(Receiver, Box_2)$ to explore where it fails.

$$\begin{array}{ll}
\underline{Select(Receiver, Box_2) \implies} & item \in Help \implies Select(Receiver, item) \\
\underline{Box_2 \in Help \implies} & Signal(item) \implies item \in Help \\
\underline{Signal(Box_2) \implies} & Mark(Sender, item) \implies Signal(item) \\
\underline{Mark(Sender, item) \implies} & \implies Mark(Sender, Box_1) \\
\implies &
\end{array} \tag{4.19}$$

Just as in section 4.6.1, we can repair this insufficiency using operation Reformation 3 from Definition 4.4 on axiom:

$$Mark(Sender, item) \implies Signal(item)$$

by renaming the right-hand-side variable $item$ to $item'$ and by adding the new precondition $item \neq item'$ to cure the resulting orphan variable.

$$Mark(Sender, item) \wedge item \neq item' \implies Signal(item') \tag{4.20}$$

Repairing an Incompatibility

Our repair has now introduced an incompatibility in the left-hand-side game: $Select(Receiver, Box_2)$ is provable, while it is in $\mathcal{F}(\mathbb{S})$. The refutation is:

$$\dfrac{\dfrac{\dfrac{\dfrac{\dfrac{Select(Receiver, Box_2) \implies \quad item \in Help \implies Select(Receiver, item)}{Box_2 \in Help \implies} \quad Signal(item) \wedge item \neq item' \implies item' \in Help}{Signal(item) \wedge item \neq Box_2 \implies} \quad Mark(Sender, item) \implies Signal(item)}{Mark(Sender, item) \wedge item \neq Box_2 \implies \implies} \quad \implies Mark(Sender, Box_1)}{Box_1 \neq Box_2 \implies} \quad \implies Box_1 \neq Box_2}{\implies} \tag{4.21}$$

We choose to break this unwanted proof at the highlighted resolution step. We will use Belief Revision 2, that is, adding another unprovable precondition to the axiom:

$$Mark(Sender, item) \wedge item \neq item' \implies Signal(item')$$

To identify a suitable precondition, note that $Side(Sender, Inner)$ has already been identified as an assertion which is an axiom in the right-hand-side game but is not an axiom in the left-hand-side game. Adding this new precondition gives:

$$Mark(Sender, item) \wedge item \neq item' \wedge Side(Sender, Inner) \implies Signal(item')$$

This new precondition will block the unwanted left-hand-side game proof (4.21) of $Select(Receiver, Box_2)$. However, to ensure that it does not block the wanted left- and right-hand-side game proofs of $Select(Receiver, Box_)$, we retain the original axiom:

$$Mark(Sender, item) \implies Signal(item) \tag{4.22}$$

Note that (4.22) is used to find helpful items regardless of inner vs outer position. The full repaired signalling convention, in line with human behaviour, is therefore:

$$\begin{aligned} Mark(Sender, item) &\implies Signal(item) \\ Mark(Sender, item) \wedge item \neq item \wedge \ Side(Sender, Inner) &\implies Signal(item') \\ Signal(item) &\implies item \in Help \end{aligned}$$

It is the unique automatic repair when modifying axiom (4.7) with the ABC system.

4.6.3 Private knowledge

One final game variant, rounding out our logical exploration, is depicted in Figure 4.5.

In this additional variant of the earlier 'avoid' condition in Figure 4.3, the receiver is allowed to possess private knowledge about one item, while the sender continues to know

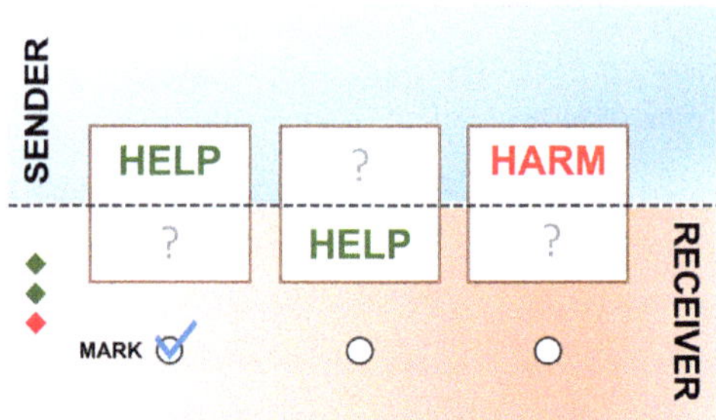

Figure 4.5 *Private vs. negotiable information. Conventions depend on negotiable information.*

(only) the contents of the other two. As a consequence, the players' previous division of game knowledge is altered: the receiver now requires the sender's assistance only for some—not all—items, and the sender no longer knows the true ratio of helpful to harmful items, removing this information from the pool of shared player knowledge.

Unlike the previous, human performance for this example has not yet been reported in the virtual bargaining literature, although equivalent scenarios have been discussed (Misyak *et al.*, 2016). However, the consequences of this arrangement for assumptions set out in this literature are vital to a fuller model of virtual bargaining.

Specifically: where players have established the signalling convention depicted in Figure 4.2 in advance of encountering this variation, *there should be no repairs at all.*

From the receiver's viewpoint, the situation is ostensibly analogous to that in Figure 4.3. However, it would be a mistake to interpret the sender's signal as warning of a scorpion in Box_1. Instead, taking into account only what is negotiable between sender and receiver, the receiver ought to interpret the situation as analogous to that in Figure 4.2, and interpret the sender's mark to mean Box_1 must contain a banana.

As formulated, virtual bargaining predicts this behaviour by assuming that any conventions players develop will depend just on what they *could have* openly negotiated (Misyak and Chater, 2014)—which does not include the receiver's private knowledge. The receiver should therefore select Box_1 as a result of a signalling convention with the sender covering Box_1 and Box_3, then open Box_2 based on their own knowledge. So the receiver is merely 'adding in' the private knowledge of Box_2 to their strategy:

$$
\begin{aligned}
Mark(Sender, item) &\implies Signal(item) \\
Signal(item) &\implies item \in Help \\
&\implies Box_2 \in Help \\
item \in Help &\implies Select(Receiver, item)
\end{aligned}
$$

Logical representation of the game (as we have used throughout) would thus allow the sender and receiver to explore the consequences of using their existing signalling convention, ranging only over Box_1 and Box_3, and observe that unlike the 'avoid' condition of Figure 4.3, there is no insufficiency or incompatibility to prompt a repair. If our analysis is correct, this repair—or lack thereof—should predict human behaviour.

4.7 Conclusion

Experiments on coordination under extreme communicative constraints have revealed a human ability to enhance weak or ambiguous signals using efficient, flexible social reasoning. Whatever the cognitive underpinnings of this ability, it is argued in this growing literature that virtual bargaining plays a vital role in efficient low-bandwidth coordination, and perhaps even society as a whole (Misyak *et al.*, 2014). In the more controlled context of lab-based coordination games, novel behaviours increasingly support this line of thinking: despite limited communication, humans readily put themselves into the shoes of cooperating partners to devise mutually consistent conventions spontaneously. As task demands change, they spontaneously and fluently adapt and combine these conventions (Misyak *et al.*, 2016).

In this chapter we have aimed to show how some of these results can be understood through the lens of logical inference, breaking down players' strategies into rules, facts, and signalling conventions. On this basis, we have then demonstrated the potential for the automated repair of these conventions, addressing logical reasoning faults relative to facts or rules such as insufficiency or incompatibility, to reproduce these behaviours. Despite being limited to spontaneous adaptation, and necessarily selective in scope, this early work demonstrates that some virtual bargaining behaviours *can* be replicated using logical representation change, from no more information than humans are given.

Moreover, we have achieved this using a purely symbolic approach. Unlike sampling-based methods, such as statistical machine learning, our logic-based method allows for efficient, one- or zero-shot revision of signalling conventions, without extensive datasets of positive and negative examples. Just like human players, our approach can create successful strategies without experimentation. It can form compound structures, in the shape of logic theories, that are intelligently adapted through representation change. Where rules from different sources are included we can intelligently *exclude* them, and the language expressing them, from automated repair.

As part of this work, we have represented the strategies for playing several selection and avoidance games as Datalog theories. This representation has the advantage that we can interpret these theories both procedurally, as logic programs whose execution will implement the strategies, and declaratively, as logical theories whose faults may then be repaired with the ABC system. The ABC system employs a combination of abduction, belief revision, and Reformation. Abduction and belief revision add/delete axioms or delete/add preconditions to rules, respectively; Reformation changes logical *concepts*, the 'C' in ABC, by modifying the *language* of the theory. It diagnoses faults by failures of reasoning. It repairs faults by blocking or unblocking appropriate proofs.

Its application to virtual bargaining has served as a driver for improving the ABC system—extending its range of repairs, while keeping within its spirit of diagnosis and repair via reasoning failures. For instance, we had not previously applied ABC to theories intended to be interpreted procedurally. Nor had we encountered theories whose correct behaviour in an old situation had to preserved, while they were adapted to deal with a new one. This called for repairs splitting rules in two, distinguished by complementary preconditions—one used in the old situation, another in the new one.

Despite this progress, the language change delivered by Reformation, in particular, remains purely syntactic in character, as to some extent do all the functions of our ABC system. It captures the repairs needed to intelligently adapt a pre-existing convention but has no semantics to call on when assigning new predicates and constants, or deciding which potential preconditions to include. Conducting theory repair in an operational domain, physical or notional, such as the item selection and avoidance game, gives us a means to explore how new concepts can be linked to those occurring in the game, as objects or as operations on them, building toward a semantic component to ABC.

More generally, we began this paper with the twofold problem of (1) constructing an initial signalling convention from players' shared knowledge of the game rules, the end goal, and other available information; and (2) adapting that convention after circumstances change. The assumption that these questions can be treated as different parts of the problem-both conceptually and based, for example, on participants forming but not always revising a convention when a less elaborate alternative would suffice (Misyak and Chater, 2017)-has enabled us to tackle spontaneous adaptation separate from the spontaneous creation of conventions. However, the objective of modelling—and of understanding—virtual bargaining is inevitably a function of both of these components.

As a result, our most pressing next step is to explore how the signalling convention, whose axioms we have been assuming as the basis for theory repair, is initially formed.

Acknowledgements

The research reported in this paper was supported by EPSRC grant EP/N014758/1.

References

Bundy, A. and Mitrovic, B. (2016). Reformation: a domain-independent algorithm for theory repair. Technical report, University of Edinburgh.

Ceri, S., Gottlob, G., and Tanca, L. (1990). *Logic Programming and Databases*. Berlin: Springer-Verlag.

Chater, N., Misyak, J. B., Watson, D., *et al.* (2014). Negotiating the traffic: Can cognitive science help make autonomous vehicles a reality? *Trends in Cognitive Sciences*, **22**(2), 93–5.

Cox, P. T. and Pietrzykowski, T. (1986). Causes for Events: their Computation and Applications, in *Lecture Notes in Computer Science: Proceedings of the 8th International Conference on Automated Deduction* (ed. J. Siekmann) Berlin: Springer-Verlag: 608–21.

Gärdenfors, P. (1992). *Belief Revision*. Cambridge: Cambridge University Press.

Kowalski, R. (1979). *Logic for Problem Solving*. Amsterdam: North Holland.

Kowalski, R. A. and Kuehner, D. (1971). Linear resolution with selection function. *Artificial Intelligence*, **2**, 227–60.

Li, X., Bundy, A., and Smaill, A. (2018). ABC repair system for Datalog-like theories, in *10th International Joint Conference on Knowledge Discovery, Knowledge Engineering and Knowledge Management*, Vol. 2. Setubel: SCITEPRESS, 335–42.

Misyak, J. B. and Chater, N. (2014). Virtual bargaining: a theory of social decision-making. *Philosophical Transactions of the Royal Society B: Biological Sciences*, **369**(1655).

Misyak, J. B., Melkonyan, T., Zeitoun, H., *et al.* (2014). Unwritten rules: virtual bargaining underpins social interaction, culture, and society. *Trends in Cognitive Sciences*, **18**(10), 512–19.

Misyak, J. B., Noguchi, T., and Chater, N. (2016). Instantaneous conventions: the emergence of flexible communicative signals. *Psychological Science*, **27**(12), 1550–61.

Misyak, J. B. and Chater, N. (2017). The spontaneous creation of systems of conventions, in *Proceedings of the 39th Annual Meeting of the Cognitive Science Society, London, UK, 16–29 July 2017.*

Mitrovic, B. (2013). Repairing inconsistent ontologies using adapted Reformation algorithm for sorted logics. UG4 Final Year Project, University of Edinburgh.

Part 2

Human-like Social Cooperation

5

Mining Property-driven Graphical Explanations for Data-centric AI from Argumentation Frameworks

Oana Cocarascu, Kristijonas Cyras, Antonio Rago, and Francesca Toni

Imperial College London, UK

5.1 Introduction

Artifical intelligence (AI) is continuing to make progress in many settings, fuelled by data availability, computational power and algorithmic and engineering advances. However, it is widely acknowledged that the adoption of systems using AI and their societal benefits are heavily dependent on human understanding of the rationale behind the systems' outputs, and that these systems' widespread inability to explain their outputs causes human mistrust and doubts regarding their regulatory compliance. For example, the UK House of Lords Select Committee report on 'AI in the UK: ready, willing and able?' (6 April 2018)[1] states that '...the development of intelligible AI systems is a fundamental necessity if AI is to become an integral and trusted tool in our society', the European Commission's Ethics Guidelines for Trustworthy AI (8 April 2019)[2] state that 'AI systems and their decisions should be explained in a manner adapted to the stakeholder concerned', and the UK Information Commissioner's Office and the Alan Turing Institute recently concluded (on 24 January 2020) a consultation on 'Explaining AI decisions guidance',[3] aimed at giving 'organisations practical advice to help explain the processes, services and decisions delivered or assisted by AI, to the individuals affected by them'.

[1] See https://publications.parliament.uk/pa/ld201719/ldselect/ldai/100/100.pdf.
[2] See https://ec.europa.eu/digital-single-market/en/news/ethics-guidelines-trustworthy-ai.
[3] See https://ico.org.uk/about-the-ico/ico-and-stakeholder-consultations/ico-and-the-turing-consultation-on-explaining-ai-decisions-guidance/.

Oana Cocarascu, Kristijonas Cyras, Antonio Rago, and Francesca Toni, *Mining Property-driven Graphical Explanations for Data-centric AI from Argumentation Frameworks* In: *Human Like Machine Intelligence*. Edited by: Stephen Muggleton and Nick Charter, Oxford University Press.
 DOI: 10.1093/oso/9780198862536.003.0005

Because of the almost universal awareness that AI-based systems need to be explainable to humans in order to be used fruitfully by them, extensive research efforts are currently devoted towards explainable AI (XAI), in both academia (e.g., see overviews in Guidotti *et al.*, 2019; Miller, 2019) and industry (for example, amongst others, at IBM, who launched AI Explainability 360 in August 2019, and at Google, with a new explainable AI service launched in November 2019).

To date, for the most part these efforts have focused on static explanations aimed at expert developers. Instead, research in (cognitive and social) psychology has identified the need for explanations with which humans can interact (e.g., see Chapter 5 in Miller, 2019). At the same time, research in psychology advocates that humans developed reasoning in order to argue (Mercier and Sperber, 2011), thus pointing to the amenability of argumentation to humans.

In this paper we use *argumentation frameworks* as understood in AI, see Atkinson *et al.*, 2017; Baroni *et al.*, 2018*a*) for recent overviews) as the scaffolding for *explanations*, amenable to *human consumption*, drawn from data-centric AI methods. Argumentation frameworks of many different kinds have been widely studied in the AI literature both in terms of formal properties they exhibit under different semantics and in terms of applications they can support. For example, several semantics have been proposed for Abstract Argumentation Frameworks (AFs) (Dung, 1995; Baroni *et al.*, 2011), Bipolar Argumentation Frameworks (BFs) (Cayrol and Lagasquie-Schiex, 2005; Cohen *et al.*, 2014; Čyras *et al.*, 2017), Quantitative Bipolar Argumentation Frameworks (QBFs) (Baroni *et al.*, 2018*b*, 2019) and structured argumentation (Besnard *et al.*, 2014), and relationships between semantics and existence properties for these semantics have been thoroughly studied, as have other properties (e.g., see Baroni *et al.*, 2018*b*; Baroni *et al.*, 2019). In addition, several practical applications of argumentation frameworks have been investigated, including of AFs in healthcare (Hunter and Williams, 2015), of BFs/QBFs in engineering design (Baroni *et al.*, 2015) and of structured argumentation for recommender systems (Briguez *et al.*, 2014) and healthcare (Fan *et al.*, 2013). These applications predominantly require the up-front definition, often by hand, of suitable argumentation frameworks of one type or another.

In this chapter, we define a variety of types of explanation as graphs fulfilling structural properties, obtained from argumentation frameworks. We then show how these (types of) explanations can be deployed with argumentation frameworks automatically mined from

- linguistic data (text), by means of relation-based argument mining (Carstens and Toni, 2015; Cocarascu and Toni, 2017; Menini *et al.*, 2018) in the context of the method of (Cocarascu and Toni, 2016; 2018),
- labelled data where combinations of features are associated with outcomes, treated as a case base in the spirit of (Čyras *et al.*, 2016*a*,*b*; Cocarascu *et al.*, 2020*b*), and
- a data-driven recommender system (where items have actual and predicted ratings, tailored to users, based on the features of the items), in the spirit of (Rago *et al.*, 2018).

For each of these three types of approaches to mining argumentation frameworks, we illustrate how they can support reasoning to outputs using existing argumentation-based approaches from the literature, and how explanation of these outputs can be generated using the approach we describe in this chapter. To ground our presentation, throughout the chapter we use illustrations from the media and entertainment industry, focusing on the music industry in particular, from the viewpoint of consumers. The illustrations in the chapter are kept simple and manually engineered, but, where possible, we indicate how these illustrations may be obtained automatically (as carried out in (Chauhan *et al.*, 2019), describing a proof-of-concept system built on this chapter).

The chapter is organized as follows. In section 5.2 we give background and illustrative context for the whole chapter. In section 5.3 we define an abstract notion of explanation and several instances of this notion, that we will use later in the chapter. In section 5.4 we study (reasoning and explaining with) BFs mined from textual data. In section 5.5 we study (reasoning and explaining with) AFs mined from labelled data. In section 5.6 we study (explaining with) QBFs mined from recommender systems. In section 5.7 we conclude.

5.2 Preliminaries

In this section we will first give the necessary background on the types of argumentation frameworks we will use in this chapter (abstract, bipolar, and quantitative bipolar argumentation frameworks) as the scaffolding for explanation (section 5.2.1), followed by a brief description of our chosen application domain for illustration throughout the chapter (section 5.2.2).

5.2.1 Background: argumentation frameworks

Abstract Argumentation frameworks (AFs) are pairs consisting of a set of arguments and a binary (attack) relation between arguments (Dung, 1995). Formally, an AF is any pair $(Args, \mathcal{R}^-)$ where $\mathcal{R}^- \subseteq Args \times Args$. Bipolar Argumentation frameworks (BFs) extend AFs by considering two binary relations: attack and support (Cayrol and Lagasquie-Schiex, 2005). Formally, a BF is any triple $(Args, \mathcal{R}^-, \mathcal{R}^+)$ where $(Args, \mathcal{R}^-)$ is an AF and $\mathcal{R}^+ \subseteq Args \times Args$. If $\mathcal{R}^+ = \{\}$, a BF $(Args, \mathcal{R}^-, \mathcal{R}^+)$ can be identified with an AF $(Args, \mathcal{R}^-)$, so we will often use the term BF to denote BFs as well as AFs.

Any $F = (Args, \mathcal{R}^-, \mathcal{R}^+)$ can be understood and visualized as a directed graph G, henceforth called *argument graph*, with nodes $Args$ and two types of edges: $\mathcal{R}^-$ and $\mathcal{R}^+$ (see e.g., Cayrol and Lagasquie-Schiex, 2005; Cohen *et al.*, 2018). In this chapter, when showing G, we will use single ($\rightarrow$) and double ($\Rightarrow$) arrows to denote $\mathcal{R}^-$ and $\mathcal{R}^+$, respectively (see illustration in Figure 5.1). A *sub-graph* of an argument graph $G = (Args, \mathcal{R}^-, \mathcal{R}^+)$ is a directed graph $\mathcal{G}_{F_s} = (Args_s, \mathcal{R}_s^-, \mathcal{R}_s^+)$, denoted $\mathcal{G}_{F_s} \sqsubseteq G$, such that $Args_s \subseteq Args$, $\mathcal{R}_s^- \subseteq \mathcal{R}^-$, $\mathcal{R}_s^+ \subseteq \mathcal{R}^+$. A *path* in F from $\mathsf{b} \in Args$ to $\mathsf{a} \in Args$, denoted $path(\mathsf{b}, \mathsf{a})$, is a sequence $s = \langle \mathsf{a}_0, \ldots, \mathsf{a}_n \rangle$ of arguments such that $\mathsf{a}_0 = \mathsf{b}$, $\mathsf{a}_n = \mathsf{a}$ and $\forall 0 \leq i < n$, $(\mathsf{a}_i, \mathsf{a}_{i+1}) \in \mathcal{R}^- \cup \mathcal{R}^+$.

Semantics of AFs/BFs amount to 'recipes' for determining 'winning' sets of arguments or the 'dialectical strength' of arguments. These semantics can be respectively defined

Figure 5.1 *The argument graphs G for the AF* $F = (\{\alpha,\beta,\gamma,\delta,\epsilon\},\{(\beta,\alpha),(\delta,\alpha),(\gamma,\epsilon)\})$ *(left) and G for the BF* $F = (\{\alpha,\beta,\gamma,\delta,\epsilon\},\{(\beta,\alpha),(\delta,\alpha),(\gamma,\epsilon)\},\{(\gamma,\alpha),(\epsilon,\delta)\})$ *(right).*

qualitatively, in terms of *extensions* (e.g., the *grounded* extension (Dung, 1995), defined below, and that we will use in this chapter), and quantitatively, in terms of a *gradual evaluation* of arguments (e.g., as in Rago *et al.*, 2016; Baroni *et al.*, 2017)—the former of which, defined below, we will use in this chapter).

Given an AF $(Args,\mathcal{R}^-)$, let $E \subseteq Args$ *defend* $\mathsf{a} \in Args$ if for all $\mathsf{b} \in Args$ attacking a there exists $\mathsf{c} \in E$ attacking b. Then, the *grounded extension* of $(Args,\mathcal{R}^-)$ is $\mathbb{G} = \bigcup_{i\geqslant 0} G_i$, where G_0 is the set of all *unattacked* arguments (i.e., the set of all arguments $\mathsf{a} \in Args$ such that there is no argument $\mathsf{b} \in Args$ with $(\mathsf{b},\mathsf{a}) \in \mathcal{R}^-$) and $\forall i \geqslant 0$, G_{i+1} is the set of all arguments that G_i defends. For any $(Args,\mathcal{R}^-)$, the grounded extension $\mathbb{G}$ always exists and is unique. As an illustration, in the simple AF in Figure 5.1, left, $\mathbb{G} = \{\beta,\delta,\gamma\}$.

On the other hand, quantitative semantics allow a *gradual evaluation* of arguments. They can be defined for BFs, as in Baroni *et al.*, 2017, or for *Quantitative Bipolar Argumentation Frameworks* (QBFs) (Baroni *et al.*, 2018*b*), of the form $(Args,\mathcal{R}^-,\mathcal{R}^+,\tau)$ where $(Args,\mathcal{R}^-,\mathcal{R}^+)$ is a BF and $\tau : Args \rightarrow I$ for some interval I (e.g. $I = [0,1]$ or $I = [-1,1]$) gives the intrinsic strength or *base score* of arguments. AFs and BFs are QBFs with special choices of τ (Baroni *et al.*, 2018*b*), so we will sometimes use the term QBF to denote AFs and BFs. The argument graph for $(Args,\mathcal{R}^-,\mathcal{R}^+,\tau)$ is the argument graph of $(Args,\mathcal{R}^-,\mathcal{R}^+)$.

Given a QBF $(Args,\mathcal{R}^-,\mathcal{R}^+,\tau)$, the strength of arguments is given by some $\sigma : Args \rightarrow I$. Several such notions have been defined in the literature (e.g., see Baroni *et al.*, 2019 for an overview). In section 5.4 we will use the notion of Rago *et al.*, (2016),[4] where $I = [0,1]$ and for $\mathsf{a} \in Args$:

$$\sigma(\mathsf{a}) = c(\tau(\mathsf{a}),\mathcal{F}'(\sigma(\mathcal{R}^-(\mathsf{a}))),\mathcal{F}'(\sigma(\mathcal{R}^+(\mathsf{a}))))$$

such that:

(1) $\mathcal{R}^-(\mathsf{a})$ is the set of all arguments attacking a and if $(\mathsf{a}_1,\ldots,\mathsf{a}_n)$ is an arbitrary permutation of the $(n \geq 0)$ elements of $\mathcal{R}^-(\mathsf{a})$, then $\sigma(\mathcal{R}^-(\mathsf{a})) = (\sigma(\mathsf{a}_1),\ldots,\sigma(\mathsf{a}_n))$ (similarly for supporters);

(2) for $v_0, v_a, v_s \in [0,1]$,

$$c(v_0,v_a,v_s) = v_0 - v_0 \cdot |v_s - v_a| \qquad \text{if } v_a \geq v_s,$$

[4] Note that several other notions could be used, as overviewed in (Baroni *et al.*, 2019). We have chosen this specific notion because it satisfies some desirable properties (Baroni *et al.*, 2019) as well as performing well in practice (Cocarascu *et al.*, 2019).

$$c(v_0, v_a, v_s) = v_0 + (1 - v_0) \cdot |v_s - v_a| \qquad \text{if } v_a < v_s;$$

and

(3) for $S = (v_1, \ldots, v_n) \in [0,1]^*$ and $f'(x,y) = x + y - x \cdot y$:
if $n = 0$: $\mathcal{F}'(S) = 0$; if $n = 1$: $\mathcal{F}'(S) = v_1$; if $n = 2$: $\mathcal{F}'(S) = f'(v_1, v_2)$;
if $n > 2$: $\mathcal{F}'(S) = f'(\mathcal{F}'(v_1, \ldots, v_{n-1}), v_n)$.

Intuitively, the strength $\sigma(\mathsf{a})$ of argument a results from the combination c of three components: the base score $\tau(\mathsf{a})$ of a, the aggregated strength $\mathcal{F}'(\sigma(\mathcal{R}^-(\mathsf{a})))$ of all arguments attacking a and the aggregated strength $\mathcal{F}'(\sigma(\mathcal{R}^+(\mathsf{a})))$ of all arguments supporting a. The combination c decreases the base score of a if the aggregated strength of the attackers is at least as high as the aggregated strength of the supporters (with the decrement proportional to the base score and to the absolute value of the difference between the aggregated strengths). The combination c increases the base score of a otherwise, if the aggregated strength of the attackers is lower than the aggregated strength of the supporters (with the increment proportional to the distance between 1 and the base score and to the absolute value of the difference between the aggregated strengths). Finally, the aggregated strengths are defined recursively (using the probabilistic sum when there are exactly two terms to aggregate—these are either strengths of attackers or of supporters).[5]

As an illustration, in the BF in Figure 5.1, right, if the base score of all arguments if 0.5, then $\sigma(\gamma) = \tau(\gamma) = 0.5$ and $\sigma(\epsilon) = c(0.5, 0.5, 0) = 0.5 - 0.5 \cdot 0.5 = 0.25$.

5.2.2 Application domain

We will consider three consumer-oriented tasks in the music setting: (1) judging whether a given album is worthwhile, based on (textual) reviews from consumers (section 5.4); (2) determining whether a given album might receive UK Platinum certification[6] based on its features, namely genre, positivity of reviews, and US Platinum certification,[7] and information about UK Platinum certification of other albums (section 5.5); (3) recommending (and how strongly) albums to consumers, based on their past ratings (section 5.6).

Concretely, we will illustrate our methodologies by: (1) judging whether the Rolling Stones' album *It's Only Rock 'n Roll* (*IOR* henceforth) is worthwhile based on several reviews from Amazon (given later in Table 5.1); (2) determining whether *IOR* might get UK Platinum certification based on the albums *Blue&Lonesome (B&L)*, *Let It Bleed* (*LiB*); and *Sticky Fingers* (*SF*) by the Rolling Stones, as well as *Time Out Of Mind* (*TooM*) and *Modern Times* (*MT*) by Bob Dylan (some of which got UK platinum certification, see Table 5.2); (3) recommending *IOR* and *MT* to a consumer who already rated some albums (see Table 5.3).

[5] Note that this recursively defined notion treats strengths of attackers and supporters as sets, but needs to consider them in sequence (thus the mention of 'an arbitrary permutation').

[6] https://www.bpi.co.uk/brit-certified/.

[7] https://www.riaa.com/gold-platinum/.

Table 5.1 *Reviews for album* It's Only Rock 'n Roll.

r_1: *It's Only Rock N' Roll is* **a very good album**. *The reason I gave it an eight-and-a-half rating is because it's* **not a masterpiece**. *The* **album does have some very strong songs**. *'If You Can't Rock Me' has a* **great guitar riff**. *'Till The Next Goodbye' is a* **beautiful love song**, *like others which they have done before.*

r_2: *It is an* **uneven album**, *some tracks are great, some bad, and some mediocre. 'Time Waits For No One' is arguably the Stones' greatest performance on record, not just Taylor's* **incredible solo** *but great playing by the entire band.*

r_3: *This is their* **worst album, the end result of deteriorating musical and lyrical standards** *that started with Goats' Head Soup. Despite this, I cannot write it off completely. Keith's* **riff machine is as good as ever'** *'If You Can't Rock Me' and the title track, the slower number 'If You Really Want To Be My Friend' contains one of their best hooks.*

r_4: *As an* **album though it's just too patchy, too mediocre'** *'Dance Little Sister' is awful, 'Luxury' sounds like a shadow of the band that recorded 'Exile' just two years earlier.*

r_5: *IORR is the* **second in a string of three mediocre mid-1970's albums**. *That's* **not to say it's bad, because an average Stones album is still pretty solid**. *All in all* **a collection of some nice songs but nothing essential**. *The Stones on cruise control rolling down the street to some pounding funk-rock R&B in a 1974 Buick Riviera.*

Table 5.2 *Albums with their features, outcomes, and representation as labelled examples.*

Album	Features ($\mathcal{F}_p$)	Outcome ($\mathcal{O}$)	Example
$B\&L$	$Blues$	UK_Plat	$(\{Blues\}, UK\_Plat)$
$TooM$	$Blues$, US_Plat	not UK_Plat	$(\{Blues, US\_Plat\}, \text{not } UK\_Plat)$
SF	$R'n'R$, US_Plat	not UK_Plat	$(\{R'n'R, US\_Plat\}, \text{not } UK\_Plat)$
LiB/MT	$Blues$, US_Plat, Pos_Rev	UK_Plat	$(\{Blues, US\_Plat, Pos\_Rev\}, UK\_Plat)$
IOR	$Blues$, $R'n'R$, US_Plat	?	$(\{Blues, R'n'R, US\_Plat\}, ?)$

Note that the aim of this chapter is to show how AFs and BFs can serve as the scaffolding for property-driven explanations rather than predictive performances. Therefore, we will not report on experiments with the datasets mentioned earlier. For experimental evaluations of the methodologies underpinning explanations in this chapter please see (Cocarascu and Toni, 2016 and 2018) for the methods in section 5.4, (Cocarascu *et al.*, 2018 and Cocarascu *et al.* 2020*b*) for the methods in section 5.5; and (Rago *et al.*, 2018) for the methods in section 5.6.

Table 5.3 *Items/Features in example RS with their associations, actual ratings (if defined) and predicted ratings.*

	Node	Item/Feature	Associations	Actual Rating	Predicted Rating
Items	i_1	*MT*	$\{a_2, g_1, p_1\}$	-	0.7
	i_2	*IOR*	$\{a_1, g_1, g_2\}$	-	0.6
	i_3	*TooM*	$\{a_2, g_1\}$	-0.3	-0.3
	i_4	*LiB*	$\{a_1, g_1, p_1\}$	1.0	1.0
	i_5	*SF*	$\{a_1, g_2\}$	-0.2	-0.2
Features	g_1	*Blues*	$\{i_1, i_2, i_3, i_4\}$	-	0.7
	g_2	$R'n'R$	$\{i_2, i_5\}$	-	-0.2
	a_1	The Rolling Stones	$\{i_2, i_4, i_5\}$	-	0.8
	a_2	Bob Dylan	$\{i_1, i_3\}$	-	-0.3
	p_1	*UK_Plat*	$\{i_1, i_4\}$	-	1.0

5.3 Explanations

We define a generic notion of *property-driven explanation* for arguments in argumentation graphs, as sub-graphs that fulfil some specified property. This notion will serve as a template to instantiate with specific properties to define and/or characterise explanation methods for various reasoning tasks later in this chapter.

Definition 5.1 *Let G be an argumentation graph, for $F = (Args, \mathcal{R}^-, \mathcal{R}^+)$. Let P be a property of graphs. A P-*driven explanation *for* a $\in$ *$Args$ is a sub-graph $\mathcal{G}_{F_s} \sqsubseteq G$ such that* a *is a node of $\mathcal{G}_{F_s}$ and $\mathcal{G}_{F_s}$ fulfils P.*

Several notions of explanation in argumentative settings considered in the literature are instances of our notion of property-driven explanation. For example, (Fan and Toni, 2015; Čyras *et al.*, 2016*b*; Schulz and Toni, 2016; Čyras *et al.*, 2019; Cocarascu *et al.*, 2020*b*) all propose or use notions of explanation based on some form of *dispute trees* (Dung *et al.*, 2006), where, given an AF $(Args, \mathcal{R}^-)$, a *dispute tree* for a $\in Args$ is a tree $\mathcal{T}$ such that: (1) every node of $\mathcal{T}$ is of the form $[L\!:\!\mathsf{x}]$, with $\mathsf{L} \in \{\mathsf{P}, \mathsf{O}\}$, $\mathsf{x} \in Args$: the node is *labelled* by argument x and assigned the *status* of either *proponent (*P*)* or *opponent (*O*)*; (2) the root of $\mathcal{T}$ is a P node labelled by a; (3) for every P node n, labelled by some b $\in Args$, and for every c $\in Args$ attacking b, there exists a child of n, which is an O node labelled by c; (4) for every O node n, labelled by some b $\in Args$, there exists at most one child of n which is a P node labelled by some c $\in Args$ attacking b; (5) there are no other nodes in $\mathcal{T}$ except those given by 1–4. Two different types of dispute trees have been advocated as explanations:

- *admissible dispute trees (ADTs)* (Dung *et al.*, 2006), namely dispute trees $\mathcal{T}$ such that (1) every O node in $\mathcal{T}$ has a child, and (2) no argument in $\mathcal{T}$ labels both P and O nodes; and
- *maximal dispute trees (MDTs)* (Čyras *et al.*, 2016*b*), namely dispute trees $\mathcal{T}$ such that for all opponent nodes [O : x] which are leaves in $\mathcal{T}$ there is no y $\in$ $Args$ attacking x.

Note that although ADTs and MDTs have been originally defined for AFs, their definition is also applicable to BFs, focusing exclusively on attacks.

Both notions (ADTs and MDTs) can be seen as instances of the notion of P-driven explanation, for P the property of being, respectively, an ADT and an MDT (but leaving the proponent/opponent status of nodes implicit). Other instances of the notion of P-driven explanation include choosing a maximal sub-graph of connected nodes and choosing some branch(es) within this maximal sub-graph. Throughout the paper we will make use of all these instances of P-driven explanation, defined as follows.

Definition 5.2 *Let G be an argumentation graph, for $F = (Args, \mathcal{R}^-, \mathcal{R}^+)$. An* ADT-driven explanation *for* a $\in$ $Args$ *is a sub-graph* $\mathcal{G}_{F_s} \sqsubseteq G$ *such that* a *is a node of* $\mathcal{G}_{F_s}$ *and* $\mathcal{G}_{F_s}$ *is an ADT for* a. *An* MDT-driven explanation *for* a $\in$ $Args$ *is a sub-graph* $\mathcal{G}_{F_s} \sqsubseteq G$ *such that* a *is a node of* $\mathcal{G}_{F_s}$ *and* $\mathcal{G}_{F_s}$ *is an MDT for* a. *A* maximally connected sub-graph-driven explanation *for* a $\in$ $Args$ *is a sub-graph* $\mathcal{G}_{F_s} \sqsubseteq G$ *such that* a *is a node of* $\mathcal{G}_{F_s}$ *and the set of nodes of* $\mathcal{G}_{F_s}$ *is* $\{N \in G : \exists path(N, \mathsf{a})\}$. *A* branch-driven explanation *for* a $\in$ $Args$ *is a sub-graph* $\mathcal{G}_{F_s} \sqsubseteq G$ *such that* a *is a node of* $\mathcal{G}_{F_s}$ *and* $\mathcal{G}_{F_s}$ *is a finite branch in the sub-graph with nodes* $\{N \in G : \exists path(N, \mathsf{a})\}$.

Note that there may be multiple ADT-, MDT- and branch-driven explanations for any given argument. Note also that these notions could be refined by imposing further restrictions, for example that a branch is of minimal length.

5.4 Reasoning and Explaining with BFs Mined from Text

In this section we focus on BFs mined from textual data by means of the method outlined in (Cocarascu and Toni, 2016, 2018) in combination with relation-based argument mining (Carstens and Toni, 2015; Cocarascu and Toni, 2017; Menini *et al.*, 2018). We show how the mined BF can be used for reasoning and for providing explanations, illustrating our methodology on reviews for the album 'It's Only Rock 'n Roll' (*IOR*) by the Rolling Stones, given in Table 5.1.

5.4.1 Mining BFs from text

In order to extract a BF from text about a product, we deploy the method of (Cocarascu and Toni, 2016 and 2018), consisting of four steps:

1. split the text into temporally ordered sentences;[8]
2. identify topics in the text and, for each topic, the sentences related to the topic;
3. for each topic, for each pair of sentences related to the topic, determine whether the most recent[9] sentence *supports*, *attacks*, or *neither supports nor attacks* the less recent sentence or, if the sentence is the least recent, the topic;[10]
4. construct a BF, including (i) a generic argument for the product being worthwhile; (ii) the topics, each treated as an argument as to whether the product is any good as far as the topic is concerned; (iii) the argumentative sentences from Step 3 (namely sentences that attack/support topics and other sentences, as well as sentences that are attacked/supported by other sentences); (iv) attacks/supports between arguments, as determined at Step 3; and (v) supports between topics and the argument at (i).

For illustration, consider the reviews in Table 5.1, assuming that they were posted in order (so r_1 was posted first and r_5 last). Here, we have indicated in bold the essential parts only (for ease of reference) of the separate sentences identified at Step 1, with a separate sentence for each bold statement. Thus, after Step 1, the arguments extracted from the reviews are (where $a_{i,j}$ represents the jth argument extracted from r_i):

$a_{1,1}$:...*a very good album*... $a_{1,2}$:...*not a masterpiece*...
$a_{1,3}$:...*album does have some very strong songs*... $a_{1,4}$:...*great guitar riff*...
$a_{1,5}$:...*beautiful love song*... $a_{2,1}$:...*uneven album*... $a_{2,2}$:...*incredible solo*...
$a_{3,1}$:...*worst album, the end result of deteriorating musical and lyrical standards*...
$a_{3,2}$:...*riff machine is as good as ever*...
$a_{4,1}$:...*album though it's just too patchy, too mediocre*...
$a_{5,1}$:...*second in a string of three mediocre mid-70's albums*...
$a_{5,2}$:...*not to say it's bad, because an average Stones album is still pretty solid*...
$a_{5,3}$:...*a collection of some nice songs but nothing essential*...

Various techniques can be used to identify topics at Step 2, from associating each encountered noun to a topic to standard topic modelling approaches such as Latent

[8] Sentences that contain specific keywords such as *whereas* and *however* are split further since, in general, the phrases before and after these keywords have different sentiment polarity and thus can potentially lead to different argumentative relations.

[9] In our approach, the arguments are ordered by the date of which the review they were extracted from was posted. Thus, more recent texts can attack or support less recent texts. If the text is a collection of online reviews then this choice is legitimate, given that users can see only arguments previously expressed. Practically, this choice allows us to reduce the number of argumentative relations whilst still capturing the debate, in the spirit of preference-based argumentation.

[10] The use of topics help reduce the number of pairs of arguments that need to be checked, as different topics are highly unlikely to be related and thus arguments about them unlikely to be in argumentative relations.

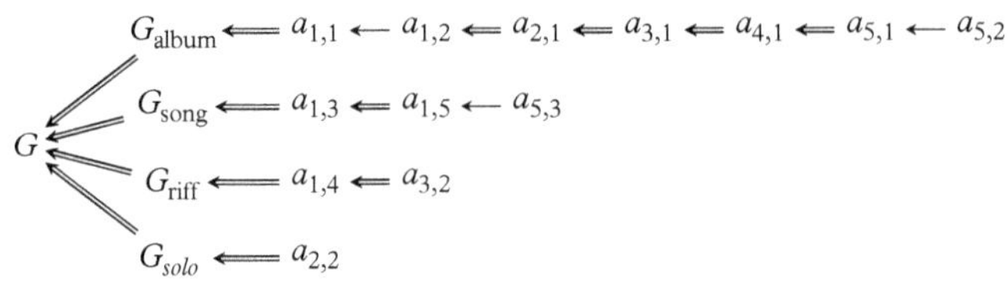

Figure 5.2 *The argument graph G for the BF* $F = (Args, \mathcal{R}^-, \mathcal{R}^+)$ *mined from the reviews in Table 5.1.*

Dirichlet Allocation (LDA) (Blei *et al.*, 2003) and Non-negative Matrix Factorization (NMF) (Lee and Seung, 1999). In our illustration, the topics obtained from the reviews may be: *album*, *song*, *riff*, and *solo*. Arguments are then associated to topics. For example, $a_{2,2}$ is the only argument with topic *solo*.

Various methods, including machine learning techniques for *Relation-based Argument Mining* (RbAM) (Carstens and Toni, 2015), as in (Cocarascu and Toni, 2017), can be used at Step 3. RbAM does not rely on a specific argument model or internal argument structure but assumes that if one text attacks or supports another, then both may be considered to be argumentative, irrespectively of their stand-alone argumentativeness. In the execution of Step 3, we compare the least recent sentence and its topic, to determine whether the sentence attacks or supports that the album in question is good as far as that topic is concerned. Thus, for example, $a_{1,1}$ supports an argument G_{album}, standing for '*IOR* is good as far as topic *album* is concerned'.

The BF obtained at Step 4 may be as in Figure 5.2, where argument G represents that the *IOR* album is worthwhile (or good).

5.4.2 Reasoning

The BF obtained at Step 4 can be used to predict whether a product (e.g., an album) is worthwhile, based on the text (e.g., reviews) provided. For example, a quantitative semantics allowing a gradual evaluation of arguments (see section 5.2.1) can be used to determine the dialectical strength of the root (G) argument or any of the topic arguments (G_{album} etc.) for the BF $F = (Args, \mathcal{R}^-, \mathcal{R}^+)$ in Figure 5.2, and this strength can be used as a measure of general worth or goodness regarding the various topics. Using the notion of strength σ of (Rago *et al.*, 2016) (see section 5.2.1), with $\tau(\text{a}) = 0.5$ for all arguments $\text{a} \in Args$, we obtain $\sigma(G) = 0.99$ and

$$\sigma(G_{\text{solo}}) = 0.75, \quad \sigma(G_{\text{riff}}) = 0.875, \quad \sigma(G_{\text{song}}) = 0.8125, \quad \sigma(G_{\text{album}}) = 0.511.$$

Based on these dialectical strengths, the user can select whether he/she should buy the album given that its highlights are specific songs and the riff (with respective topics having strengths higher than 0.8), rather than the album (with respective topic having a 0.5 strength). The strength of G_{album} in particular will serve as a feature for prediction, as we will see later in section 5.5.

$$G \Longleftarrow G_{\text{riff}} \Longleftarrow a_{1,4} \Leftarrow a_{3,2}$$

Figure 5.3 *Explanation as to why IOR is worthwhile based on the 'riff' topic.*

$$G \Longleftarrow G_{\text{album}} \Longleftarrow a_{1,1} \leftarrow a_{1,2} \Leftarrow a_{2,1} \Leftarrow a_{3,1} \Leftarrow a_{4,1} \Leftarrow a_{5,1} \leftarrow a_{5,2}$$

Figure 5.4 *Explanation as to why IOR is not worthwhile based on the 'album' topic.*

5.4.3 Explaining

The BF obtained at Step 4 can also be used to explain the output of reasoning, for example as to whether a product is worthwhile, based on the text provided. For example, Figure 5.3 gives an explanation as to why the *IOR* album is worthwhile for a user deeming the riff to be important. This explanation is a *branch-driven explanation* (see Definition 5.2) for G, focusing on the branch for the strongest topic argument (since $\sigma(G_{\text{riff}}) = 0.875$ is the highest). Figure 5.4 shows an explanations as to why the *IOR* album is not good 'as an album'. This explanation is again a *branch-driven explanation* for G, focusing on the branch for the weakest topic argument (since $\sigma(G_{\text{album}}) = 0.511$ is the lowest).

5.5 Reasoning and Explaining with AFs Mined from Labelled Examples

In this section we use AA-CBR[11] (Čyras *et al.*, 2016*a* and 2016*b*; Cocarascu *et al.*, 2018; Cocarascu *et al.*, 2020*b*) to mine AFs from labelled data points, called examples, represented as sets of features together with an outcome. In a nutshell, AA-CBR affords methods for: (1) reasoning about examples, understood as arguments in favour of a specific outcome (i.e. their label), to determine the outcome of any new example; (2) explaining that reasoning via debates that employ examples as arguments. Both tasks are supported by the mined AFs, which can be used for both reasoning for predicting outcomes for new, unlabelled examples, represented solely as sets of features, and for explaining the prediction.

In our application domain, examples are music albums related via subset inclusion of (sets of) features and with, as outcome, whether they obtained UK Platinum certification, which is also the *focus* of the model. AA-CBR uses grounded argumentation semantics to determine the acceptability of the (argument representing the) focus, and consequently to determine the outcome of a new example. AA-CBR provides explanations for outcomes as sub-graphs (specifically, dispute trees) of (the graph representing) the AF. The explanations can be seen as debates between two parties arguing about the focus. They also indicate how the parties could change the examples to obtain different outcomes.

[11] The formalism's name uses acronyms of abstract argumentation (AA) and case-based reasoning (CBR).

In our application domain of music albums, AA-CBR assumes that albums are identified with sets of features and a classification of having a UK Platinum album certification. It treats albums with more specific features as exceptions to albums with less specific features, as long as they have diverging classifications. For instance, a non-UK Platinum album with certain features may have as an exception a UK Platinum album with the same and some additional features. The exceptions among albums are the interactions that AA-CBR models.

We next give formally, and illustrate specifically within our application domain the methods for reasoning and explaining in AA-CBR (see Čyras *et al.*, 2016*a* and 2016*b*, Cocarascu *et al.*, 2018 Cocarascu *et al.*, 2020*b*) for additional information—our purpose here is to give sufficient details to illustrate property-driven graphical explanations in section 5.5.3).

5.5.1 Mining AFs from examples

Consider a fixed but otherwise arbitrary (possibly infinite) set $\mathcal{F}_p$ of *features* and a set $\mathcal{O} = \{\varphi, \overline{\varphi}\}$ of two (distinct) *outcomes*, with φ called the *focus* (this is the outcome expected for the empty set of features). Then:

- an *example* is a pair (X, o) with $X \subseteq \mathcal{F}_p$ and $o \in \mathcal{O}$;
- a *dataset* is a finite set $DS \subseteq \wp(\mathcal{F}_p) \times \mathcal{O}$ such that for any $(X, \mathrm{O}_X), (Y, \mathrm{O}_Y) \in DS$ with $X = Y$ we have $\mathrm{O}_X = \mathrm{O}_Y$.

For illustration, we can use: as features, the genres of albums ($Blues$ and $R'n'R$), existence of highly positive album review (Pos_Rev), e.g. $\sigma(G_{album}) > 0.75$ for σ as in section 5.4, and the albums' US Platinum certification status (US_Plat); and as outcomes, whether or not albums have UK Platinum certification, the former case (UK_Plat) being the focus. So $\mathcal{F}_p = \{Blues, R'n'R, US\_Plat, Pos\_Rev\}$ and $\mathcal{O} = \{UK\_Plat, \text{not } UK\_Plat\}$. Then our DS may be as in the top four rows in Table 5.2.[12] The albums with their features and outcomes are given in the following table:

We can then use AA-CBR to mine an AF, forming a model of the dataset, given the focus: the *AF corresponding to DS and* $\varphi \in \mathcal{O}$, denoted $aaf(DS, \varphi)$, is $(Args, \mathcal{R}^-)$ with

- $Args = DS \cup \{(\{\}, \varphi)\}$ (with $(\{\}, \varphi)$ the *focus argument*)
- for $(X, \mathrm{O}_X), (Y, \mathrm{O}_Y) \in Args$, it holds that $((X, \mathrm{O}_X), (Y, \mathrm{O}_Y)) \in \mathcal{R}^-$ iff

  1. $\mathrm{O}_X \neq \mathrm{O}_Y$, and (different outcomes)
  2. $Y \subsetneq X$, and (specificity)
  3. $\nexists (Z, \mathrm{O}_X) \in DS$ with $Y \subsetneq Z \subsetneq X$. (concision)

The construction of $(Args, \mathcal{R}^-)$ singles out conflicts between labelled examples. In particular, a conflict arises if features of one example form a proper subset of the features

[12] Note well that LiB and MT have the same features (and the same outcome), so they make up the same example, which we call LiB/MT.

Figure 5.5 *Graphs G (left) and $\mathcal{G}_{F_N}$ (right) for $F = (Args, \mathcal{R}^-)$, $F_N = (Args_N, \mathcal{R}_N^-)$ in the running illustration.*

of another (specificity) and the two examples have different outcomes. AA-CBR models this with an attack from the latter to the former, provided that the latter is the most concise such case. In the case of our running illustration, the argument graph G for the corresponding $F = (Args, \mathcal{R}^-)$ is given in Figure 5.5, left.

5.5.2 Reasoning

Given an unlabelled example $(N, ?)$, with $N \subseteq \mathcal{F}_p$ and ? indicating the as-yet-unknown outcome, AA-CBR expands $aaf(DS, \varphi)$ thus: the *AF corresponding to DS*, $\varphi \in \mathcal{O}$ *and an unlabelled example* $(N, ?)$, denoted $aaf(DS, \varphi, N)$, is $(Args_N, \mathcal{R}_N^-)$ with

- $Args_N = Args \cup \{(N, ?)\}$, and
- $\mathcal{R}_N^- = \mathcal{R}^- \cup \{((N, ?), (Y, O_Y)) \ : \ (Y, O_Y) \in Args \text{ and } Y \nsubseteq N\}$.

Put simply, $aaf(DS, \varphi, N)$ extends $aaf(DS, \varphi)$ so that $(N, ?)$ attacks all arguments containing 'irrelevant' features, that is features not in N.

As an illustration, let $(Args_N, \mathcal{R}_N^-) = aaf(DS, \varphi, N)$ be the AF corresponding to DS and unlabelled example $(N, ?)$ in Table 5.2 (where $(N, ?)$ is given in the bottom row), and focus UK_Plat. Then $F_N = (Args_N, \mathcal{R}_N^-)$ yields argument graph $\mathcal{G}_{F_N}$ in Figure 5.5, right.

In this setting, reasoning amounts to predicting the outcome of the unlabelled example, which in turns boils down to establishing whether the focus argument belongs to the grounded extension $\mathbb{G}$ of $(Args_N, \mathcal{R}_N^-)$ (Čyras *et al.*, 2016*a* and 2016*b*). Formally, the *AA-CBR outcome* of $(N, ?)$ is:

- φ, if $(\{\}, \varphi) \in \mathbb{G}$;
- $\overline{\varphi}$, otherwise, if $(\{\}, \varphi) \notin \mathbb{G}$.

To continue our illustration, the grounded extension of $(Args_N, \mathcal{R}_N^-)$ in Figure 5.5, right, is $\mathbb{G} = \{IOR, TooM, SF\}$, so the AA-CBR outcome of $(\{Blues, R'n'R, US\_Plat\}, ?)$ is not UK_Plat. That is, IOR is not deemed UK Platinum by AA-CBR (matching reality). This prediction can in turn be used as a feature for other reasoning (see section 5.6).

5.5.3 Explaining

In AA-CBR, explanations for why the AA-CBR outcome of $(N, ?)$ is φ / $\overline{\varphi}$ are defined in terms of admissible / maximal dispute trees, respectively (Čyras *et al.*, 2016*b*). These

[O:*IOR*] ⟶ [P:*LiB*/*MT*] ⟶ [O:*TooM*] ⟶ [P:({}, *UK_Plat*)]

Figure 5.6 *MDT-driven explanation for the AA-CBR outcome of* $(\{Blues, R'n'R, US\_Plat\}, ?)$.

notions can be trivially recovered as instances of our notions of ADT- / MDT-driven explanations (see Definition 5.2) as follows: an explanation for why the AA-CBR outcome of $(N, ?)$ is φ / $\overline{\varphi}$ is an ADT- / MDT- driven explanation, respectively, for $(\{\}, \varphi)$.

Thus, explanations in AA-CBR are debates between two parties arguing about the focus. (Note that there is always an explanation for why the AA-CBR outcome of $(N, ?)$ is φ or $\overline{\varphi}$ (Čyras *et al.*, 2016*b*).)

To conclude our illustration, an MDT-driven explanation for why the AA-CBR outcome of $(\{Blues, R'n'R, US\_Plat\}, ?)$ is not UK_Plat is given in Figure 5.6 (with the proponent P/opponent O status of nodes left explicit). Here, the focus (argued for by P) is the album being UK Platinum. O has an argument concerning the non-UK Platinum album $TooM$. P tries to put forward the argument that LiB/MT are UK Platinum, but these albums have highly positive reviews, which album IOR in question lacks. So O defends against P's argument by using the argument drawn from the unlabelled example.

Note that the sub-graph of G from Figure 5.5 consisting of the nodes from the explanation in Figure 5.6 is *not* maximally connected sub-graph-driven, because it does not contain node SF; but it is branch-driven. The deployment of these other types of graph-driven explanations (Definition 5.2) and the relations amongst them in AA-CBR is an interesting direction of future work.

AA-CBR explanations exhibit a degree of interactivity in the sense that they indicate what arguments could be sought by either the proponent or the opponent to establish a different outcome for the new case. For instance, the explanations may indicate what features an album under consideration lacks or possesses that result in a particular classification: if IOR in the above example had highly positive reviews, it would have the feature Pos_Rev and its AA-CBR outcome would be UK_Plat, with two explanations (in linear notation) $[\mathsf{P}: LiB/MT] \rightarrow [\mathsf{O}: TooM] \rightarrow [\mathsf{P}: (\{\}, UK\_Plat)]$ and $[\mathsf{P}: IOR] \rightarrow [\mathsf{O}: SF] \rightarrow [\mathsf{P}: (\{\}, UK\_Plat)]$.

5.6 Reasoning and Explaining with QBFs Mined from Recommender Systems

Recommender systems estimate a user's sentiment on items (e.g., products) using factors such as their features and similar users' sentiments in order to determine item recommendations. We show how recommender systems (of a certain type) can be mapped onto QBFs representing a single user's viewpoint, in the spirit of (Rago *et al.*, 2018). The mapping is such that the user's actual ratings of items give the arguments' base scores and the predicted ratings for the user give the strength of arguments. The construction of the QBF from the ratings guarantees desirable properties. In this section, argumentation is used solely for explaining, as recommender systems are naturally equipped with reasoning (recommending) capabilities.

5.6.1 Mining QBFs from recommender systems

We consider recommender systems giving a single user predicted ratings for items based on the associations between items and their features and the actual ratings from the user:[13] a *Recommender System* (RS) *for a given user* is a tuple $\langle \mathcal{I}, \mathcal{F}_r, \mathcal{A}, \mathcal{P} \rangle$ where:

- $\mathcal{I}$ is a finite, non-empty set of *items*;
- $\mathcal{F}_r$ is a finite, non-empty set of *features*;[14]
- the sets $\mathcal{I}$ and $\mathcal{F}_r$ are pairwise disjoint and each $i \in \mathcal{I}$ is associated with $\mathcal{F}_{r_i} \subseteq \mathcal{F}_r$;
- $\mathcal{A} : \mathcal{I} \cup \mathcal{F}_r \to [-1,1]$ is a total function of *actual ratings* from the user;
- $\mathcal{P} : \mathcal{I} \cup \mathcal{F}_r \to [-1,1]$ is a partial function of *predicted ratings* for the user.

We assume that ratings, when defined, are real numbers in [−1,1] where, straightforwardly, a positive/negative rating indicates positive/negative sentiment (respectively). For simplicity, in the remainder we focus on simple RSs where:

(1) $\forall f \in \mathcal{F}_r$, $\mathcal{A}(f)$ is undefined (i.e., the user can only rate items, as in many real RSs);

(2) $\forall f \in \mathcal{F}_r$, $\mathcal{P}(f) = max(\{\mathcal{A}(i_+) | i_+ \in \mathcal{I} \wedge f \in \mathcal{F}_{r_{i_+}} \wedge \mathcal{A}(i_+) > 0\}) - max(\{-\mathcal{A}(i_-) | i_- \in \mathcal{I} \wedge f \in \mathcal{F}_{r_{i_-}} \wedge \mathcal{A}(i_-) < 0\})$ (namely, the predicted rating of a feature is given in terms of the most positive and the most negative from the feature's associated items' actual ratings);

(3) $\forall i \in \mathcal{I}$, if $\mathcal{A}(i)$ is defined then $\mathcal{P}(i) = \mathcal{A}(i)$; otherwise, $\mathcal{P}(i) = max(\{\mathcal{P}(f_+) | f_+ \in \mathcal{F}_{r_i} \wedge \mathcal{P}(f_+) > 0\}) - max(\{-\mathcal{P}(f_-) | f_- \in \mathcal{F}_{ri} \wedge \mathcal{P}(f_-) < 0\})$[15] (namely, for an item without an actual rating from the user, the item's predicted rating is calculated using the most positive and the most negative from the item's associated features' predicted ratings).

Table 5.3 shows an example RS (also visualized in Figure 5.7a) in the domain of album recommendation (note that we have included a feature p_1 predicted as in section 5.5).[16] Actual ratings from the user are only available for three of the albums and RS may be used to recommend to the user i_1 over i_2 based on their predicted ratings (note that without the associations of p_1 the predicted ratings would be $\mathcal{P}(i_1) = 0.4$ and $\mathcal{P}(i_2) = 0.6$ and therefore i_2 would be recommended instead).

We define a class of functions from the set $\mathbb{R}$ of all RSs as defined earlier to the set $\mathbb{Q}$ of all QBFs as follows:

[13] This recommender system is more abstract and considerably simpler than real systems, which take, in particular, several users into accounts. It is a simplification of our recommender system in (Rago *et al.*, 2018), used here for simplicity of presentation, given that our focus is on illustrating how our explanations can be deployed.

[14] Note that the set of features $\mathcal{F}_r$ used here in the context of recommendations may be different, in general, from the set of features $\mathcal{F}_p$ used in Section 5.5 for the purpose of prediction with AA-CBR.

[15] We assume that for any $S \subseteq [0,1]^*$, if $S = \{\}$ then $max(S) = 0$; otherwise, $max(S) = argmax_{s \in S}\ s$.

[16] Here, $B\&L$ has been omitted since its associated features $\{a_1, g_1, p_1\}$ in RS_1 are identical to those of LiB.

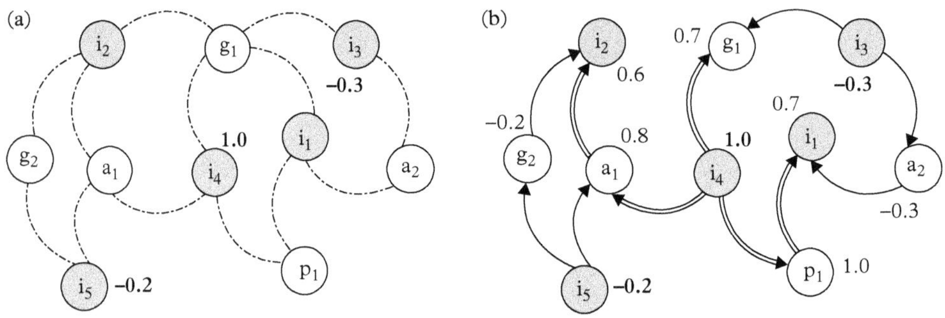

Figure 5.7 *a. Example RS and b. corresponding QBF $\mu_1(RS)$ with non-zero base scores (actual ratings) in bold and strength (predicted ratings) in normal font.*

Definition 5.3 *An* Argumentative Recommender Reading *function $\mu : \mathbb{R} \to \mathbb{Q}$ is such that, for any $\mathcal{R} \in \mathbb{R}$, $\mu(\mathcal{R}) = \langle \mathcal{I} \cup \mathcal{F}_r, \mathcal{R}^-, \mathcal{R}^+, \tau \rangle$ such that $\mathcal{R}^- : \mathcal{I} \times \mathcal{F}_r \cup \mathcal{F}_r \times \mathcal{I}$, $\mathcal{R}^+ : \mathcal{I} \times \mathcal{F}_r \cup \mathcal{F}_r \times \mathcal{I}$, $\mathcal{R}^-$ and $\mathcal{R}^+$ are pairwise disjoint, and for any $x \in \mathcal{I} \cup \mathcal{F}_r$, $\tau(x) = \mathcal{A}(x)$ if this is defined, and $\tau(x) = 0$ otherwise.*

Intuitively, items and features are treated as arguments: if a user rates an item highly/lowly then this item can be seen as an argument for/against, respectively, the argument that the user likes the features associated with the item and, similarly, if a user rates a feature highly/lowly then this feature can be seen as an argument for/against, respectively, the argument that the user likes items which hold this feature. Also, the actual ratings, when available, are treated as base scores. If, in addition, we understand the predicted ratings $\mathcal{P}$ as a dialectical strength function (see section 5.2.1) $\mathcal{P} : \mathcal{I} \cup \mathcal{F}_r \to [-1, 1]$ for the QBF, then it is natural to refine μ as follows:

Definition 5.4 *A* $\mathcal{P}$-induced Argumentative Recommender Reading function *$\mu_1 : \mathbb{R} \to \mathbb{Q}$, is an Argumentative Recommender Reading function such that, for $i \in \mathcal{I}, f \in \mathcal{F}_r$:*

- $\mathcal{R}^-(f) = argmax_{i_- \in \mathcal{I}}\{-\mathcal{A}(i_-) | f \in \mathcal{F}_{r_{i_-}} \wedge \mathcal{A}(i_-) < 0\}$;
- $\mathcal{R}^+(f) = argmax_{i_+ \in \mathcal{I}}\{\mathcal{A}(i_+) | f \in \mathcal{F}_{r_{i_+}} \wedge \mathcal{A}(i_+) > 0\}$.[17]

If $\mathcal{A}(i)$ is defined then $\mathcal{R}^-(i) = \mathcal{R}^+(i) = \{\}$, otherwise:

- $\mathcal{R}^-(i) = argmax_{f_- \in \mathcal{F}_r}\{-\mathcal{P}(f_-) | f_- \in \mathcal{F}_{r_i} \wedge \mathcal{P}(f_-) < 0\}$;
- $\mathcal{R}^+(i) = argmax_{f_+ \in \mathcal{F}_r}\{\mathcal{P}(f_+) | f_+ \in \mathcal{F}_{r_i} \wedge \mathcal{P}(f_+) > 0\}$.

Figure 5.7b shows $\mu_1(RS)$, for RS illustrated earlier.

$\mu_1(RS)$ and $\mathcal{P}$, treated as a strength function, together satisfy the following intuitive properties, variants of monotonicity (Baroni *et al.*, 2018*b* and 2019) and stating that

[17] $argmax_{s \in S}\{\alpha(s) | \pi(s)\}$, for a function α and a property π, stands for the set $\{s \in \mathcal{S} | \alpha(s) = v_{max}\}$, where $v_{max} = max\{\alpha(s) | s \in S \wedge \pi(s)\}$.

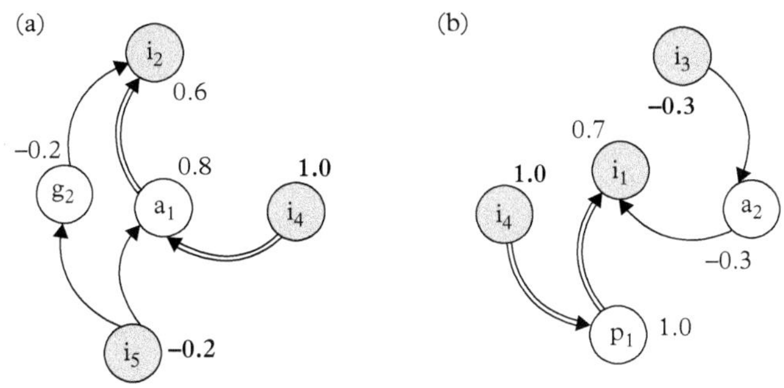

Figure 5.8 *The argumentation explanations for items i_1 (**a**) and i_2 (**b**).*

increasing the strength of an attacker (supporter) will reduce (increase, respectively) the strength of an argument:

Definition 5.5 (Property) *For any $\alpha, \beta \in Args$ in a QBF $\langle Args, \mathcal{R}^-, \mathcal{R}^+, \tau\rangle$, with $\tau(\alpha) = \tau(\beta)$:*

- *if $\mathcal{R}^+(\alpha) = \mathcal{R}^+(\beta)$, $\gamma \in \mathcal{R}^-(\alpha)$, $\delta \in \mathcal{R}^-(\beta)$ and $\mathcal{R}^-(\beta)\backslash\{\delta\} = \mathcal{R}^-(\alpha)\backslash\{\gamma\}$, then $\sigma(\beta) < \sigma(\alpha)$ whenever $\sigma(\delta) > \sigma(\gamma)$;*
- *if $\mathcal{R}^-(\alpha) = \mathcal{R}^-(\beta)$, $\gamma \in \mathcal{R}^+(\alpha)$, $\delta \in \mathcal{R}^+(\beta)$ and $\mathcal{R}^+(\beta)\backslash\{\delta\} = \mathcal{R}^+(\alpha)\backslash\{\gamma\}$, then $\sigma(\beta) > \sigma(\alpha)$ whenever $\sigma(\delta) > \sigma(\gamma)$.*

It is easy to see that $\mathcal{P}$ satisfies Property 5.5, for QBF $\mu_1(RS_1)$.

5.6.2 Explaining

This argumentative reading of predicted ratings on items and their features facilitates the extraction of explanations for recommendations, based on Definition 5.2, with which users could potentially interact (based on properties such as Property 5.5) to provide feedback and improve future predictions. In particular, the notion of maximally connected sub-graph-driven explanation is useful in this setting.

As an illustration, if the user wonders why item i_1, instead of i_2, is recommended by RS in Table 5.3, maximally connected sub-graph-driven explanations for each item, shown in Figures 5.8a and b, respectively, will provide relevant information.

5.7 Conclusions

We have explored (reasoning and) explanation using argumentation frameworks of different kinds, mined from different types of data. We have advanced general templates of argument graph-driven explanations and have given instances of these from three different approaches, in the setting of media and entertainment industry.

An actual (albeit preliminary) proof-of-concept music recommender system integrating explanations as described in this chapter has been implemented (Chauhan *et al.*, 2019). In the future we plan to explore how users can interact with various explanations to obtain different reasoning outcomes, and to evaluate experimentally the usefulness of these interactions.

Several works define methods for determining explanations for the (non-)acceptability of arguments in argumentation (see e.g., García *et al.*, 2013; Fan and Toni, 2015; Schulz and Toni, 2016). These works use trees as the underlying mechanism for computing explanations, that we have adapted for some of our purposes in this paper. In some works (e.g., García *et al.*, 2013), explanations are branches of the grounded extension of an acyclic AA framework. Study of formal relationships between our branch-driven explanations and these works is left for future work.

The differences and relationships between arguments and explanations as pieces of text are explored in (Bex and Walton, 2016). Also, the links between arguments, explanations and causal models are studied (e.g., in Timmer *et al.*, 2017), where a support graph is constructed from a Bayesian network and arguments are built from that support graph to facilitate the correct interpretation and explanation of the relation between hypotheses and evidence that is modelled in the Bayesian network. Study of formal relationships between our and these works is also left for future explorations.

Other work in argumentation (e.g., Cerutti *et al.*, 2014), investigates the usefulness of explanation in argumentation with users. Similar explorations for our approach are also left for the future, in particular to ascertain whether our property-driven explanations can provide a basis for machines that humans want to use, fully in the spirit of human-like computing.

Acknowledgements

The authors were partially supported by the EPSRC project EP/P029558/1 ROAD2H. Rago and Toni were also partially supported by the Human-Like Computing EPSRC Network of Excellence.

References

Atkinson, K., Baroni, P., Giacomin, M. *et al.* (2017). Towards artificial argumentation. *AI Magazine*, **38**(3), 25–36.

Baroni, P., Caminada, M., and Giacomin, M. (2011). An introduction to argumentation semantics. *Knowledge Engineering Review*, **26**(4), 365–410.

Baroni, P., Comini, G., Rago, A. *et al.* (2017). Abstract games of argumentation strategy and game-theoretical argument strength, in *International Conference on Principles and Practice of Multi-Agent Systems*. Cham: Springer, 403–19.

Baroni, P., Gabbay, D., Giacomin, M. *et al.* (eds) (2018*a*). *Handbook of Formal Argumentation*. Rickmansworth, UK: College Publications.

Baroni, P., Rago, A., and Toni, F. (2018*b*). How many properties do we need for gradual argumentation?, in S. A. McIlwraith and K. Q. Weinberger, eds, *32nd AAAI Conference on Artificial Intelligence*. New Orleans, LO: AAI Press, 1736–43.

Baroni, P., Rago, A., and Toni, F. (2019). From fine-grained properties to broad principles for gradual argumentation: a principled spectrum. *International Journal of Approximate Reasoning*, **105**, 252–86.

Baroni, P., Romano, M., Toni, F. *et al.* (2015). Automatic evaluation of design alternatives with quantitative argumentation. *Argument & Computation*, **6**(1), 24–49.

Besnard, P., García, A. J., H. *et al.* (2014). Introduction to structured argumentation. *Argument & Computation*, **5**(1), 1–4.

Bex, F. and Walton, D. (2016). Combining explanation and argumentation in dialogue. *Argument & Computation*, 7(1), 55–68.

Blei, D. M., Ng, A. Y., and Jordan, M. I. (2003, March). Latent dirichlet allocation. *Journal of Machine Learning Research*, **3**, 993–1022.

Briguez, C. E., Budán, M. C., Deagustini, C. *et al.* (2014). Argument-based mixed recommenders and their application to movie suggestion. *Expert Systems with Applications*, **41**(14), 6467–82.

Carstens, L. and Toni, F. (2015). Towards relation based argumentation mining, in *Proceedings of the 2nd Workshop on Argumentation Mining*. Denver, CO: Association of Computational Linguistics, 29–34.

Cayrol, C. and Lagasquie-Schiex, M.-C. (2005). On the acceptability of arguments in bipolar argumentation frameworks, in *European Conference on Symbolic and Quantitative Approaches to Reasoning and Uncertainty*. Berlin: Springer, 378–89.

Cerutti, F., Tintarev, N., and Oren, N. (2014). Formal arguments, Preferences, and natural language interfaces to humans: an empirical evaluation, in *Proceedings of the European Conference on Artificial Intelligence*, Prague, 207–12.

Chauhan, R., Anicai, A.-E., Gavrielov, D. *et al.* (2019). Explainable automated decisions for consumer-oriented tasks in the music industry. Master's thesis, MSc Group Projects (Computing Science), Department of Computing, Imperial College London.

Cocarascu, O., Čyras, K., and Toni, F. (2018). Explanatory predictions with artificial neural networks and argumentation, in *2nd Workshop on XAI at the 27th IJCAI and the 23rd ECAI*, Stockholm, Sweden.

Cocarascu, O., Rago, A., and Toni, F. (2019). Dialogical Explanations for review aggregations with argumentative dialogical agents, in *Proceedings of the 18th International Conference on Autonomous Agents and MultiAgent Systems*, 13–17 May, Montreal.

Cocarascu, O., Stylianou, A., Čyras, K. *et al.* (2020*b*). Data-empowered argumentation for dialectically explainable predictions, in *ECAI 2020—24th European Conference on Artificial Intelligence, Santiago de Compostela, Spain, 10–12 June 2020*.

Cocarascu, O. and Toni, F. (2016). Detecting deceptive reviews using argumentation, *Proceedings of the 1st International Workshop on AI for Privacy and Security*, 1–8.

Cocarascu, O. and Toni, F. (2017). Identifying attack and support argumentative relations using deep learning, in *Proceedings of the 2017 Conference on Empirical Methods in Natural Language Processing*, 1374–9.

Cocarascu, O. and Toni, F. (2018). Combining deep learning and argumentative reasoning for the analysis of social media textual content using small datasets. *Computational Linguistics*, **44**(4), 833–58.

Cohen, A., Gottifredi, S., García, A. *et al.* (2014). A survey of different approaches to support in argumentation systems. *The Knowledge Engineering Review*, **29**(5), 513–50.

Cohen, A., Parsons, S., Sklar, E. I. *et al.* (2018). A characterization of types of support between structured arguments and their relationship with support in abstract argumentation. *International Journal of Approximate Reasoning*, **94**, 76–104.

Čyras, K., Birch, D., Guo, Y. *et al.* (2019). Explanations by arbitrated argumentative dispute. *Expert Systems with Applications*, **127**, 141–56.

Čyras, K., Satoh, K., and Toni, F. (2016*a*). Abstract argumentation for case-based reasoning, in *Proceedings of the 15th International Conference on the Principles of Knowledge Representation and Reasoning (KR 2016)*, 549–52.

Čyras, K., Satoh, K., and Toni, F. (2016*b*). Explanation for case-based reasoning via abstract argumentation, in *Proceedings of Computational Models of Argument*, 243–54.

Čyras, K., Schulz, C., and Toni, F. (2017). Capturing bipolar argumentation in non-flat assumption-based argumentation, in *International Conference on Principles and Practice of Multi-Agent Systems*. Cham: Springer, 386–402.

Dung, P. M. (1995). On the acceptability of arguments and its fundamental role in nonmonotonic reasoning, logic programming and n-person games. *Artificial Intelligence*, 77, 321–57.

Dung, P. M., Kowalski, R., and Toni, F. (2006). Dialectic proof procedures for assumption-based, admissible argumentation. *Artificial Intelligence*, **170**(2), 114–59.

Fan, X., Craven, R., Singer, R. *et al.* (2013). Assumption-based argumentation for decision-making with preferences: a medical case study in *International Workshop on Computational Logic in Multi-Agent Systems*. Berlin: Springer, 374–90.

Fan, X. and Toni, F. (2015). On computing explanations in argumentation, in *Twenty-Ninth Association for the Advancement of Artificial Intelligence Conference on Artificial Intelligence*, Austin. 1496–502.

García, A. J., Chesñevar, C., Rotstein, N. *et al.* (2013). Formalizing dialectical explanation support for argument-based reasoning in knowledge-based systems. *Expert Systems with Applications*, **40**, 3233–47.

Guidotti, R., Monreale, A., Ruggieri, S. *et al.* (2019). A survey of methods for explaining black box models. *Association of Computing Machinery Computing Surveys*, **51**(5), 1–42.

Hunter, A. and Williams, M. (2015). Aggregation of clinical evidence using argumentation: A tutorial introduction, in *Foundations of Biomedical Knowledge Representation—Methods and Applications*, pp. 317–37.

Lee, D. D. and Seung, H. S. (1999). Learning the parts of objects by nonnegative matrix factorization. *Nature*, **401**, 788–91.

Menini, S., Cabrio, E., Tonelli, S. *et al.* (2018). Never retreat, never retract: argumentation analysis for political speeches in *32nd Advancement of Artificial Intelligence Conference*, New Orleans, 4889–96.

Mercier, H. and Sperber, D. (2011). Why do humans reason? arguments for an argumentative theory. *Behavioral and Brain Sciences*, **34**, 57–111.

Miller, T. (2019). Explanation in artificial intelligence: insights from the social sciences. *Artificial Intelligence*, **267**, 1–38.

Rago, A., Cocarascu, O., and Toni, F. (2018). Argumentation-based recommendations: fantastic explanations and how to find them in *Proceedings of the Twenty-Seventh International Joint Conference on Artificial Intelligence*, Stockholm, Sweden, 1949–55.

Rago, A., Toni, F., Aurisicchio, M. *et al.* (2016). Discontinuity-free decision support with quantitative argumentation debates. In: *Principles of Knowledge Representation and Reasoning: Proceedings of the Fifteenth International Conference (KR 2016)*, Cape Town, AAAI Press, 25–29.

Schulz, C. and Toni, F. (2016). Justifying answer sets using argumentation. *Theory and Practice of Logic Programming*, **16**(1), 59–110.

Timmer, S. T., Meyer, J.-J. Ch., Prakken, H. *et al.* (2017). A two-phase method for extracting explanatory arguments from Bayesian networks. *International Journal of Approximate Reasoning*, **80**, 475–94.

6

Explanation in AI systems

Marko Tesic and Ulrike Hahn
Department of Psychological Sciences, Birkbeck, University of London, UK

In this chapter, we consider recent work aimed at guiding the design of algorithmically generated explanations. The chapter proceeds in four parts. First, we introduce the general problem of machine-generated explanation and illustrate different notions of explanation with the help of Bayesian belief networks. Secondly, we introduce key theoretical perspectives on what constitutes an explanation, and more specifically a 'good' explanation, from the philosophical literature. We compare these theoretical perspectives and the criteria they propose with a case study on explaining reasoning in Bayesian belief networks and present implications for AI. Thirdly, we consider the pragmatic nature of explanation with the focus on its communicative aspects that are manifested in considerations of trust. Finally, we present conclusions.

6.1 Machine-generated Explanation

Recent years have seen a groundswell of interest in machine-generated explanation for AI systems (DARPA, 2016; Doshi-Velez and Kim, 2017; Samek *et al.*, 2017; Montavon *et al.*, 2018; Rieger *et al.*, 2018). Multiple factors exert pressure for supplementing AI systems with explanations of their outputs. Explanations provide transparency for what are often black-box procedures. Hence transparency is viewed as critical for fostering the acceptance of AI systems in real-world practice (Bansal *et al.*, 2014; Chen *et al.*, 2014; Mercado *et al.*, 2016; Hayes and Shah, 2017; Wachter *et al.*, 2017; Fallon and Blaha, 2018), last but not least, because transparency might be a necessary ingredient for dealing with legal liability (Goodman and Flaxman, 2016; Doshi-Velez *et al.*, 2017; Wachter *et al.*, 2017*a*; Felzmann *et al.*, 2019). At the same time, decades of research in AI make plausible the claim that AI systems genuinely able to navigate real-world challenges are likely to involve joint human–system decision-making, at least for the foreseeable future. This however, requires AI systems to communicate outputs in such a way as to allow humans to make informed decisions.

The challenge of developing adequate, machine-generated explanation is a formidable one. For one thing, it requires an accessible model of how AI system's outputs or conclusions were arrived at. This poses non-trivial challenges for many of the presently most

Marko Tesic and Ulrike Hahn, *Explanation in AI systems* In: *Human Like Machine Intelligence.* Edited by: Stephen Muggleton and Nick Charter, Oxford University Press. © Oxford University Press 2021. DOI: 10.1093/oso/9780198862536.003.0006

successful types of AI systems, such as convolutional, deep learning networks (Collobert *et al.*, 2011; Graves and Schmidhuber, 2005; Krizhevsky *et al.*, 2012; Goodfellow *et al.*, 2016) which are already notoriously opaque to external observers, let alone offering up representations of a system's internal processes that could be used to drive explanation generation. As such, the automated generation of explanation is arguably easier to achieve with AI systems that operate with models that are formulated at acceptable user levels of engagement; at least here, the step of translating low-level representations into a suitable higher-level representations accessible to us is, in a large number of cases, already taken care of.

Bayesian Belief Networks (BNs) are an AI technique that has been viewed as significantly more interpretable and transparent than deep neural networks (Gunning and Aha, 2019), while still possessing a notable predictive power and being applied to various contexts ranging from defence and military (Laskey and Mahoney, 1997; Falzon, 2006; Lippmann *et al.*, 2006) and cyber security (Xie *et al.*, 2010; Chockalingam *et al.*, 2017), over medicine (Wiegerinck *et al.*, 2013; Agrahari *et al.*, 2018), and law and forensics (Fenton *et al.*, 2013; Lagnado *et al.*, 2013), to agriculture (Drury *et al.*, 2017) as well as psychology, philosophy, and economics (see below). As such, BNs seem to serve well one of the goals of this chapter which is to bring and overlay insights on explanations from different areas of research: they are a promising meeting point connecting the research on machine-generated explanation in AI and the research on human understanding of explanation in psychology and philosophy. We thus use BNs as the focal point of our analysis in this chapter. Given the increasing popularity of BNs within AI (Friedman *et al.*, 1997; Ng and Jordan, 2002; Pernkopf and Bilmes, 2005; Roos *et al.*, 2005), including their relation to deep neural networks (Wang and Yeung, 2016; Rohekar *et al.*, 2018; Choi *et al.*, 2019) and efforts to explain deep neural networks via BNs (Harradon *et al.*, 2018), this should be intrinsically interesting. Furthermore, we take the kinds of issues we identify here to be indicative of the kinds of problems and distinctions that are likely to emerge in *any* attempt at machine-generated explanation.

6.1.1 Bayesian belief networks: a brief introduction

Bayesian belief networks provide a simple graphical formalism for summarising and simplifying joint probability distributions in such a way as to facilitate Bayesian computations (Pearl, 1988; Neapolitan, 2003). Specifically, BNs use independence relations between variables to simplify the computation of joint probability distributions in cases of multivariate problems. As Bayesian models, they have a clear normative foundation: 'being Bayesian', that is, assigning degrees of belief in line with the axioms of probability (the Dutch Book argument, (Ramsey, 2016; Vineberg, 2016)), and the use of Bayes' rule to update believes in light of new evidence (also known as 'conditionalization') which uniquely minimizes the inaccuracy of an agent's beliefs across all possible worlds (i.e., regardless of how the world turns out), on the condition that inaccuracy is measured with the Brier score and those worlds are finite (see e.g., the formal results outlined in (Pettigrew, 2016)). In other words, Bayesian computations specify how agents *should* change their beliefs, if they wish those beliefs to be accurate. As a result, the Bayesian framework has seen widespread application not just in AI but also in philosophy, economics, and psychology (Bovens and Hartmann, 2003; Hahn and Oaksford, 2006; Howson and Urbach, 2006; Hahn and Oaksford, 2007; Neil *et al.*, 2008; Dizadji-Bahmani *et al.*, 2011; Fenton *et al.*, 2013; Rehder, 2014; Rottman and Hastie, 2014; Hahn and Hornikx, 2016; Harris *et al.*, 2016; Spiegler, 2016; Dewitt *et al.*, 2018; Liefgreen *et al.*, 2018; Madsen *et al.*, 2018; Phillips *et al.*, 2018; Pilditch *et al.*, 2018; Dardashti *et al.*, 2019; Tešić, 2019; Tešić and Hahn, 2019; Tešić *et al.*, 2020). In particular, BNs are helpful in spelling out the implications of less intuitive interactions between variables. This is readily illustrated with the example of "explaining away", a phenomenon that has received widespread psychological investigation (Morris and Larrick, 1995; Fernbach and Rehder, 2013; Rehder, 2014; Rottman and Hastie, 2014, 2016; Davis and Rehder, 2017; Rehder and Waldmann, 2017; Liefgreen *et al.*, 2018; Pilditch *et al.*, 2019; Sussman and Oppenheimer, 2011; Tešić *et al.*, 2020). Figure 6.1 illustrates a simple example of explaining away. There are two potential causes, a physical abuse and haemophilia (a genetic bleeding disorder), of a single effect, bruises on a child's body. Before finding out anything about whether there are bruises on a body, the two causes are independent: learning that a child is suffering from haemophilia will not change our beliefs about whether the child is physically abused. However, if we learn that the child has bruises on its body, then the two causes become dependent: additionally learning that the child is suffering from haemophilia will change (decrease) the probability that it has been physically abused since haemophilia alone is sufficient to explain away the bruises.

The example illustrates not just BNs ability to model explaining away situations and provide us with both qualitative and quantitative normative answers, but also their advantage over classical logic and rule-based expert systems. A rule-based expert system consisting of a set of IF-THEN rules and a set of facts (see (Grosan and Abraham, 2011)) may carry out an incorrect chaining in situations representing explaining away. For instance, a rule-based system may combine plausible-looking rules 'If the child is suffering from haemophilia, then it is likely the child has bruises' with 'If the child has bruises, then it is likely the child is physically abused' to get 'If the child is suffering from haemophilia, then it is likely the child is physically abused'. However, we know that actually the opposite is true: learning about haemophilia makes physical abuse less likely (Pearl, 1988). The application of rule-based expert systems to legal and medical contexts (Grosan and Abraham, 2011) where explaining away and other

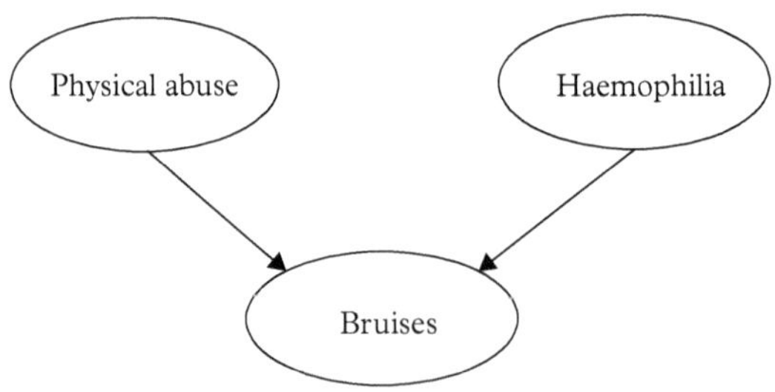

Figure 6.1 *A BN model of explaining away.*

causal-probabilistic relationships can be found highlights the importance of accurately capturing these relationships in computational terms.

6.1.2 Bayesian belief networks: explaining evidence

Another important feature of BNs is that they can be used for both predictive reasoning and diagnostic reasoning (often referred to as an abduction, see e.g., (Korb and Nicholson, 2010)). Consider the BN in Figure 6.1. An example of predictive reasoning would be inferring a probability that the child has bruises given that its suffering from haemophilia (i.e., inference from causes (h) to effects (e)); whereas diagnostic reasoning would be inferring a probability of physical abuse from learning that the child has bruises (i.e., inference from effects (e) to causes (h)). Often, diagnostic reasoning (abduction) is used to find the most probable explanations (causes) of observed evidence (effects), that is to find the configuration h with the maximum $p(\mathrm{h} \mid \mathrm{e})$ (Pearl, 1988). Similarly, Shimony's 1991 partial abduction approach first marginalizes out variables that are not part of explanations (x) and then searches for the most probable h: that is, find h with the maximum $\sum_x p(\mathrm{h}, \mathrm{x} \mid \mathrm{e})$. More recently, Yuan *et al.* (2011) introduced a method they call 'Most Relevant Explanation' (MRE) which chooses the explanation that has the highest likelihood ratio compared to all other explanations: that is, find h with the maximum $p(\mathrm{e} \mid \mathrm{h})/p(\mathrm{e} \mid \overline{\mathrm{h}})$, where $\overline{\mathrm{h}}$ denotes all other alternative explanations to h. Nielsen *et al.* (2008) introduced a 'Causal Explanation Tree' (CET) method which uses the post-intervention distribution of variables (Pearl, 2000) in selecting explanations, which is in contrast to all previous methods since they use non-interventional distribution of variables in a BN. Drawing on their definition of causation, Halpern and Pearl (2005*b*) develop a definition of explanation to address a question of why certain evidence holds given users epistemic state. Their definition of explanation states that (1) a user should consider evidence to hold, (2) an explanation (h) is a sufficient cause of evidence, (3) h is minimal (i.e., it does not contain irrelevant or redundant elements), and (4) h is not known at the beginning, but it is considered as a possibility. This is an improvement on other accounts. However, their account of causation has as an output again a set of variables in a BN model which are deemed as causes of evidence in the model. Yap *et al.* (2008) employ Markov blanket to determine which variables should feature in an explanation. A Markov blanket is of a node X includes all nodes that are direct parents, children, or children's parents of node X. A powerful property of Markov blanket is that knowing the sates of all the variables in a Markov blanket of X would uniquely determine the probability distribution of X: additionally learning the states of other variables outside the Markov blanket of X would not affect the probability distribution of X. Yap *et al.*'s 'Explaining BN Inferences' procedure identifies Markov nodes of evidence (i.e., nodes in a Markov blanket of the evidence node) and learns context-specific independences in Markov nodes with a decision tree to exclude irrelevant nodes in an explanation of the evidence.

Despite the difference among the methods, they all share at least one commonality: explanation of evidence is exhausted by a set of variables in a BN that these methods have pointed to. In other words, evidence provided a justification in terms of a set of variables. This is undoubtedly useful, but in certain contexts (e.g., high-cost domains, see (Herlocker *et al.*, 2000)) it is arguably not enough to meet the demands of user transparency. In contrast to the notion of explanation as a justification of evidence is that of explanation of *reasoning processes* in BNs, and expert systems in general (Wick and Thompson, 1992; Lacave and Díez, 2002; Sørmo *et al.*, 2005). Here one is interested in how evidence propagates in a BN rather than in selecting a set of variables that would account for evidence. Explaining reasoning processes in BNs been a research focus amongst researchers for some time (see (Lacave and Díez, 2002) for an overview). We next describe one more recent attempt in the context of the Bayesian Argumentation via Delphi (BARD) project.

6.1.3 Bayesian belief networks: explaining reasoning processes

The BARD project (Dewitt *et al.*, 2018; Liefgreen *et al.*, 2018; Phillips *et al.*, 2018; Pilditch *et al.*, 2018, Pilditch et al., 2019; Cruz *et al.*, 2020) set as its goal the development of assistive technology that could facilitate group decision-making in an intelligence context. To this end, BARD provides a graphical user interface enabling intelligence analysts to represent arguments as BNs and allowing them to examine the impact of different pieces of evidence on arguments as well as to bring groups of analysts to a consensus via an automated Delphi method. An essential component of the system is the algorithm for generating natural language explanations of inference in a BN, or more specifically, an explanation of evidence propagation in a BN. This algorithm builds on earlier work by Zukerman and colleagues that have sought to use BNs to generate arguments (Zukerman *et al.*, 1998, 1999). The algorithm uses an evidence-to-goal approach to generate explanations for a BN. An explanation starts with the given pieces of evidence and traces paths that describe their influence on intervening nodes until the goal is reached. In essence, the algorithm adopts a causal interpretation of the links between the connected nodes, finds a set of rules that describe causal relations in a BN, and calculates all paths between evidence nodes and target nodes and builds corresponding trees in order to determine the impact of evidence on target nodes. Figure 6.2 provides an example. There we have four pieces of evidence, Emerson Report, Quinns Report, and AitF Sawyer Report all stating that 'The Spider' is in the facility and Comms Analyst Winter Report stating that 'The Spider' is not in the facility. The goal is to explain the impact of these four pieces of evidence on two target variables, namely *'Is The Spider in the facility?'* and *'Are logs true?' ('Are Emerson & Quinn spies?')*. The algorithm first finds all relevant paths between evidence and target nodes, builds a corresponding tree and calculates the impact of evidence on the target, which is simply a difference between the probability of the target node *before* learning particular piece(s) of evidence and *after* learning particular piece(s) of evidence. This way the algorithm can find *HighImpSet*—nodes that have the highest impact on the target node, *NoImpSet*—nodes that, in light of the other evidence nodes, have no impact on the target node, and *OppImpSet*—nodes that have the opposite impact to that of *HighImpSet*. Finally, the algorithm realizes the explanations in English language using sentences, clauses and

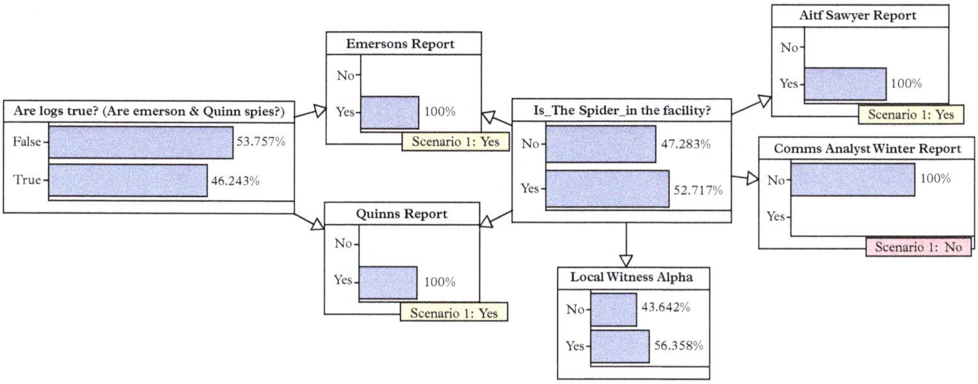

Figure 6.2 *A BN of a fictional scenario used in BARD testing phase. Four pieces of evidence are available:* Emerson Report=*Yes,* Quinns Report=*Yes,* AitF Sawyer Report=*Yes, and* Comms Analyst Winter Report=*No.*

Is The Spider in the facility?		
HighImpSet *MinHIS* *CombMinSet* *NoImpSet* *OppImpSet*	{{*ASR*}} {{*ASR*}} Ø Ø {{*ER*}, {*QR*}}	In the absence of evidence, the probability of *Is The Spider in the facility?* = Yes is 10% (very unlikely). Observing *Emerson's Report* = No and *Quinn's Report* = No reduces the probability of *Is The Spider in the facility?* = Yes. However, adding the evidence *AitF Sawyer's Report* = Yes increases the probability of *Is The Spider in the facility?* = Yes. The final probability of *Is The Spider in the facility?* = Yes is 5.3% (very unlikely).

Are the logs true? (Are Emerson and Quinn spies?)		
HighImpSet *MinHIS* *CombMinSet* *NoImpSet* *OppImpSet*	{{*ER*}, {*QR*}} {{*ER*}, {*QR*}} {*ER, QR*} Ø {{*ASR*}}	In the absence of evidence, the probability of *Are the logs true?* = True is 10% (very unlikely). Observing either *Emerson's Report* = No or *Quinn's Report* = No reduces the probability of *Are the logs true* = True. However, adding the evidence *AitF Sawyer's Report* = Yes increases the probability of *Are the logs true?* = Yes. The final probability of *Are the logs true?* = Yes is 4.8% (almost no chance).

Figure 6.3 *A summary report generated by the BARD algorithm applied on the BN from Figure 6.2. In addition to the natural language explanation, it provides sets with nodes that are* HighImpSet, NoImpSet, *and* OppImpSet. *For the purposes of this chapter we can ignore* MinHIS *and* CombMinSet.

phrases devised and combined by means of a semantic grammar (Burton, 1976). The output of the algorithm is presented in Figure 6.3.

As can be seen, the output in Figure 6.3 provides significantly more information to the user than just a single verdict on whether or not the variable *'Is The Spider in the facility?'* is part of the explanation of the evidence, as would be the output of methods looking for a justification of evidence. In addition to the impact sets, it provides a natural language explanation on how different pieces of evidence influence the probability of the target variable. Nevertheless, there remain continued challenges with this approach.

First, the algorithm retains difficulties in coping adequately with soft evidence, namely evidence that we do not learn with probability 1. For instance, imagine that in the BN in Figure 6.2 we additionally learn that the logs are most likely true, but we are not absolutely convinced. To reflect that we would set *p(Are logs true? (Are Emerson & Quinn spies)* = True) to equal 0.95 for instance. Thus the probability of *Are logs true? (Are Emerson & Quinn spies)* = True has changed from 0.46243 to 0.95, but it didn't go all the way to 1. The current version of the algorithm is not able to calculate the impact of such change. Second, the explanations generated by the algorithm are not aimed specifically at what a human user *might find hard to understand*. To make matters worse, it is arguably the interactions between variables and the often counterintuitive effects of these, that users will most struggle with (for psychological evidence to this effect see e.g. (Dewitt *et al.*, 2018; Liefgreen *et al.*, 2018; Phillips *et al.*, 2018; Pilditch *et al.*, 2018, 2019; Tešić *et al.*, 2020)). In other words, the system generates an (accurate) explanation, but not necessarily a good explanation. For further guidance on what might count as a good explanation we consult research on this topic within the philosophy of science and epistemology, where the topic has raised decades of interest.

6.2 Good Explanation

6.2.1 A brief overview of models of explanation

The historic point of departure for thinking about the nature of explanation in philosophy is the covering law model (Hempel and Oppenheim, 1948), also known as the 'deductive-nomological model' of scientific explanation (where nomological means pertaining to the laws of nature). This model construes explanation as a deductive argument with true premises that has the phenomenon to be explained (the so-called explanandum) as its conclusion. Specifically, this conclusion is derived from general laws and particular facts. For example, an explanation of a position of a planet at a point in time consists of a derivation of that position from the Newtonian laws governing gravity (general law), and information about the mass of the sun, the mass of the planet, and position at a particular time and velocity of each (particular facts) (Woodward, 2017). A key feature of this model is that it views explanation and prediction as essentially two sides of the same coin. In the same way that Newtonian laws and information about the mass of the sun and the planet, etc., can be used to predict the position of the planet at some future time the inference can also be used to explain the position of the planet after we observe it. In other words, we see here the same tight coupling between diagnostic reasoning and predictive reasoning that we mentioned earlier in the context of BNs. However, while this coupling works in BNs across the range of possible probabilities, it becomes forced in the covering law model when dealing with probabilistic explanations, in particular when dealing with cases where the probability of observing the conclusion is low. Not only do probabilistic contexts move the inference from deduction to an ampliative inference where the conclusion is no longer certain, the symmetry between explanation and prediction also becomes forced. We might, for example, readily explain someone being struck by the lightning by appealing to stormy weather conditions and the fact that they were out in the open. But we would nevertheless hasten to predict that someone will be struck by the lightning even if they are out in the open and there is a storm as it is a low probability event. This limits the utility of the covering law model within the social sciences where deduction is not commonplace and where low probability events are often found. Hempel himself was aware of these difficulties to the extent that he proposed two versions of the model the deductive-nomological model and an inductive-statistical one, and thought that the inductive statistical model applied only when the explanatory theory involves high probabilities. Even this restriction, however, does not deal appropriately with the asymmetries involved in explanation. These can be observed even in purely deductive context as is illustrated by the following example from Salmon (1992). Imagine there is a flagpole with a shadow of 20 metres (m) and someone asks why that shadow is 20 m long. In this context, it seems appropriate to explain the length of the shadow by appealing to the height of the flagpole, the position of the sun, and the laws of trigonometry. These together adequately explain the shadows' length. But note that this inference can be reversed when one can also use the sun's position, the laws of trigonometry, and the length of the shadow to explain the height of the flagpole. This, however, seems wrong; an adequate explanation of the height of that flagpole presumably involves an appeal to the maker of the flagpole in some form or other. Examples such as these serve to illustrate not just the limits of Hempel's account but the limits of deductive approaches in the context of explanation more generally.

The asymmetric relations involved in explanation prompted alternative accounts of scientific explanation within the subsequent literature. Chief among these are causal accounts which assert that to explain something is to give a specification of its causes. The standard explication of cause in this context is that of factors without which something could not be the case (i.e., *conditio sine qua non*). This deals readily even with low probability events, and causes can be identified through a process of 'screening off'. If one finds that $p(M \mid N, L) = p(M \mid N)$, then N screens off L from M and that M is causally irrelevant to L. For example, a reading of a barometer (B) and whether there is a storm (S) are correlated. However, knowing the atmospheric pressure (A) will make these two independent: $p(B \mid A, S) = p(B \mid A)$, suggesting no causal relationship between B and S. However, the notion of cause in itself is notoriously fraught as is evidenced by J. L. Mackie's convoluted (Mackie, 1965) definition whereby a cause is defined as an 'insufficient but necessary part of an unnecessary but sufficient condition'. This rather tortured definition reflects the difficulties with the notion of causation when multiple causes are present which is giving rise to overdetermination (e.g., decapitation and arsenic in the blood stream can both be the causes of death), the difficulties created by causal chains (e.g., tipping over the bottle which hits the floor which releases the toxic liquid), and the impact of background conditions (e.g., putting yeast in the dough causes it to rise, but only if it is actually put in the oven, the oven works, the electrical bills have been paid, and so on). It is a matter of ongoing research to what extent causal Bayes nets, that is BNs supplemented with the *do*-calculus Pearl (2000), provide a fully satisfactory account of causality and these difficulties (see also Halpern and Pearl, 2005*a*). At the same time, the difficulty of picking out a single one out of multiple potential causes points to the second main alternative to Hempel's covering law model, namely so-called pragmatic accounts of explanation.

According to van Fraassen 1977 an explanation always has a pragmatic component: specifically what counts as an explanation in any given context depends on the possible contrasts the questioner has in mind. For example, consider the question 'why did the dog bury the bone?' Different answers are required for different prosodic contours: 'why did the *dog* (i.e., not some other animal) bury the bone?'; 'why did the dog *bury* the bone?' (say, rather than eat it); 'why did the dog bury the *bone*?' (say, rather than the ball). In short, pragmatic accounts bring into the picture the recipient of an explanation while rejecting a fundamental connection between explanation and inference assumed by Hempel's model.

6.2.2 Explanatory virtues

Philosophy has not only tried to characterize the nature of explanation, it has also sought to identify the so-called explanatory virtues. Of the many things that might count as an explanation according to a particular theoretical account of explanation, not all may seem equally good or compelling. Among 'explanations', we might ask what distinguishes better ones from poorer ones. In search of explanatory virtues that characterize good explanations, a number of factors have been identified: explanatory power, unification, coherence, and simplicity are chief among these. Explanatory power often relates to the ability of an explanation to decrease the degree to which we find the explanandum surprising; the less surprising the explanadum in light of an explanation, the more powerful the explanation. For instance, a geologist may find a prehistoric earthquake as explanatory of deformation in layers of bedrock to the extent that these deformations would be less surprising given the occurrence of such an earthquake (Schupbach and Sprenger, 2011). Unification refers to explanations' ability to provide a unified account of a wide range of phenomena. For example, Maxwell's theory (explanation) managed to unify electricity and magnetism (phenomena). Coherence renders explanations that better fit our already established beliefs to be preferred to those that do not (Thagard, 1989). Explanations can also have internal coherence, namely how parts of an explanation fit together. An often motioned explanatory virtue is simplicity. According to Thagard (1978), simplicity is related to the size and nature of auxiliary assumptions needed by an explanation to explain evidence. For instance, the phlogiston theory of combustion needed a number of auxiliary assumptions to explain facts that are easily explained by Lavoisier's theory: it assumed existence of a fire-like element 'phlogiston' that is given away in combustion and that had 'negative weight' since bodies undergoing combustion increase in weight. Others operationalize simplicity as a number of causes invoked in an explanation: the more causes the less simple an explanation (Lombrozo, 2007).

While all of these factors seem intuitive, debate persists about their normative basis. In particular, there is ongoing debate within the philosophy of science about whether these factors admit of adequate probabilistic reconstruction (Glymour, 2014). At the same time, there is now a sizeable program within psychology that seeks to examine the application of these virtues to every day lay explanation. This body of work probes the extent to which lay reasoners endorse these criteria when distinguishing better from worse explanations (Pennington and Hastie, 1992; Sloman, 1994; Lombrozo, 2007; Williams and Lombrozo, 2010; Bonawitz and Lombrozo, 2012; Johnson *et al.*, 2014*a*,*b*; Lombrozo, 2016; Bechlivanidis *et al.*, 2017; Zemla *et al.*, 2017). To date, researchers found some degree of support for these factors, but also seeming deviations in practice.

Finally, there is a renewed interest in both philosophy and psychology in the notion of inference to the best explanation (Harman, 1965; Lipton, 2003). Debate here centres around the question of whether the fact that an explanation seems in some purely non-evidential way better than its rivals should provide grounds for thinking that explanation is more probable. In other words, the issue is whether an explanation exhibiting certain explanatory considerations that other explanations do not should be considered more likely to be true (Harman, 1967; Thagard, 1978; Lipton, 2003; Douven, 2013; Henderson, 2013). Likewise this has prompted psychological research into whether such probability boosts can actually be observed in reasoning contexts (Douven and Schupbach, 2015). The research on explanatory virtues in both philosophy and psychology is still very active.

6.2.3 Implications

What if anything can be inferred for the project of machine-generated explanation from this body of research? First, the notion of explanation that emerges is a potentially very different one across different parts of this literature. An explanation is variously a hypothesis or a variable, an inference, or an answer to a question. From afar, in the context of inference, we may distinguish explanation as a product from explanation as a process (Lombrozo, 2012). From a product perspective, an explanation is a hypothesis or a claim that accounts for evidence when prompted to do so. In contrast, explanation can also be viewed as a cognitive activity (process) that has as its goal to generate explanation 'products'. This distinction nicely corresponds to what we have found in the literature on explanations in BNs reported in section 6.1. There, explanations are also viewed as products that consist of nodes in a BN that aim to account for other nodes in a BN (i.e., evidence nodes) as well as reasoning processes that include not just the final products but also how one arrives at these products. This nicely illustrates how different disciplines can come to similar conceptual distinctions without these distinctions being communicated from one to another. Establishing the necessary communication channels will arguably help connect the research areas working on closely related questions thus bringing different perspectives and inputs into these areas.

This brings us to our second point. It seems clear that BNs provide a potential tool that is compatible with present thinking about the explanation at least in principle. They can capture the asymmetry in explanation as arcs are directed and can have a causal interpretation (Pearl, 2000), whilst at the same time being able to make predictions. This is in contrast to, for instance, a rule-based expert system with IF-THEN rules and a set of facts which would be susceptible to the symmetry 'error' in explanation illustrated by the flagpole example from section 6.2.1. A BN on the other hand would be able to account for the asymmetry given a causal interpretation and directional representation of arrows. However, it is neither clear how explanations in BNs can capture the pragmatic component that van Fraassen raises nor how to operationalize explanatory virtues in the context of BNs. These are all potential avenues for further research.

Another point that can be drawn is that the debates about the nature of explanation and explanatory virtues have been conducted at very high levels of abstraction. They have also typically focused on philosophy of science and issues tightly related to it. This is true even for psychological research on explanation, to the extent that it has tried to model psychological investigations more or less directly on philosophical distinctions. However, for the purposes of developing suitable AI algorithms, it also seems important to work in the opposite direction, as it were from the bottom up. In other words, it seems important to start simultaneously with simple applications of BNs to multiple variable problems, and consider what kinds of explanations a human (expert) would produce. This would shed light on the kinds of explanations that seem natural and appropriate human users as well as provide guidelines on possible theories of explanation. A similar point has been made in AI literature with additional emphasison the importance of human-generated explanations serving as a baseline for comparison with machine-generated explanations (Doshi-Velez and Kim, 2017). To explore these ideas further, we conducted a case study on explanation in BNs which we describe next.

6.2.4 A brief case study on human-generated explanation

The main motivation of the study was to find out what kinds of explanations a human (expert) would produce upon being presented with evidence in a BN. This is interesting from both the psychological and the AI perspective as, on the one hand, it could give us further inputs into human explanatory intuitions and preferences and, on the other hand, it could inform the AI researcher that aims to build algorithms for an automated generation of explanations.

In the study we used four BNs of different complexity found on a publicly available BN repository https://www.norsys.com/netlibrary/index.htm. The number of nodes in the BNs ranged from 4 to 18 and the number of arcs raged from 4 to 20. Figure 6.4 includes a BN used in the study.

Three independent raters, all of whom were experts in probabilistic reasoning, were then given access to implementations of the respective model in order to probe the BNs in more detail, and asked to then provide answers to questions that prompted them to consider how learning evidence changed the probabilities of the target nodes. Below are sample questions:

- Given evidence: {*Neighbours grass* = wet}
  Question: How does the probability of '*Our Sprinkler* = was on' change compared to when there was no evidence and why?
- Given Evidence: {*Our grass* = wet, *Wall* = wet}
  Question: How does the probability of '*Rain* = rained' change compared to when the only available evidence was '*Our Grass* = wet' and why?

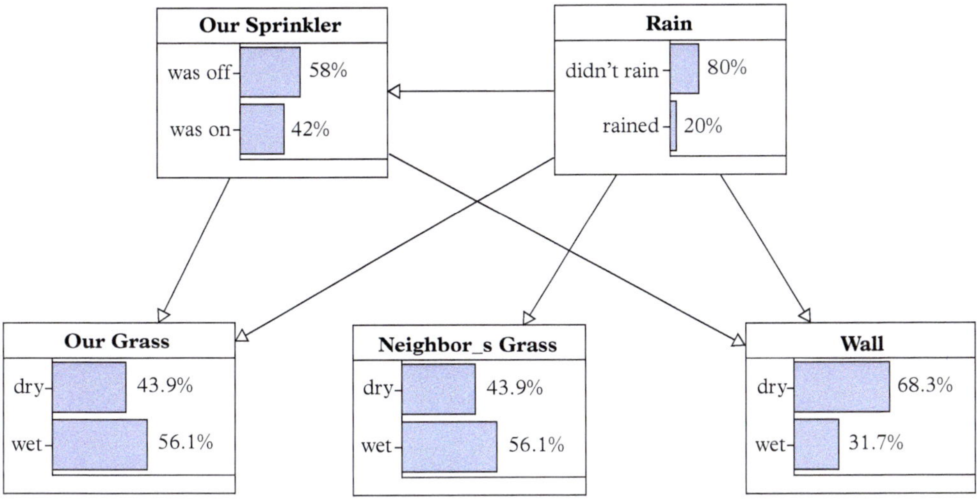

Figure 6.4 *A BN used in the case study.*

Subsequently, the three independent sets of answers were subjected to an analysis by a fourth person in order to identify both commonalities and differences across the answers. This then formed the basis of the subsequent evaluation of those answers.

We describe the full set of results elsewhere (Tešić and Hahn, prep*a*), restricting ourselves here to an initial summary of the results. First, we observed high levels of agreement across answers. Differences were typically more presentational than substantive. For example, the following three statements all seek to describe the same state of affairs:

- 'As A is true C is more likely to be true if B is true and less likely to be true if B is false. As we do not know B these alternatives essentially cancel themselves out and leave the probability of C unchanged.'
- 'It does not change. $P(\mathrm{C} \mid \mathrm{A})$ is equal to $P(\mathrm{C})$ if $P(\mathrm{C} \mid \mathrm{A}, \mathrm{B}) = P(\mathrm{C} \mid {\sim}\mathrm{A}, {\sim}\mathrm{B})$ and $P(\mathrm{C} \mid \mathrm{A}, {\sim}\mathrm{B}) = P(\mathrm{C} \mid {\sim}\mathrm{A}, \mathrm{B})$ (assuming $P(\mathrm{B}) = 0.5$), which here is the case.'
- 'According to model parameters: If A and B both true or both false, then C has probability .75. If A true but B false, or vice-versa, then C has probability .25. When we know A is true, and prior for B is 50%, there is a 50% that probability of C is 75% and a 50% that probability of C is 2%, therefore overall probability of C is 50%.'

Second, all appeal to *hypothetical reasoning* as a way of unpacking interactions of evidence variables:

- '*Wall* = wet is a lot more likely if the sprinkler was on than if it rained (as a matter of fact, if it rained, the wall is more likely to be dry than wet). Since, *Our sprinkler* = was on went down, *Wall* = wet went down.'

Third, causal explanations are prevalent, and, where present, typically appeal to the underlying target system being modelled (i.e., sprinkler, wall, rain) as opposed to the model itself:

- 'The probability of rain decreases because, although the sprinkler and rain can both cause our grass to be wet, the wet wall is more likely to happen when the sprinkler is on rather than rain.'

Notably, in appealing to causes, it is the most probable cause that seems to be highlighted as an explanation:

- 'There is a decrease [in probability] because the most likely cause of our grass being wet is the sprinkler and since the wall is dry the sprinkler is unlikely to be on.'

Finally, these data seem to suggest that the structure of the BNs is exploited in order to zero in on 'the explanation' as a subset of all variables described in the problem. Specifically, explanations seemed to make use of the Markov blanket: a set of nodes consisting of the another node's parents, its children, and its children's parents and making that node conditionally independent of the rest of the network (Korb and Nicholson, 2010). In addition to Markov blanket, the rater's descriptions mostly followed the direction of evidence propagation, that is followed the directed paths in a BN:

- 'The probability of *Battery voltage* = dead increases because failure of the car to start could be explained by the car not cranking and the likely cause of this is a faulty starter system. A dead battery is one possible explanation for a faulty starter system.'

This suggests that the explanatory virtue of 'simplicity' might, in a BN context, be conceptualized in terms terms of a Markov blanket and path direction.

In summary, we see multiple features of the general philosophical literature reflected in these explanations of everyday situations expressed with a BN model: a focus on an inference or a reasoning process; the use of causal explanation for a probabilistic system; the directional nature of explanation (its asymmetry); indications of pragmatic sensitivity in that hypotheticals are used to express relevant 'contrasts'; and, finally, an emerging notion of simplicity in the use of the Markov blanket.

These results are, however, still very much preliminary and further research is needed, but hopefully they give us some sense of the kinds of explanations a human (expert) may produce and, potentially, prefer.

6.3 Bringing in the user: bi-directional relationships

One of the themes of this chapter is that an explanation is an explanation for someone. We have already encountered this in pragmatic theories of explanation within the philosophical literature, but it seems particularly pertinent to the AI context. As illustrated by the BARD assistive reasoning tool example, if explanations for AI systems are to be effective, they must take the user into account. On the one hand, this will require detailed research into what it is that the users of a given system do and do not readily understand. These concerns will largely be specific to the context of the AI system in question. But there are also further general considerations, which apply across potential systems, that have not yet been adequately discussed.

6.3.1 Explanations are communicative acts

Only very recently has it been noted that the provision of an explanation, whether from human or machine, is a *communicative act*. In order to understand the impact of explanations, one must consequently consider the pragmatics of (natural language) communication. Pragmatics, that is the part of language that deals with meaning in context, tells us much about how users will come to interpret explanations. The need for AI researchers to consider pragmatics in the context of machine-generated explanation has recently been highlighted by Miller (2019) who very persuasively argues that the research in explainable AI has largely neglected insights from social sciences, one of which is a very important commonplace regarding the social aspect of explanations where it is argued that explanations are often presented relative to the explainer's beliefs about the explainee's beliefs. In particular, Miller argues that explanations go beyond search for and presentation of associations and causes of evidence, but that they are also contextual: explainer and explainee may have different background beliefs regrading certain observed pieces of evidence and an explainable AI should address this.

We briefly consider some further potential implications for the recipients of explanations here. One general feature of communicative acts is that they provide information about the speaker, intended or otherwise. Recent work concerned with the question of trust has highlighted the interplay between culture and its contents and the perceived reliability of the speaker (Bovens and Hartmann, 2003; Olsson and Vallinder, 2013). One upshot of this is that receiving an explanation is likely to change perceptions of the reliability of the explanation's source. Here, it is not only characteristics of the explanation such as its perceived cogency, how articulately it is framed, or how easily to process it that are likely to influence perceived source reliability, there are also likely to be effects of the specific content. In particular, the extent to which the content of the message fits with our present (uncertain) beliefs about the world has been shown to affect beliefs about the issue at hand as well as beliefs about the perceived reliability of the speaker (Collins *et al.*, 2018; Collins and Hahn, 2020).

6.3.2 Explanations and trust

Because message content and source reliability/trustworthiness jointly determine the impact of the communication on our beliefs, these interactions are likely to be consequential for the extent to which the machine conclusion being explained is itself perceived to be true. The literature in AI, in particular recommender systems, has long recognized relationship between trust and explanation (Zhang and Chen, 2020). The majority of research suggests that providing an explanation improves user's trust in an AI system (Herlocker *et al.*, 2000; Sinha and Swearingen, 2002; Symeonidis *et al.*, 2009). However, the situation seems more intricate as more transparent systems do not always lead to increase in trust (Cramer *et al.*, 2008), and sometimes poor explanations can lead to reduced acceptance of the AI systems (Herlocker *et al.*, 2000). To explore the interactions between explanations and trust, in addition to manipulation transparency of AI systems, one would also need to manipulate experimentally a level of trust users have in them. Recently, we have conducted empirical work on the relationship between reliability (which is related to the notion of trust as understood in AI) and explanation in non-AI context (Tešić and Hahn, prep*b*). We have performed three experiments where we used simple dialogues between two people (in the condition where an explanation was provided these were then explainer and explainee) on five different issues to show that (1) providing an explanation for a claim increases not just people's ratings of convincingness in the claim but also their reliability of the person providing an explanation compared when there is no such explanation, and (2) providing an explanation has a significantly greater impact on the convincingness and reliability when people's initial (prior) reliability of the source is low compared to when that reliability is high. In the context of AI, these results suggest that providing a (good) explanation of AI system's decisions will arguably increase people's ratings of convincingness in/acceptance of these decisions as well as people's perceived the reliability/trust of the system. In particular, the impact of providing an explanation will be greater (and most useful) if people's initial perceived reliability/trust of an AI system is low.

6.3.3 Trust and fidelity

The recent surge of model agnostic post-hoc explanations of black-box deep learning models has significantly pushed the horizons of explainable AI, but at the same time it has also introduced *fidelity* problems, namely unlike explanations of BN where original BN models could be used to generate explanations (either as justification of evidence or as explanation of reasoning processes), deep learning models are not transparent enough for either a lay or an expert human user to be able to explain the models' outputs. Instead, one resorts to explanatory models that are independent of deep learning models to generate explanations of these black-box models' decisions after these decisions have been made, that is post-hoc (Ribeiro *et al.*, 2016; Zhang and Chen, 2020). The explanations models are often model agnostic as they should be able to explain decisions of any (black-box) model. Post-hoc model agnostic explanation models have certainly furthered the work on explanation in AI, but they have also prompted questions regarding the degree to which the explanations generated by models are reflecting the real mechanisms that generated decisions of a deep learning model: that is, they have raised questions regarding the fidelity of explanation models (Sørmo *et al.*, 2005; Ribeiro *et al.*, 2016). In the literature, the tradeoff between fidelity and interpretability of explanations models is often acknowledged: the higher the fidelity of an explanation model to the black-box model the lower the interpretability of that model and its transparency to a human user (Ribeiro *et al.*, 2016). This however brings trust into the consideration. On the one hand, if higher interpretability is to increase trust, then trust may be negatively affected by higher fidelity. On the other hand, if users expect higher fidelity explanation models, then lower fidelity may now negatively affect trust. This potentially interesting relationship between fidelity and trust is another open issue related to the interplay between a user and the system that could be addressed in the future research.

6.3.4 Further research avenues

The effect of the communication on the reliability of the source/trust and possibly fidelity, however, are unlikely to be the only way in which explanations alter beliefs about what it is that is being explained. For example, does providing an explanation constrain and/or make less ambiguous the underlying (causal) structure of the world that the explainee had in mind before receiving the explanation (or in the case of a BN, does providing an explanation restrict the number of potential BN structures that the explainee has mind)? How does providing an explanation of an (ab)normal event in a causal chain of events reflect on our perceptions of that explanation? Does a detailed explanation of a usual and obvious succession of events make that explanation less preferred or worse, compared to a less detailed explanation? All these questions call for further investigation and can have implications for the explainable AI project.

6.4 Conclusions

We have seen multiple ways to build different notions of what counts as an explanation. One of these involves explanation as identification of the variables that mattered in generating certain outcomes. In the context of computational models of explanation in BNs this corresponds to the usual focus on explaining observed evidence via unobserved nodes within the network (Pacer *et al.*, 2013). In other words, the explanation identifies a justification/hypothesis. This is the notion of explanation that has figured prominently in work on computer-generated explanations as well as in the psychological and philosophical literature on explanation. The second notion of explanation we considered includes explanation of the inference that links evidence and hypothesis. In the context of BNs this means explaining the inferences that lead to a change (or no change) in the probabilities of the query nodes. In other words, the explanation involves a target hypothesis plus information about the incremental reasoning process that identifies that hypothesis. Finally, we considered the notion of explanation in terms of providing understanding of what is hard to understand and or surprising to the system's user. Explanation in this widest sense requires not just identification of a best hypothesis and explanation of the incremental steps that lead to the identification of that hypothesis but also to a user model that tells us what it is that human users find difficult. For this widest sense of explanation, psychological research is essential. Explanation so understood constitutes a fundamental problem of human computer interaction and only empirical research that seeks to understand the human user can lead to fully satisfactory answers.

Choosing among these three different notions of explanation also directly affects the answer to the question of what counts as a good explanation. As we saw in this chapter, there is some guidance on the notion of good explanation that can be drawn from both the philosophy and psychology literature, but it is also clear that more specific work is required. In particular, most recent research suggests that such work will need to take into account that providing an explanation is a communicative act that changes perceptions of the communicator. In other words, explanation will not merely translate an extant result into a language understood by the user, it is likely to affect how the user interprets the output of the system and the reliability of the system itself. This means the provision of explanation will likely affect what the user considers to be the verdict of the system in the first place, which could lead to further intricate relationships between trust and concepts such as fidelity. It is thus essential that future work on explanation within AI engage more fully with the pragmatic consequences of communicating explanation.

Acknowledgements

This work was supported by the Intelligence Advanced Research Projects Activity (IARPA) as part of the Bayesian Argumentation via Delphi (BARD) project, and by the Humboldt Foundation.

References

Agrahari, R., Foroushani, A., Docking, T. R. *et al.* (2018). Applications of Bayesian network models in predicting types of hematological malignancies. *Scientific Reports*, 8(1), 6951.

Bansal, A., Farhadi, A., and Parikh, D. (2014). Towards transparent systems: semantic characterization of failure modes, in *European Conference on Computer Vision*. New York, NY: Springer, 366–81.

Bechlivanidis, C., Lagnado, D. A., Zemla, J. C. *et al.* (2017). Concreteness and abstraction in everyday explanation. *Psychonomic Bulletin & Review*, **24**(5), 1451–64.

Bonawitz, E. B. and Lombrozo, T. (2012). Occam's rattle: children's use of simplicity and probability to constrain inference. *Developmental Psychology*, **48**(4), 1156.

Bovens, L. and Hartmann, S. (2003). *Bayesian Epistemology*. Oxford: Oxford University Press.

Burton, R. R. (1976). *Semantic Grammar: an Engineering Technique for Constructing Natural Language Understanding Systems*. Cambridge, MA: Bolt, Beranek and Newman, inc.

Chen, J. Y., Procci, K., Boyce, M. *et al.* (2014). Situation awareness-based agent transparency. Technical report, Army research lab Aberdeen proving ground MD human research and engineering, *Theoretical Issues in Ergonomics Science*, (19)3, 259–82, DOI: 10.1080/1463922X.2017.1315750.

Chockalingam, S., Pieters, W., Teixeira, A. *et al.* (2017). Bayesian network models in cyber security: a systematic review, in *Nordic Conference on Secure IT Systems*. New York, NY: Springer, 105–22.

Choi, A., Wang, R., and Darwiche, A. (2019). On the relative expressiveness of Bayesian and neural networks. *International Journal of Approximate Reasoning*, **113**, 303–23.

Collins, P. J. and Hahn, U. (2020). We might be wrong, but we think that hedging doesn't protect your reputation. *Journal of Experimental Psychology. Learning, Memory, and Cognition*, **46**(7), 1328.

Collins, P. J., Hahn, U., von Gerber, Y. *et al.* (2018). The bi-directional relationship between source characteristics and message content. *Frontiers in Psychology*, **9**, 18.

Collobert, R., Weston, J., Bottou, L. *et al.* (2011). Natural language processing (almost) from scratch. *Journal of Machine Learning Research*, **12**(Aug), 2493–537.

Cramer, H., Evers, V., Ramlal, S. *et al.* (2008). The effects of transparency on trust in and acceptance of a content-based art recommender. *User Modeling and User-Adapted Interaction*, **18**(5), 455.

Cruz, N., Desai, S. C., Dewitt, S. *et al.* (2020). Widening access to Bayesian problem solving. *Frontiers in Psychology*, **11**, 660.

Dardashti, R., Hartmann, S., Thébault, K. *et al.* (2019). Hawking radiation and analogue experiments: a Bayesian analysis. *Studies in History and Philosophy of Science Part B: Studies in History and Philosophy of Modern Physics*, **67**, 1–11.

DARPA (2016). Explainable artificial intelligence (XAI) program. Available at: https://www.darpa.mil/program/xplainable-artificial-intelligence.

Davis, Z. and Rehder, B. (2017). The causal sampler: a sampling approach to causal representation, reasoning, and learning, in *Proceedings of the Cognitive Science Society*. Red Hook, NY: Curran Associates, 1896–901.

Dewitt, S., Lagnado, D. A., and Fenton, N. E. (2018). Updating Prior Beliefs Based on Ambiguous Evidence, in *Proceedings of the 40th Annual Conference of the Cognitive Science Society*. Red Hook, NY: Curran Associates, 306–11.

Dizadji-Bahmani, F., Frigg, R., and Hartmann, S. (2011). Confirmation and reduction: a Bayesian account. *Synthese*, **179**(2), 321–38.

Doshi-Velez, F. and Kim, B. (2017). Towards a Rigorous Science of Interpretable Machine Learning. *arXiv preprint arXiv:1702.08608.*

Doshi-Velez, F., Kortz, M., Budish, Ryan, B. *et al.* (2017). Accountability of AI under the law: the role of explanation. *arXiv preprint arXiv:1711.01134.*

Douven, I. (2013). Inference to the best explanation, Dutch books, and inaccuracy minimisation. *The Philosophical Quarterly*, **63**(252), 428–44.

Douven, I. and Schupbach, J. N. (2015). The role of explanatory considerations in updating. *Cognition*, **142**, 299–311.

Drury, B., Valverde-Rebaza, J., Moura, M.-F. *et al.* (2017). A survey of the applications of Bayesian Networks in Agriculture. *Engineering Applications of Artificial Intelligence*, **65**, 29–42.

Fallon, C. K. and Blaha, L. M. (2018). Improving automation transparency: addressing some of machine learning's unique challenges, in *International Conference on Augmented Cognition*, 245–254.

Falzon, L. (2006). Using Bayesian network analysis to support centre of gravity analysis in military planning. *European Journal of Operational Research*, **170**(2), 629–43.

Felzmann, H., Villaronga, E. F., Lutz, C. *et al.* (2019). Transparency you can trust: transparency requirements for artificial intelligence between legal norms and contextual concerns. *Big Data & Society*, **6**(1), 2053951719860542.

Fenton, N., Neil, M., and Lagnado, D. A. (2013). A general structure for legal arguments about evidence using Bayesian networks. *Cognitive Science*, **37**(1), 61–102.

Fernbach, P. M. and Rehder, B. (2013). Cognitive shortcuts in Causal Inference. *Argument & Computation*, **4**(1), 64–88.

Friedman, N., Geiger, D., and Goldszmidt, M. (1997). Bayesian network classifiers. *Machine Learning*, **29**(2–3), 131–63.

Glymour, C. (2014). Probability and the explanatory virtues. *British Journal for the Philosophy of Science*, **66**(3), 591–604.

Goodfellow, I., Bengio, Y., and Courville, A. (2016). *Deep Learning*. Cambridge, MA: MIT Press.
Goodman, B. and Flaxman, S. (2016). EU regulations on algorithmic decision-making and a 'right to explanation', in *ICML Workshop on Human Interpretability in Machine Learning (WHI 2016), New York, NY. http://arxiv. org/abs/1606.08813 v1.*
Graves, A. and Schmidhuber, J. (2005). Framewise phoneme classification with bidirectional lstm and other neural network architectures. *Neural Networks*, **18**(5–6), 602–10.
Grosan, C. and Abraham, A. (2011). *Intelligent Systems*. New York, NY: Springer.
Gunning, D. and Aha, D. W. (2019). Darpa's explainable artificial intelligence program. *AI Magazine*, **40**(2), 44–58.
Hahn, U. and Hornikx, J. (2016). A normative framework for argument quality: argumentation schemes with a Bayesian foundation. *Synthese*, **193**(6), 1833–73.
Hahn, U. and Oaksford, M. (2006). A Bayesian approach to informal argument fallacies. *Synthese*, **152**(2), 207–36.
Hahn, U. and Oaksford, M. (2007). The rationality of informal Argumentation: a Bayesian approach to reasoning fallacies. *Psychological Review*, **114**(3), 704.
Halpern, J. Y. and Pearl, J. (2005*a*). Causes and explanations: A structural-model approach. Part I: Causes. *The British Journal for the Philosophy of Science*, **56**(4), 843–87.
Halpern, J. Y. and Pearl, J. (2005*b*). Causes and explanations: A structural-model approach. Part II: Explanations. *The British Journal for the Philosophy of Science*, **56**(4), 889–911.
Harman, G. (1965). The inference to the best explanation. *The Philosophical Review*, **74**(1), 88–95.
Harman, G. (1967). Detachment, probability, and maximum likelihood. *Nous*, 401–11.
Harradon, M., Druce, J., and Ruttenberg, B. (2018). Causal learning and explanation of deep neural networks via autoencoded activations. *arXiv preprint arXiv:1802.00541.*
Harris, A. J. L., Hahn, U., Madsen, J. K. *et al.* (2016). The appeal to expert opinion: quantitative support for a Bayesian network approach. *Cognitive Science*, **40**(6), 1496–533.
Hayes, B. and Shah, J. A. (2017). Improving robot controller transparency through autonomous policy explanation, in *2017 12th ACM/IEEE International Conference on Human–Robot Interaction*. New York, NY: Institute of Electrical and Electronics Engineers, 303–12.
Hempel, C. G. and Oppenheim, P. (1948). Studies in the logic of explanation. *Philosophy of Science*, **15**(2), 135–75.
Henderson, L. (2013). Bayesianism and inference to the best explanation. *The British Journal for the Philosophy of Science*, **65**(4), 687–715.
Herlocker, J. L., Konstan, J. A., and Riedl, J. (2000). Explaining collaborative filtering recommendations, in *Proceedings of the 2000 ACM Conference on Computer Supported Cooperative Work*. Strasbourg, PA: Association for Computational Linguistics, 241–50.
Howson, C. and Urbach, P. (2006). *Scientific Reasoning: The Bayesian Approach*. LaSalle, IL: Open Court Publishing.
Johnson, S., Jin, A., and Keil, F. (2014*a*). Simplicity and goodness-of-fit in explanation: the case of intuitive curve-fitting, in *Proceedings of the Annual Meeting of the Cognitive Science Society*, Vol. 36. Red Hook, NY: Curran Associates, 2453–8.
Johnson, S., Johnston, A., Toig, A. *et al.* (2014*b*). Explanatory scope informs causal strength inferences, in *Proceedings of the Annual Meeting of the Cognitive Science Society*, Vol. 36.Red Hook, NY: Curran Associates, 2453–8.
Korb, K. B. and Nicholson, A. E. (2010). *Bayesian Artificial Intelligence*. Boca Raton, FL: CRC Press.
Krizhevsky, A., Sutskever, I., and Hinton, G. E. (2012). Imagenet classification with deep convolutional neural networks, in F. Pereira, C. J. C. Burges, L. Bottou et al., eds, *Advances in Neural Information Processing Systems*. Red Hook, NY: Curran Associates, 1097–105.

Lacave, C. and Díez, F. J. (2002). A review of explanation methods for Bayesian networks. *The Knowledge Engineering Review*, **17**(2), 107–27.

Lagnado, D. A., Fenton, N., and Neil, M. (2013). Legal idioms: a framework for evidential reasoning. *Argument & Computation*, **4**(1), 46–63.

Laskey, K. B. and Mahoney, S. M. (1997). Network fragments: Representing knowledge for constructing probabilistic models, in *Proceedings of the Thirteenth Conference on Uncertainty in Artificial Intelligence*, San Francisco, CA: Morgan Kaufmann Publishers Inc, 334–41.

Liefgreen, A., Tešić, M., and Lagnado, D. (2018). Explaining away: significance of priors, diagnostic reasoning, and structural complexity, in *Proceedings of the 40th Annual Conference of the Cognitive Science Society*. Red Hook, NY: Curran Associates, 2047–52.

Lippmann, R., Ingols, K., Scott, C. *et al.* (2006). Validating and restoring defense in depth using attack graphs, in *MILCOM 2006–2006 IEEE Military Communications Conference*. New York, NY: Institute of Electrical and Electronics Engineers, 1–10.

Lipton, P. (2003). *Inference to the Best Explanation*. London: Routledge.

Lombrozo, T. (2007). Simplicity and probability in causal explanation. *Cognitive Psychology*, **55**(3), 232–57.

Lombrozo, T. (2012). Explanation and abductive inference. *Oxford Handbook of Thinking and Reasoning*, Oxford: Oxford University Press, 260–76.

Lombrozo, T. (2016). Explanatory preferences shape learning and inference. *Trends in Cognitive Sciences*, **20**(10), 748–59.

Mackie, J. L. (1965). Causes and conditions. *American Philosophical Quarterly*, **2**(4), 245–64.

Madsen, J. K., Hahn, U., and Pilditch, T. D. (2018). Partial source dependence and reliability revision: the impact of shared backgrounds, in *Proceedings of the 40th Annual Conference of the Cognitive Science Society*. Red Hook, NY: Curran Associates, 720–5.

Mercado, J. E., Rupp, M. A., Chen, J. Y. C. *et al.* (2016). Intelligent agent transparency in human–agent teaming for multi-uxv management. *Human Factors*, **58**(3), 401–15.

Miller, T. (2019). Explanation in artificial intelligence: insights from the social sciences. *Artificial Intelligence*, **267**, 1–38.

Montavon, G., Samek, W., and Müller, K.-R. (2018). Methods for interpreting and understanding deep neural networks. *Digital Signal Processing*, **73**, 1–15.

Morris, M. W. and Larrick, R. P. (1995). When one cause casts doubt on another: a normative analysis of discounting in causal attribution. *Psychological Review*, **102**(2), 331–55.

Neapolitan, R. E. (2003). *Learning Bayesian Networks*. Upper Saddle River, NJ: Prentice Hall.

Neil, M. and Fenton, N. et al. (2008). Using Bayesian networks to model the operational risk to information technology infrastructure in financial institutions *Journal of Financial Transformation*, **22**, 131–38.

Ng, A. Y. and Jordan, M. I. (2002). On discriminative vs. generative classifiers: A comparison of logistic regression and naive Bayes, in *Advances in Neural Information Processing Systems*. Cambridge, MA: MIT Press, 841–48.

Nicholson, A. E., Korb, K. B., Nyberg, E. P. *et al.* (2020). BARD: a structured technique for group elicitation of Bayesian networks to support analytic reasoning. *arXiv preprint arXiv:2003.01207*.

Nielsen, U., Pellet, J.-P., and Elisseeff, A. (2008). Explanation trees for causal Bayesian networks. *Proceedings of the 24th Annual Conference on Uncertainty in Artificial Intelligence (UAI-08)*. Corvallis, OR: AUAI Press, 427–34.

Olsson, E. J. and Vallinder, A. (2013). Norms of assertion and communication in social networks. *Synthese*, **190**(13), 2557–71.

Pacer, M., Williams, J., Chen, X. *et al.* (2013). Evaluating computational models of explanation using human judgments, in *Proceedings of the 29th Conference on Uncertainty in Artificial Intelligence*. Corvallis, OR: AUAI Press.

Pearl, J. (1988). *Probabilistic Reasoning in Intelligent Systems: Networks of Plausible Inference*. San Francisco, CA: Morgan Kauffman.

Pearl, J. (2000). *Causality: Models, Reasoning and Inference*, Vol. 29. New York, NY: Springer.

Pennington, N. and Hastie, R. (1992). Explaining the Evidence: Tests of the story model for juror decision making. *Journal of Personality and Social Psychology*, **62**(2), 189–206.

Pernkopf, F. and Bilmes, J. (2005). Discriminative versus generative parameter and structure learning of Bayesian network classifiers, in *Proceedings of the 22nd International Conference on Machine Learning*. New York, NY: ACM, 657–64.

Pettigrew, R. (2016). *Accuracy and the Laws of Credence*. Oxford: Oxford University Press.

Phillips, K., Hahn, U., and Pilditch, T. D. (2018). Evaluating Testimony from multiple witnesses: single cue satisficing or integration? in *Proceedings of the 40th Annual Conference of the Cognitive Science Society*. Red Hook, NY: Curran Associates, 2247–52.

Pilditch, T. D., Fenton, N., and Lagnado, D. (2019). The zero-sum fallacy in evidence evaluation. *Psychological Science*, **30**(2), 250–60.

Pilditch, T. D., Hahn, U. and Lagnado, D. A. (2018). Integrating dependent evidence: naïve reasoning in the face of complexity, in *Proceedings of the 40th Annual Conference of the Cognitive Science Society*. Red Hook, NY: Curran Associates,882–7.

Ramsey, F. P. (2016). Truth and probability, in H. Arló-Costa, V. F. Hendricks, J. van Benthem et al., eds, *Readings in Formal Epistemology*. New York, NY: Springer. 21–45.

Rehder, B. (2014). Independence and dependence in human causal reasoning. *Cognitive Psychology*, **72**, 54–107.

Rehder, B. and Waldmann, M. R. (2017). Failures of explaining away and screening off in described versus experienced causal learning scenarios. *Memory & Cognition*, **45**(2), 245–60.

Ribeiro, M. T., Singh, S., and Guestrin, C. (2016). 'Why should I trust you?', Explaining the predictions of any classifier, in *Proceedings of the 22nd ACM SIGKDD International Conference on Knowledge Discovery and Data Mining*. New York, NY: ACM, 1135–44.

Rieger, L., Chormai, P., Montavon, G. *et al.* (2018). Structuring neural networks for more explainable predictions, in *Explainable and Interpretable Models in Computer Vision and Machine Learning*. New York, NY: Springer, 115–31.

Rohekar, R. Y., Nisimov, S., Gurwicz, Y. *et al.* (2018). Constructing deep neural networks by Bayesian network structure learning, in *Advances in Neural Information Processing Systems*. Cambridge, MA: MIT Press, 3047–058.

Roos, T., Wettig, H., Grünwald, P. *et al.* (2005). On discriminative Bayesian network classifiers and logistic regression. *Machine Learning*, **59**(3), 267–96.

Rottman, B. M. and Hastie, R. (2014). Reasoning about causal relationships: inferences on causal networks. *Psychological Bulletin*, **140**(1), 109–39.

Rottman, B. M. and Hastie, R. (2016). Do people reason rationally about causally related events? Markov violations, weak inferences, and failures of explaining away. *Cognitive Psychology*, 87, 88–134.

Salmon, W. C. (1992). Scientific Explanation, in M. H. Salmon, J. Earman, C. Gilmore (eds), *Introduction to the Philosophy of Science* , Englewood Cliffs, NJ: Prentice-Hall, 7–41.

Samek, W., Wiegand, T., and Müller, K.-R. (2017). Explainable artificial intelligence: understanding, visualizing and interpreting deep learning models. *arXiv preprint arXiv:1708.08296*.

Schupbach, J. N. and Sprenger, J. (2011). The logic of explanatory power. *Philosophy of Science*, **78**(1), 105–27.

Shimony, S. E. (1991). Explanation, irrelevance and statistical independence, in *Proceedings of the Ninth National Conference on Artificial intelligence, Vol. 1*. Palo Alto, CA: AAAI Press, 482–7.

Sinha, R. and Swearingen, K. (2002). The role of transparency in recommender systems, in *CHI'02 Extended Abstracts on Human Factors in Computing Systems*, New York, NY: ACM, 830–31.

Sloman, S. A. (1994). When explanations compete: the role of explanatory coherence on judgements of likelihood. *Cognition*, **52**(1), 1–21.

Sørmo, F., Cassens, J., and Aamodt, A. (2005). Explanation in case-based reasoning—perspectives and goals. *Artificial Intelligence Review*, **24**(2), 109–43.

Spiegler, R. (2016). Bayesian networks and boundedly rational expectations. *The Quarterly Journal of Economics*, **131**(3), 1243–90.

Sussman, A. B. and Oppenheimer, D. M. (2011). A causal model theory of Judgment, in *Proceedings of the 33rd Annual Conference of the Cognitive Science Society*. Red Hook, NY: Curran Associates, 1703–8.

Symeonidis, P., Nanopoulos, A., and Manolopoulos, Y. (2009). Moviexplain: a recommender system with explanations, in *Proceedings of the Third ACM Conference on Recommender Systems*. New York, NY: ACM, 317–20.

Tešić, M. (2019). Confirmation and the generalized Nagel–Schaffner model of reduction: a Bayesian analysis. *Synthese*, **196**(3), 1097–129.

Tešić, M. and Hahn, U. (2019). Sequential diagnostic reasoning with independent causes, in *Proceedings of the 41th Annual Conference of the Cognitive Science Society*. Red Hook, NY: Curran Associates, 2947–53.

Tešić, M. and Hahn, U. (in prep.*a*). Human-generated explanations of inferences in Bayesian networks: a case study.

Tešić, M. and Hahn, U. (in prep.*b*). The impact of explanation on explainee's beliefs and explainer's perceived reliability.

Tešić, M., Liefgreen, A., and Lagnado, D. (2020). The propensity interpretation of probability and diagnostic split in explaining away. *Cognitive Psychology*, **121**, 101–293.

Thagard, P. (1989). Explanatory coherence. *Behavioral and Brain Sciences*, **12**(3), 435–67.

Thagard, P. R. (1978). The best explanation: criteria for theory choice. *The Journal of Philosophy*, **75**(2), 76–92.

Van Fraassen, B. C. (1977). The pragmatics of explanation. *American Philosophical Quarterly*, **14**(2), 143–50.

Vineberg, S. (2016). Dutch book arguments, in E. N. Zalta (ed.), *The Stanford Encyclopedia of Philosophy* (Spring 2016 edn) Metaphysics Research Lab, Stanford University.

Wachter, S., Mittelstadt, B., and Floridi, L. (2017*a*). Why a right to explanation of automated decision-making does not exist in the general data protection regulation. *International Data Privacy Law*, 7(2), 76–99.

Wachter, S., Mittelstadt, B., and Russell, C. (2017*b*). Counterfactual explanations without opening the black box: automated decisions and the GDPR. *Harvard Journal of Law & Technology*, **31**, 841.

Wang, H. and Yeung, D.-Y. (2016). Towards Bayesian deep learning: a survey. *arXiv preprint arXiv:1604.01662.*

Wick, M. R. and Thompson, W. B. (1992). Reconstructive expert system explanation. *Artificial Intelligence*, **54**(1–2), 33–70.

Wiegerinck, W., Burgers, W., and Kappen, B. (2013). Bayesian networks, introduction and practical applications, in Bianchini, M., Maggini, M., and Jain, L. C., eds, *Handbook on Neural Information Processing*. New York, NY: Springer, 401–31.

Williams, J. J. and Lombrozo, T. (2010). The role of explanation in discovery and generalization: Evidence from category learning. *Cognitive Science*, **34**(5), 776–806.

Woodward, J. (2017). Scientific explanation, in E. N. Zalta (ed.), *The Stanford Encyclopedia of Philosophy* (Fall 2017 edn) Stanford, CA: Metaphysics Research Lab, Stanford University.

Xie, P., Li, J. H., Ou, X. *et al.* (2010). Using Bayesian networks for cyber security analysis, in *2010 IEEE/IFIP International Conference on Dependable Systems & Networks (DSN)*, 211–20. IEEE, Chicago, IL, June 28–July 1, 2010.

Yap, G.-E., Tan, A.-H., and Pang, H.-H. (2008). Explaining inferences in Bayesian networks. *Applied Intelligence*, **29**(3), 263–78.

Yuan, C., Lim, H., and Lu, T.-C. (2011). Most relevant explanation in Bayesian networks. *Journal of Artificial Intelligence Research*, **42**, 309–52.

Zemla, J. C., Sloman, S., Bechlivanidis, C. *et al.* (2017). Evaluating everyday explanations. *Psychonomic Bulletin & Review*, **24**(5), 1488–500.

Zhang, Y. and Chen, X. (2020). Explainable recommendation: a survey and new perspectives. *Foundations and Trends in Information Retrieval*, **14**(1), 1–101.

Zukerman, I., McConachy, R., and Korb, K. B. (1998). Bayesian Reasoning in an abductive mechanism for argument generation and analysis, in *AAAI/IAAI*, July 26–30, 1998; Madison, Wisconsin, pp. 833–38.

Zukerman, I., McConachy, R., Korb, K. B. *et al.* (1999). Exploratory interaction with a Bayesian argumentation system, in *IJCAI*, Palo Alto, CA: AAAI Press, 1294–99.

7

Human-like Communication

Patrick G. T. Healey

Queen Mary, University of London, UK

> *'And the words I uttered myself, and which must nearly always have gone with an effort of the intelligence, were often to me as the buzzing of an insect. And this is perhaps one of the reasons I was so untalkative, I mean this trouble I had in understanding not only what others said to me, but also what I said to them. It is true that in the end, by dint of patience, we made ourselves understood, but understood with regard to what I ask of you, and to what purpose?'*
>
> (Samuel Beckett, 1951, 50)

7.1 Introduction

Human-like communication is both the best known and arguably most underestimated challenge for machine intelligence. The challenge was set by Alan Turing 1950 who wanted to sidestep the relatively ill-defined question of whether a machine can think by substituting it with a relatively well-defined behavioural test: whether a machine is distinguishable from a human in conversation. However, he adopted a form of interaction that suppresses some of the hardest challenges posed by human-like communication.

Turing's starting point was a party game in which an interrogator must decide which of two people is a woman: a woman and a man who is pretending to be a woman.[1] The players are seated in different rooms and questions are asked and answered in writing. Turing suggested a variation on this game in which the man is substituted with a computer that is also pretending to be a woman. If interrogators 'decide wrongly as often' which is the woman then the computer has passed the test (Turing, 1950, 434). Turing's test thus replaces fundamental questions about the nature of human thought with a practical test of performance in conversation.

The test foregrounds the verbal content of the conversation but pushes the details of delivery into the background. Turing proposed the use of typed communication via a teleprinter to remove cues to identity such as voice characteristics, appearance, or

[1] If Turing had formulated his test today he would probably have prefixed *cis* to each use of 'man' and 'woman'.

Patrick G. T. Healey, *Human-like Communication* In: *Human Like Machine Intelligence*. Edited by: Stephen Muggleton and Nick Charter, Oxford University Press. © Oxford University Press 2021. DOI: 10.1093/oso/9780198862536.003.0007

behaviour that might incidentally give away a person's or machine's identity. He also envisaged an alternating question–answer format for the interrogation as illustrated in the examples below. This assumes transmission of complete turns and also simplifies the types of turn or speech act that are likely to be used. In combination with the teleprinter this removes many of the incremental and collaborative processes typical of normal conversation.

> Q: Please write me a sonnet on the subject of the Forth Bridge.
> A: Count me out on this one. I never could write poetry.
>
> Q: Add 34957 to 70764.
> A: (Pause about 30 seconds and then give as answer) 105621.
>
> (Turing, 1950, 434)

These simplifications have been preserved in subsequent versions of the test including the Loebner Prize (Loebner, 1990) and the 2014 Royal Society competition where it was claimed that the test was passed for the first time (Warwick and Shah, 2016). Performance of gender has been dropped from these competitions in favour of open-ended conversations but they all use typed communication with alternating transmission of whole turns. When Stuart Shieber criticized the Loebner Prize it was for the topics of the conversations, not its format. For Shieber, even text transcripts of a conversation are a useful instance of the Turing test (Shieber, 1994).

Text-based interaction is now so familiar that it is easy to forget that it is not the native habitat for human communication. We first encounter, learn, and use language in face-to-face interaction. Even digital natives still spend an estimated 32–47% of their waking hours in spoken conversation (Mehl *et al.*, 2007; Milek *et al.*, 2018). Text-based communication is only possible because it piggybacks on our prior capacity for face-to-face interaction.

This chapter argues that the differences between text chat and natural conversation are not just incidental differences in format. Rather, they raise basic questions about how human communication works and about what capabilities a machine would need to be able to engage in human-like interaction.

The most obvious difference is practical; text chat is strictly unimodal and sequential whereas natural conversation is multimodal, concurrent, and incremental. The rich variety of verbal and non-verbal resources that can be recruited in human communication is illustrated below. The claim is not that these processes pose insurmountable technical challenges for a machine. They are active research areas in the development of embodied conversational agents and human–robot interaction. However, their complexity is still well beyond the state of the art in machine intelligence.

The second difference relates to our understanding of how human communication works. The Turing test is designed to interrogate knowledge of gender norms (or other topics) but assumes that the interrogation is conducted in a shared language. Failure to understand what is said could be diagnostic—especially if it involves a gender specific word or usage—but in general the test presupposes some shared language competence through which a participant's knowledge is probed. This is an assumption that is built into most models of communication in the cognitive sciences. Communication is assumed to be underwritten by a shared language, and individual variation is modelled by differences in knowledge, context, or other pragmatic factors. While this can be a useful idealization, it leads to systematic discounting of one of the distinguishing characteristics of human communication: misunderstanding.

People have different interests, experience, and expertise, and they do not all mean the same things when using the same words (Clark, 1996). As a result, misunderstandings are a common, recurrent problem. Models of communication often marginalize misunderstandings as noise or performance errors. Sometimes this is because of an explicit idealization to a standard average language competence. Sometimes it is a side effect of the construction of a single language model from large corpora. Often the phenomena associated with misunderstanding are not actually in the data. Machine-learning corpora are dominated by monologue (e.g., news articles or Wikipedia entries) and rarely contain either dialogue or examples of misunderstandings. Psycholinguistic experiments often study individual language processing in situations where dialogue and the possibility of misunderstandings are not encountered.

The second part of this chapter argues that the ability to detect and adjust to misunderstandings on the fly is fundamental to successful human communication and underpins our ability to develop new conventions and new meanings on a conversation by conversation basis (Healey, 2008; Healey *et al.*, 2018*a*; Healey *et al.*, 2018*b*; Clark, 2020). Machine intelligence that aims to achieve human-like communication needs to embrace the detection and resolution of misunderstandings as a fundamental feature of natural interaction. This is necessary in order to engage constructively with ordinary human diversity; not just for ethical reasons but also to pass the Turing test.

7.2 Face-to-face Conversation

Human communication is multimodal, it exploits a rich variety of verbal and non-verbal resources including: voice characteristics, speech, gaze, gesture, facial expression, body position, orientation, touch, and even the structure and composition of the shared environment. The remarkably complex and fine-grained organization of these resources began to be documented in the 1950s when video was first used to analyse the non-verbal dynamics of conversation at millisecond resolutions (Condon and Ogston, 1966; Scheflen, 1973*a*). These studies uncovered some remarkably fine-grained phenomena. For example, Scheflen's detailed observations of the coordination of speech and body movements in psychotherapeutic interaction led him to speculate that eye blinks are significant communicative signals (Scheflen, 1973*b*). Recent experimental evidence shows that blinks are systematically related to turn-taking in conversation (Hömke *et al.*, 2017; Hömke *et al.*, 2018).

The use of behaviours such as blinking as part of turn-taking illustrates the diverse range of non-verbal resources that can be deployed in conversation. It also highlights how people use parallel channels of interaction to provide rich concurrent feedback during conversation. Addressees use verbal backchannels such as 'mmm' and 'aha' as well as gaze, nods, smiles, shrugs, eyebrow raises, blinks, posture changes etcetera to signal their ongoing response to what is being said (Yngve, 1970; Bavelas *et al.*, 1992; Bavelas *et al.*, 2000). Speakers monitor these signals carefully and will edit or amend their turns mid-utterance in response (Bavelas *et al.*, 1992; Clark and Wilkes-Gibbs, 1986; Bavelas *et al.*, 2000). Goodwin (1979) described this reciprocal process as the 'interactive construction of a sentence'. Speakers and listeners jointly build up each turn at conversation increment by increment.

Scheflen (1973*a*) argued that conversation should not be understood as an alternation of turns but rather as a continuous system of concurrent, parallel streams of verbal and non-verbal behaviours in which active participants hold facial expressions and postures and move together throughout an interaction (Scheflen, 1973a, 6). The importance of these interactive social processes for understanding the significance of individual behaviours can be illustrated by considering the use of facial expressions, gesture, and voice.

7.2.1 Facial expressions

Darwin 1872 characterized facial expressions as genetically determined displays of universal human emotions triggered by specific mental states. The basic categories of human facial expression and the muscle configurations involved in producing them have been extensively studied (Ekman and Friesen, 1971). This has provided the foundations for cognitive and neurocognitive models developed of the processing of facial expressions (e.g., Calder and Young, 2005) and sophisticated computational tools that can automatically detect and classify basic facial expressions (e.g., Ruf *et al.*, 2011).

Importantly, people's tendency to display a particular facial expression is more strongly predicted by social context than by emotional state. For example, smiles are not simply a reflection of how happy someone is. Smiling is less strongly associated with events that make people happy—including even winning a gold medal—than it is with whether people are engaged in a social encounter (Fernández-Dols and Ruiz-Belda, 1995). Kraut and Johnston (1979) found a strong positive correlation between smiling and social contact, with smiles approximately 10 times more likely when people were in social contact and independently of whether a positive or negative event had occurred. Similarly Provine (1996) estimated laughter is about 30 times more likely during social contact than when social contact is absent.

These studies show that facial expressions are *recipient* or *audience designed* (Sacks *et al.*, 1974; Clark and Murphy, 1982). People use facial expressions strategically to provide others present with a signal of how they construe a situation. Moreover, they are used to signal a wide range of things. Smiles can be used to display happiness, task completion, sympathy, gratitude, confusion, awkwardness, irony, and sarcasm, among other things. Knowing which of these things is being signalled depends on understanding the context of the interaction and the current orientation and activities of others present.

Recipient design matters because the datasets for both experimental testing and machine learning of facial expressions are typically photographs of faces with posed expressions in a neutral context i.e. without any physical or social context (e.g., Kanade *et al.*, 2000; Pantic *et al.*, 2005; Qu *et al.*, 2017). It also matters for engineering human-like emotional responses for robots that implement basic emotion categories as a direct window on the internal state of the machine (Breazeal, 2003). As Jung (2017) argues, human–robot interaction needs to take a much more interactive, conversational approach to the deployment of multi-modal signals. Similar problems attend attempts to use people's facial expressions as an automatic index of their emotional states (e.g., Gunes and Pantic, 2010; Theodorou *et al.*, 2019). Human-computer interaction often involves dealing with unsocial materials such as to-do lists and spreadsheets. In these contexts our typical facial expression is blank.

7.2.2 Gesture

Social processes also play an important role in the production and comprehension of gesture. David McNeill's seminal work on gesture is primarily based on videos of people gesturing as they recount stories to camera rather than a live addressee (McNeill, 1992; McNeill, 2006). McNeil's primary interest is in the timing and content of gestures, their relationship to speech, and what this can tell us about the underlying cognitive representations that generate them. As McNeill (2006) acknowledges, this focus on processes of production by an individual speaker obsucres the different classes of gesture that become visible when people are engaged in a social encounter. For example, (Bavelas *et al.*, 1995) document a number of dialogue-specific interactive gestures with functions such as offering a turn, referring back to a previous turn, eliciting help with finding a word and soliciting agreement.

Ostensibly non-interactive gestures also change in type, form, and position depending on what an addressee can see (Bavelas and Healing, 2013). The position and delivery of content specific gestures, such as iconics, is sensitive to the number and arrangement of interlocutors (Özyürek, 2002). Some gestures are even jointly produced. People sometimes reach into each other's gesture spaces to manually modify, elaborate or rephrase each other's gestures (Tabensky, 2001; Furuyama, 2002). This also underlines how shared three-dimensional space provides a specific resource for human communication that is absent from modalities like text chat and video mediated communication.

Research on sensing and generating gestures has predominantly focused on non interactive gestures such as iconics and diectics and almost exclusively explores the relationship between the speaker's words and their hand movements. Gesture generation for embodied conversational agents typically focuses on temporal and semantic congruence between the speaking agent's words and their hand movements. Gesture perception normally focuses on sensing and modelling speaker's hand shapes (Wagner *et al.*, 2014; Rautaray and Agrawal, 2015). It is instructive that the attempt to build virtual embodied conversational agents reinforced Scheflen's 1973*b* point about the dynamics of the conversational system since it made it clear that non-speaking addressees should not just stop moving and wait for their next turn. Addressee's gestures, posture and nods are also systematically organized (Maatman *et al.*, 2005; Morency *et al.*, 2010; Vinciarelli *et al.*, 2011; Healey *et al.*, 2013).

7.2.3 Voice

Non-speech voice characteristics provide a third illustration of the use of non-verbal signals in face-to-face interaction. Turing removed voice characteristics because they would give away too much about someone's identity, particularly gender, but this also removes a rich variety of other communicative signals. Laughter, sighs, pauses, filled pauses ('umms' and 'errs'), in-breaths, and snorts all have communicative uses. Research on speech recognition and speech synthesis machines was still in its infancy in the 1950s (although work started much earlier, e.g., Dudley, 1939). Progress since has been remarkable and recent work on voice morphing has reached the point where gender transformation might be good enough to make a spoken version of the Turing test possible (Ye and Young, 2006).

The cues that voice characteristics contribute to the real-time management of conversation are technically challenging. The average offset between the end of a question and its answer is around 250 milliseconds (ms), approximately the length of a syllable, and there is surprisingly little cross-linguistic variation in this timing (Stivers *et al.*, 2009). This is much faster than speech production. It takes around 600–1200 ms to produce the name of an object from the moment of seeing it. Human turn-taking thus displays a close cross-person co-ordination that depends on people's ability to project when they should speak next (Sacks *et al.*, 1974). Machines are currently unable to achieve this speed of response in conversation partly because of the complexity of the phonetic, semantic, and syntactic cues that appear to be involved in projecting turn endings (canned responses can be produced at much shorter latencies) (de Ruiter *et al.*, 2006; Riest *et al.*, 2015; Levinson, 2016).

Speed of response matters because deviations from the 250 ms average are systematically interpreted. For example when a nominated next speaker takes longer than 250 ms this is interpreted as a signal that they are in some way unwilling or unable to answer (Jefferson, 1989; Levinson, 2016). Turing's second Q&A example illustrates this intuition as he suggests in the computer should pause before answering a calculation question that would be trivial for the machine (see above). A difficult question should be answered more slowly than an easy one. The same applies to socially weighted exchanges; people accede to requests more quickly than they refuse them (Levinson and Torreira, 2015).

In addition to different kinds of silence people are also sensitive to different kinds of filled pause. Clark and Fox Tree (2002) provide corpus evidence that a long filled pause ('uhh') and a short filled pause ('um') systematically signal whether there will be a short or long delay in speaking and are used to implicate, among other things, whether the speaker wants to keep the turn or hand it over (see also Fox Tree and Clark, 1997). These differences in filled pauses also affect listener's comprehension (Fox Tree, 2001).

Breathing is a similarly subtle cue. Audible in-breaths are interpreted as a signal of the intention to take a turn, especially in muti-party settings where there might be competition for the floor (Mondada, 2007). The length of an in-breath predicts how much someone is about to say and, like a filled pause, can also signal a possible problem with what they are about to say (Torreira *et al.*, 2015). Breathing has been simulated for (virtual) embodied conversational agents but only as an on/off as parameter for behavioural realism not as a signal for turn-taking (Novick *et al.*, 2018).

7.3 Coordinating Understanding

The multimodal signals people use in ordinary conversation are complex, fine-grained, incremental, and have a clear interactional organization. Most of these signals are not yet recognized, modelled, or produced by machines, and those that are fall short of the sophistication needed for human-like conversational intelligence.[2] These phenomena do not present an insoluble technical problem for machine intelligence. Signal processing techniques can capture audible in-breaths, 'uhhs' and 'ums', laughter, silences, blinks, gesture, gaze, posture, and head movements at very fine resolutions. Many social context effects could be captured by, in effect, a wider camera angle such as that offered by full three-dimensional reality capture. The difficulty lies more in recognizing how the collaborative and interactive nature of these signals matters for communication (Vinciarelli *et al.*, 2011; Wagner *et al.*, 2014). As noted above, the relevant phenomena are typically absent from the datasets and models used to analyse or recreate human interaction.

The Turing test and the multi-modal phenomena discussed above are characterized in essentially behavioural terms. The second, more fundamental challenge to human-like machine intelligence relates to way in which we manage shared meaning.

7.3.1 Standard average understanding

Cognitive psycholinguistics focuses on the processes that underpin individual production and comprehension. Typical experimental paradigms involve individuals in lexical decision tasks or eye-movement studies that test effects such as priming where, for example, processing of words or images is facilitated or inhibited by prior exposure to related words or images. Often supplemented by neurocognitive studies, the resulting models make proposals about the cognitive representations and processes that are necessary to account for the data, for example the separation of lexical, syntactic, and semantic processing of language or the separation of facial expressions and personal identity in visual cognition.

The simplest way to generalize these individualistic models to communication is to assume multiple copies of these processing capabilities and then investigate what is needed to bring them into alignment. This is the basic intuition behind the application of Shannon and Weaver's (1949) model of machine communication to human interaction (not its original purpose). This assumes that communicating agents have identical processes for encoding and decoding the meaning of signals transmitted between them (Cherry, 1966; Healey *et al.*, 2018*a*).

A similar intuition informs explanations of communication in terms of mirror neurons, concepts of resonance, or concepts of automatic, unconscious, cross-person priming

[2] So-called deepfakes might seem like a counter-example, however these are post-processed, canned monologues that are currently incapable of live interaction.

that cause direct coordination of behaviour (Chartrand and Bargh, 1999; Pickering and Garrod, 2004; Chartrand and Lakin, 2013). These models are inherently conservative and assume that coordination consists of reciprocal activation of the same representations e.g. words or syntax or a non-verbal behaviour through interaction. This requires some qualification since if all we do is automatically mutually prime each other our interactions would rapidly become deadlocked. To avoid this difficulty priming effects are normally moderated by factors such as degree of affiliation, type of interaction or social situation (Leander *et al.*, 2012).

These models are underwritten by a shared language, a shared basic set of representations and processes or an idealised basic competence as a starting point for communication. If we don't share this basic capacity we don't, by definition, speak the same language.

> [A]s actual users of one and the same language code... A common code is their communication tool, which actually underlies and makes possible the exchange of messages.
>
> (Jakobson, 1961, 247)

> [A] language is a code which pairs phonetic and semantic representations of sentences.
>
> (Sperber and Wilson, 1986, 9)

The problem for a theory of communication, on this view, is to account for how these shared processes become aligned in different processors. Communication can only be successful where the same, or sufficiently similar, representations are already available to participants.

The idea of an idealized average competence has deep roots in the cognitive sciences. It can be traced back to early formal models of syntax and semantics for natural languages (Chomsky, 1957; Montague, 1973). These models focus on characterizing the structure of sentences and relationships between sentences in a natural language *in toto*. They inspired many attempts at developing computational models of cognitive processing of theses structures and were subsequently co-opted as the foundations for pragmatic theory including communication. In these models individual deviation from the average competence is treated as noise or performance error (Healey *et al.*, 2018*a*). The same idealization is incorporated into contemporary machine-learning (ML) algorithms, including Transformers and deep neural networks (DNNs) (Schmidhuber, 2015; Devlin *et al.*, 2018). These systems are typically trained on very large text corpora and model statistcial global average usage across thousands and sometimes millions of examples. The models produced are static and relatively inflexible and necessarily idealize away from individual variation.

7.3.2 Misunderstandings

The idealization to a standard average competence has impeded progress both in understanding how people communicate and in building machines capable of human-like communication; conversation is both harder and more creative than models of average competence imply.

The principle behavioural prediction of models such as communication under-pinned by automatic mutual priming is that we should see frequent repetition of words, syntax, and gestures. Moreover, mutual priming predicts not just that people should repeat each other but that use of different words and syntax should be inhibited. Significant amounts of repetition are found in laboratory-based studies of priming but these effects are not found in ordinary unscripted conversation. Even on a generous measure of repetition that includes closed-class words ('the', 'a', 'of', 'and') people repeat less than 5% of each other's words on a turn-by-turn basis in ordinary conversation and actively diverge from each other in choice of syntactic structures (Howes *et al.*, 2010; Healey *et al.*, 2014).

A second difficulty is that misunderstandings are not occasional performance problems or noise in the signal; they are ubiquitous and highly structured (Sacks *et al.*, 1974; Schegloff, 1984, 1992). One of the most commonly used words in dialogue and one that appears to be universal across human languages is 'huh?' (Dingemanse *et al.*, 2013). In ordinary English conversation it is used about once every 84 seconds (Enfield, 2017). This is just one, especially explicit, signal of misunderstanding. Hand-coded corpus studies estimate some form of problem with understanding occurs once every three turns or once every 20 words (Brennan and Schober, 2001; Colman and Healey, 2011).

The potential sources of misunderstanding in conversation are diverse. Schegloff (1987) includes misarticulations, malapropisms, use of a 'wrong' word, unavailability of a word when needed, failure to hear or to be heard, trouble on the part of the recipient in understanding, incorrect understandings by recipients. Not all of these relate to content; sometimes it is the speakers intention that is being queried as in 'what do you mean by saying that'. Nonetheless, in controlled experimental tasks it is clear that even for native speakers of English simple terms such as 'box', 'row', and 'column' mean different things especially when applied in new contexts (Garrod and Anderson, 1987; Healey, 1997; Healey *et al.*, 2018*b*).

The importance of misunderstanding in ordinary conversation is especially vivid in the case of healthcare where it affects the delivery of diagnosis and treatment (McDonald and Sherlock, 2016; NHS Improvement, 2018). In the NHS in England there are over 1 million spoken interactions between members of the public and the NHS every 36 hours. These interactions are time-pressured and depend on communication between people with different backgrounds, knowledge, expertise, and concerns. As a result, misunderstandings are common and can be highly consequential, leading to worse health status, poorer understanding of diagnosis and treatment, increased hospitalization rates, and a waste of resources (Williams *et al.*, 2002; Watson and McKinstry, 2009; McDonald and Sherlock, 2016; NHS Improvement, 2018). Clinicians estimate that 50% of what is said in consultations is not fully understood (Williams *et al.*, 2002; Watson and McKinstry, 2009).

7.4 Real-time Adaptive Communication

The misunderstandings that follow from individual differences are not marginal or incidental phenomena, they are a fundamental, recurrent, and critically important feature of ordinary human communication. The challenge for machine intelligence is to switch from models of human communication that focus almost on understanding in a shared language to models that focus on detecting and dealing with misunderstanding.

Conversation analysts provide some clues about how this might be achieved. They have described a variety of turn-based mechanisms, termed *repairs*, that are associated with detecting and dealing with problems with understanding. These are locally organized procedures that operate over short sequences of turns. Different types of repair are specified in terms of how a problem is signalled, the scope of the problem (e.g., a generic 'huh?' versus a specific repetition of a problematic word) and how these signals are systematically related to different forms of response (Sacks *et al.*, 1974; Schegloff, 1987; Schegloff, 1992; Dingemanse *et al.*, 2013). These different parameters are structured within a local repair space that operates like a moving window over sequences of four turns or *positions* and is remarkably stable across languages and cultures (Stivers *et al.*, 2009; Dingemanse *et al.*, 2015).

The challenge for the cognitive sciences is to develop models that connect these descriptions of repair procedures with local, turn-by-turn, and word-by-word updates of meaning. Some formal semantic models of dialogue such as Ginzburg's KoS (Ginzburg, 2012) are providing frameworks in which different repair phenomena can be formally modelled as meaning updates. Kempson's development of dynamic syntax (Kempson *et al.*, 2016) provides incremental ways to build and revise semantic structures as a dialogue proceeds. There are also experimental methodologies emerging that enable studies of the causal connections between repairs and changes in meaning (Healey *et al.*, 2003, 2018*b*). The challenge for machine intelligence is to develop learning algorithms that are capable of local, real-time updates to a language model without sacrificing the power of machine learning. This will require a new generation of hybrid machine learning approaches that enable content addressable forms of online meaning updates without requiring thousands of examples.

7.5 Conclusion

The Turing test was designed to avoid a hard question about what really constitutes human-like computation by asking the apparently simpler question of what is needed for human-like communication. It turns out that this is a hard question too. It has taken more than 60 years before a chatbot has been able to fool even some of the people some of the time i.e., 33% of the judges for a maximum of 5 minutes (Warwick and Shah, 2016). Even in its sequential text-chat form human-like communication is a difficult test. The incremental, concurrent, multimodal, and multiparty nature of ordinary human conversations poses much more complex technical challenges that are beyond the current state of the art in machine intelligence.

The technical challenges involved in human-like communication appear ultimately tractable but achieving this also requires rethinking how human communication works. Modelling communication as the problem of coordinating an idealized average competence has two key limitations. It fails to address the evidence that misunderstandings are a ubiquitous and structured feature of natural communication. It also fails to account for how people can adapt their communication in response on a person-by-person and turn-by-turn basis. Misunderstandings are not marginal phenomena they provide a structured way in which people can adapt a variety of verbal and non-verbal resources in the service of their constantly changing communicative needs.

References

Bavelas, J. and Healing, S. (2013). Reconciling the effects of mutual visibility on gesturing: A review. *Gesture*, **13**(1), 63–92.

Bavelas, J. B., Chovil, N., Coates, L. *et al.* (1995). Gestures specialized for dialogue. *Personality and Social Psychology Bulletin*, **21**(4), 394–405.

Bavelas, J. B., Chovil, N., Lawrie, D. A. *et al.* (1992). Interactive gestures. *Discourse Processes*, **15**(4), 469–89.

Bavelas, J. B., Coates, L., and Johnson, T. (2000). Listeners as co-narrators. *Journal of Personality and Social Psychology*, **79**(6), 941.

Breazeal, C. (2003). Emotion and sociable humanoid robots. *International Journal of Human-computer Studies*, **59**(1-2), 119–55.

Brennan, S. E. and Schober, M. F. (2001). How listeners compensate for disfluencies in spontaneous speech. *Journal of Memory and Language*, **44**(2), 274–96.

Calder, A. J. and Young, A. W. (2005). Understanding the recognition of facial identity and facial expression. *Nature Reviews Neuroscience*, **6**(8), 641–51.

Chartrand, T. L. and Bargh, J. A. (1999). The chameleon effect: the perception–behavior link and social interaction. *Journal of Personality and Social Psychology*, **76**(6), 893.

Chartrand, T. L. and Lakin, J. L. (2013). The antecedents and consequences of human behavioral mimicry. *Annual Review of Psychology*, **64**, 285–308.

Cherry, C. (1966). *On human communication.* Cambridge, MA: MIT Press.

Chomsky, N. (1957). Syntactic Structures. Mouton and Co, The Hague.

Clark, E. V. (2020). Conversational repair and the acquisition of language. *Discourse Processes*, 1–19.

Clark, H. H. (1996). Communities, commonalities. *Rethinking Linguistic Relativity* (17), 324.

Clark, H. H. and Fox Tree, J. E. (2002). Using uh and um in spontaneous speaking. *Cognition*, **84**(1), 73–111.

Clark, H. H. and Murphy, G. L. (1982). Audience design in meaning and reference, in *Advances in Psychology*, Vol. 9. pp. 287–99. Amsterdam: Elsevier.

Clark, H. H. and Wilkes-Gibbs, D. (1986). Referring as a collaborative process. *Cognition*, **22**(1), 1–39.

Colman, M. and Healey, P. (2011). The distribution of repair in dialogue, in *Proceedings of the Annual Meeting of the Cognitive Science Society*, in L. Carlson and T. F. Shipley eds, *Proceedings of the 33rd Annual Meeting of the Cognitive Science Society Boston, Massachusetts*, Vol. 33. Cognitive Science Society. pp. 1563–1568. July 20–23.

Condon, W. S. and Ogston, W. D. (1966). Sound film analysis of normal and pathological behavior patterns. *Journal of Nervous and Mental Disease.*

Darwin, C. (1872). *The Expression of the Emotions in Man and Animals.* London: Murray.

De Ruiter, J.-P., Mitterer, H., and Enfield, N. J. (2006). Projecting the end of a speaker's turn: A cognitive cornerstone of conversation. *Language*, **82**(3), 515–535.

Devlin, J., Chang, M.-W., Lee, K. *et al.* (2018). BERT: pre-training of deep bidirectional transformers for language understanding. *CoRR*, **abs/1810.04805**.

Dingemanse, M., Roberts, S. G., Baranova, J. *et al.* (2015). Universal principles in the repair of communication problems. *PLoS ONE*, **10**(9), e0136100.

Dingemanse, M., Torreira, F., and Enfield, N. (2013). Is 'Huh?' a universal word? *PLoS ONE*, 8(11), e78273.

Dudley, H. (1939). The automatic synthesis of speech. *Proceedings of the National Academy of Sciences of the United States of America*, **25**(7), 377.

Ekman, P. and Friesen, W. V. (1971). Constants across cultures in the face and emotion. *Journal of Personality and Social Psychology*, **17**(2), 124.

Enfield, N. J. (2017). *How We Talk: The Inner Workings of Conversation*. London: Basic Books.

Fernández-Dols, J.-M. and Ruiz-Belda, M.-A. (1995). Are smiles a sign of happiness? gold medal winners at the olympic games. *Journal of Personality and Social Psychology*, **69**(6), 1113.

Fox Tree, J. E. (2001). Listeners' uses ofum anduh in speech comprehension. *Memory & cognition*, **29**(2), 320–326.

Fox Tree, J. E. and Clark, H. H. (1997). Pronouncing the as thee to signal problems in speaking. *Cognition*, **62**(2), 151–67.

Furuyama, N. (2002). Prolegomena of a theory of between-person coordination of speech and gesture. *International Journal of Human-Computer Studies*, **57**(4), 347–74.

Garrod, S. and Anderson, A. (1987). Saying what you mean in dialogue: A study in conceptual and semantic co-ordination. *Cognition*, **27**(2), 181–218.

Ginzburg, J. (2012). *The Interactive Stance: Meaning for Conversation*. Oxford: Oxford University Press.

Goodwin, C. (1979). The interactive construction of a sentence in natural conversation. *Everyday language: Studies in Ethnomethodology*, 97–121.

Gunes, H. and Pantic, M. (2010). Automatic, dimensional and continuous emotion recognition. *International Journal of Synthetic Emotions (IJSE)*, **1**(1), 68–99.

Healey, P. G. T. (2008). Interactive misalignment: The role of repair in the development of group sub-languages, in C. R. Kempson and R. Kempson eds, *Language in Flux: Relating Dialogue Coordination to Language Variation, Change and Evolution*. London: Palgrave-Macmillan, 341–54.

Healey, P., Lavelle, M., Howes, C. *et al.* (2013). How listeners respond to speaker's troubles, in *Proceedings of the Annual Meeting of the Cognitive Science Society*, in M. Knauff, N. Sebanz, M. Pauen and I. Wachsmuth eds, *Proceedings of the 35th Annual Meeting of the Cognitive Science Society*. Berlin, Germany: Cognitive Science Society. July 31-August 3, 2013, pp. 2506–22511. Vol. 35.

Healey, P. G. T. (1997). Expertise or expertese?: The emergence of task-oriented sub-languages, in *Proceedings of the Ninteenth Annual Conference of The Cognitive Science Society*, in M. G. Shafto and Langley, P. eds. August 7th–10th, Stanford University, CA. pp. 301–306. Stanford University Stanford, CA. Publisher is LEA, Mahwah, NJ.

Healey, P. G. T., De Ruiter, J. P., and Mills, G. J. (2018*a*). Editors' introduction: Miscommunication. *Topics in Cognitive Science*, **10**(2), 264–78.

Healey, P. G. T., Mills, G. J., Eshghi, A. *et al.* (2018*b*). Running repairs: Coordinating meaning in dialogue. *Topics in Cognitive Science*, **10**(2), 367–88.

Healey, P. G. T., Purver, M., and Howes, C. (2014). Divergence in dialogue. *PloS ONE*, **9**(6).

Healey, P. G. T., Purver, M., King, *et al.* (2003). Experimenting with clarification in dialogue. In *Proceedings of the Annual Meeting of the Cognitive Science Society*, in Alterman, R. and Kirsh, D.

eds, *Proceedings of the 25th Annual Conference of the Cognitive Science Society.* July 31st-August 2nd, Park Plaza Hotel, Boston. Vol. 25, Mahwah, N.J.: LEA. pp. 539–544.

Hömke, Paul, Holler, J., and Levinson, S. C. (2017). Eye blinking as addressee feedback in face-to-face conversation. *Research on Language and Social Interaction*, **50**(1), 54–70.

Hömke, P., Holler, J., and Levinson, S. C. (2018). Eye blinks are perceived as communicative signals in human face-to-face interaction. *PloS ONE*, **13**(12).

Howes, C., Healey, P. G. T., and Purver, M. (2010). Tracking lexical and syntactic alignment in conversation. In *Proceedings of the Annual Meeting of the Cognitive Science Society*, in In S. Ohlsson and R. Catrambone eds, Proceedings of the 32nd Annual Conference of the Cognitive Science Society. Portland, Oregon, 11–14th August. Austin, TX: Cognitive Science Society pp. 2004–2009. Vol. 32.

Jakobson, R. (1961). *Structure of language and its mathematical aspects.* Volume 12. American Mathmatical Society. 190 Hope Street, Providence, Rhode Island. ISBN 0–8218–1312–9.

Jefferson, G. (1989). Preliminary notes on a possible metric which provides for a 'a standard maximum silence' of approximately one second in conversation in Roger, D., and Bull, P. eds, *Conversation: An Interdisciplinary Perspective.* Vol. 3. Philadhelphia, PA: Multilingual Matters LTD, 166–96.

Jung, M. F. (2017). Affective grounding in human-robot interaction. In *2017 12th ACM/IEEE International Conference on Human-Robot Interaction HRI*, pp. 263–273. IEEE. 6–9th March, 2017. Vienna, Austria: The publisher is the Institute of Electrical and Electronics Engineers (IEEE). ISBN: 978–1–4503–4336–7.

Kanade, T., Cohn, J. F., and Tian, Y. (2000). Comprehensive database for facial expression analysis. In *Proceedings Fourth IEEE International Conference on Automatic Face and Gesture Recognition (Cat. No. PR00580).* Grenoble, France: Institute of Electrical and Electronics Engineers (IEEE). 28–30 March. pp. 46–53. IEEE.

Kempson, R., Cann, R., Gregoromichelaki, E. *et al.* (2016, October). Language as mechanisms for interaction. *Theoretical Linguistics*, **42**(3-4), 203–76.

Kraut, R. E., and Johnston, R. E. (1979). Social and emotional messages of smiling: an ethological approach. *Journal of Personality and Social Psychology*, **37**(9), 1539.

Leander, N. P., Chartrand, T. L., and Bargh, J. A. (2012). You give me the chills: Embodied reactions to inappropriate amounts of behavioral mimicry. *Psychological Science*, **23**(7), 772–9.

Levinson, S. C. (2016). Turn taking in human communication–origins and implications for language processing. *Trends in Cognitive Sciences*, **20**(1), 6–14.

Levinson, S. C., and Torreira, F. (2015). Timing in turn-taking and its implications for processing models of language. *Frontiers in Psychology*, **6**, 731.

Loebner, H. G. (1990). The loebner prize—the first turing test. *Cambridge Center for Behavioral Studies, Massachusetts, United States.*

Maatman, R. M., Gratch, J., and Marsella, S. (2005). Natural behavior of a listening agent. IN *International workshop on intelligent virtual agents*, pp. 25–36. *Intelligent Virtual Agents: 5th International Working Conference, IVA 2005*, Kos, Greece, September 12-14, 2005. Themis Panayiotopoulos, Jonathan Gratch, Thomas Rist, Ruth Aylett and Daniel Ballin (eds) *Proceedings: 3661 (Lecture Notes in Computer Science)* Springer, Berlin, Heidelberg.

McDonald, A. and Sherlock, J. (2016). A long and winding road. improving communication with patients in the NHS. *Marie Curie: Care and Support through Terminal Illness.*

McNeill, D. (1992). *Hand and Mind: What Gestures Reveal about Thought.* Chicago, IL: University of Chicago Press.

McNeill, D. (2006). Gesture: a psycholinguistic approach. *The encyclopedia of language and linguistics*, 58–66.

Mehl, M. R., Vazire, S., Ramírez-Esparza, N. *et al.* (2007). Are women really more talkative than men? *Science*, **317**(5834), 82–82.

Milek, A., Butler, E. A., Tackman, A. M. *et al.* (2018). "eavesdropping on happiness" revisited: A pooled, multisample replication of the association between life satisfaction and observed daily conversation quantity and quality. *Psychological Science*, **29**(9), 1451–62.

Mondada, L. (2007). Multimodal resources for turn-taking: Pointing and the emergence of possible next speakers. *Discourse Studies*, **9**(2), 194–225.

Montague, R. (1973). The proper treatment of quantification in ordinary english, in Hintikka K. J. J., Moravcsik J. M. E., Suppes P. eds, *Approaches to Natural Language*. Synthese Library (Monographs on Epistemology, Logic, Methodology, Philosophy of Science, Sociology of Science and of Knowledge, and on the Mathematical Methods of Social and Behavioral Sciences), vol 49. pp. 221–242. Springer, Dordrecht. https://doi.org/10.1007/978-94-010-2506-5_10

Morency, L.-P., de Kok, I., and Gratch, J. (2010). A probabilistic multimodal approach for predicting listener backchannels. *Autonomous Agents and Multi-Agent Systems*, **20**(1), 70–84.

NHS Improvement (2018). Spoken communication and patient safety in the NHS. Annex: Much more than words. https://improvement.nhs.uk/resources/improving-safety-critical-spoken-communication/.

Novick, D., Afravi, M., and Camacho, A. (2018). Paolachat: A virtual agent with naturalistic breathing, in *International Conference on Virtual, Augmented and Mixed Reality*. In Jessie Y.C. ChenGino Fragomeni eds, *10th International Conference, VAMR 2018, Held as Part of HCI International 2018*, Las Vegas, NV, USA, July 15–20, 2018, Proceedings, Part I. Springer Lecture Notes in Computer Science (LNCS, volume 10909), Springer, Berlin, Heidelberg. pp. 351–60.

Özyürek, A. (2002). Do speakers design their cospeech gestures for their addressees? the effects of addressee location on representational gestures. *Journal of Memory and Language*, **46**(4), 688–704.

Pantic, M., Valstar, M., Rademaker, R. *et al.* (2005). Web-based database for facial expression analysis. In *2005 IEEE International Conference on Multimedia and Expo*, Amsterdam, Netherlands: Institute of Electrical and Electronics Engineers (IEEE). pp. 5, doi: 10.1109/ICME.2005.1521424.

Pickering, M. J. and Garrod, S. (2004). Toward a mechanistic psychology of dialogue. *Behavioral and Brain Sciences*, **27**(2), 169–90.

Provine, R. R. (1996). Laughter. *American Scientist*, **84**(1), 38–45.

Qu, F., Wang, S.-J., Yan, W. J. *et al.* (2017). Cas (me) ˆ2: A database for spontaneous macro-expression and micro-expression spotting and recognition. *IEEE Transactions on Affective Computing*, **9**(4), 424–36.

Rautaray, S. S. and Agrawal, A. (2015). Vision based hand gesture recognition for human computer interaction: a survey. *Artificial Intelligence Review*, **43**(1), 1–54.

Riest, C., Jorschick, A. B., and de Ruiter, J. P. (2015). Anticipation in turn-taking: mechanisms and information sources. *Frontiers in Psychology*, **6**, 89.

Ruf, T., Ernst, A., and Küblbeck, C. (2011). Face detection with the sophisticated high-speed object recognition engine (shore), in *Microelectronic Systems: Circuits, Systems and Applications*. Albert Heuberger, Günter Elst and Randolf Hanke eds. Springer, Berlin, Heidelberg, pp. 243–252.

Sacks, H., Schegloff, E. A., and Jefferson, G. (1974). A simplest systematics for the organization of turn-taking for conversation. *Language*, **50**(4), 696–735.

Scheflen, A. E. (1973*a*). *Communicational Structure: Analysis of a Psychotherapy Transaction.* Bloomington, IN: Indiana University Press.

Scheflen, A. E. (1973*b*). *How Behavior Means.* Aronson, New York.

Schegloff, E.A. (1987). Some sources of misunderstanding in talk-in-interaction. *Linguistics*, **25**, 201–18.

Schegloff, E. A. (1984). On some gestures' relation to talk, in Atkinson, J. M., Heritage, J., and Oatley, K. eds. *Structures of Social Action.* Cambridge University Press, Cambridge.

Schegloff, E. A. (1992). Repair after next turn: The last structurally provided defense of intersubjectivity in conversation. *American Journal of Sociology*, **97**(5), 1295–345.

Schmidhuber, J. (2015). Deep learning in neural networks: An overview. *Neural Networks*, **61**, 85–117.

Shannon, C. E. and Weaver, W. (1949). *The Mathematical Theory of Communication.* Urbana, IL: University of Illinois Press.

Shieber, S. M. (1994). Lessons from a restricted turing test. *arXiv preprint cmp-lg/9404002.*

Sperber, D. and Wilson, D. (1986). *Relevance: Communication and Cognition.* Vol. 142. Cambridge, MA: Harvard University Press.

Stivers, T., Enfield, N. J., Brown, P. et al. (2009). Universals and cultural variation in turn-taking in conversation. *Proceedings of the National Academy of Sciences*, **106**(26), 10587–92.

Tabensky, A. (2001). Gesture and speech rephrasings in conversation. *Gesture*, **1**(2), 213–35.

Theodorou, L., Healey, P. G. T., and Smeraldi, F. (2019). Engaging with contemporary dance: What can body movements tell us about audience responses? *Frontiers in Psychology*, **10**, 71.

Torreira, F. J., Bögels, S., and Levinson, S. C. (2015). Breathing for answering: the time course of response planning in conversation. Frontiers in psychology, 6, 284. https://doi.org/10.3389/fpsyg.2015.00284

Turing, A. (1950). Computing machinery and intelligence. *Mind*, **59**(236), 433.

Vinciarelli, A., Pantic, M., Heylen, D. *et al.* (2011). Bridging the gap between social animal and unsocial machine: A survey of social signal processing. *IEEE Transactions on Affective Computing*, **3**(1), 69–87.

Wagner, P., Malisz, Z. and Kopp, S. (2014). Gesture and speech in interaction: An overview. Speech Communication, Vol. 57, pp. 209–232, ISSN 0167–6393, https://doi.org/10.1016/j.specom.2013.09.008.

Warwick, K. and Shah, H. (2016). Can machines think? a report on turing test experiments at the Royal Society. *Journal of Experimental & Theoretical Artificial Intelligence*, **28**(6), 989–1007.

Watson, P. and McKinstry, B. (2009). A systematic review of interventions to improve recall of medical advice in healthcare consultations. *Journal of the Royal Society of Medicine*, **102**(6), 235–43.

Williams, M., Davis, T., Parker, R. et al. (2002). The role of health literacy in patient-physician communication. *Family Medicine*, **34**(5), 383–9.

Ye, H. and Young, S. (2006). Quality-enhanced voice morphing using maximum likelihood transformations. *IEEE Transactions on Audio, Speech, and Language Processing*, **14**(4), 1301–12.

Yngve, V. H. (1970). On getting a word in edgewise, in *Chicago Linguistics Society, 6th Meeting, 1970*, pp. 567–78. April 16–18th. Chicago: Chicago Linguistic Society.

8

Too Many cooks: Bayesian inference for coordinating Multi-agent Collaboration

Rose E. Wang[1], Sarah A. Wu[1], James A. Evans[2], David C. Parkes[3], Joshua B. Tenenbaum[1], and Max Kleiman-Weiner[1,3]

[1]*Massachusetts Institute of Technology,* [2]*University of Chicago, and*
[3]*Harvard University, USA*

8.1 Introduction

*Working together enables a group of agents to achieve together what no individual could achieve on their own (Tomasello, 2014; Henrich, 2015). However, collaboration is challenging as it requires agents to coordinate their behaviours. In the absence of prior experience, social roles, and norms, we still find ways to negotiate our joint behaviour in any given moment to work together with efficiency (Tomasello, Carpenter, Call, Behne and Moll, 2005; Misyak, Melkonyan, Zeitoun and Chater, 2014). Whether we are writing a scientific manuscript with collaborators or preparing a meal with friends, core questions we ask ourselves are: how can I help out the group? What should I work on next, and with whom should I do it with? Figuring out how to flexibly coordinate a collaborative endeavor is a fundamental challenge for any agent in a multi-agent world.

Central to this challenge is that agents' reasoning about what they should do in a multi-agent context depends on the future actions and intentions of others. When agents, like people, make independent decisions, these intentions are unobserved. Actions can reveal information about intentions, but predicting them is difficult because of uncertainty and ambiguity—multiple intentions can produce the same action. In humans, the ability to understand intentions from actions is called theory-of-mind (ToM). Humans rely on this ability to cooperate in coordinated ways, even in novel situations (Tomasello, Carpenter, Call, Behne and Moll, 2005; Shum, Kleiman-Weiner, Littman and Tenenbaum, 2019).

* Rose Wang and Sarah Wu contributed equally to this chapter

Rose E. Wang, Sarah A. Wu, James A. Evans, David C. Parkes, Joshua B. Tenenbaum, and Max Kleiman-Weiner, *Too Many cooks: Bayesian inference for coordinating Multi-agent Collaboration* In: *Human Like Machine Intelligence.* Edited by: Stephen Muggleton and Nick Charter, Oxford University Press. DOI: 10.1093/oso/9780198862536.003.0008

We aim to build agents with theory-of-mind and use these abilities for coordinating collaboration.

In this work, we study these abilities in the context of multiple agents cooking a meal together, inspired by the video game *Overcooked* (Ghost Town Games, 2016). These problems have hierarchically organized sub-tasks and share many features with other object-oriented tasks such as construction and assembly. These sub-tasks allow us to study agents that are challenged to coordinate in three distinct ways: (A) Divide and conquer: agents should work in parallel when sub-tasks can be efficiently carried out individually, (B) Cooperation: agents should work together on the same sub-task when most efficient or necessary, (C) Spatio-temporal movement: agents should avoid getting in each other's way at any time.

To illustrate, imagine the process required to make a simple salad: first chopping both tomato and lettuce and then assembling them together on a plate. Two people might collaborate by first dividing the sub-tasks up: one person chops the tomato and the other chops the lettuce. This doubles the efficiency of the pair by completing sub-tasks in parallel (challenge A). On the other hand, some sub-tasks may require multiple to work together. If only one person can use the knife and only the other can reach the tomatoes, then they must cooperate to chop the tomato (challenge B). In all cases, agents must coordinate their low-level actions in space and time to avoid interfering with others and be mutually responsive (challenge C).

Our work builds on a long history of using cooking tasks for evaluating multi-agent coordination across hierarchies of sub-tasks (Grosz and Kraus, 1996; Cohen and Levesque, 1991; Tambe, 1997). Most recently, environments inspired by *Overcooked* have been used in deep reinforcement learning studies where agents are trained using self-play and human data (Song, Wang, Lukasiewicz, Xu and Xu, 2019; Carroll, Shah, Ho, Griffiths, Seshia, Abbeel and Dragan, 2019). In contrast, our approach is based on techniques that dynamically learn while interacting rather than requiring large amounts of pre-training experience for a specific environment, team configuration, and sub-task structure. Instead our work shares goals with the ad-hoc coordination literature, where agents must adapt on the fly to variations in task, environment, or team (Chalkiadakis and Boutilier, 2003; Stone, Kaminka, Kraus and Rosenschein, 2010; Barrett, Stone and Kraus, 2011). However, prior work is often limited to action coordination (e.g,. chasing or hiding) rather than coordinating actions across and within sub-tasks. Our approach to this problem takes inspiration from the cognitive science of how people coordinate their cooperation in the absence of communication (Kleiman-Weiner, Ho, Austerweil, Littman and Tenenbaum, 2016). Specifically, we build on recent algorithmic progress in Bayesian theory-of-mind (Ramırez and Geffner, 2011; Nakahashi, Baker and Tenenbaum, 2016; Baker, Jara-Ettinger, Saxe and Tenenbaum, 2017; Shum, Kleiman-Weiner, Littman and Tenenbaum, 2019) and learning statistical models of others (Barrett, Stone, Kraus and Rosenfeld, 2012; Melo and Sardinha, 2016), and extend these works to decentralized multi-agent contexts.

Our strategy for multi-agent hierarchical planning builds on previous work linking high-level coordination (sub-tasks) to low-level navigation (actions) (Amato, Konidaris, Kaelbling and How, 2019). In contrast to models which have explicit communication mechanism or centralized controllers (McIntire, Nunes and Gini, 2016; Brunet, Choi and How, 2008), our approach is fully decentralized and our agents are never trained together. Prior work has also investigated ways in which multi-agent teams can mesh inconsistent plans (e.g. two agents doing the same sub-task by themselves) into consistent plans (e.g. the agents perform different sub-tasks in parallel) (Cox and Durfee, 2004, 2005), but these methods have also been centralized. We draw more closely from decentralized multi-agent planning approaches in which agents aggregate the effects of others and best respond (Claes, Robbel, Oliehoek, Tuyls, Hennes and Van der Hoek, 2015; Claes, Oliehoek, Baier and Tuyls, 2017). These prior works focus on tasks with spatial sub-tasks called *Spatial Task Allocation Problems* (SPATAPs). However, there are no mechanisms for agents to cooperate on the same sub-task as each sub-task is spatially distinct.

We develop *Bayesian Delegation*, an algorithm for decentralized multi-agent coordination that rises to the challenges described above. Bayesian Delegation leverages Bayesian inference with inverse planning to rapidly infer the sub-tasks others are working on. Our probabilistic approach allows agents to predict the intentions of other agents under uncertainty and ambiguity. These inferences allow agents to efficiently delegate their own efforts to the most high-value collaborative tasks for collective success. We quantitatively measure the performance of Bayesian Delegation in a suite of novel multi-agent environments. First, Bayesian Delegation outperforms existing approaches, completing all environments in less time than alternative approaches and maintaining performance even when scaled up to larger teams. Finally, we show Bayesian Delegation is an ad-hoc collaborator. It performs better than baselines when paired with alternative agents.

8.2 Multi-Agent MDPs with Sub-Tasks

A multi-agent Markov decision process (MMDP) with sub-tasks is described as a tuple $\langle n, \mathcal{S}, \mathcal{A}_{1\ldots n}, T, R, \gamma, \mathcal{T} \rangle$ where n is the number of agents, $s \in \mathcal{S}$ are object-oriented states specified by the locations, status and type of each object and agent in the environment (Boutilier, 1996; Diuk, Cohen and Littman, 2008). $\mathcal{A}_{1\ldots n}$ is the joint action space with $a_i \in \mathcal{A}_i$ being the set of actions available to agent i; each agent chooses its own actions independently. $T(s, a_{1\ldots n}, s')$ is the transition function which describes the probability of transitioning from state s to s' after all agents act $a_{1\ldots n}$. $R(s, a_{1\ldots n})$ is the reward function shared by all agents and γ is the discount factor. Each agent aims to find a policy $\pi_i(s)$ that maximizes expected discounted reward. The environment state is fully observable to all agents, but agents do not observe the policies $\pi_{-i}(s)$ ($-i$ refers to all other agents except i) or any other internal representations of others agents.

Unlike traditional MMDPs, the environments we study have a partially ordered set of sub-tasks $\mathcal{T} = \{\mathcal{T}_0 \ldots \mathcal{T}_{|\mathcal{T}|}\}$. Each sub-task $\mathcal{T}_i$ has preconditions that specify when a sub-task can be started, and postconditions that specify when it is completed. They provide structure when R is very sparse. These sub-tasks are also the target of high-level coordination between agents. In this work, all sub-tasks can be expressed as `Merge(X,Y)`, that is, to bring `X` and `Y` into the same location. Critically, unlike in SPATAPs, this location is not fixed or predetermined if both `X` and `Y` are movable. In the cooking environments we study here, the partial order of sub-tasks refers to a "recipe". Figure 8.1 shows an example of sub-task partial orders for a recipe.

The partial order of sub-tasks ($\mathcal{T}$) introduces two coordination challenges. First, `Merge` does not specify how to implement that sub-task in terms of efficient actions nor

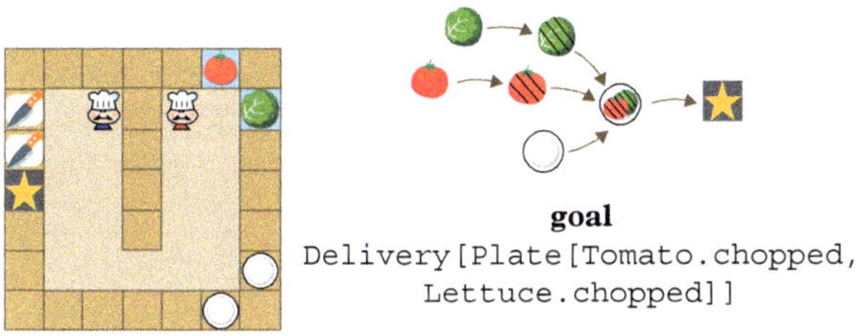


```
Merge(Tomato.unchopped,Knife)
Merge(Lettuce.unchopped,Knife)
Merge(Tomato.chopped,
Lettuce.chopped)
Merge([Tomato.chopped
Lettuce.chopped],Plate[])
Merge(Plate[Tomato.chopped,
Lettuce.chopped],Delivery)
```

(a) Partial-Divider. (b) Salad. (c) Example sub-task order for Salad.

Figure 8.1 *The Overcooked environment. (a) The Partial-Divider kitchen offers many counters for objects, but forces agents to move through a narrow bottleneck. (b) The Salad recipe in which two chopped foods must be combined on a single plate and delivered, and (c) one of the many possible orderings for completing this task. All sub-tasks are expressed in the* `Merge` *operator. Different recipes are possible in each kitchen, allowing for variation in high-level goals while keeping the low-level navigation challenges fixed.*

which agent(s) should work on it. Second, because the ordering of sub-tasks is partial, the sub-tasks can be accomplished in many different orders. For instance, in the *Salad* recipe (Figure 8.1b), once the tomato and lettuce are chopped, they can: (a) first combine the lettuce and tomato and then plate, (b) the lettuce can be plated first and then add the tomato, or (c) the tomato can be plated first and then add the lettuce. These distinct orderings make coordination more challenging since to successfully coordinate, agents must align their ordering of sub-tasks.

The partially ordered set of sub-tasks $\mathcal{T}$ is given in the environment and generated by representing each recipe as an instance of STRIPS, an action language (Fikes and Nilsson, 1971). Each instance consists of an initial state, a specification of the goal state, and a set of actions with preconditions that dictate what must be true/false for the action to be executable, and postconditions that dictate what is made true/false when the action is executed. For instance, for the STRIPS instance of the recipe *Tomato*, the initial state is the initial configuration of the environment (i.e. all objects and their states), the specification of the goal state is `Delivery[Plate[Tomato.chopped]]`, and the actions are the `Merge` sub-tasks.. A plan for a STRIPS instance is a sequence of actions that can be executed from the initial state and results in a goal state. To generate these partial orderings, we construct a graph for each recipe in which the nodes are the states of the environment objects and the edges are valid actions. We then run breadth-first-search starting from the initial state to determine the nearest goal state, and return all shortest "recipe paths" between the two states.

8.2.1 Coordination Test Suite

We now describe the Overcooked inspired environments we use as a test suite for evaluating multi-agent collaboration. Each environment is a 2D grid-world kitchen. Figure 8.1a shows an example layout. The kitchens are built from counters that contain both movable food and plates and immovable stations (e.g. knife stations). The state is

Table 8.1 *State representation and transitions for the objects and interactions in the Overcooked environments. The two food items (tomato and lettuce) can be in either chopped or unchopped states. Objects with status [] are able to "hold" other objects. For example, an Agent holding a Plate holding an unchopped tomato would be denoted Agent[Plate[Tomato.unchopped]]. Once combined, these nested objects share the same $\{x, y\}$ coordinates and movement. Interaction dynamics occur when the two objects are in the same $\{x, y\}$ coordinates.*

Object state representation:

Type	Location	Status
Agent	{x, y}	[]
Plate	{x, y}	[]
Counter	{x, y}	[]
Delivery	{x, y}	[]
Knife	{x, y}	N/A
Tomato	{x, y}	{chopped, unchopped}
Lettuce	{x, y}	{chopped, unchopped}

Interaction dynamics:

Food.unchopped + Knife $\rightarrow$ Food.chopped + Knife

Food1 + Food2 $\rightarrow$ [Food1, Food2]

X + Y[] $\rightarrow$ Y[X]

represented as a list of entities and their type, location, and status (Diuk, Cohen and Littman, 2008). See Table 8.1 for a description of the different entities, the dynamics of object interactions, and the statuses that are possible. Agents (the chef characters) can move north, south, east, west or stay still. All agents move simultaneously. They cannot move through each other, into the same space, or through counters. If they try to do so, they remain in place instead. Agents pick up objects by moving into them and put down objects by moving into a counter while holding them. Agents chop foods by carrying the food to a knife station. Food can be merged with plates. Agents can only carry one object at a time and cannot directly pass to each other.

The goal in each environment is to cook a recipe in as few time steps as possible. The environment terminates after either the agents bring the finished dish specified by the recipe to the star square or 100 time steps elapse.

8.3 Bayesian Delegation

We now introduce *Bayesian Delegation*, a novel algorithm for multi-agent coordination that uses inverse planning to make probabilistic inferences about the sub-tasks other agents are performing. Bayesian Delegation models the latent intentions of others in order to dynamically decide whether to divide-and-conquer or to cooperate, and an action planner finds approximately optimal policies for each sub-task. Note that planning is decentralized at both levels, i.e., agents plan and learn for themselves without any access to each other's internal representations.

Inferring the sub-tasks others are working on enables each agent to select the right sub-task when multiple are possible. Agents maintain and update a belief state over the possible sub-tasks that all agents (including itself) are likely working on based on a history of observations that is commonly observed by all. Formally, Bayesian Delegation maintains a probability distribution over task allocations. Let $\mathbf{ta}$ be the set of all possible allocations of agents to sub-tasks where all agents are assigned to a sub-task. For example, if there are two sub-tasks ($[\mathcal{T}_1, \mathcal{T}_2]$) and two agents ($[i,j]$), then $\mathbf{ta} = [(i:\mathcal{T}_1, j:\mathcal{T}_2), (i:\mathcal{T}_2, j:\mathcal{T}_1), (i:\mathcal{T}_1, j:\mathcal{T}_1), (i:\mathcal{T}_2, j:\mathcal{T}_2)]$ where $i:\mathcal{T}_1$ means that agent i is "delegated" to sub-task $\mathcal{T}_1$. Thus, $\mathbf{ta}$ includes both the possibility that agents will divide and conquer (work on separate sub-tasks) and cooperate (work on shared sub-tasks). If all agents pick the same $ta \in \mathbf{ta}$, then they will easily coordinate. However, in our environments, agents cannot communicate before or during execution. Instead Bayesian Delegation maintains uncertainty about which ta the group is coordinating on, $P(ta)$.

At every time step, each agent selects the most likely allocation $ta^* = \arg\max_{ta} P(ta|H_{0:T})$, where $P(ta|H_{0:T})$ is the posterior over ta after having observed a history of actions $H_{0:T} = [(s_0, \mathbf{a_0}), \ldots (s_T, \mathbf{a}_T)]$ of T time steps and $\mathbf{a}_t$ are all agents' actions at time step t. The agent then plans the next best action according to ta^* using a model-based reinforcement learning algorithm described below. This posterior is computed according by Bayes rule:

$$P(ta|H_{0:T}) \propto P(ta)P(H_{0:T}|ta) = P(ta)\prod_{t=0}^{T} P(\mathbf{a_t}|s_t, ta) \tag{8.1}$$

where $P(ta)$ is the prior over ta and $P(\mathbf{a_t}|s_t, ta)$ is the likelihood of actions at time step t for all agents. Note that these belief updates do not explicitly consider the private knowledge that each agent has about their own intention at time $T-1$. Instead each agent performs inference based only on the history observed by all, i.e., the information a third-party observer would have access to (Sugden, 2003; Bacharach, 1999; Nagel, 1986). The likelihood of a given ta is the likelihood that each agent i is following their assigned task ($\mathcal{T}_i$) in that ta.

$$P(\mathbf{a_t}|s_t, ta) \propto \prod_{i:\mathcal{T} \in ta} exp(\beta * Q^*_{\mathcal{T}_i}(s, a_i)) \tag{8.2}$$

where $Q^*_{\mathcal{T}_i}(s, a_i)$, is the expected future reward of a towards the completion of sub-task $\mathcal{T}_i$ for agent i. The soft-max accounts for non-optimal and variable behavior as is typical in Bayesian theory-of-mind (Kleiman-Weiner, Ho, Austerweil, Littman and Tenenbaum, 2016; Baker, Jara-Ettinger, Saxe and Tenenbaum, 2017; Shum, Kleiman-

Weiner, Littman and Tenenbaum, 2019). β controls the degree to which an agent believes others are perfectly optimal. When $\beta \to 0$, the agent believes others are acting randomly. When $\beta \to \infty$, the agent believes others are perfectly maximizing. Since the likelihood is computed by planning, this approach to posterior inference is called inverse planning. Note that even though agents see the same history of states and actions, their belief updates will not necessarily be the same because updates come from $Q_{\mathcal{T}_i}$, which is computed independently for each agent and is affected by stochasticity in exploration.

The prior over $P(ta)$ is computed directly from the environment. First, $P(ta) = 0$ for all ta that have sub-tasks without satisfied preconditions. We set the remaining priors to $P(ta) \propto \sum_{\mathcal{T} \in ta} \frac{1}{V_{\mathcal{T}(s)}}$, where $V_{\mathcal{T}(s)}$ is the estimated value of the current state under sub-task $\mathcal{T}$. This gives ta that can be accomplished in less time a higher prior weight. Priors are reinitialized when new sub-tasks have their preconditions satisfied and when others are completed. Figure 8.2 shows an example of the dynamics of $P(ta)$ during agent interaction. The figure illustrates how Bayesian delegation enables agents to dynamically align their beliefs about who is doing what (i.e., assign high probability to a single ta).

Action planning transforms sub-task allocations into efficient actions and provides the critical likelihood for Bayesian Delegation (see Equation 8.1). Action planning takes the ta selected by Bayesian Delegation and outputs the next best action while modeling the movements of other agents. In this work, we use bounded real-time dynamic programming (BRTDP) extended to a multi-agent setting to find approximately optimal Q-values and policies (McMahan, Likhachev and Gordon, 2005). Each simulation was run in parallel (3 recipes, 3 environments, 5 models, 20 seeds) each on 1 CPU core, which took up to 15 GB of memory and roughly 3 hours to complete. We next describe the details of our BRTDP implementation: (McMahan, Likhachev and Gordon, 2005).

$$V^b_{\mathcal{T}_i}(s) = \min_{a \in \mathcal{A}_i} Q^b_{\mathcal{T}_i}(s,a), \quad V^b_{\mathcal{T}_i}(g) = 0$$

$$Q^b_{\mathcal{T}_i}(s,a) = C_{\mathcal{T}_i}(s,a) + \sum_{s' \in S} T(s'|s,a) V^b_{\mathcal{T}_i}(s')$$

where C is cost and $b = [l,u]$ is the lower and upper bound respectively. Each time step is penalized by 1 and movement (as opposed to staying still) by an additional 0.1. This cost structure incentivizes efficiency. The lower-bound was initialized to the Manhattan distance between objects (which ignores barriers). The upper-bound was the sum of the shortest-paths between objects which ignores the possibility of more efficiently passing objects. While BRTDP and these heuristics are useful for the specific spatial environments and subtask structures we develop here, it could be replaced with any other algorithm for finding an approximately optimal single-agent policy for a given sub-task. For details on how BRTDP updates on V and Q, see (McMahan, Likhachev and Gordon, 2005). BRTDP was run until the bounds converged ($\alpha = 0.01, \tau = 2$) or for a maximum of 100 trajectories each with up to 75 roll-outs for all models. The softmax during inference used $\beta = 1.3$. At each time step, agents select the action with the highest value for their sub-task. When agents do not have any valid sub-tasks, i.e. sub-task is None, they take a random action (uniform across the movement and

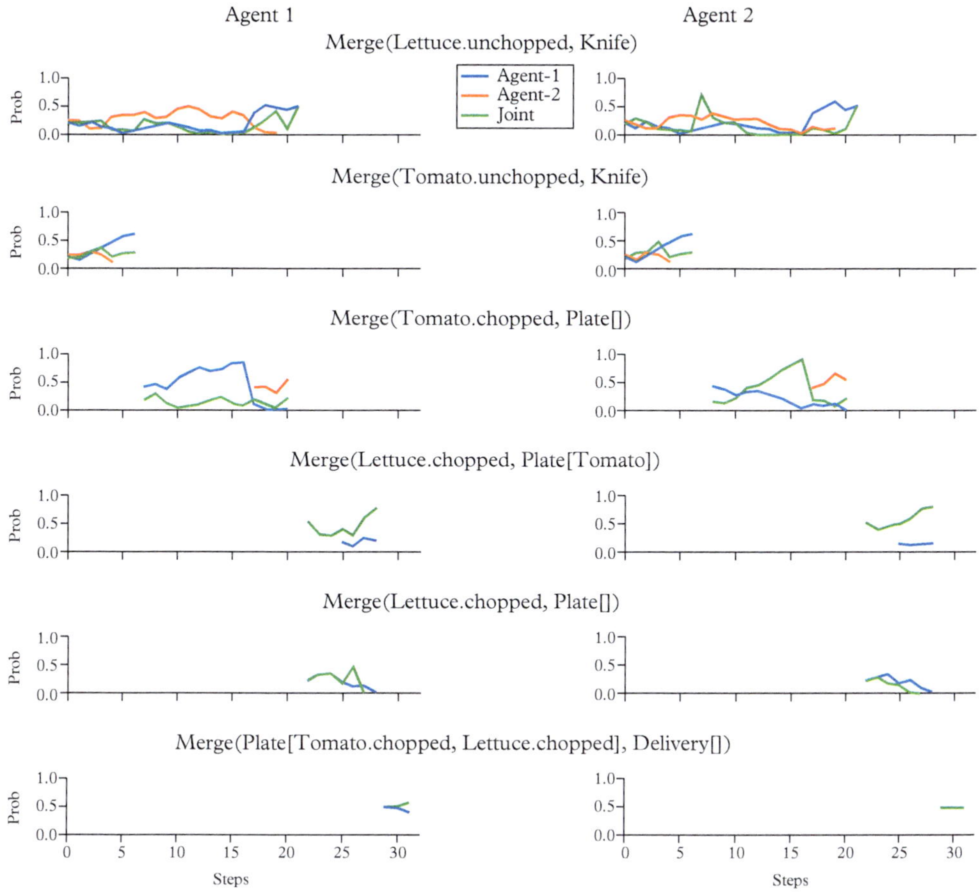

Figure 8.2 *Dynamics of the belief state, P(ta) for each agent during Bayesian delegation with the Salad recipe on Partial-Divider (Figure 8.1). During the first 7 time steps, only the* `Merge(Lettuce.unchopped, Knife)` *and* `Merge(Tomato.unchopped, Knife)` *sub-tasks are nonzero because their preconditions are met. These beliefs show alignment across the ordering of sub-tasks as well as within each sub-task. Salad can be completed in three different ways (see Figure 8.5), yet both agents eventually drop* `Merge(Lettuce.unchopped, Plate[])` *in favor of* `Merge(Tomato.unchopped, Plate[])` *followed by* `Merge(Lettuce.chopped, Plate[Tomato])`*. Agents' beliefs also converge over the course of each specific sub-task. For instance, while both agents are at first uncertain about who should be delegated to* `Merge(Lettuce.unchopped, Knife)`*, they eventually align to the same relative ordering. This alignment continues, even though there is never any communication or prior agreement on what sub-task each agent should be doing or when.*

stay-in-place actions). This greatly improves the performance of the alternative (lesioned) models: without this noise, they often get stuck and block each other from completing the recipe. It has no effect on Bayesian Delegation.

Agents use ta^* from Bayesian Delegation to address two types of low-level coordination problems: (1) avoiding getting in each others way while working on distinct sub-tasks, and (2) cooperating efficiently when working on a shared sub-task. ta^* contains agent i's best guess about the sub-tasks carried out by others, $\mathcal{T}_{-i}$. In the first case, $\mathcal{T}_i \neq \mathcal{T}_{-i}$. Agent i first creates models of the others performing $\mathcal{T}_{-i}$ assuming others agents are stationary ($\pi^0_{\mathcal{T}_{-i}}(s)$, level-0 models). These level-0 models are used to reduce the multi-agent transition function to a single agent transition function T' where the transitions of the other agents are assumed to follow the level-0 policies, $T'(s'|s,a_{-i}) = \sum_{a_i} T(s'|s,a_{-i},a_i)\prod_{A\in -i}\pi^0_{\mathcal{T}_A}(s)$. Running BRTDP on this transformed environment finds an approximately optimal level-1 policy $\pi^1_{\mathcal{T}_i}(s)$ for agent i that "best responds" to the level-0 models of the other agents. This approach is similar to level-K or cognitive hierarchy (Wright and Leyton-Brown, 2010; Kleiman-Weiner, Ho, Austerweil, Littman and Tenenbaum, 2016; Shum, Kleiman-Weiner, Littman and Tenenbaum, 2019).

When $\mathcal{T}_i = \mathcal{T}_{-i}$, agent i attempts to work together on the same sub-task with the other agent(s). The agent simulates a fictitious centralized planner that controls the actions of all agents working together on the same sub-task (Kleiman-Weiner, Ho, Austerweil, Littman and Tenenbaum, 2016). This transforms the action space: if both i and j are working on $\mathcal{T}_i$, then $\mathcal{A}' = a_i \times a_j$. Joint policies $\pi^J_{\mathcal{T}_i}(s)$ can similarly be found by single-agent planners such as BRTDP. Agent i then takes the actions assigned to it under $\pi^J_{\mathcal{T}_i}(s)$. Joint policies enable emergent decentralized cooperative behavior—agents can discover efficient and novel ways of solving sub-tasks as a team such as passing objects across counters. Since each agent is solving for their own $\pi^J_{\mathcal{T}_i}(s)$, these joint policies are not guaranteed to be perfectly coordinated due to stochasticity in the planning process. Note that although we use BRTDP, any other model-based reinforcement learner or planner could also be used.

8.4 Results

We evaluate the performance of Bayesian Delegation across two different experimental paradigms. First, we test the performance of each agent type when all agents are the same type with both two and three agents (self-play). Second, we test the performance of each agent type when paired with an agent of a different type (ad-hoc).

We compare the performance of Bayesian Delegation (BD) to four alternative baseline agents: Uniform Priors (UP), which starts with uniform probability mass over all valid ta and updates through inverse planning; Fixed Beliefs (FB), which does not update $P(ta)$ in response to the behavior of others; Divide and Conquer (D&C) (Ephrati and Rosenschein, 1994), which sets $P(ta) = 0$ if that ta assigns two agents to the same sub-task (this is conceptually similar to Empathy by Fixed Weight Discounting (Claes, Robbel, Oliehoek, Tuyls, Hennes and Van der Hoek, 2015) because agents cannot share sub-tasks and D&C discounts sub-tasks most likely to be attended to by other agents

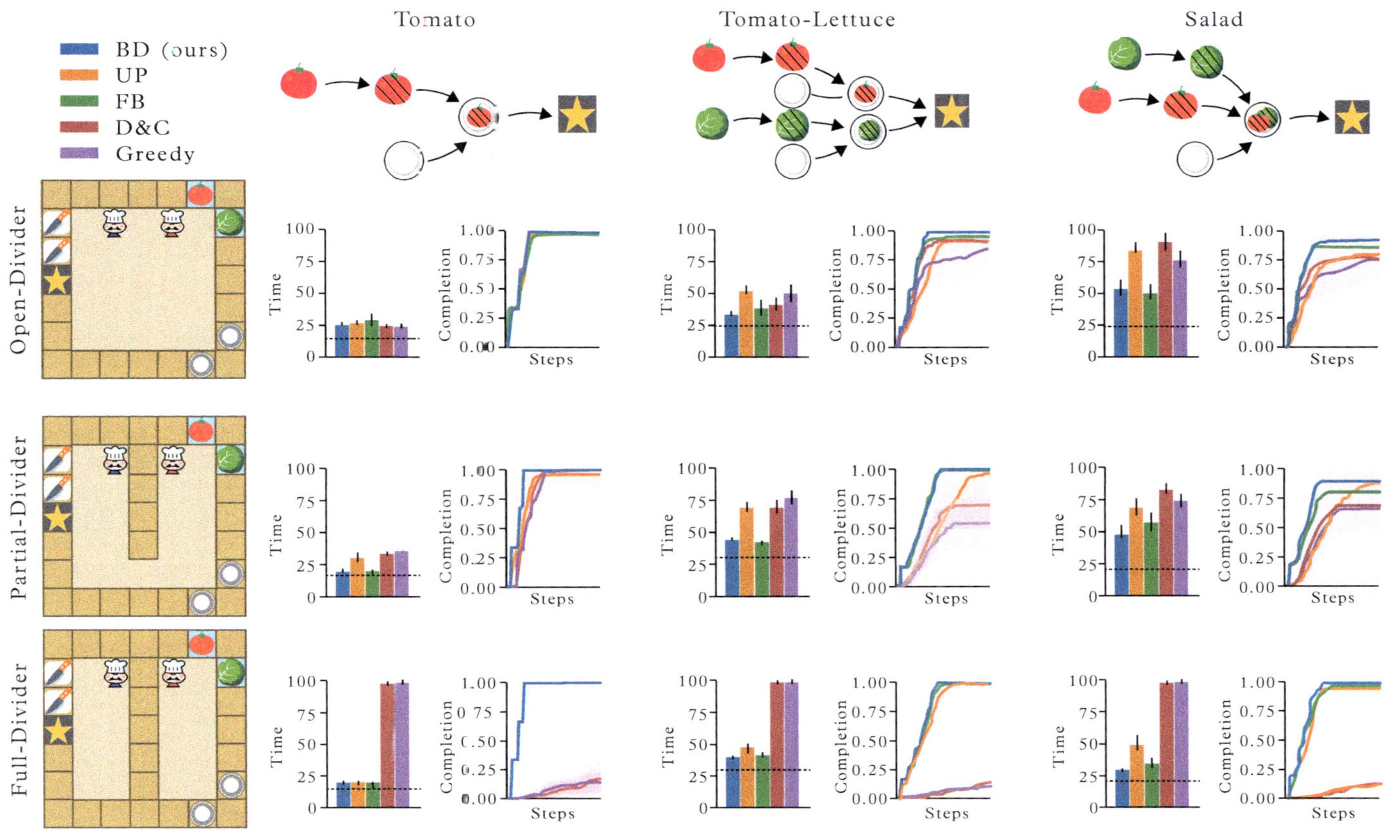

Figure 8.3 *Performance results for each kitchen-recipe composition (lower is better) for two agents in self-play. The row shows the kitchen and the column shows the recipe. Within each composition, the left graph shows the number of time steps needed to complete all sub-tasks. The dashed lines on the left graph represent the optimal performance of a centralized team. The right graph shows the fraction of sub-tasks completed over time. Bayesian Delegation completes more sub-tasks and does so more quickly compared to baselines.*

based on $P(ta|H)$); Greedy, which selects the sub-task it can complete most quickly without considering the sub-tasks other agents are working on. All agents take advantage of the sub-task structure because end-to-end optimization of the full recipe using techniques such as DQN (Mnih, Kavukcuoglu, Silver, Graves, Antonoglou, Wierstra and Riedmiller, 2013) and Q-learning (Watkins and Dayan, 1992) never succeeded under our computational budget.

To highlight the differences between our model and the alternatives, let us consider an example with two possible sub-tasks ($[\mathcal{T}_1, \mathcal{T}_2]$) and two agents ($[i, j]$). The prior for Bayesian Delegation puts positive probability mass on $\mathbf{ta} = [(i : \mathcal{T}_1, j : \mathcal{T}_2), (i : \mathcal{T}_2, j : \mathcal{T}_1), (i : \mathcal{T}_1, j : \mathcal{T}_1), (i : \mathcal{T}_2, j : \mathcal{T}_2)]$ where $i : \mathcal{T}_1$ means that agent i is assigned to sub-task $\mathcal{T}_1$. The UP agent proposes the same $\mathbf{ta}$, but places uniform probability across all elements, i.e., $P(ta) = \frac{1}{4}$ for all $ta \in \mathbf{ta}$. FB would propose the same $\mathbf{ta}$ with the same priors as Bayesian Delegation, but would never update its beliefs. The D&C agent does not allow for joint sub-tasks, so it would reduce to $\mathbf{ta} = [(i : \mathcal{T}_1, j : \mathcal{T}_2), (i : \mathcal{T}_2, j : \mathcal{T}_1)]$. Lastly, Greedy makes no inferences so each agent i would propose $\mathbf{ta} = [(i : \mathcal{T}_1), (i : \mathcal{T}_2)]$. Note that j does not appear.

In the first two computational experiments, we analyze the results in terms of three key metrics. The two pivotal metrics are the number of time steps to complete the full recipe and the total fraction of sub-tasks completed. We also analyze average number of shuffles, a measure of uncoordinated behavior. A *shuffle* is any action that negates the previous action, such as moving left and then right, or picking an object up and then putting it back down (see Figure 8.4a for an example). All experiments show the average performance over 20 random seeds. Agents are evaluated in 9 task-environment combinations (3 recipes $\times$ 3 kitchens).

8.4.1 Self-play

Table 8.2 quantifies the performance of all agents aggregated across the 9 environments. Bayesian Delegation outperforms all baselines and completes recipes with less time step and fewer shuffles. The performance gap was even larger with three agents. Most other agents performed worse with three agents than they did with two, while the performance of Bayesian Delegation did not suffer. Figure 8.3 breaks down performance by kitchen and recipe. All five types of agents are comparable when given the recipe *Tomato* in *Open-Divider*, but when faced with more complex situations, Bayesian Delegation outperform the others. For example, without the ability to represent shared sub-tasks, D&C and Greedy fail in *Full-Divider* because they cannot explicitly coordinate on the same sub-task to pass objects across the counters. Baseline agents were also less capable of low-level coordination resulting in more inefficient shuffles (Figure 8.4). A breakdown of three agent performance is shown in Figure 8.7.

Learning about other agents is especially important for more complicated recipes that can be completed in different orders. In particular, FB and Greedy, which do not learn, have trouble with the *Salad* recipe on *Full Divider*. There are two challenges in this composition. One is that the *Salad* recipe can be completed in three different orders: once the tomato and lettuce are chopped, they can be (a) first combined together and

Table 8.2 *Self-play performance of our model and alternative models with two versus three agents. All metrics are described in the text. See Figure 8.3 for more detailed results on Time Steps and Completion for two agents in self-play, and see Figure 8.4 for more detailed results on shuffles. Averages ± standard error of the mean.*

		Time Steps (↓ better)	Completion (↑ better)	Shuffles (↓ better)
Two agents	BD (ours)	**35.29 ± 1.40**	**0.98 ± 0.06**	**1.01 ± 0.05**
	UP	50.42 ± 2.04	0.94 ± 0.05	5.32 ± 0.03
	FB	37.58 ± 1.60	0.95 ± 0.04	2.64 ± 0.03
	D&C	71.57 ± 2.40	0.61 ± 0.07	13.08 ± 0.05
	Greedy	71.11 ± 2.41	0.57 ± 0.08	17.17 ± 0.06
Three agents	BD (ours)	**34.52 ± 1.66**	**0.96 ± 0.08**	1.64 ± 0.05
	UP	56.84 ± 2.12	0.91 ± 0.22	5.02 ± 0.12
	FB	41.34 ± 2.27	0.92 ± 0.08	**1.55 ± 0.05**
	D&C	67.21 ± 2.31	0.67 ± 0.15	4.94 ± 0.09
	Greedy	75.87 ± 2.32	0.62 ± 0.22	12.04 ± 0.13

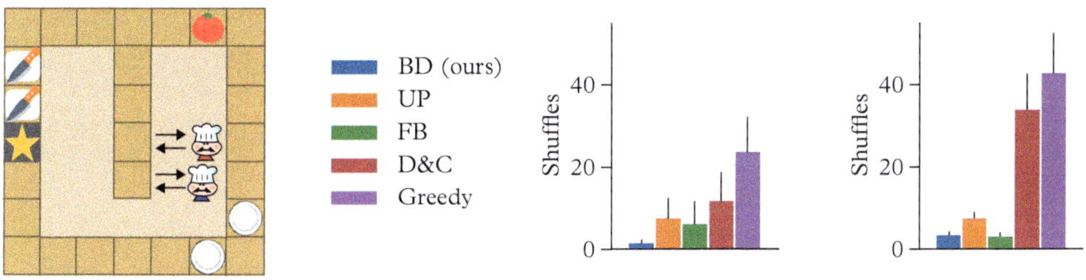

(a) Example "shuffle" (b) *Open-Divider* (c) *Partial-Divider*

Figure 8.4 *Shuffles observed for recipe Tomato+Lettuce. (a) Example of a shuffle, where both agents simultaneously move back and forth from left to right, over and over again. This coordination failure prevents them from passing each other. Note that they are not colliding. Average number of shuffles by each agent in the (b) Open-Divider and (c) Partial-Divider environments. Error bars show the standard error of the mean. Bayesian Delegation and Joint Planning help prevent shuffles, leading to better coordinated behavior.*

then plated, (b) the lettuce can be plated first and then the tomato added or (c) the tomato can be plated first and then the lettuce added. The second challenge is that neither agent can perform all the sub-tasks by themselves, thus they must converge to the same order. Unless the agents that do not learn coordinate by luck, they have no way of recovering. Figure 8.5 shows the diversity of orderings used across different runs of Bayesian Delegation. Another failure mode for agents lacking learning is that FB and Greedy frequently get stuck in cycles in which both agents are holding objects that must be merged (e.g., a plate and lettuce). They fail to coordinate their actions such that one puts their object down in order for the other to pick it up and merge. Bayesian

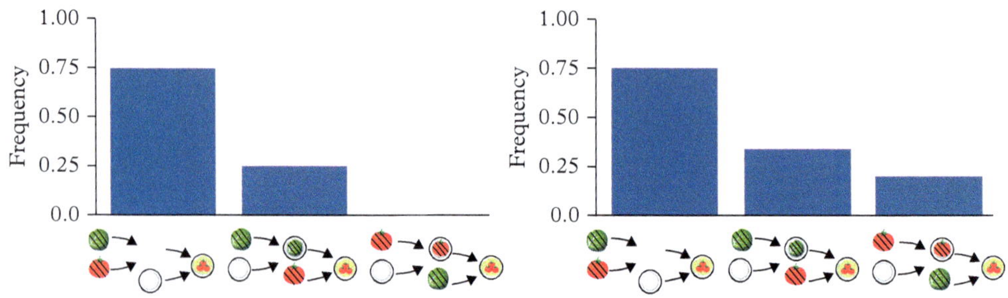

(a) *Open-Divider* (b) *Partial-Divider*

Figure 8.5 *Frequency that three orderings of Salad are completed by our model agents, in (a) Open-Divider and (b) Partial-Divider.*

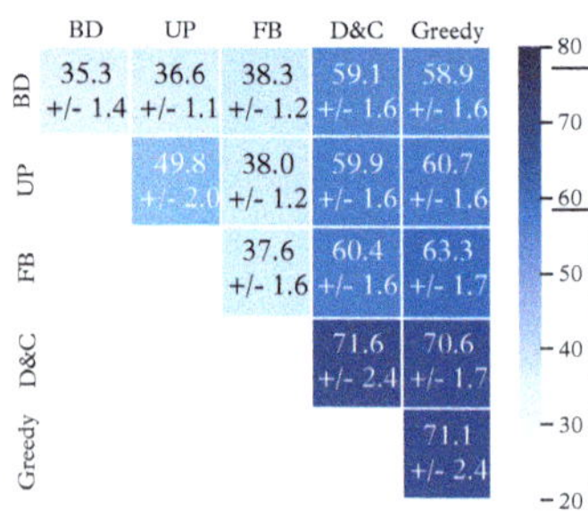

	Times Steps (↓ better)	Completion (↑ better)	Shuffles (↓ better)
BD (ours)	**48.25 ± 0.75**	**0.90 ± 0.01**	**3.96 ± 0.33**
UP	48.84 ± 0.77	0.89 ± 0.01	4.17 ± 0.34
FB	50.00 ± 0.78	0.87 ± 0.01	5.11 ± 0.42
D&C	62.49 ± 0.83	0.77 ± 0.01	6.84 ± 0.43
Greedy	63.40 ± 0.84	0.76 ± 0.01	6.61 ± 0.41

Figure 8.6 *Ad-hoc performance of different agent pairs in time steps (the lower and lighter, the better). (Left) Rows and columns correspond to different agents. Each cell is the average performance of one the row agent playing with the column agent. (Right) Mean performance (± SE) of agents when paired with the others.*

Delegation can break these symmetries by yielding to others so long as they make net progress towards the completion of one of the sub-tasks. For these reasons, only Bayesian Delegation performs on par (if not more efficiently) with three agents than with two agents. As additional agents join the team, aligning plans becomes even more important in order for agents to avoid performing conflicting or redundant sub-tasks.

8.4.2 Ad-hoc

Next, we evaluated the ad-hoc performance of the agents. We show that Bayesian Delegation is a successful ad-hoc collaborator. Each agent was paired with the other agent types. None of the agents had any prior experience with the other agents. Figure 8.6 shows the performance of each agent when matched with each other and in aggregate across all recipe-kitchen combinations. Bayesian Delegation performed well even when matched with baselines. When paired with UP, D&C, and Greedy, the dyad performed better than when UP, D&C, and Greedy were each paired with their own type. Because

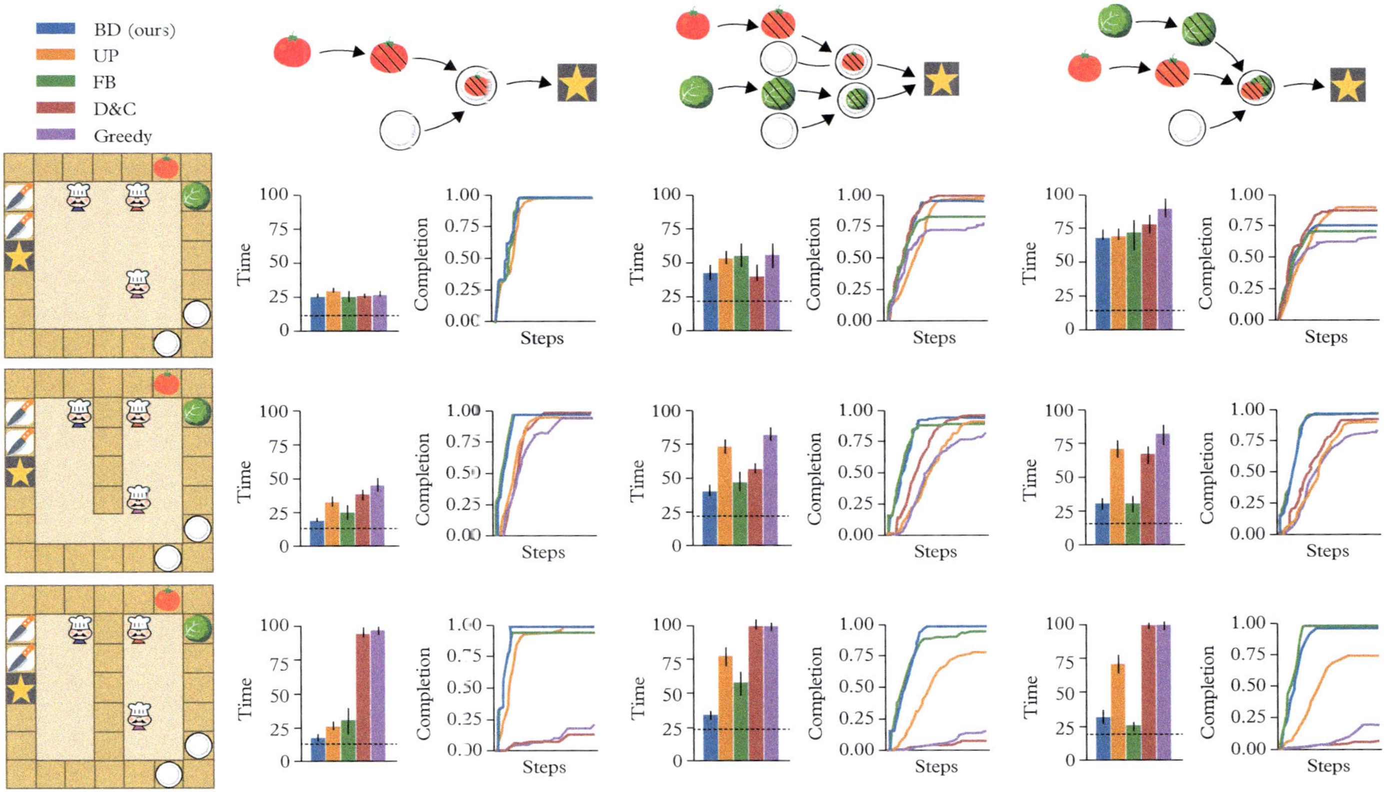

Figure 8.7 *Performance results for each kitchen-recipe composition (lower is better) for* ***three*** *agents. The row shows the kitchen and the column shows the recipe. Within each composition, the left graph shows the number of time steps needed to complete all sub-tasks. The dashed lines on the left graph represent the optimal performance of a centralized team. The right graph shows the fraction of sub-tasks completed over time. The full agent completes more sub-tasks and does so more quickly compared to the alternatives.*

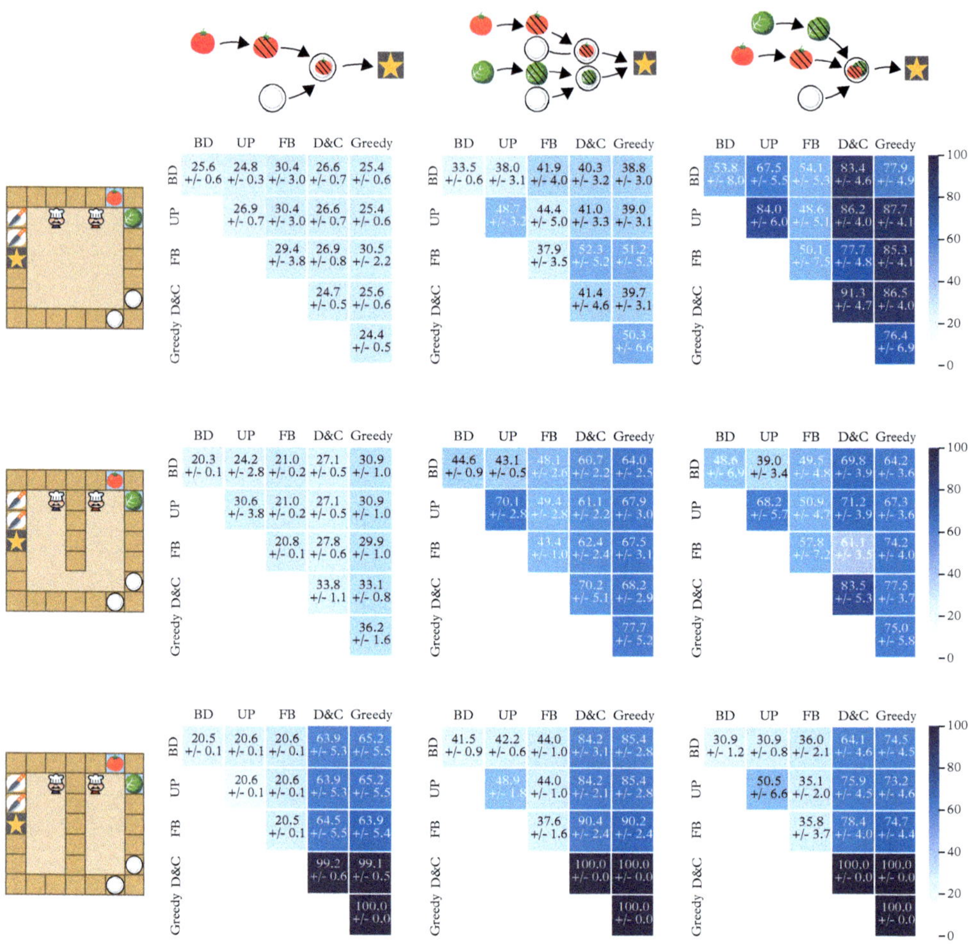

Figure 8.8 *Heat map performance for each kitchen-recipe composition (lower is better) for two agents using different models. The figure breaks down the aggregate performance from Figure 8.6 by kitchen (row) and recipe(column). In most compositions, as models perform as better adhoc coordinators as they become more "Bayesian Delegation"-like (going bottom to top by row, right to left by column).*

Bayesian Delegation can learn in-the-moment, it can overcome some of the ways that these agents get stuck. UP performs better when paired with Bayesian Delegation or FB compared to self-play, suggesting that as long as one of the agents is initialized with smart priors, it may be enough to compensate for the other's uninformed priors. D&C and Greedy perform better when paired with Bayesian Delegation, FB, or UP. Crucially, these three agents all represent cooperative plans where both agents cooperate on the same sub-task. Figure 8.8 breaks down the ad-hoc performance of each agent pairing by recipe and kitchen.

8.5 Discussion

We developed Bayesian Delegation, a new algorithm inspired by and consistent with human theory-of-mind. Bayesian Delegation enables efficient ad-hoc coordination by rapidly inferring the sub-tasks of others. Agents dynamically align their beliefs about who is doing what and determine when they should help another agent on the same sub-task and when they should divide and conquer for increased efficiency. It also enables them to complete sub-tasks that neither agent could achieve on its own. Our agents reflect many natural aspects of human cooperation, such as the emergence of joint behavior when joint planning is deemed better than planning alone (Tomasello, 2014).

The environments studied here are highly challenging from a coordination perspective. There are multiple ways to complete each goal and spatial movement is relatively constrained leading to a high probability of miscoordination. Furthermore, there are no channels for communication. If communication were possible in these environments, many of these coordination problems could be reasoned about directly. Instead, Bayesian Delegation is a kind of implicit mechanism for coordinating group behavior. One might hypothesize that implicit coordination mechanisms such as Bayesian Delegation were important for collaborative hunting and other kinds of early coordinated behavior (Tomasello, 2014). Indeed, these kinds of implicit, pre-linguistic mechanisms for coordinating the mental states of other may have been important for the emergence and acquisition of language (Misyak, Melkonyan, Zeitoun and Chater, 2014).

While Bayesian Delegation reflects progress towards human-like coordination, there are still limitations which we hope to address in future work. One challenge is that when agents jointly plan for a single sub-task, they currently have no way of knowing when they have completed their individual "part" of the joint effort. Consider a case where one agent needs to pass lettuce and tomato across the divider for the other to chop it, after dropping off the lettuce, the first agent is currently unable to reason that it has fulfilled its role in that joint plan and can move on, i.e., that the rest of the sub-task depends only on the actions of the other agent. Currently, our agents considers sub-tasks active as long as their post-conditions remain unsatisfied. If agents were able to recognize when their sub-tasks were finished with respect to themselves, then they would be able to coordinate even more efficiently and flexibly. This opens the possibility of looking ahead to future sub-tasks that will need to be done even before their preconditions are satisfied. For example, once an agent passes off a tomato to another to chop, the first agent can go and get a plate in anticipation of also passing that over even before the chopping has begun.

At some point, as one scales up the number of agents, there can be "too many cooks" in the kitchen! The algorithms presented here scale poorly with the number of agents. In some sense this is a natural trade-off, as Bayesian Delegation through inverse planning requires computing policies not just for oneself but also for each other agent. Other less flexible but more efficient mechanisms may also play a crucial role. Over time, people build up and establish behavioral norms and conventions which yield coordination without sophisticated agent modeling (Young, 1993; Bicchieri, 2006; Lewis, 1969). Roles often emerge between people that spend significant time together (Misyak, Melkonyan,

Zeitoun and Chater, 2014). For instance in *Partial-Divider* a pair of agents could break the symmetry by converging on a norm where one person always yields to the other or in *Open-Divider* a pair of agents might decide to always move in a clockwise direction to minimize the probability of collisions (Lerer and Peysakhovich, 2019; Carroll, Shah, Ho, Griffiths, Seshia, Abbeel and Dragan, 2019). Models that allow for these kinds of subtle norms and roles to emerge are needed for agents to form longer term collaborations that persist beyond a single short interaction. Such representations are essential for building AI agents that are capable of partnering with human teams and with each other.

Acknowledgements

We thank Alex Peysakhovich, Barbara Grosz, DJ Strouse, Leslie Kaelbling, Micah Carroll, and Natasha Jaques for insightful ideas and comments. This work was funded by the Harvard Data Science Initiate, Harvard CRCS, Templeton World Charity Foundation, The Future of Life Institute, DARPA Ground Truth, and The Center for Brains, Minds and Machines (NSF STC award CCF-1231216).

References

Amato, Christopher, Konidaris, George, Kaelbling, Leslie Pack, and How, Jonathan P. (2019). Modeling and planning with macro-actions in decentralized pomdps. *Journal of Artificial Intelligence Research*, **64**, 817–859.

Bacharach, Michael (1999). Interactive team reasoning: A contribution to the theory of co-operation. *Research in economics*, **53**(2), 117–147.

Baker, Chris L, Jara-Ettinger, Julian, Saxe, Rebecca, and Tenenbaum, Joshua B (2017). Rational quantitative attribution of beliefs, desires and percepts in human mentalizing. *Nature Human Behaviour*, **1**, 0064.

Barrett, Samuel, Stone, Peter, and Kraus, Sarit (2011). Empirical evaluation of ad hoc teamwork in the pursuit domain. In *The 10th International Conference on Autonomous Agents and Multi-agent Systems-Volume 2*, pp. 567–574. International Foundation for Autonomous Agents and Multiagent Systems.

Barrett, Samuel, Stone, Peter, Kraus, Sarit, and Rosenfeld, Avi (2012). Learning teammate models for ad hoc teamwork. In *AAMAS Adaptive Learning Agents (ALA) Workshop*, pp. 57–63.

Bicchieri, Cristina (2006). *The grammar of society: The nature and dynamics of social norms*. Cambridge University Press.

Boutilier, Craig (1996). Planning, learning and coordination in multiagent decision processes. In *Proceedings of the 6th conference on Theoretical aspects of rationality and knowledge*, pp. 195–210. Morgan Kaufmann Publishers Inc.

Brunet, Luc, Choi, Han-Lim, and How, Jonathan (2008). Consensus-based auction approaches for decentralized task assignment. In *AIAA guidance, navigation and control conference and exhibit*, p. 6839.

Carroll, Micah, Shah, Rohin, Ho, Mark, Griffiths, Thomas, Seshia, Sanjit, Abbeel, Pieter, and Dragan, Anca (2019). On the utility of learning about humans for human-ai coordination. In *Advances in Neural Information Processing Systems*.

Chalkiadakis, Georgios and Boutilier, Craig (2003). Coordination in multiagent reinforcement learning: A bayesian approach. In *Proceedings of the second international joint conference on Autonomous agents and multiagent systems*, pp. 709–716.

Claes, Daniel, Oliehoek, Frans, Baier, Hendrik, and Tuyls, Karl (2017). Decentralised online planning for multi-robot warehouse commissioning. In *Proceedings of the 16th Conference on Autonomous Agents and MultiAgent Systems*, pp. 492–500. International Foundation for Autonomous Agents and Multiagent Systems.

Claes, Daniel, Robbel, Philipp, Oliehoek, Frans A, Tuyls, Karl, Hennes, Daniel, and Van der Hoek, Wiebe (2015). Effective approximations for multi-robot coordination in spatially distributed tasks. In *Proceedings of the 2015 International Conference on Autonomous Agents and Multiagent Systems*, pp. 881–890. International Foundation for Autonomous Agents and Multiagent Systems.

Cohen, Philip R and Levesque, Hector J (1991). Teamwork. *Noûs*, **25**(4), 487–512.

Cox, Jeffrey S and Durfee, Edmund H (2004). Efficient mechanisms for multiagent plan merging. In *Proceedings of the Third International Joint Conference on Autonomous Agents and Multiagent Systems, 2004. AAMAS 2004.*, pp. 1342–1343. IEEE.

Cox, Jeffrey S and Durfee, Edmund H (2005). An efficient algorithm for multiagent plan coordination. In *Proceedings of the fourth international joint conference on Autonomous agents and multiagent systems*, pp. 828–835.

Diuk, Carlos, Cohen, Andre, and Littman, Michael L (2008). An object-oriented representation for efficient reinforcement learning. In *Proceedings of the 25th international conference on Machine learning*, pp. 240–247. ACM.

Ephrati, Eithan and Rosenschein, Jeffrey S (1994). Divide and conquer in multi-agent planning. In *AAAI*, Volume 1, p. 80.

Fikes, Richard E. and Nilsson, Nils J. (1971). Strips: A new approach to the application of theorem proving to problem solving. *Artificial Intelligence*, **2**(3), 189 – 208.

Ghost Town Games (2016). Overcooked.

Grosz, Barbara J and Kraus, Sarit (1996). Collaborative plans for complex group action. *Artificial Intelligence*, **86**(2), 269–357.

Henrich, Joseph (2015). *The secret of our success: how culture is driving human evolution, domesticating our species, and making us smarter.* Princeton University Press.

Kleiman-Weiner, Max, Ho, Mark K, Austerweil, Joseph L, Littman, Michael L, and Tenenbaum, Joshua B (2016). Coordinate to cooperate or compete: abstract goals and joint intentions in social interaction. In *Proceedings of the 38th Annual Conference of the Cognitive Science Society*.

Lerer, Adam and Peysakhovich, Alexander (2019). Learning existing social conventions via observationally augmented self-play. In *Proceedings of the 2019 AAAI/ACM Conference on AI, Ethics, and Society*, pp. 107–114. ACM.

Lewis, David (1969). *Convention: A philosophical study*. John Wiley & Sons.

McIntire, Mitchell, Nunes, Ernesto, and Gini, Maria (2016). Iterated multi-robot auctions for precedence-constrained task scheduling. In *Proceedings of the 2016 International Conference on Autonomous Agents & Multiagent Systems*, pp. 1078–1086.

McMahan, H Brendan, Likhachev, Maxim, and Gordon, Geoffrey J (2005). Bounded real-time dynamic programming: Rtdp with monotone upper bounds and performance guarantees. In *Proceedings of the 22nd international conference on Machine learning*, pp. 569–576. ACM.

Melo, Francisco S and Sardinha, Alberto (2016). Ad hoc teamwork by learning teammates' task. *Autonomous Agents and Multi-Agent Systems*, **30**(2), 175–219.

Misyak, Jennifer B, Melkonyan, Tigran, Zeitoun, Hossam, and Chater, Nick (2014). Unwritten rules: virtual bargaining underpins social interaction, culture, and society. *Trends in cognitive sciences*.

Mnih, Volodymyr, Kavukcuoglu, Koray, Silver, David, Graves, Alex, Antonoglou, Ioannis, Wierstra, Daan, and Riedmiller, Martin (2013). Playing atari with deep reinforcement learning. *arXiv preprint arXiv:1312.5602*.

Nagel, Thomas (1986). *The view from nowhere*. Oxford University Press.

Nakahashi, Ryo, Baker, Chris L, and Tenenbaum, Joshua B (2016). Modeling human understanding of complex intentional action with a bayesian nonparametric subgoal model. In *AAAI*, pp. 3754–3760.

Ramırez, Miquel and Geffner, Hector (2011). Goal recognition over pomdps: Inferring the intention of a pomdp agent. In *IJCAI*, pp. 2009–2014. IJCAI/AAAI.

Shum, Michael, Kleiman-Weiner, Max, Littman, Michael L, and Tenenbaum, Joshua B (2019). Theory of minds: Understanding behavior in groups through inverse planning. In *Proceedings of the Thirty-Third AAAI Conference on Artificial Intelligence (AAAI-19)*.

Song, Yuhang, Wang, Jianyi, Lukasiewicz, Thomas, Xu, Zhenghua, and Xu, Mai (2019). Diversity-driven extensible hierarchical reinforcement learning. In *Proceedings of the AAAI Conference on Artificial Intelligence*, Volume 33, pp. 4992–4999.

Stone, Peter, Kaminka, Gal A, Kraus, Sarit, and Rosenschein, Jeffrey S (2010). Ad hoc autonomous agent teams: Collaboration without pre-coordination. In *Twenty-Fourth AAAI Conference on Artificial Intelligence*.

Sugden, Robert (2003). The logic of team reasoning. *Philosophical explorations*, **6**(3), 165–181.

Tambe, Milind (1997). Towards flexible teamwork. *Journal of artificial intelligence research*, 7, 83–124.

Tomasello, Michael (2014). *A natural history of human thinking*. Harvard University Press.

Tomasello, Michael, Carpenter, Malinda, Call, Josep, Behne, Tanya, and Moll, Henrike (2005). Understanding and sharing intentions: The origins of cultural cognition. *Behavioral and brain sciences*, **28**(05), 675–691.

Watkins, Christopher JCH and Dayan, Peter (1992). Q-learning. *Machine learning*, **8**(3-4), 279–292.

Wright, James R and Leyton-Brown, Kevin (2010). Beyond equilibrium: Predicting human behavior in normal-form games. In *AAAI*.

Young, H Peyton (1993). The evolution of conventions. *Econometrica: Journal of the Econometric Society*, 57–84.

9

Teaching and Explanation: Aligning Priors between Machines and Humans

José Hernández-Orallo and Cèsar Ferri

Universitat Politècnica de València, Spain

9.1 Introduction

'*Intelligenti pauca*', says the somewhat self-referential abbreviation of the Latin proverb, '*intelligenti pauca sufficiunt*'. Along with many other formulations, from Plautus's *Persa* (Plautus, BCE) to Chomsky's *Poverty of the Stimulus* (Chomsky, 1992), it reflects the phenomenon that the intelligent do not need much. The power of efficiently exploring many possible explanations to account for an observation is key to finding the right one, but as more explanations are considered, the relevance of priors, to select among the explanations, becomes more important.

From this perspective, data-hungry machine learning algorithms should not be considered very intelligent, even if they are able to find patterns and make good predictions in situations where humans are clearly overwhelmed. Of course, there is a positive side of this story: humans and machines are complementary nowadays. Humans are good at learning from very few examples, using context and priors very advantageously. Machines, on the other hand, are good at learning from huge amounts of raw data, without the need for background knowledge.

We expect this to change in the future, especially as renewed efforts to make machine learning more human-like crystallize. One important approach in this direction is known as *machine teaching* (Zhu, 2015). Instead of expecting that a machine learns a concept—or pattern—from a set of examples randomly sampled from a distribution, the examples are chosen wisely by a teacher. This more closely resembles the way humans learn, as social—and cultural—animals. It is also related to the way humans communicate, finding the right signs or words such that the receiver understands the meaning of the message (Goodman and Frank, 2016). This duality between teacher and learner (sender and receiver in communication) is illuminating about the way humans interact and learn.

José Hernández-Orallo and Cèsar Ferri, *Teaching and Explanation: Aligning Priors between Machines and Humans* In: *Human Like Machine Intelligence.* Edited by: Stephen Muggleton and Nick Charter, Oxford University Press.
 DOI: 10.1093/oso/9780198862536.003.0009

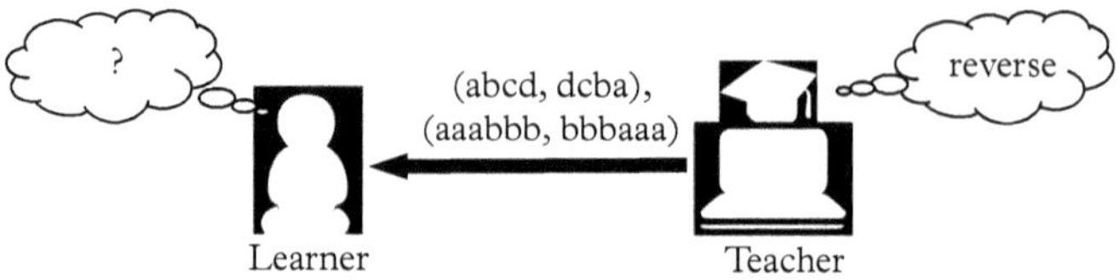

Figure 9.1 *Machine-teaching procedure. The teacher wants to teach the concept of the* reverse *of a string. The teacher chooses two examples carefully and shows them to the learner: the input string* abcd *being mapped into* dcba, *and the input string* aaabbb *being mapped into* bbbaaa. *The learner must figure out the concept from only these two examples. In this chapter we assume a unidirectional batch teaching session.*

Figure 9.1 shows a situation where the teacher has the concept of 'reverse'. The teacher could try to transmit the concept, but the languages used by the learner and teacher might be different. Instead, as happens with human learners very often, a few examples may be more effective. In the image, the teacher sends a couple of input-output pairs to the learner, thinking that this would be sufficient for the learner to build and identify the concept.

The field of machine teaching has usually placed machines as learners and humans (or other machines) as teachers. In this setting, there are interesting connections with areas such as learning from demonstrations (Schaal, 1997; Argall *et al.*, 2009), programming by example (Gulwani, 2016), curriculum learning (Bengio *et al.*, 2009), active learning (Winston and Horn, 1975), query (Angluin, 1988), inductive logic programming (Muggleton, 1991), and, more generally, inductive programming (Gulwani *et al.*, 2015).

It is only very recently that attention has been directed to the case where humans are learners and machines are teachers. There seems to be some reluctance to call this *machine teaching* as well, but it is even more natural than in the original setting. We now deal with *machines that teach* in the same way that we employ *machine learning* to refer to the area dealing with *machines that learn.*[1] In addition, this perspective of machine teaching is closely connected with an increasingly important area: explainable artificial intelligence (XAI). XAI, as covered in other chapters in this volume, deals with making machine behaviour understandable to humans. In particular, when machine-learning models are to be explained, this can be done with examples, known as exemplar-based explanations (Molnar, 2019). Then, the problem of choosing these examples becomes extremely similar to the problem of choosing examples in machine teaching. In both cases, the question the teacher (a machine) has to solve is: what are the best examples so that the learner (a human) can identify the concept? A key element in machine teaching and exemplar-based explanations is that the explanation (a concept, a model, or a proxy for it) is not given, but the learner must build it from the examples. And in this game,

[1] We would welcome a different term such as 'computational' or 'theoretical' teaching to cover humans teaching machines, machines teaching humans, and machines teaching machines, and use more specialized terms for each of these three situations. In this chapter, we will use the term 'machine teaching' for the three of them, in order to keep the connection and terminology with the existing literature.

the realization of the priors that both teacher and learner have is critical for this process to be efficient and successful.

In recent work, Telle *et al.* (2019) and Hernández-Orallo and Telle (2020) consider two priors in machine teaching: the sampling prior and the learning prior. The sampling prior is the probability that each concept is to be taught, that is it reflects the expectation of what concepts the teacher is going to teach. The learning prior is the expectation the learner has about the concepts being taught. In both cases, they represent the probability of each concept a priori. Ideally, these two priors should be aligned to ensure that both parts identify the same concept when shown the same evidence. For instance, in a mathematics class, if a student sees numbers $\{2, 7, 17\}$ she will expect the prime numbers to be the more likely concept than the numbers on the jerseys of a group of selected football players.

Even for universal Turing-complete languages, the use of strong priors can make teaching very efficient in expectation, in terms of the number of examples (Hernández-Orallo and Telle, 2020) or their size (Telle *et al.*, 2019). A natural choice for both priors is simplicity, formally defined in terms of the size of the programs in the concept language. Finally, there may be other priors, especially in a Bayesian setting, such as the example prior, which is the expectation of examples, unconditional or conditional (likelihood) given the concept.[2] For instance, a learner would expect number 7 to be more likely as an exemplar of prime numbers than the prime number 15,485,863. Interestingly, if the teaching process aims at minimizing the information that is exchanged for the sake of efficiency, this actually imposes a simplicity example prior.

When the machine-teaching setting is applied to human learners, we rarely have a precise articulation of their priors. Even if we assume simplicity as a prior for teacher and learner, the representational languages and coding may differ and the machine-teaching process may be affected. We can only partly rely on the invariance theorem (Solomonoff, 1964), which states that the difference in sizes of the shortest descriptions of an object between two (universal) representations is bounded by a constant that only depends on the two representations. However, this constant may be huge, so it may be more relevant to know whether the 'order' or 'preference' of objects given by the prior is similar or not. Also, in this teaching scenario we have a double simplicity prior (on examples and programs): the teacher sends the first set of examples—according to their *size*—such that the learner identifies the concept using an enumeration—ordered by the *size* of the programs. If alignment is perfect (the priors are the same between teacher and learner), teaching is efficient in terms of the length of the *teaching message*, and the learner will look for the shortest representation of the concept at hand. However, even if the alignment is not perfect, the schema can still work as most preference orders in the priors may be preserved. This presumed *teaching invariance* is one of the questions we will explore in this chapter, because effective teaching and communication depends on the teacher

[2] If teacher and learner can update their beliefs about each other using Bayes rule, the situation resembles a Bayesian game. In this chapter, we leave belief update for future work and assume that learner and teacher do not modify their priors during the teaching (batch) session.

considering some imperfection in the alignment of priors between teacher and learner (because of the different representations). Ultimately, the better the teacher knows the priors the learner is using, the more information-efficient teaching and communication can be. The more robust the desired exchange, the more redundancy is needed.

The rest of the chapter is organized as follows. Section 9.2 gives a quick introduction to machine teaching and the key concepts of teaching dimension and teaching size. Section 9.3 discusses related work and connections between exemplar-based explanations and machine teaching. Section 9.4 introduces our setting, a new teaching algorithm with exceptions, and the explanation protocol. Section 9.5 shows how this is applied to universal languages, and perform simple experiments to analyse the alignment of priors with humans. Section 9.6 instantiates the setting to feature-value languages and also uses two simple examples to investigate the alignment between humans and machines. Finally, section 9.7 analyses the way this setting should be extended in the future to allow for differences between representations used by learner and teacher.

9.2 Teaching Size: Learner and Teacher Algorithms

Machine teaching is an area dating back several decades (Goldman and Kearns, 1995; Zhao *et al.*, 2011), which has recently become more popular in artificial intelligence (Zhu, 2015; Zhu *et al.*, 2018; Telle *et al.*, 2019). Machine teaching is sometimes considered as an inverse problem to machine learning, but it is better characterized as one edge in a space of learning settings where teacher and learner may or may not control the examples to be provided. In batch and incremental learning, examples are randomly sampled from a distribution, and the learner has no control over them. There is no teacher to help either. In active learning, the examples can be selected by the learner to refine its hypotheses accordingly. The teacher is passive, and just labels examples, at the learner's request. In query learning, the learner can even ask the teacher (usually called the 'oracle') about the validity of statements or (partial) concepts. Finally, in machine teaching, the teacher is responsible for the selection of examples, either as a batch (traditional machine teaching setting), incrementally (also known as curriculum learning), or interactively (such as Bayesian teaching). We will focus on the batch situation for the rest of the paper.

Most research in machine teaching has been based on a simple, abstract idea: the *teaching dimension* (TD), for a concept, defined as the minimum number of examples that are required so that the learner identifies the concept. When extended to a whole concept class, we refer to the—maximum or expected—teaching dimension of a concept class. The teaching dimension has been used to establish important connections with complexity notions in computational learning theory, such as the Vapnik-Chervonenkis (VC) dimension or Probably Approximately Correct (PAC) learning (Chen *et al.*, 2016). Recently, however, the model has been extended in different ways, including preferences or priors (Balbach, 2008; Gao *et al.*, 2017; Hernández-Orallo and Telle, 2020). In all these cases, the teacher has a perfect model of the learner. More recent, and complex, models of teaching consider that the teacher does not know the behaviour of the learner;

referred to as 'black-box learner' (Dasgupta *et al.*, 2019). We will relate to this at the end of the chapter.

Let us now introduce some notation to define some of these notions properly. We adapt the notation in Telle *et al.* (2019) for our purposes, including the definitions of teaching dimension, teaching size, and the learner and teacher algorithms. We consider a possibly infinite example (or instance) space X. From now on, we will consider examples as pairs $\langle i,o \rangle$ and concepts as functions mapping inputs to outputs. We consider a possibly infinite concept class C. An example set $S = \{\langle i_1, o_1 \rangle, ..., \langle i_k, o_k \rangle\}$ is just a finite set of i/o pairs, used as *witness* for the teaching process. Given a concept $c \in C$ and an example set $S \subseteq X$, we say that c satisfies S, denoted by $c \models S$, if c is consistent with all the examples in S. In the functional presentation of examples, we can also express that c satisfies S if $c(i) = o$ for all the pairs $\langle i,o \rangle$ in S. All concepts satisfy the empty set. Given a language L, the same concept can be represented by one or more programs in L. We say that a program $p \in L$ satisfies the example set S, denoted by $p \models S$, if it has the i/o-behaviour specified by all i/o pairs in S. The equivalence class of programs compatible with concept c is $Class_L(c) = \{p : \forall S, p \models S \iff c \models S\}$.

For any concept $c \in C$, the goal of the teacher is to find a *small* witness set $w \subseteq X$ such that the learner uniquely identifies the concept from it. The classical notion of teaching dimension for a concept c is simply the cardinality of the smallest set (in number of examples) that allows the learner to unequivocally identify c (Zhu, 2015). While this is appropriate in situations where all examples have the same size (e.g., fixed dimension vectors, such as Boolean functions), in other concept classes and example domains, some examples may be larger than others, as in Figure 9.1. Actually, as shown in Telle *et al.* (2019), there exist concepts that may be taught with a single example, but this example may be arbitrarily large.

Accordingly, we generalize the notion of teaching dimension by choosing an encoding function δ so that $\delta(S)$ is the number of bits needed to encode the example set S. This function can be derived from some probability distribution on examples (using fewer bits for frequent examples) or it can be just defined in terms of the size of the examples. In either case, we will refer to $\delta(S)$ as the size of S.

9.2.1 Uniform-prior teaching size

With the above definition of size, we can now define the teaching size (TS) of a concept c as follows:

$$TS(c) = \min_S \{\delta(S) : \{c\} = \{c' \in C : c' \models S\}\}$$

If the minimum does not exist, the teaching size is infinite. This is the size of the smallest example set S (in terms of the δ encoding) that allows a learner to uniquely identify c. This is expressed by the set of compatible concepts being a singleton $\{c\}$, only containing c. The set S is known as a witness set for the concept c and the learner can use it to infer a program in $Class_L(c)$. The teaching size is a generalization of the teaching dimension, TD, where $\delta(S) = |S|$.

Let us see a few examples, one with Boolean formulas, as usual in the machine-teaching literature, one where we see why the teaching size is important (taken from Dasgupta *et al.* 2019) and one where teaching needs preferences or priors.

Example 9.1 Consider the concept space of all Boolean functions of m variables. Clearly, there are 2^m combinations of such variables, and 2^{2^m} assignments to those combinations, hence concepts in that space. In this case, we need 2^m examples to identify univocally the right concept, as we have to tell all the assignments in order to rule out all alternative concepts, because no preference or prior is assumed. As a result, the $TD = 2^m$. With a simple δ calculated as the number of bits for each example ($m+1$ to account for the variables and the class), the teaching size would be $TS = (m+1)2^m$. Note that both the teaching dimension and the teaching size depend on the number of variables m.

Example 9.2 Consider the concept space composed of all threshold concepts c_t over real numbers $x \in T \subseteq \mathbb{R}$ such that $c_t(x) = 1$ if $x \geq t$ and 0 otherwise. For finite T, we only have $|T|$ concepts.[3] The two single points closest to the threshold are enough to identify one unique concept consistent and rule all the rest. Namely, $S = \{x_0, t\}$, with x_0 being the rightmost nearest value $x_0 < t$ in T, which will necessarily make $c_t(x) = 0$, and t itself, which will necessarily make $c_t(t) = 1$. In this case the teaching dimension is 2, independently of how large X is, provided X is finite. We see that even if T grows larger and larger, the teaching dimension remains constant. This is somewhat counterintuitive. If a learner has to guess a threshold among a million, this should take more effort than guessing a threshold among a dozen. The teaching *size* gives an answer to this problem. By using a simple coding δ for the examples, indexing T, we would have that the cost of referring to an element in T would be simply given by $\forall x : \delta(x) = \log_2 |T|$. This is just the bits[4] that are needed to determine which one has been chosen from T. With this, the teaching size would just be $TS(c_t) = 2\log_2 |T|$ for every concept c_t, the coding cost of the two examples using δ.

Example 9.3 Consider the following setting for regular expressions. Let X be composed of all pairs $\langle s, y \rangle$, where s is a string formed with the alphabet $\Sigma = \{\epsilon, a, b\}$ and $y \in [0, 1]$. Let L be the language composed by the regular expressions that can be formed by Σ and the operators $\{(, *, |,)\}$. As the alphabet is binary and the classes are too, let $\delta(\langle s, y \rangle)$ be just[5] defined as the length of string s plus one for coding 0 or 1. Consider we want to teach the concept $a*$ and we employ $S = \{\langle aa, 1 \rangle, \langle b, 0 \rangle\}$. Given that the concepts $a*$ and aa are consistent with S, a learner could never learn the target concept. Similarly, if we add ab to the set, we can still not distinguish $a*$ from $aa|ab$. In the end, as there are infinitely many regular expressions consistent with any finite number of examples, given

[3] Dasgupta *et al.* (2019) derives $|T| + 1$ as the number of concepts but here we consider the set of possible thresholds and examples to be the same set T.

[4] For simplicity of the argument, we just assume here that T is a power of 2.

[5] A more appropriate encoding should use delimiters or self-delimiting codes between examples.

the uniform-prior assumption, $TD(a*) = \infty$ and $TS(a*) = \infty$, and so it is for any other concept.

Note that things become even worse as the concept class becomes richer, especially when Turing-complete languages are considered, as the example in Figure 9.1. With a uniform prior (where there is no way to distinguish between all consistent hypotheses), both the teaching dimension and size would be infinite for most rich languages. Even if this is the traditional setting for the teaching dimension, these negative results have suggested that the model should be changed or extended. One option is to assume that we still do not know the learner preferences in the beginning but try to adapt to them (Dasgupta *et al.*, 2019), and another option is to assume that the teacher knows the learner perfectly, and the preferences it has. This is simply introduced as priors, leading to models such as the preference-based teaching dimension (Balbach, 2008; Gao *et al.*, 2017). We can generalize all this by the use of priors on concepts, or more precisely on programs. In the following subsection, we see how we can apply a simplicity prior to get a more powerful notion of teaching size.

In these approaches, we assume that the teacher has a model of the learner. So it is useful to make the learner model explicit. We can understand the learner as a *partial function* Φ mapping sets to programs: $\Phi(S) = p$. Consequently, we can express the teaching size as, $TS(c) = \min_S \{\delta(S) : \Phi(S) = p \in Class_L(c)\}$. In other words, the teaching size of a concept is the minimal set (in terms of δ), such that the learner recovers a program expressing the desired concept. Note that if more than one concept is consistent with a set S, and the learner has no preference, then it will not return any of them. This happens especially when the learner's concept prior (learning prior) is uniform so that the learner cannot tell between two equally consistent concepts.

9.2.2 Simplicity-prior teaching size

Consider a language L as a set of strings over an instruction alphabet representing programs. For each p in L, its length in bits is denoted by $\ell(p)$, using some appropriate encoding for programs. Let $\prec$ be a total order on programs, ordered by ℓ; thus shorter programs precede longer programs, and for programs with equal value of ℓ, $p_1 \prec p_2$ if p_1 goes before p_2 lexicographically (programs are just strings).

We can now define the learner as returning the first program for an example set S according to ℓ and $\prec$ as follows:

Definition 9.1 *The ℓ-optimal learner algorithm takes a set of examples S and outputs the simplest program (according to ℓ) that is compatible with S:*

$$\Phi_\ell(S) = \underset{p}{\operatorname{argmin}}^{\prec} \{\ell(p) : p \models S\}$$

This is our **definition of the ℓ-optimal learner algorithm**. The Kolmogorov complexity of an example set is simply given by $K(S) = \ell(\Phi_\ell(S))$, the length of the shortest program having the i/o-behaviour specified by S.

Finally, the teaching size of a concept c based on ℓ is defined as:

$$TS_\ell(c) = \min_S\{\delta(S) : \Phi_\ell(S) \in Class_L(c)\}$$

In plain words, the teaching size for a concept c using the order given by ℓ and $\prec$ is the size of the smallest witness set S, under δ encoding, such that the first program in the order that satisfies S is in the equivalence class of c:

$$TS_\ell(c) = \min_S\{\delta(S) : p \models S \wedge (p' \prec p \Rightarrow p' \not\models S) \wedge p \in Class_L(c)\}$$

The learner $\Phi(S)$ can be thought as an enumeration procedure that, given S, tries all programs in increasing size-lexicographic order, that is, following $\prec$, until the first program that is compatible with S is found.

Example 9.4 Consider the setting of Example 9.3, where L was the language of regular examples. We can assume $\prec$ is defined by the number of symbols in the regular expression, and in case of same length, the lexicographical order given by this symbol preference $* \prec a \prec b \prec (\prec) \prec |$. Now, when given $S = \{\langle aa, 1\rangle, \langle b, 0\rangle\}$, the following concepts are consistent with S: $a*$, $*a$, aa, as infinitely many longer concepts. In this case $\Phi_\ell(S) = *a$ because of the symbol preference. As a result, since we cannot find a witness set that works with one example or with shorter examples, we have that $TD(*a) = 2$ and $TS(*a) = 5$. Note that the larger the expression of a concept is, the larger the teaching dimension and the teaching size are expected to be. Consider concept $*abbbaababaaa$ for instance.

Finally, we can think of a teacher as filling a Teaching Book, a list of *entries* in the form of pairs of program and witness $\langle w, p\rangle$, with p the smallest program compatible with w, and w the smallest witness for which p is compatible. The teacher could use a nested loop to fill this book, generating example sets in some order of non-decreasing size, breaking ties in a deterministic manner (using an order, denoted by $\triangleleft$), and, for each example set, running the learner algorithm $\Phi(S)$.

Sometimes the teacher does not want or need to build a book, especially when looking for only one witness set for a particular concept, but the effort is very similar, as the teacher still needs to complete a nested loop over witness sets and programs until it certifies that the desired witness set does not lead to any other program, which means building the book until it finds the program. In this case, the procedure is still a nested loop using the learner algorithm, stopping the first time a p-equivalent program is found.

Definition 9.2 *The (ℓ, δ)-optimal teacher algorithm, $\Omega_{\ell,\delta}$, takes a concept c and returns a witness set. It works as follows:*

```
for all witness sets w in increasing δ(w) and ordered by ◁ for equal size do
    if Φℓ(w) is equivalent to c then
        return w
    end if
end for
```

Example 9.5 Following with Example 9.4, we would start with witness sets with $\delta = 0$, that is the empty sample set $S_0 = \emptyset$, whose $\Phi_\ell(S_0) = *$ because of the symbol preference. Then, there are no syntactically correct witness sets with $\delta = 1$. Then, we would try witness sets with $\delta = 2$, leading to four sets with the shortest example, $S_1 = \{\langle a, 0\rangle\}$, $S_2 = \{\langle a, 1\rangle\}$, $S_3 = \{\langle b, 0\rangle\}$ and $S_4 = \{\langle b, 1\rangle\}$, with $\Phi_\ell(S_1) = b$, $\Phi_\ell(S_2) = *$, $\Phi_\ell(S_0) = a$, $\Phi_\ell(S_0) = *$. Note that some concepts are repeated ($*$), and would not be added to the teaching book, as they are already there. The next iteration would work with witness sets with $\delta = 3$, meaning that they still have to contain one example but they can use strings of size 2. Then, with witness sets with $\delta = 4$, we could either work with one example with strings of size 3, or with sets with two examples with strings of size 1. This process would continue until we find a witness set that gives the desired regular expression (concept).

This formulation makes it explicit that in this teaching model the teacher uses a (perfect) model of the learner in order to think what witness set to provide.

9.3 Teaching and Explanations

Explainable AI (XAI) is a term used to capture a series of domains and approaches aiming at explaining the decisions and actions of AI systems (Swartout *et al.*, 1991). Machine learning is an important type and component of AI systems, and therefore XAI focuses ons explaining machine-learning models, covering supervised and unsupervised learning, and reinforcement learning (Hernández-Orallo, 2019; Samek and Müller, 2019).

Explainable AI must usually face several tradeoffs, such as the tension between fidelity[6] and comprehensibility (Guidotti *et al.*, 2018) or between these two elements and actionability (Bella *et al.*, 2011). In general, being able to make successful explanations among these tensions requires a great deal of abstraction, ultimately modelling machine behaviour (Rahwan *et al.*, 2019) in a way that is comprehensible to humans, possibly using machine learning itself (Fabra-Boluda *et al.*, 2017).

There are two main families of XAI approaches. In one, the goal is to extract an abstract representation of what the AI system is doing and provide this as an explanation to a human, for example extracting comprehensible rules from models (Domingos, 1998). This is what we would call a model-based explanation. In the other, the goal is to give examples such that humans can build their explanation themselves, known as exemplar-based explanations, for example using partial anchors (Ribeiro *et al.*, 2018). We will therefore explore the connection between connection between (exemplar-based) machine teaching and exemplar-based explanations. But first, it is important to understand what interpretability is when dealing with humans as receivers of explanations.

[6] Level of coincidence between the extracted comprehensible model (surregated model) and the original black-box model.

9.3.1 Interpretability

Interpretability in the context of explainable AI typically refers to the degree in which some machine-generated representation can be understood by a human. It is no surprise that this has much to do with psychology, neuroscience, and many other disciplines dealing with humans. We will not summarize the vast literature here (see, e.g., Miller 2019), but we will highlight two important factors in human interpretability: simplicity bias and confirmation bias. Simplicity has been recognized as a fundamental cognitive principle (Chater and Vitányi, 2003), but we have to consider that simplicity must always be understood in terms of a representational language and background knowledge; what is simple for a given person (because it can be expressed rather concisely and easily given their previous knowledge) may be complicated for another person. Similarly, confirmation bias determines how easily supporting or contradicting facts are assimilated. What matters here is that the biases, or priors, are so relevant for interpretability that it is astonishing that some techniques in AI could even consider generating one optimal explanation for all, such as generating the same set of rules from a black box. This would only work if the users, humans, share the same priors.

Still, because humans share genes and culture to some degree, some representations have been shown to be better than others for explanations. For instance, as humans are usually educated in number arithmetic and (basic) logic, explanations based on simple equations or rules are usually deemed more interpretable than other representations, but not all variants are equally good (Lakkaraju *et al.*, 2016). For instance, Ribeiro *et al.* (2018) claim that 'compared to other interpretable options, rules fare well; users prefer, trust and understand rules better than alternatives (Lim *et al.*, 2009), in particular rules similar to anchors'. We will revisit 'anchors' later on.

In any case, interpretable models can directly be learned in the first place, or a black-model could be interrogated to extract examples from which an interpretable model is built (Domingos, 1998). Many rule learning methods argue that a trade-off must be expressed in terms of loss plus a regularization term (Angelino *et al.*, 2017) or, more directly, in terms of the original MDL principle (Aoga *et al.*, 2018; Proença and van Leeuwen, 2020).

Simplicity as a regularization term does not necessarily imply interpretability. Samek and Müller (2019, section 1.5) focus mostly on quality evaluations that are not performed with human intervention, called 'objective' or 'automatic', such as a measure of simplicity. On the other hand, 'subjective' or 'human' evaluations of explanations usually ask humans to assess the interpretability of explanations or patterns (Ribeiro *et al.*, 2016; Doshi-Velez and Kim, 2017; Nguyen, 2018; Lage *et al.*, 2019). In this chapter we see simplicity as a mental prior to align the hypothesis space.

9.3.2 Exemplar-based explanation

A completely different approach to XAI considers that humans may have very different internal knowledge representations, and a one-size-fits-all expression of an explanation would not work. Consequently, instead of *giving* the explanation, exemplar-based explanation techniques provide the elements—usually in the form of facts, that is examples—such that humans can build their own explanations.

But this does not mean that representations are irrelevant for exemplar-based explanations; examples must also be represented in some way. Molnar (2019) puts it clearly: 'Exemplar-based explanations only make sense if we can represent an instance of the data in a humanly understandable way. This works well for images, because we can view them directly... It is more challenging to represent tabular data in a meaningful way, because an instance can consist of hundreds or thousands of (less structured) features... It works well if there are only a handful of features or if we have a way to summarize an instance.'

In this chapter, we actually explore this possibility of *summarizing an instance*. We will see that even for feature-value examples, we can find partial representations, also known as anchors, which may vary in *size*. It is not the same to express an example with all instantiated variables (e.g., $X_1 = v_1, X_2 = v_2, ..., X_m = v_m$) than by just using a small subset of them (e.g., $X_3 = v_3, X_7 = v_7$).

In the XAI literature there are different kinds of exemplars that can be key for understanding a concept (Molnar, 2019):

- Foils: a foil or counterfactual[7] is a full example e' derived as a small perturbation of a given example e such that e and e' have different outputs.
- Anchors: an anchor is a *partial* example for which some features are fixed, and all the others are not specified, as they are very unlikely to change the output produced by the fixed values.
- Prototypes: a prototype is a full example that is central or representative for a region or large group of examples.
- Criticisms: a criticism is a full instance that is not well represented by the prototypes (Kim *et al.*, 2016), but it is not necessary an error or an exception. They are sometimes related to outliers.

It is insightful to consider the duality between those examples that work as positive exemplars (what the model does), such as anchors and prototypes, and those that work as negative exemplars (what the model does not). This reminds us of a common duality between general patterns and exceptions. Sometimes, it is easier to explain something with a general rule without noise or outliers, plus a separate list of anomalies. This is behind coding and learning schemes such as the minimum message length (MML) (Wallace and Dowe, 1999) and minimum description length (MDL) (Wallace and Boulton, 1968; Wallace and Dowe, 1999; Rissanen, 1983). Of course, this has not been ignored by the literature of explainable AI (e.g., Proença and van Leeuwen, 2020).

In this chapter, we aim at finding the examples in a different, more principled way. We will build on representations that are similar to anchors, and exceptions that are similar

[7] Lipton (1990) uses the term foil instead of counterfactual, as we are actually talking about a change of the outcome, not a situation where the cause did not happen. Some recent sources are also favouring foil over counterfactual (see, e.g., Google, 2020).

to criticisms, but we identify these examples using a machine-teaching setting based on the teaching size. Before that, we need to briefly review the relation between machine teaching and explainable AI.

9.3.3 Machine teaching for explanations

Machine teaching has been used, in the first place, to understand or model how humans teach. For instance, Khan *et al.* (2011) analyse how humans teach one-dimensional concepts (intervals) to machines, comparing a machine-teaching setting with a curriculum learning setting. The question in both cases is whether humans give the examples at the boundaries to help the learner recreate these boundaries or give those examples in clear areas so that the user can interpolate (Basu and Christensen, 2013).

However, here we are interested in using machine teaching for explaining concepts to humans. Some models are interactive, where the teacher can ask queries of the learner (e.g., Liu *et al.* 2017). A few approaches, however, have tried to bring or extend the machine-teaching setting using examples to explainable AI. Some proposals deviate significantly from the primitive machine-teaching (and teaching dimension) formulation, such as using carefully chosen demonstrations in inverse reinforcement learning (Ho *et al.*, 2016), or in a cooperative framework (Hadfield-Menell *et al.*, 2016).

Yang and Shafto (2017) use a Bayesian approach where teacher and learner interact and converge on the likelihood of the data, given the model on the teacher's side, and the posterior of the model, given the data on the learner's side. In the traditional machine-teaching setting, we assume that the teacher works in a batch mode and sends a witness set that is optimized for some kind of efficiency (number or size of examples), not the one that maximizes the probability of identification only. Other variations of the teaching paradigm for XAI include the decomposition of the learner's hypothesis into an attention function and a decision function (Chen *et al.*, 2018).

A Bayesian inference algorithm of regular expressions from examples is presented by Ouyang (2018). The proposed teaching paradigm is also related to how humans communicate and how the speaker selects the appropriate word according to the listener. Given a concept, how do speakers choose from the myriad of referring expressions at their disposal? This *referring* process is analysed by Degen *et al.* (2019) considering speakers as agents that rationally trade off cost and informativeness of sentences. This work is based on the the rational speech act (RSA) framework for pragmatic reasoning. The RSA model provides a social cognition approach to utterance understanding based on the considering the speaker as a utility-maximizing agent and then the listener estimates his beliefs via Bayesian inference (Goodman and Frank, 2016).

9.4 Teaching with Exceptions

We are now ready to adapt the teaching size setting introduced in section 9.2 for the purpose of explanation in a more realistic situation considering exceptions, as described above. We start from a concept c that teacher A has learned or programmed into a language L_A, represented by a program p_A. It is this p_A that A wants to explain. For instance, p_A could be a program in Java or Python that implements the actions of a robot according to its observations, or it may be a classifier implemented with a neural network. For our purposes the teacher assumes a learner B with another language L_B, (e.g., decision lists, Python, or $P3$, as used in Telle *et al.*, 2019) trying to find a program p_B in L_B that is equivalent to p_A, and hence c. The teacher wants to achieve this by sending a small witness example set w to B.

However, the problem is that we do not have an easy (sometimes not even decidable) mechanism to determine whether p_A and p_B are equivalent. Accordingly, what we are going to do is to determine equivalence modulo a set of examples D (usually much larger than the witness set w). We denote this equivalence as $p_A \equiv_D p_B$, meaning that for every $\langle i, o \rangle$ in D, $p_A(i) = p_B(i)$ or, similarly, $p_A \models D$ and $p_B \models D$.

In a general situation, because of the presence of noise when building p_A or simply because A and B are very different, some examples in D are left as exceptions so that the whole can be coded more compactly. Also, we want to ensure that a witness set S leads to the right concept covering D and not merely the given examples in the witness set. This leads us to an MML/MDL variation of our teaching setting, as follows.

We extend the definition of δ to code the exceptions by just adding the index of the example in D. If D has n examples and E is the set of the $m \leq n$ exceptions in D, then the cost of coding the m exceptions is exactly $m \log_2 n = |E| \log_2 |D|$.

Our learner is the same as in Definition 9.1, but the teacher has to be modified:

Definition 9.3 *The (ℓ, δ)-optimal teacher algorithm considering exceptions (strategy 1), $\Omega_{\ell,\delta}$, takes a set of examples D and returns a witness set and the exceptions to D:*

for all $\langle w, E \rangle$ *in increasing* $\delta(\langle w, E \rangle)$ *and ordered by* $\triangleleft$ *for equal size* ***do***
  $p \leftarrow \Phi_\ell(w)$
  if $p \models D \setminus E$ ***then***
    return $\langle w, E \rangle$
  end if
end for

The previous definition of a teacher may suggest that we need to look through the combined enumeration of witness sets w and exception sets E. However, for some w, the learner will identify programs p for which some of the examples in E are not really exceptions for p. In other words, not all pairs $\langle w, E \rangle$ really make sense. An alternative strategy is as follows:

Definition 9.4 *The (ℓ, δ)-optimal teacher algorithm considering exceptions (strategy 2), $\Omega_{\ell,\delta}$, takes a set of examples D and returns a witness set and the exceptions to D:*

  $s_b \leftarrow \infty, w_b \leftarrow \emptyset, E_b \leftarrow \emptyset$
for all w *in increasing* $\delta(w)$ *and ordered by* $\triangleleft$ *for equal size* ***do***
  if $\delta(w) > s_b$ ***then***
    return $\langle w_b, E_b \rangle$
  end if
  $p \leftarrow \Phi_\ell(w)$

$E \leftarrow \{\langle i,o\rangle \in D : p(i) \neq o\}$
$s \leftarrow \delta(\langle w,E\rangle) = \delta(w) + |E|\log_2|D|$
if $s < s_b$ ***then***
  $s_b \leftarrow s, w_b \leftarrow w, E_b \leftarrow E$
else if $s = s_b$ ***and*** $\langle w_b, E_b\rangle \lhd \langle w,E\rangle$ ***then***
  $s_b \leftarrow s, w_b \leftarrow w, E_b \leftarrow E$
end if
end for

Proposition 9.1 *The algorithm in Definition 9.4 (strategy 2) is equivalent to the algorithm in Definition 9.3 (strategy 1).*

Proof: If there is a pair $\langle w,E\rangle$ found by strategy 1, then it is the shortest pair, of size s. Then strategy 2 goes on exploring pairs until the size of w exceeds s_b, the size of the best solution so far. If that is the case no shorter solution could be found (as exceptions take space coding), and then the solution that is found must be the right one. Note that the best solution for strategy 2 is always the shortest one, and in case of ties the first one in lexicographical order is kept, as strategy 1 does. So we see that both strategies find the first pair in lexicographical order. □

The advantage of strategy 2 is that it does not have to check the redundancy or even the consistency of the exceptions as it just enumerates on the witness sets. The disadvantage of strategy 2 is that it explores longer witness sets than necessary. However, we can adapt some existing (approximate) algorithms for strategy 2, which would be more difficult to adapt for strategy 1. This is important because, as we will see, both strategies are intractable in general.

In principle, we do not impose any condition that $w \subset D$, but we will find reasons for connecting these two sets, also by efficiency. For instance, in the feature-value case, we will consider that the values are taken from D, especially when there is an infinite number of possible values (e.g., real numbers), to avoid making the search infinite and the size coding of exceptions ill-fitted for this. In the following sections, we will explore some of these options depending on the kinds of concepts to be explained.

Given all this, we are now ready to use the witness set computed by the algorithm $\Omega_{\ell,\delta}(D)$ in Definition 9.3 (or 9.4) for explanation, as per the following protocol:

Protocol 9.5 (Explanation) *Given some labelled data D, we explain how labels were generated by an unknown algorithm as follows:*

1. $\langle w,E\rangle \leftarrow \Omega_{\ell,\delta}(D)$ *using the algorithm in Definition 9.3 (or 9.4)*
2. *We show w (but not E) to the human.*
3. *We ask the human to think of a concept and use it to label* $D \setminus E$.
4. *(Optionally) We can check how well the human scores for* $D \setminus E$ *to see whether the human has understood the concept.*
5. *(Optionally) We show E to the human as exceptions.*

Note that we expect D to be much larger than w. This means that w must consist of very significant examples that allow the learner to generalize and cover much of D. The use of exceptions allows for some slack, especially in cases where there are some patterns only affecting a small area or number of examples, and of course to deal with the presence of noise. We can consider weights for the two components of $\delta(\langle w, E\rangle)$. They could easily be introduced as $\alpha\delta(w) + (1-\alpha)|E|\log_2|D|$, with $\alpha \in [0,1]$. For a range of values of α, we could account for situations where more or less noise is expected. If α is near 0 that means that exceptions are strongly penalized against the witness set, and only (nearly) perfect solutions would be considered. For values of α closer to 1, this would mean that we are more lenient with exceptions, not only because there might be noise, but also because the concept may be complex and we want to teach it up to a level of accuracy (α can be defined in terms of the size of $|D|$).

9.5 Universal Case

We will instantiate the algorithms and protocol for several concept languages, starting with a Turing-complete language, P3. This language is a variant of P″ (Böhm, 1964), a primitive programming language that was shown to be Turing-complete. Our P3 variant is also universal, but has only seven instructions: `<>+-[]o`. We consider a binary alphabet and the special symbol dot '.', so that the tape alphabet has three symbols $\Sigma = \{0, 1, .\}$. P3 programs work on an input/memory tape of cells with values in Σ, by moving a pointer left (instruction `<`) and right (instruction `>`), similar to a Turing Machine. The instruction `o` is used to print the symbol where the pointer is onto a sequential output tape. The tape alphabet has the cyclic ordering ('0' $<$ '1' $<$ '.' $<$ '0') and there are two instructions that modify the symbol in the pointer cell (+ transforms it to the following symbol in the ordering, and - tranforms it to the previous one). Finally, the bracket instruction `[` loops to the corresponding bracket `]`, that is the instructions that are placed between the two brackets are executed repeatedly. The loop exits to the instruction following `]` when the content of the pointer cell is '.'. This value is also considered the end of the input and output strings. Consequently, the output of a program is given by all the 0s or 1s found on the output tape until a '.' is found. Consequently, if a '.' is output the program stops. The coding schema for programs and examples is the same of Telle *et al.* (2019). Basically, for programs it is simply $N\log_2 7$ bits, where N is the number of instructions. For the examples, we use Elias Delta Coding (Elias, 1975). We are now going to show a few cases. In what follows we consider $\alpha = 0$, so no exceptions are allowed.

9.5.1 Example 1: Non-iterative concept

First we start with a non-iterative example. Consider that the teacher has the concept c: 'output the first symbol of a string and then append a zero after it'. By following the teacher procedure in Definition 9.2, the teacher finds the shortest witness set such that an enumeration of P3 programs returns a program compatible with c.

Table 9.1 *A non-iterative concept and associated witness set, its size, and shortest program.*

$\langle$Input, Output$\rangle$	Size	Program
$\{\langle 0,00\rangle\langle 10,10\rangle\}$	19	`o<+o`

Table 9.1 shows the witness set w and a P3 program that is returned by $\Phi(w)$, as calculated by Telle *et al.* (2019).[8] In this case, the program prints the content on the cell (initially on the first position, so printing the first symbol of the string), then it moves the pointer left, to a cell that is '.' (as all strings are left-padded by '.'), then increases the content's cell ('.'), converting it into a 0, and outputs it. So it actually implements the concept.

The teacher knows for certain that if a learner uses P3 and simplicity, these two examples are sufficient to identify the concept. However, the teacher hopes that these two examples would also be sufficient for any learner with a strong simplicity criterion, even if they are not using P3 and their coding is different.

In order to check this with humans as learners, we did some simple experiments[9] considering that humans are also bound by simplicity priors, as we discussed earlier in the chapter. The subjects, after seeing the witness set of a new string, should figure out a concept to infer the output string (in a free text box). This is done for five fresh input strings (chosen with a simplicity prior too).

The results are quite poor in this case, with accuracies of 20%, 5%, 0%, 0%, and 15%. After interviews with some of the subjects, they clearly do not identify the concept in this case, but most would have succeeded with three or four examples, or with only two *larger* examples. In this case, despite the invariance theorem of representations seen before, we see that the minimal witness set is insufficiently informative to compensate for the difference in representation language. These two short examples do not seem enough for this concept.

9.5.2 Example 2: Iterative concept

As a second example, imagine the teacher considers the 'copy' concept, that is a function that just maps inputs into outputs. Table 9.2 shows the witness set associated with this concept. Only one example suffices. In this case, the program `[o>]` starts a loop with `[`, outputs the character of the pointer, and moves the cursor to the right one position. This is done until the special character '.' is reached, which is found at the end of

[8] See the whole teaching book on https://github.com/ceferra/The-Teaching-Size-with-P3/blob/master/filteredprogsTS7.txt.

[9] For all the experiments that follow, we had the same human sample consisting of 20 university graduates. In the questionnaire, only the examples were given, with no information at all about the concept or the language P3. The form we used can be found at http://tiny.cc/921qiz.

Table 9.2 *An iterative concept and associated witness set, its size, and shortest program.*

⟨Input, Output⟩	Size	Program
$\{\langle 010, 010\rangle\}$	16	`[o>]`

the strings, which are right-padded with '.'. This behaviour, hence, corresponds to the 'copy' concept.

We follow the same experimental procedure with humans as before, and we now get accuracies 75%, 90%, 50%, 90%, and 60%. Even with one single example, the accuracies are much higher. Note that the size is not much lower than in the previous case.

9.6 Feature-value Case

Feature-value representations are probably the most common way of expressing data in machine learning, especially in supervised scenarios. In order to apply Protocol 9.5 to feature-value representations, we need to define how we code examples and programs.

In classification, for instance, 'complete' examples are again pairs $\langle i, o\rangle$, where i is a vector of m features (or attributes) X_j, each one specifying either a quantitative (numeric) or a qualitative (nominal) value v depending on the type of the attribute, and o is a value one of the n_C possible classes. For instance, given three attributes, an example would be $\langle\{X_1 = v_1, X_2 = v_2, X_3 = v_3\}, o\rangle$.

Note that in cases where there are many attributes, coding a complete example may use many bits that are not informative at all, as many attributes may be useless or redundant. As we are working with teaching size and not teaching dimension, it also makes sense to use bits (and hence witness size) for valuable information only. Consequently, we consider the coding of partial examples too. In explainable AI, the term anchor is given to an example where some features are specified, but others are not, because their influence is small. For instance, given three attributes, a partial version of the example above could be $\langle\{X_2 = v_2\}, o\rangle$. Sometimes we will just use the term 'example' indistinctly for partial and complete examples.

Definition 9.6 *In the feature-value representation, a witness set is a set of consistent examples (complete or partial).*

We now introduce δ, the coding for witness sets for this representation. In the case of numeric variables, we could choose to code any possible value, but this would require an infinite number of bits for unlimited real number resolution. Instead, as we have a dataset D, we can look up how many different values there are for each attribute. To code a value we simply select one of those, which requires $\log_2 n_j$, with n_j is the number of different values of attribute X_j in D. All together:

Definition 9.7 *Given a dataset with n_c classes, m features and n_j different values for each attribute $X_j : j = 1 \dots m$, the code size δ of a witness set is defined as:*

$$\delta(w) = \log_2 |w| + \sum_{x \in w} \{\log_2 n_c + \sum_{j=1}^{m} I(j)\{\log_2(m+1) + \log_2 n_j\}\}$$

where $I(j) = 1$ if attribute X_j is specified in the example and 0 otherwise.

We first code how many examples we have. Then, for each example x in the witness set w, we code it as follows. We first code the class ($\log_2 n_c$) and then, for each included attribute, we have to select which one it is (or whether it is the last one, that is why we use $m+1$ instead of m), and then code the value for that attribute.[10]

Let us now choose the language L that will be used for expressing concepts by the learner algorithm. We look for a language that uses elements that are familiar for humans and that can be coded effectively. Humans typically understand the operators $=$ and $\neq$ for nominal attributes and $\leq$ and $>$ for numeric attributes. It is no surprise that many languages that are used for interpretable models are based on conditions using these operators and some logical operators with them.

In particular, we will work with decision lists, a common representation from the early days of concept learning (Rivest, 1987), as we discussed in section 9.3.1. A decision list is a sequence of if-then-else-if conjunctive rules, with a default class rule at the end. Each rule has one or more conditions and one class. We define the coding function ℓ for decision lists as follows:

Definition 9.8 *The coding function ℓ of a decision list p is defined as:*

$$\ell(p) = \log_2 n_c + \sum_{k=1}^{K} \left[\log_2 n_c + 1 + \sum_{j=1}^{m} I(j,k)\{\log_2 2 + \log_2(m+1) + \log_2 n_j\} \right]$$

where K is the number of rules (excluding the default rule) and $I(j,k) = 1$ if attribute X_j appears in the condition of rule k and 0 otherwise.

Basically, we need $\log_2 n_c$ for the class of the default rule; for each non-default rule we need $\log_2 n_c$ to account for the class and the extra 1 is to determine whether more rules are needed or not. For each condition $X_j \odot v$ we require 1 bit to determine the operator $\odot$ ($=$ or $\neq$ for nominal attributes and $\leq$ or $>$ for numeric attributes), $\log_2(m+1)$ to determine the variable j (or the end of coding variables) and $\log_2 n_j$ to determine v.

Given the way in which witness sets and programs are constructed and coded, we can now run the algorithm in Definition 9.4, and eventually Protocol 9.5. The input is a dataset D and the output will be a witness set w and exceptions E. Humans will only receive w, and then will be asked to classify some elements in $D \setminus E$.

[10] As we cannot code the same attribute more than once, this coding can be improved slightly, but we will keep this schema for simplicity.

9.6.1 Example 1: Concept with nominal attributes only

We start with one of the Monk's problems, popular Boolean concepts frequently used in machine learning (Wnek *et al.*, 1990). The problems have six input attributes (all nominal) describing features of robots, and the output is Boolean. In particular, Monk1 captures the pattern where the class is true if $X_1 = X_2$ or $X_5 = 1$, and false otherwise.[11] In many available versions of the dataset, true is represented with value 2 and false is represented with value 1. We will use true and false for clarity.

We build an inscrutable model with a neural network (multilayer perceptron) using Weka (Hall *et al.*, 2009), using the default parameters, with 50 examples chosen randomly as the training dataset. From the rest of examples, we select 50 randomly, which we use as the test set. The neural network model has three errors for the test dataset. This means that the neural network has not captured the relational characteristics of Monk1. Instead of the true test set, we take the labellings of the NN as ground truth, as it is the behaviour of NN that we want to explain.

Variables X_1, X_2, and X_4 have three possible values each, X_5 has four, while X_3 and X_6 have two each, which means that conditions require different amounts of bits depending on the attribute. We use the 50 test examples labelled by the NN as D, and run $\Omega_{\ell,\delta}(D)$ as per Definition 9.4. We get the following witness set:

$$\langle\{X_5 = 1\}, \text{True}\rangle \quad \langle\{X_4 = 3, X_5 = 2\}, \text{True}\rangle \quad \langle\{X_4 = 1, X_5 = 2\}, \text{False}\rangle$$
$$\langle\{X_4 = 2, X_5 = 2\}, \text{False}\rangle \quad \langle\{X_5 = 3\}, \text{False}\rangle \quad \langle\{X_5 = 4\}, \text{False}\rangle$$

Note that all examples are partial. The coding cost $\delta(w) = 50.61$. We see that the only original pattern that remains is that $X_5 = 1$ is true. The other partial examples are there to justify the decision list that is generated (as we will see below). Both the witness set and the decision list try to compress the 50 examples with the possibility of leaving exceptions out.

To understand clearly how the teacher gets this, we need to look at the coding. Each extra condition in an example with variable X_5 (with four possible values) requires $\log_2(6+1) + \log_2 4 = 4.81$ bits, whereas each exception takes $\log_2(50) = 5.64$ bits. Adding a condition to an example is only justified if it removes an exception but does not add another. Typically, however, trying to cover a positive example will also imply that some negative examples are covered, and this is why exceptions are common. With this analysis, a large set of exceptions are expected. In particular, E is:

$$\langle\{X_1 = 2, X_2 = 2, X_3 = 1, X_4 = 1, X_5 = 2, X_6 = 2\}, \text{True}\rangle$$
$$\langle\{X_1 = 3, X_2 = 3, X_3 = 2, X_4 = 2, X_5 = 2, X_6 = 1\}, \text{True}\rangle$$
$$\langle\{X_1 = 3, X_2 = 3, X_3 = 2, X_4 = 3, X_5 = 3, X_6 = 2\}, \text{True}\rangle$$
$$\langle\{X_1 = 1, X_2 = 1, X_3 = 1, X_4 = 3, X_5 = 3, X_6 = 2\}, \text{True}\rangle$$
$$\langle\{X_1 = 1, X_2 = 1, X_3 = 2, X_4 = 2, X_5 = 3, X_6 = 1\}, \text{True}\rangle$$
$$\langle\{X_1 = 3, X_2 = 3, X_3 = 2, X_4 = 3, X_5 = 3, X_6 = 2\}, \text{True}\rangle$$
$$\langle\{X_1 = 2, X_2 = 2, X_3 = 1, X_4 = 3, X_5 = 3, X_6 = 2\}, \text{True}\rangle$$

[11] Note that this pattern cannot be expressed with only two rules in L since it does not consider direct comparisons between variables.

$\langle\{X_1 = 1, X_2 = 1, X_3 = 2, X_4 = 1, X_5 = 3, X_6 = 2\}, \text{True}\rangle$
$\langle\{X_1 = 3, X_2 = 3, X_3 = 1, X_4 = 2, X_5 = 4, X_6 = 1\}, \text{True}\rangle$
$\langle\{X_1 = 1, X_2 = 1, X_3 = 2, X_4 = 2, X_5 = 4, X_6 = 2\}, \text{True}\rangle$
$\langle\{X_1 = 3, X_2 = 3, X_3 = 1, X_4 = 3, X_5 = 4, X_6 = 2\}, \text{True}\rangle$
$\langle\{X_1 = 2, X_2 = 3, X_3 = 2, X_4 = 3, X_5 = 2, X_6 = 1\}, \text{False}\rangle$

The coding cost $\delta(E) = 12\log_2 50 = 67.73$.

Following our approach, we are able to find the simplest decision list p that the learner finds for this witness set:

if $X_5 = 1$ then True
elsif $X_5 = 2$ and $X_4 = 3$ then True
else (default rule) False

with coding cost $\ell(p) = 22.01$ bits.

The cost of the witness set (without exceptions) is 50.61 bits, much larger than the cost of the decision list (without exceptions), which is 22.01 bits. The decision list becomes very small but this is again due to how costly covering some exceptions is for this problem. For instance, a rule with two conditions of four-valued variables requires 13.61 bits, which only makes sense if three or more exceptions are removed. This decision list is only approximately similar to the original concept, but with the exceptions it captures the neural network perfectly. It may seem that giving more cost to exceptions could be a good thing, but this would generate more spurious rules.

We performed a similar experiment to those conducted with strings in the previous section, also using Protocol 9.5 with the same 20 human subjects. Humans only received w (no concept, no exceptions) and were asked to classify five fresh examples (taken randomly from $D \setminus E \setminus w$). The results are quite good, with accuracies 95%, 90%, 90%, 90%, and 90%.

9.6.2 Example 2: Concept with numeric attributes

We can do similarly for numeric attributes. We will illustrate it with the Iris flower dataset, a very popular classification problem.

In our case, we again use a neural network (multilayer perceptron) using Weka (Hall *et al.*, 2009) with the default parameters, using 75 examples as training dataset. The NN model has 4 errors (out of 75) for the test dataset.[12] Again, we take the NN as the ground truth, the concept we want to explain.

Note that the four variables have different numbers of values, so some attributes are cheaper to code than others. The cheapest one is X_4, and it is also the one that seems

[12] Example input $\{X_1 = 6.3, X_2 = 2.8, X_3 = 5.1, X_4 = 1.5\}$ is predicted by the NN as Versicolor, and should be Virginica. Example input $\{X_1 = 6.0, X_2 = 2.7, X_3 = 5.1, X_4 = 1.6\rangle$, is predicted by the NN as Virginica, and should be Versicolor. Example input $\{X_1 = 7.2, X_2 = 3.0, X_3 = 5.8, X_4 = 1.6\rangle$, is predicted by the NN as Versicolor, and should be Viriginica. Example input $\{X_1 = 5.9, X_2 = 3.2, X_3 = 4.8, X_4 = 1.8\rangle$, is predicted by the NN as Virginica, and should be Versicolor.

to be most discriminating as well. We use the 75 test examples labelled by the NN as D, and run $\Omega_{\ell,\delta}(D)$ as per Definition 9.4. We get the following witness set:

$\langle\{X_4 = 0.5\}, \text{Setosa}\rangle$ $\quad$ $\langle\{X_4 = 1.0\}, \text{Versicolor}\rangle$
$\langle\{X_4 = 1.6\}, \text{Versicolor}\rangle$ $\quad$ $\langle\{X_4 = 1.8\}, \text{Virginica}\rangle$

As D only has values 0.1, 0.2, 0.3, 0.4, 0.5, 1, 1.1, 1.2, 1.3, 1.4, 1.5, 1.6, 1.8, 1.9, 2, 2.1, 2.2, 2.3, 2.4, and 2.5 for X_4, the coding cost $\delta(w) = 34.92$. Note that all examples are partial. And E, the exceptions, are:

$\langle\{X_1 = 6.1, X_2 = 2.6, X_3 = 5.6, X_4 = 1.4\}, \text{Virginica}\rangle$
$\langle\{X_1 = 6.0, X_2 = 2.7, X_3 = 5.1, X_4 = 1.6\}, \text{Virginica}\rangle$

The coding cost $\delta(E) = 2\log_2 75 = 12.46$.

The simplest decision list p that the learner finds for this witness set is:

if $X_4 \leq 0.5$ then Setosa
elsif $X_4 \leq 1.6$ then Versicolor
else (default rule) Virginica

which only considers one attribute. The coding cost $\ell(p) = 40.92$.

It is interesting to compare that $\ell(p) = 40.92 > \delta(w) = 34.92$ when ignoring exceptions. Note that exceptions should be considered in both sides or none of them.

These simple rules show that learning with examples is not only meaningful and effective, but requires less information to be transmitted than if an 'interpretable model' were produced by the teacher. However, we have to check what happens when the simplicity priors are agreed but the representation languages are different. We conduct a similar experiment with our 20 human subjects with this new problem. The results are excellent in this case, with accuracies 95%, 100%, 100%, 100%, and 95%.

9.7 Discussion

In this chapter we have analysed how machine teaching, and the concept of teaching size, can be used to explore the way in which examples may be chosen for explaining a concept. We are not claiming that we have an operative explanation procedure yet, but we have some important insights that could lead to effective exemplar-based explanation mechanisms in the future.

In our protocol, the teacher, in order to find the witness set, assumes a particular learner $\Phi_\ell(w)$. This becomes explicit in the Algorithms 9.3 and 9.4, where the teacher calls *a model of the learner* repeatedly in order to find the optimal witness set. Consequently, what matters is a good model of the learner, which boils down to knowing its priors. Note that this model is constant, unlike interactive Bayesian approaches (Goodman and Frank, 2016; Yang and Shafto, 2017; Melo *et al.*, 2018). In our setting and experiments we have assumed some languages (P3 in one scenario and decision lists in the other scenario), and then a strong simplicity prior over this representation. Also, example sets are encoded (and hence enumerated) using another simplicity prior.

This contrasts with the classical view of having different kinds of exemplar-based explanations, looking for particular kinds of examples (counterfactuals, anchors, prototypes and criticisms) as shown in section 9.3.2. The principles under machine teaching are general, and are the same for all representation languages. Given a representation language for examples (including some possible summarizations or partial patterns), the teacher finds the shortest witness for the learner (the explainee) to grasp the concept. Because of the use of examples that might be of different size (or coding cost), the use of the teaching size instead of teaching dimension becomes necessary.

However, we are interested in the general case where the actual learner may have some other internal representation mechanisms and codings, even if the learner follows a simplicity prior. From the experiments with humans, we see that even with the minimal information, identification is possible in some cases. In other cases, larger than minimal witness sets are needed for a likely correct identification. This probabilistic view of machine teaching was recently introduced in Hernández-Orallo and Telle (2020), and it is a vision of machine teaching that has to be further developed in the future, understanding the explanation tradeoff accordingly. Also, there is an additional reason for this. By using some slack in the minimality we could convert the typically intractable teacher algorithms into approximate ones, possibly reusing some efficient algorithms for decision lists (Angelino *et al.*, 2017) or other representations. This, jointly with more experimental research with humans, opens an avenue of research questions to understand in full the prior alignment between teachers and learners using different representational languages, and the particular case when teachers are machines and learners are humans in the context of explainable AI.

Acknowledgements

We would like to thank the reviewers for their thoughtful comments. We thank Jan Arne Telle for insightful comments and discussions before and during the elaboration of this chapter. We also thank the human participants of the experimental setting at UPV. This work was funded by the EU (FEDER) and Spanish MINECO under RTI2018–094403-B-C32, and Generalitat Valenciana under PROMETEO/2019/098.

References

Angelino, E., Larus-Stone, N., Alabi, D. *et al.* (2017). Learning certifiably optimal rule lists for categorical data. *Journal of Machine Learning Research*, 18(1), 8753–830.

Angluin, D. (1988). Queries and concept learning. *Machine Learning*, 2(4), 319–342.

Aoga, J. O. R., Guns, T., Nijssen, S. *et al.* (2018). Finding probabilistic rule lists using the minimum description length principle, in *International Conference on Discovery Science*. Limassol, Cyprus: Springer, Cham, 66–82.

Argall, B. D., Chernova, S., Veloso, M. *et al.* (2009). A survey of robot learning from demonstration. *Robotics and Autonomous Systems*, 57(5), 469–83.

Balbach, F. J. (2008). Measuring teachability using variants of the teaching dimension. *Theoretical Computer Science*, **397**(1–3), 94–113.

Basu, S. and Christensen, J. (2013). Teaching classification boundaries to humans, in *Twenty-Seventh AAAI Conference on Artificial Intelligence*. Washington: Bellevue, 109–15.

Bella, A., Ferri, C., Hernández-Orallo, J. *et al.* (2011). Using negotiable features for prescription problems. *Computing*, **91**(2), 135–68.

Bengio, Y., Louradour, J., Collobert, R. *et al.* (2009). Curriculum learning, in *International Conference on Machine Learning*. Montreal, Canada, 41–8.

Böhm, C. (1964). On a family of Turing machines and the related programming language. *ICC Bulletin*, **3**(3), 187–94.

Chater, N. and Vitányi, P. (2003). Simplicity: a unifying principle in cognitive science? *Trends in Cognitive Sciences*, 7(1), 19–22.

Chen, Xi, Chen, Xi, Cheng, Yu, and Tang, Bo (2016). On the recursive teaching dimension of VC classes, in *Proceedings of the 30th International Conference on Neural Information Processing Systems*. Barcelona Spain, 2171–79.

Chen, Y., Mac Aodha, O., Su, S. *et al.* (2018). Near-optimal machine teaching via explanatory teaching sets, in *International Conference on Artificial Intelligence and Statistics*. Playa Blanca, Lanzarote, 1970–78.

Chomsky, N. (1992). On cognitive structures and their development: a reply to Piaget, in B. Beakley, P. Ludlow, P. J. Ludlow *et al.*, eds, *Philosophy of Mind: Classical Problems/Contemporary Issues*. Cambridge, MA: MIT Press, 751–755.

Dasgupta, S., Hsu, D., Poulis, S. *et al.* (2019). Teaching a black-box learner, in *International Conference on Machine Learning*. Long Beach, CA, 1547–55.

Degen, J., Hawkins, R. D., Graf, C. *et al.* (2019). When redundancy is useful: A Bayesian approach to 'overinformative' referring expressions. arXiv: 1903.08237.

Domingos, P. (1998). Knowledge discovery via multiple models. *Intelligent Data Analysis*, **2**(1–4), 187–202.

Doshi-Velez, F. and Kim, B. (2017). Towards a rigorous science of interpretable machine learning. *arXiv preprint arXiv:1702.08608*.

Elias, P. (1975). Universal codeword sets and representations of the integers. *IEEE Transactions on Information Theory*, **21**(2), 194–203.

Fabra-Boluda, R., Ferri, C., Hernández-Orallo, J. *et al.* (2017). Modelling machine learning models, in *3rd Conference on Philosophy and Theory of Artificial Intelligence*, University of Leeds, UK. Cham: Springer, 175–86.

Gao, Z., Ries, C., Simon, H. *et al.* (2017). Preference-based teaching. *Journal of Machine Learning Research*, **18**, 31:1–31:32.

Goldman, S. A. and Kearns, M. J. (1995). On the complexity of teaching. *Journal of Computer and System Sciences*, **50**(1), 20–31.

Goodman, N. D. and Frank, M. C. (2016). Pragmatic language interpretation as probabilistic inference. *Trends in Cognitive Sciences*, **20**(11), 818–29.

Google (2020). Explainable AI report. https://sites.google.com/view/www20-explainable-ai-tutorial

Guidotti, R., Monreale, A., Ruggieri, S. *et al.* (2018). A survey of methods for explaining black box models. *ACM Computing Surveys*, **51**(5), 93.

Gulwani, S. (2016). Programming by examples: applications, algorithms, and ambiguity resolution, in *International Joint Conference on Automated Reasoning*. Cham: Springer, 9–14.

Gulwani, S., Hernández-Orallo, J., Kitzelmann, E. *et al.* (2015). Inductive programming meets the real world. *Communications of the ACM*, **58**(11).

Hadfield-Menell, D., Russell, S. J., Abbeel, P. *et al.* (2016). Cooperative inverse reinforcement learning, in *Proceedings Neural Information Processing Systems 29 (NIPS 2016)*, 3909–17.

Hall, M., Frank, E., Holmes, G. *et al.* (2009). The WEKA data mining software: an update. *ACM SIGKDD Explorations Newsletter*, **11**(1), 10–18.

Hernández-Orallo, J. (2019). Gazing into clever hans machines. *Nature Machine Intelligence*, **1**(4), 172.

Hernández-Orallo, J. and Telle, J. A. (2020). Finite and confident teaching in expectation: sampling from infinite concept classes, in *24th European Conference on Artificial Intelligence (ECAI2020)*.

Ho, M. K., Littman, M., MacGlashan, J. *et al.* (2016). Showing versus doing: teaching by demonstration. *Advances in Neural Information Processing Systems*, **29**, 3027–35.

Khan, F., Mutlu, B. and Zhu, J. (2011). How do humans teach: on curriculum learning and teaching dimension. *Advances in Neural Information Processing Systems*, 1449–57.

Kim, B., Khanna, R., and Koyejo, O. O. (2016). Examples are not enough, learn to criticize! Criticism for interpretability. *Advances in Neural Information Processing Systems*, **29**, 2280–88.

Lage, I., Chen, E., He, J. *et al.* (2019). An evaluation of the human-interpretability of explanation. *arXiv preprint arXiv:1902.00006*.

Lakkaraju, H., Bach, S. H., and Leskovec, J. (2016). Interpretable decision sets: A joint framework for description and prediction, in *Proceedings of the 22nd ACM SIGKDD International Conference on Knowledge Discovery and Data Mining*, 1675–84.

Lim, Brian Y, Dey, A. K., and Avrahami, D. (2009). Why and why not explanations improve the intelligibility of context-aware intelligent systems, in *Proceedings of the SIGCHI Conference on Human Factors in Computing Systems*, 2119–28.

Lipton, P. (1990). Contrastive explanation. *Royal Institute of Philosophy Supplements*, **27**, 247–66.

Liu, W., Dai, B., Li, X. *et al.* (2017). Towards black-box iterative machine teaching. *arXiv preprint arXiv:1710.07742*.

Melo, F. S., Guerra, C., and Lopes, M. (2018). Interactive optimal teaching with unknown learners, in *IJCAI*, 2567–73.

Miller, T. (2019). Explanation in artificial intelligence: insights from the social sciences. *Artificial Intelligence*, **267**, 1–38.

Molnar, C. (2019). Interpretable machine learning: a guide for making black box models explainable. *christophm.github.io/interpretable-ml-book/*.

Muggleton, S. (1991). Inductive logic programming. *New Generation Computing*, **8**(4), 295–318.

Nguyen, D. (2018). Comparing automatic and human evaluation of local explanations for text classification, in *Proceedings of the 2018 Conference of the North American Chapter of the Association for Computational Linguistics: Human Language Technologies*, Vol. 1 (Long Papers), 1069–78.

Ouyang, L. (2018). Bayesian inference of regular expressions from human-generated example strings. arXiv:1805.08427.

Plautus, T. M. (2005). *Rome and the Mysterious Orient: Three Plays by Plautus*. University of California Press.

Proença, H. M. and van Leeuwen, M. (2020). Interpretable multiclass classification by MDL-based rule lists. *Information Sciences*, **512**, 1372–93.

Rahwan, I., Cebrian, M., Obradovich, N. *et al.* (2019). Machine behaviour. Nature, **568**(7753), 477–86.

Ribeiro, M. T., Singh, S., and Guestrin, C. (2016). Why should I trust you? Explaining the predictions of any classifier, in *Proceedings of the 22nd ACM SIGKDD International Conference on Knowledge Discovery and Data Mining*, 1135–44.

Ribeiro, M. T., Singh, S., and Guestrin, C. (2018). Anchors: High-precision model-agnostic explanations, in *Proceedings of the AAAI Conference on Artificial Intelligence*, Vol. 32, No. 1.

Rissanen, J. (1983). A universal prior for integers and estimation by minimum description length. *Annals of Statistics*, **11**(2), 416–31.

Rivest, R. L. (1987). Learning decision lists. *Machine Learning*, **2**(3), 229–46.

Samek, W. and Müller, K.-R. (2019). Towards explainable artificial intelligence, in W. Samek, ed., *Explainable AI: Interpreting, Explaining and Visualizing Deep Learning*. Cham: Springer, 5–22.

Schaal, S. (1997). Learning from demonstration. *Advances in Neural Information Processing Systems*, 1040–46.

Solomonoff, R. J. (1964). A formal theory of inductive inference. Part I. *Information and Control*, 7(1), 1–22.

Swartout, W., Paris, C., and Moore, J. (1991). Explanations in knowledge systems: Design for explainable expert systems. *IEEE Expert*, **6**(3), 58–64.

Telle, J. A., Hernández-Orallo, J., and Ferri, C. (2019). The teaching size: computable teachers and learners for universal languages. *Machine Learning*, **108**(8–9), 1653–75.

Wallace, C. S. and Boulton, D. M. (1968). An information measure for classification. *Computer Journal*, **11**(2), 185–94.

Wallace, C. S. and Dowe, D. L. (1999). Minimum message length and Kolmogorov complexity. *Computer Journal*, **42**(4), 270–83.

Winston, P. H. and Horn, B. (1975). *The Psychology of Computer Vision*. McGraw-Hill Companies.

Wnek, J., Sarma, J., Wahab, A. A. *et al.* (1990). Comparing learning paradigms via diagrammatic visualization. *Methodologies for Intelligent Systems*, **5**, 428–37.

Yang, S. C.-H. and Shafto, P. (2017). Explainable artificial intelligence via Bayesian teaching, in *NIPS 2017 workshop on Teaching Machines, Robots, and Humans*, 127–37.

Zhao, H., Sinha, A. P., and Bansal, G. (2011). An extended tuning method for cost-sensitive regression and forecasting. *Decision Support Systems*, **51**(3), 372–383.

Zhu, X. (2015). Machine teaching: an inverse problem to machine learning and an approach toward optimal education, in *Proceedings of the AAAI Conference on Artificial Intelligence*, Vol. 29, No. 1.

Zhu, X., Singla, A., Zilles, S. *et al.* (2018). An overview of machine teaching. *arXiv preprint arXiv:1801.05927*.

Part 3

Human-like Perception and Language

10

Human-like Computer Vision

Stephen Muggleton and Wang-Zhou Dai

Department of Computing, Imperial College London

10.1 Introduction

Galileo's *Siderius Nuncius* (Galilei, 1610) describes the first ever telescopic observations of the moon. Using sketches of shadow patterns Galileo conjectured the existence of mountains containing hollow areas (i.e., craters) on a celestial body previously thought perfectly spherical. His reasoned description, derived from a handful of observations, relies on a knowledge of (1) classical geometry, (2) straight line movement of light, and (3) the Sun as an out-of-view light source. This chapter investigates the use of Inductive Logic Programming (ILP) to derive logical hypotheses, related to those of Galileo, from a small set of real-world images. Figure 10.1 illustrates part of the generic background knowledge used by ILP for interpreting object convexity (see section 10.3.2). Humans regularly use logical reasoning in science and mathematics (Bronkhorst *et al.*, 2020) and are known to learn effectively from small numbers of examples (Tenenbaum *et al.*, 2011). This chapter indicates that Human-like Computer Vision based on logical reasoning, of the form demonstrated by Galileo, enables efficient and accurate perception and learning from small numbers of training images.

Logical versus statistical vision. Statistical machine learning is widely used in image classification. However, most techniques (1) require many images to achieve high accuracy and (2) do not provide support for reasoning below the level of classification, and so are unable to support secondary reasoning, such as the existence and position of light sources and other objects outside the image. This chapter describes an Inductive Logic Programming (Muggleton, 1991; Muggleton and Raedt, 1994; Muggleton *et al.*, 2011) approach called Logical Vision (LV) (Muggleton *et al.*, 2018) and an abduction-based learning approach called Abductive Learning (ABL) (Zhou, 2019; Dai *et al.*, 2019) to overcome some of these limitations. The idea of Human-Like Computer Vision is to model explicitly the secondary reasoning paradigm of human vision, in which *logical reasoning* is in charge of analysing and learning high-level relationships between abstract visual concepts (e.g., points and edges), which are obtained from *low-level perception*. More importantly, the two aspects of vision work simultaneously and keep affecting each other during the visual process.

Stephen Muggleton and Wang-Zhou Dai, *Human-like Computer Vision* In: *Human Like Machine Intelligence.* Edited by: Stephen Muggleton and Nick Charter, Oxford University Press. © Oxford University Press 2021. DOI: 10.1093/oso/9780198862536.003.0010

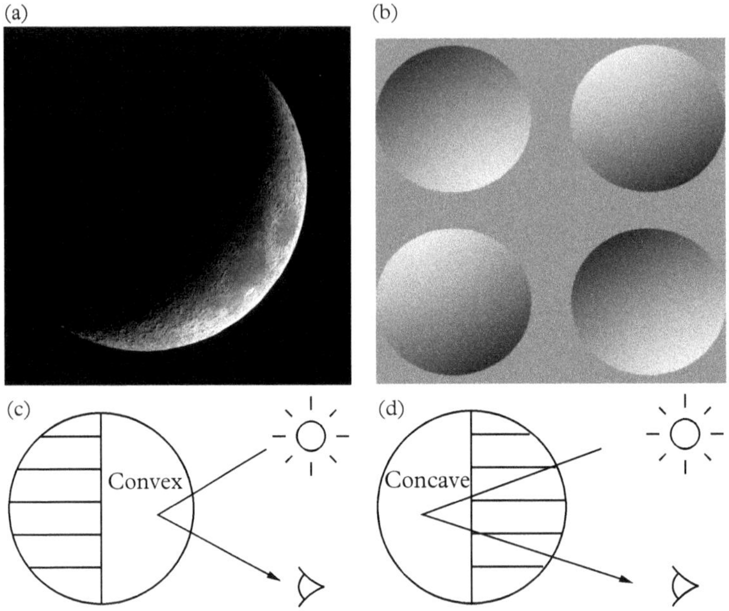

(e)
```
light(X, X).
light(X, Y): -reflect (X, Z),
light(Z, Y).
```

Figure 10.1 *Interpretation of light source direction: (a) waxing crescent moon (Credit: UC Berkeley), (b) concave/convex illusion, (c) concave and (d) convex photon reflection models, (e) Prolog recursive model of photon reflection.*

This chapter summarizes and integrates recent published work on LV by the authors. LV uses Meta-Interpretive Learning (MIL) (Muggleton *et al.*, 2014; Muggleton *et al.*, 2015; Muggleton, 2017) combined with low-level extraction of high-contrast points sampled from the image to learn recursive logic programs describing the image. In published work (Dai *et al.*, 2015) LV has been demonstrated capable of high-accuracy prediction of classes such as *regular polygon* from small numbers of images where Support Vector Machines and Convolutional Neural Networks gave near random predictions in some cases. In this early work LV was only applicable to noise-free, artificially generated images. In more recent work (Dai *et al.*, 2015), LV was extended by (1) addressing *classification noise* using a new noise-telerant version of the MIL system Metagol, (2) addressing *attribute noise* using primitive-level statistical estimators to identify sub-objects in real images, (3) using a wider class of background models representing classical two-dimensional (2D) shapes such as circles and ellipses, (4) providing richer learnable background knowledge in the form of a simple but generic recursive theory of light reflection. In the experiments we consider noisy images in natural science settings, which involve identification of the position of the light source in telescopic and microscopic images. Our results indicate that with real images the new noise-robust version of LV using a single example (i.e., one-shot LV) converges to an accuracy at least comparable to a 30-shot statistical machine learner on prediction of hidden light sources. Moreover, we demonstrate that a general background recursive theory of light can itself be invented using LV and used to identify ambiguities in the convexity/concavity of objects such as craters.

Instead of using a pretrained statistical model to extract primitive visual concepts, ABL learns a low-level perception model with a high-level reasoning model jointly. In published work (Dai *et al.*, 2019), the perception model is a neural net that learns to extract primitive logical facts from raw pixels, and the reasoning model performs logical inference based on first-order logical background knowledge and the perceived facts to obtain the final output. A key difficulty in training ABL models lies in the integration and simultaneous optimization of sub-symbolic and symbolic reasoning models. In particular, (1) we do not not have ground-truth for the primitive logic facts for training the machine learning model; (2) without accurate primitive logic facts, the reasoning model cannot deduce the correct output or learn the an appropriate logical theory. ABL addresses these challenges with logical abduction (Kakas *et al.*, 1992; Flach and Kakas, 2000) and consistency optimization. Given a training sample associated with a label for the target concept, logical abduction can conjecture the missing information—for example, candidate primitive facts in the example, or logic clauses that can complete the background knowledge—to establish a consistent proof from the sample to its final output. The abduced primitive facts and logic clauses are then used for training the machine learning model and stored as symbolic knowledge, respectively. Consistency optimization is used for maximizing the consistency between the conjectures and the background knowledge. To solve this highly complex problem, we transform it into a task that searches for a function guessing about possibly mistaken primitive facts. We verified the effectiveness of ABL with handwritten equation decipherment puzzles: the task is to learn image recognition (perception) and mathematical operations for calculating the equations (reasoning) simultaneously. Experimental results show that ABL generalise better than state-of-the-art deep learning models and can leverage learning and reasoning in a mutually beneficial way.

Structure of chapter. The structure of the chapter is as follows. In section 10.2 we describe related work on computer vision paradigms, use of machine learning, and other approaches which employ logical reasoning and learning for perception. The LV approach is described in section 10.3 and its effectiveness is demonstrated in experiments involving analysis of scientific images. Section 10.4 extends the LV paradigm through use of logical abduction to extract lower-level features, and provides experimental results based on Mayan character analysis. Lastly, section 10.5 concludes and describes directions for future work.

10.2 Related Work

The human vision paradigm defines cognition as manipulation of symbolic representations according to the rules of a formal syntax (Newell and Simon, 1976; Marr, 1982).

By contrast, computer vision based on low-level feature extraction, for example, end-to-end models such as deep neural networks (DNNs) (Goodfellow *et al.*, 2016) and statistical machine learning models (Poppe, 2010) has achieved above-human performance on many pattern recognition tasks, especially image classification problems (Le *et al.*, 2011; Krizhevsky *et al.*, 2012). State-of-the-art computer vision models are usually trained from large-scale datasets, such as (Le *et al.*, 2011; Krizhevsky *et al.*, 2012). For small-scale tasks, DNN descriptors are learned from large-scale data based on a feature space for learning and recognition. It has been shown that the DNN features with standard statistical learning techniques still achieved state-of-the-art performance in these kinds of tasks (Sermanet *et al.*, 2014; Simonyan and Zisserman, 2015).

Some approaches attempt to parse images hierarchically, in a way comparable to human cognition (Ohta *et al.*, 1978). For example, grammar-like models can be used for hierarchical object recognition (Hartz and Neumann, 2007; Felzenszwalb, 2011; Porway *et al.*, 2010). Other approaches extend the hierarchical image models with probability (Jin and Geman, 2006). Other methods use hierarchical features to encode high-level features (Liu *et al.*, 2014; Borenstein and Ullman, 2008; Zheng *et al.*, 2007). However, hierarchical information is embedded in the objective functions or introduced as functional constraints in the statistical learning procedure. It is difficult for these approaches to utilize symbolic representation in a human-like way. In an attempt to do so, some approaches use symbolic background knowledge to constrain the statistical learning process (Maclin *et al.*, 2005; Dai and Zhou, 2017). Others directly learn vision models in a high-level feature space, in which first-order logic background knowledge can be naturally applied (Andrzejewski *et al.*, 2011; Mei *et al.*, 2014; Lake *et al.*, 2015).

On the other hand, high-level vision, involving interpretation of objects and their relations in the external world, is still relatively poorly understood (Cox, 2014). Since the 1990s *perception-by-induction* (Gregory, 1998) has been the dominant model within computer vision, where human perception is viewed as inductive inference of hypotheses from sensory data. The idea originated in the work of the nineteenth-century physiologist Hermann von Helmholtz (von Helmholtz, 1962). The approach described in this paper is in line with *perception-by-induction* in using ILP for generating high-level perceptual hypotheses by combining sensory data with a strong bias in the form of explicitly encoded background knowledge. Whilst Gregory (1974) was one of the earliest to demonstrate the power of the Helmholtz's perception model for explaining human visual illusion, recent experiments (Heath and Ventura, 2016) show Deep Neural Networks fail to reproduce human-like perception of illusion.

Shape-from-shading (Horn, 1989; Zhang *et al.*, 1999) is a key computer vision technology for estimating low-level surface orientation in images. Unlike our approach for identifying concavities and convexities, shape-from-shading generally requires observation of the same object under multiple lighting conditions. By using background knowledge as a bias we reduce the number of images for accurate perception of high-level shape properties such as the identification of convex and concave image areas.

ILP has previously been used for learning concepts from images. For instance, in (Needham *et al.*, 2005; Dubba *et al.*, 2015) object recognition is carried out using existing low-level computer vision approaches, with ILP being used for learning general relational concepts from this already symbolized starting point. Farid (Farid and Sammut, 2014) adopted a similar approach, extracting planar surfaces from a three-dimensional (3D) image of objects encountered by urban search and rescue robots, then using ILP to learn relational descriptions of those objects. By contrast, LV (Dai *et al.*, 2015; Dai *et al.*, 2017) uses ILP to provide a bridge from very low-level features, such as high contrast points, to high-level interpretation of objects. The present chapter extends the earlier work on LV by implementing a noise-proofing technique, applicable to real images, and extending the use of generic background knowledge to allow the identification of objects, such as light sources, not directly identifiable within the image itself.

10.3 Logical Vision

In this section we introduce LV (Dai *et al.*, 2015; Dai *et al.*, 2017; Muggleton *et al.*, 2018), which incorporates background knowledge and utilises modern ILP techniques to learn visual concepts in terms of high-level relations.

10.3.1 Learning geometric concepts from synthetic images

As the first step, LV is applied to tasks involving learning simple geometrical concepts from synthetic images of triangles, quadrilaterals, regular polygons, and so on. Owing to its symbolic representation, LV can be fully implemented in Prolog given low-level image feature extraction primitives as the initial background knowledge.

According to (Marr, 1982), the human vision process can be postulated as a hierarchical architecture with different intermediate representations and processing levels. At each stage of recognition, the representations (symbols) obtained from previous stages play the role of background knowledge.

As an analogy to the hierarchical human vision, LV divides the visual concept learning process into three stages: low-level perception, mid-level concept extraction, and high-level concept learning. Given an input image, LV first alternately conjectures about mid-level visual objects and samples low-level primitives to support or revise those conjectures. After obtaining mid-level concepts, a meta-interpreter can be executed for learning the target visual concepts.

Here low-level perception is based on the features referred to local visual metrics such as colour information, gradients, SIFT and SURF descriptors, etc. The term 'mid-level feature/symbol' is a relative concept: these features/symbols are logical facts that represent possible sub-parts or components of the higher-level concepts. For example, a low-level feature 'colour gradient' can be useful for describing mid-level features such as edge discovery or contour extraction. However, 'edge' can also be regarded as mid-level concepts sub-parts of other higher-level concepts like 'shape' and 'region'.

In the geometric concept learning tasks, the low-level primitive 'point' is defined as a pixel which has large gradient in its local region, and 'edge' is a line segment consisting of points. Based on the notion of points, LV can learn more complex objects such as edges, polygons and combinations of polygons stage by stage.

The mid-level symbol extraction in LV is implemented by repeatedly executing a 'conjecturing and sampling' procedure. It uses the conjectures of edges to guide the sampling of points during the low-level perception; they are then used to revise previously constructed conjectures.

The intuition of the 'conjecturing and sampling' process is also an analogy to the human vision process. Suppose a man stands in front of a huge wall painting, which is very large so he can only clearly observe a small region at a time. To see the entire painting, he can try to move his eyes around to see different small regions in the painting and guess what the totality looks like. During the observation, he either can sample more details to support his conjecture, or can revise them by doing more sampling. After doing enough samples, he will believe that his final conjecture is the ground truth.

Briefly speaking, LV uses mid-level symbolic conjectures to guide the sampling of low-level features, then uses the sampled results to revise previously obtained conjectures. The low-level features themselves like pixel colours, local colour variances, or gradient directions are usually redundant and inappropriate for representing higher-level visual concepts. After the background-knowledge-guided extraction, they can be compactly abduced into logical symbols such as edges, regions, textures, etc., to serve as bases for higher-level concepts learning. An example of LV's edge discovering is illustrated in Figure 10.2.

After obtaining the logical facts about the mid-level symbols, LV uses Meta-Interpretive Learning (MIL) (Muggleton *et al.*, 2015) to learn a logic program defining the target visual concepts.

The experiments are conducted on a synthetic image dataset of geometric shapes (e.g., triangles, quadrilaterals, etc.), regular polygons, and right-angle triangles. For simplicity, the images are binary-coloured, each image contains one polygon, part of the dataset is shown in Figure 10.3. In the experiments, LV is compared to statistics-based computer vision, which learns a support vector machine model from different types of widely used computer vision features including deep convolutional neural network

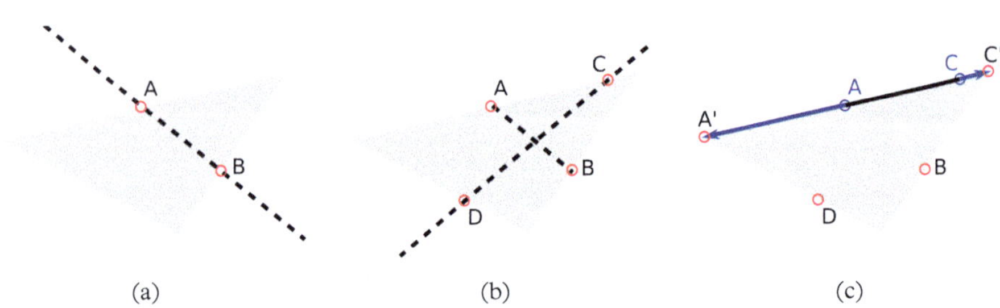

Figure 10.2 *(a) Two points A and B are sampled; (b) Edge AB is conjectured but it is invalid because if the midpoint of AB is not an edge point. So a random line that crosses AB is sampled, two new points C and D have been discovered; (c) Edge AC is conjectured and it satisfies the definition of edge, so AC is extended until no continuous edge points were found.*

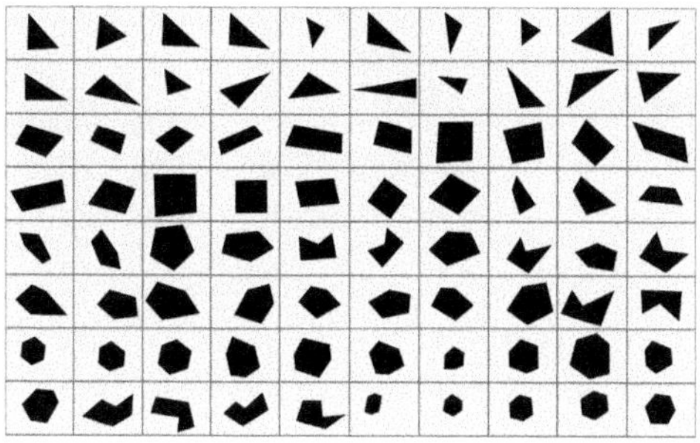

(a) Learning triangles, quadrilaterals, etc.

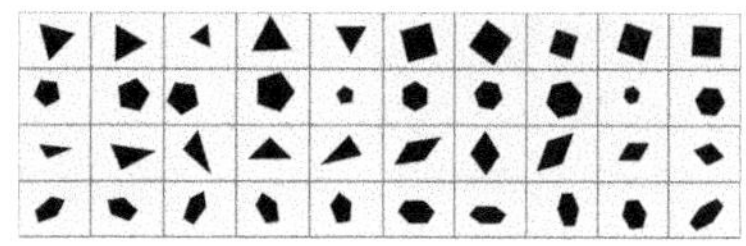

(b) Learning regular polygons

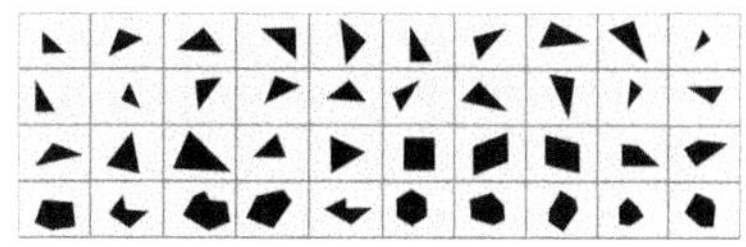

(c) Learning right-angle triangles

Figure 10.3 *Part of the image datasets for learning concepts of polygons.*

features from VGG net (Simonyan and Zisserman, 2015) and combinations of them. The results verified that LV is able to learn explainable hypotheses about visual concepts with a handful of examples, which is beyond the ability of the statistics-base computer vision methods.

10.3.2 One-shot learning from real images

Humans are good at learning from limited numbers of noisy observations. A good example of this type of situation is scientific discovery. In order to do this, humans utilize complex background knowledge as well as noise-tolerant cognition.

In this section, we show that the LV framework can also be extended to be noise robust and is able to learn logic theory in real images from noisy and data-insufficient domains such as microbiology and astronomy. More concretely, LV provided with basic generic background knowledge about radiation and reflection of photons can inform the generation of hypotheses in the form of logic programs based on evidence sampled from a single real image, that is, perform one-shot learning. Moreover, benefiting from its symbolic formalization, LV can discover theories that could be used for explaining ambiguity.

To perform one-shot learning from real images, the noise-robust version LV is extended by using: (1) richer background knowledge enabling secondary reasoning from raw images, such as a simple but generic recursive theory of light reflection for resolving visual ambiguities which cannot be easily modelled using pure statistical approaches, (2) a wider class of background models representing classical 2D shapes such as circles and ellipses, (3) primitive-level statistical estimators to handle noise in real images and demonstrating that the extended LV can learn well-performed models from only one training example (i.e., one-shot LV).

Examples of microscopic and telescopic images are shown in Figure 10.4. The two real image datasets are: (1) **Protists**, drawn from a microscope video of a *Protist*

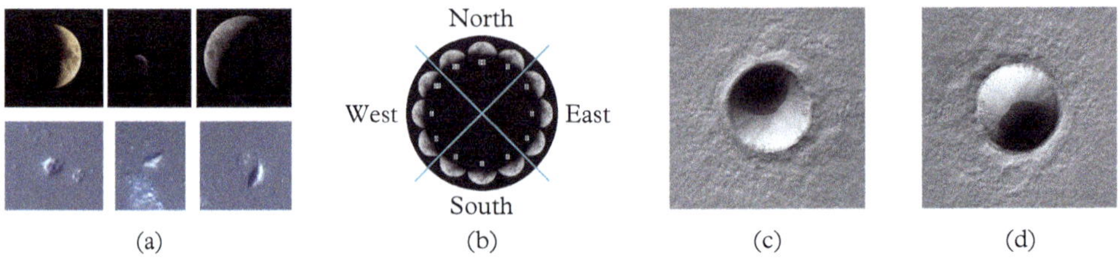

Figure 10.4 *Illustrations of data: (a) examples of the datasets, (b) four classes for twelve light source positions, (c) crater on Mars (Credit: NASA/JPL/University of Arizona), (d)* 180° *rotated crater.*

micro-organism, and (2) **Moons**, a collection of images of the moon drawn from Google images. As we can see from the figure, there is high variance in the image sizes and colours. The task is to learn a model to predict the correct category of light-source angle from real images, which is not presented in the images.

To deal with the noise among real images, LV enhances the low-level perception by employing pretrained statistical models. For example, the sampling of 'points'—which decides if an image pixel belongs to the edge of an target object—is based on a statistical image background model which can categorize pixels into foreground and background points (Zhang *et al.*, 2008). Compared to the gradient-based point sampler, statistical models are more robust in natural images. Moreover, the geometric background knowledge of LV includes shapes such as ellipse and circle to serve as mid-level concepts for representing objects of interest in microscopic and telescopic images that are often composed of curves. Another important mid-level concept for figuring out the light source position is 'highlight', which is determined by dividing the extracted shape into two halves and comparing their brightness. Finally, LV employs a noise-tolerant Meta-Interpretive Learner to perform one-shot learning from the extracted facts of mid-level concepts. The noise-tolerant version of MIL is implemented as a wrapper around the generalized MIL and returns the highest score hypothesis learned from randomly sampled examples after multiple iterations.

Experiments are conducted on LV versus statistics-based approaches. The performance is measured by predictive accuracy given different. The results in Figure 10.5 demonstrate that LV can learn an accurate model using a single training example.[1] By comparison, the statistics-based approaches require 40 or even 100 more training examples to reach similar accuracy. Examples of the learned hypotheses are shown with Figure 10.6, which is an abductive theory that explains the angle of the highlight observed on the object of interest.

During the learning process, LV also invents a predicate `clock_angle2` to represent the property of the object of interest `obj`, which can be interpreted as *convexity* or *concavity*. By including them in the background knowledge, LV is able to interpret ambiguity

[1] LV can use multiple examples but was limited to one-shot learning in the experiment, as described in (Muggleton *et al.*, 2018).

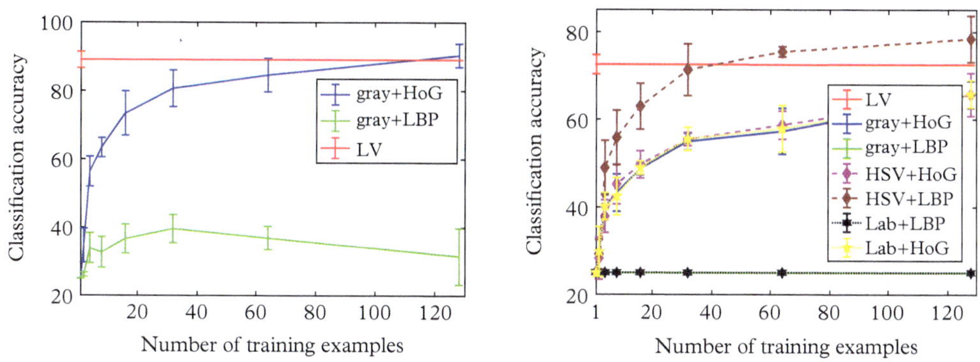

(a) Moons

(b) Protists

Figure 10.5 *Classification accuracy on the two real image datasets.*

```
clock_angle(A,B,C):-
    clock_angle1(A,B,D),
    light_source_angle(A,D,C).
clock_angle1(A,B,C):-
    highlight(A,B),
    clock_angle2(A), clock_angle3(C).
clock_angle2(Obj).
clock_angle3(Light).
```

(a)

```
clock_angle(A,B,C):-
    clock_angle1(A,B,D),
    clock_angle4(A,D,C).
clock_angle1(A,B,C):-
    highlight(A,B),
    clock_angle2(A),clock_angle3(C).
clock_angle4(A,B,C):-
    light_source_angle(A,B,D),
    opposite_angle(D,C).
clock_angle2(Obj).
clock_angle3(Light).
```

(b)

Figure 10.6 *Program learned by* LV: *(a) Hypothesis learned when training data only contains convex objects. (b) Hypothesis learned when training data only contains concave objects.* `clock_angle/3` *denotes the clock angle from* `B` *(highlight) to* `A` *(object).* `highlight/2` *is a built-in predicate meaning* `B` *is the brighter half of* `A`. `light_source_angle/3` *is an abducible predicate and the learning target. With background knowledge about lighting and comparing the two programs, we can interpret the invented predicate* `clock_angle2` *as* `convex`, `clock_angle3` *as* `light_source_name`.

like humans can. For example, Figure 10.4c and 10.4d shows two images of a crater on Mars, where Figure 10.4d is a 180° rotated image of Figure 10.4c. Human perception often confuses the convexity of the crater in such images. This phenomenon, called the *crater/mountain illusion*, occurs because human vision usually interprets pictures under the default assumption that the light is from the top of the image.

When we input Figure 10.4c to LV, it abduces four different hypotheses to explain the image, as shown in Figure 10.7. From the first two results we see that by considering different possibilities of light source direction, LV can predict that the main object (which is the crater) is either convex or concave, which shows the power of learning ambiguity.

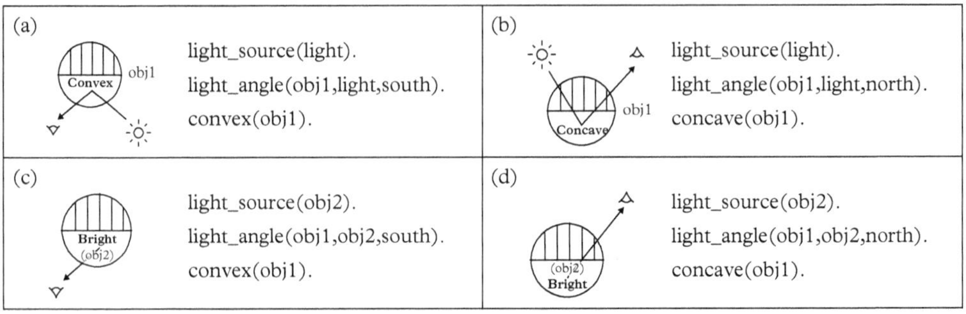

Figure 10.7 *Depiction and output hypotheses abduced from Figure 10.4c.*

The last two results are even more interesting: they suggest that `obj2` (the brighter half of the crater) might be the light source as well, which indeed is possible, though seems unlikely.

Hence, by applying a logic-based learning paradigm, LV is able to reuse the learned models in image processing. In this way, the paradigm approximates the human reasoning process in such scientific discovery tasks more closely than techniques unable to employ logical reasoning and learning.

10.4 Learning Low-level Perception through Logical Abduction

LV assumes that the model for low-level perception (e.g., gradient-based point detector and background/foreground segmentation model) are pretrained. However, in many cases, such a model is not available, and the logical theory that defines mid-level concepts with low-level features is also missing in the background knowledge. A typical example of this situation in archaeology is the discovery of the Mayan head-variant hieroglyphs (Houston *et al.*, 2001).

Mayan scripts were a mystery to modern humanity until its numerical systems and calendars were successfully deciphered in the late nineteenth century. As described by historians, the number recognition was derived from a handful of images that show mathematical regularity. The decipherment was not trivial because the Mayan calendars are composed of a sequence of unknown hieroglyphs, as shown in Figure 10.8a and Figure 10.8c.

Nevertheless, historians still managed to solve these puzzles. An example of the decipherment made by Charls Bowditch on the middle tablet in Figure 10.8a is illustrated with Figure 10.8b (Bowditch, 1910). In the tablets of Figure 10.8a, the large hieroglyph in rows 1–2 represent the mythical creation day, rows 3–6 are time spans represented with Mayan long count calendars, and rows 8–9 are Mayan Tzolk'in and Haab' calendars that encode the date computed from the creation date and long count. In rows 3–9, the odd columns represent numbers and the even columns represent time units. Column II shows the standard representations for the units; columns IV and VI are identical unit

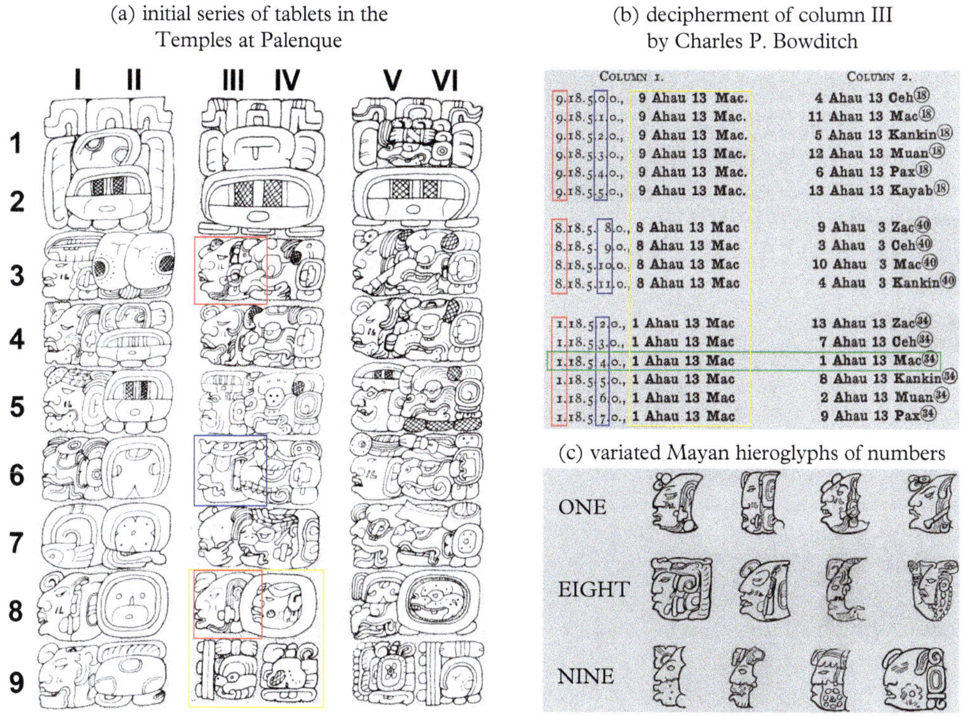

Figure 10.8 *Illustration of Mayan hieroglyph decipherment [Credits: (A) is reproduced from (Stuart, 2006); (B) and (C) are reproduced from (Bowditch, 1910)].*

representations but in different drawings. The hieroglyphs marked by boxes correspond to the numbers and units in the same coloured boxes in subfigure Figure 10.8b, in which the Column 1 lists possible interpretations from Column III of Figure 10.8a, and Column 2 lists the results calculated from the numbers in Column 1. Bowditch first identified the hieroglyphs at III4, III5, III7, and III9 in Figure 10.8a and then confirmed that III3 and III9 represent the same numbers. Initially, he conjectured that III3 is 9 based on his past experience with Mayan calendars; however, that was impossible because the calculated results were inconsistent with the dates on the tablet. Then, he tried substituting those positions with numbers that have similar hieroglyphs. Finally, he confirmed that the interpretation '1.18.5.4.0 1 Ahau 13 Mac' should be correct by validating its consistency with subsequent passages in the same tablet.

By using the approach above, Bowditch cracked most of the head-variant numbers, that is, successfully learned the low-level perception model mapping from images (Figure 10.8c) to natural numbers (mid-level symbols). The deciphering procedure succeeds because it utilizes the high-level reasoning based on mathematical background knowledge during perception, and these two abilities function at the same time and affect each other: in this case, a tunnel between perception and reasoning was established through a trial-and-error process of the hieroglyphic interpretations. The trial step perceives, interprets the picture, and passes the interpreted symbols for consistency checking, while the error step evaluates the consistency, uses reasoning to find errors in the interpretation, and provides error feedback to correct the perception.

This problem-solving process was called 'abduction' by Charles S. Peirce (1955) and termed 'retro-production' by Herbert A. Simon (Simon and Newell, 1971); it refers to the process of selectively inferring certain facts and hypotheses that explain phenomena and observations based on background knowledge (Magnani, 2009). In Bowditch's Mayan number decipherment, the background knowledge involved arithmetic and some basic facts about Mayan calendars; the hypotheses involved a recognition model for mapping hieroglyphs to meaningful symbols and a more complete understanding of the Mayan calendar system (e.g., the partial vigesimal number system). Finally, the validity of the hypotheses was ensured by trial-and-error searches and consistency checks.

The *Aductive Learning* (ABL) framework models the joint reasoning and perception, from which the logical reasoning uses structured and symbolic background knowledge to help learning the low-level perception models (Zhou, 2019; Dai *et al.*, 2019). The structure of this framework is illustrated in Figure 10.9, in which the input data such as images only have noisy low-level representations, the final concept is symbolic and constructed from complex relations among the mid-level symbols with background knowledge. As human abductive problem-solving, ABL trains the low-level perception model—which maps the low-level input to the mid-level symbols—by optimizing the consistency between perceived symbols and the reasoning. When the perception model is under-trained, the perceived mid-level symbols are likely to be wrong, so ABL uses trial-and-error to maximize the consistency, as the historians did.

Figure 18.1 shows an example of synthetic vision task, handwritten equation decipherment,[2] sharing the same idea of the Mayan hieroglyph decipherment: The equations for

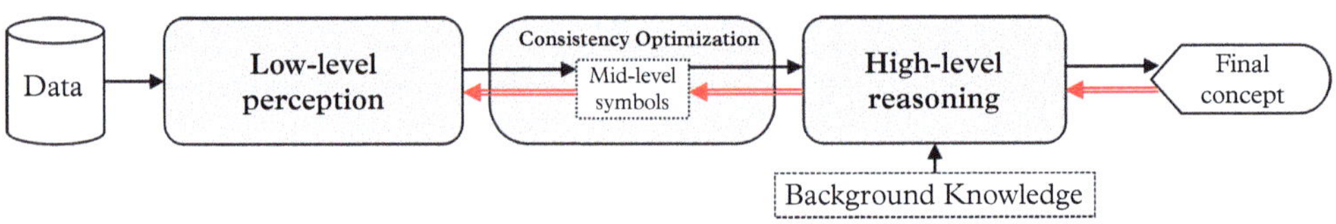

Figure 10.9 *The structure of ABL framework.*

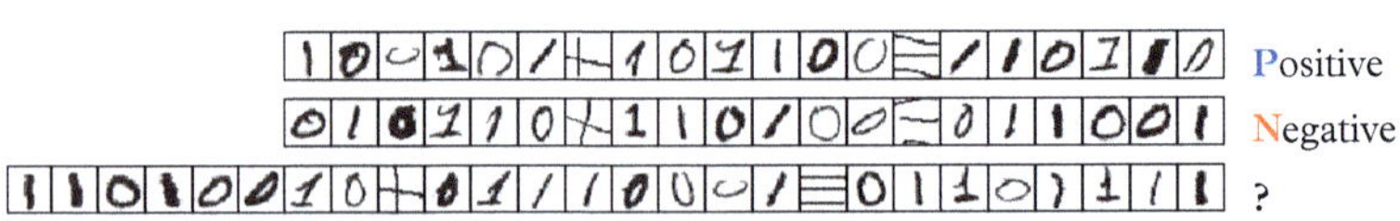

Figure 10.10 *Handwritten equation decipherment puzzle: a computer should learn to recognize the symbols and figure out the unknown operation rules ('xnor' in this example) simultaneously.*

[2] The dataset and codes are available at https://github.com/AbductiveLearning/ABL-HED. We are using synthetic datasets because it is difficult to obtain sufficient sample images of Mayan hieroglyphs for training the perceptual deep convolutional neural net.

the decipherment tasks consist of sequential pictures of characters in the form of low-level pixels. The equations are constructed from images of some mid-level symbols ('0', '1', '+' and '='), while the correspondence between image and the symbols are unknown. Moreover, the equations are generated with *unknown* operation rules, and each example is associated with a label of the final concept that indicates whether the equation is correct. A machine is tasked with learning from a training set of labelled equations, and the trained model is expected to predict unseen equations correctly. Thus, this task demands the same ability as a human jointly utilizing low-level perceptual and high-level reasoning abilities in Figure 10.8.

Before training, the domain knowledge—written as a logic program—is provided as the high-level background knowledge, which involves only the *structure* of the equations and a *recursive* definition of *bit-wise operations*. As shown in Table 10.1, the background knowledge about equation structures is a set of definite clause grammar (DCG) rules recursively define that a digit is a sequence of '0' and '1', and each equation shares the structure of X+Y=Z, although the length of X, Y, and Z may be varied. The knowledge about *bit-wise operations* is a recursive logic program that calculates X+Y in reverse, that is, it operates on X and Y digit by digit and from the last digit to the first. Meanwhile, the specific rules for calculating the operations are *undefined* in B, that is, results of '0+0', '0+1' and '1+1' could be '0', '1', '00', '01' or even '10'. The missing calculation rules are also required to be *learned* from the data.

Table 10.1 *Background knowledge about bit-wise calculation.*

```
% Abductive bit-wise calculation with given pseudo-labels,
% this procedure abduces missing pseudo-labels together with
% unknown operation rules.
calc(Rules, Pseudo) :-
    calc([], Rules, Pseudo).
calc(Rules0, Rules1, Pseudo) :-
    parse_eq(Pseudo, eq(X,Y,Z)),
    bitwise_calc(Rules0, Rules1, X, Y, Z).

% Bit-wise calculation that handles carrying
bitwise_calc(Rules, Rules1, X, Y, Z) :-
    reverse(X, X1), reverse(Y, Y1), reverse(Z, Z1),
    bitwise_calc_r(Rules, Rules1, X1, Y1, Z1),
    maplist(digits, [X,Y,Z]).
% Recursively calculate back-to-front
bitwise_calc_r(Rs, Rs, [], Y, Y).
bitwise_calc_r(Rs, Rs, X, [], X).
bitwise_calc_r(Rules, Rules1, [D1|X], [D2|Y], [D3|Z]) :-
    % Abduces Δ_C (my_op/3) during the calculation.
    abduce_op_rule(my_op([D1],[D2],Sum), Rules, Rules2),
    % Handling carry
    ((Sum = [D3], Carry = []); (Sum = [C,D3], Carry = [C])),
    bitwise_calc_r(Rules2, Rules3, X, Carry, X_carried),
    bitwise_calc_r(Rules3, Rules1, X_carried, Y, Z).
```

After training starts, the low-level perception model will try to interpret the images to the symbolic equations constructed by mid-level symbols '0', '1', '+' and '='. Because the perception model is untrained, the perceived symbols are typically wrong. In this case, the reasoning part cannot abduce any consistent hypothesis, that is, no calculation rules can establish a proof from the perceived mid-level symbols to the associated labels. In order to find the consistent assignments of the symbols to the images and crack the calculation rules, ABL tries to mark possible incorrectly perceived mid-level symbols like the historians.

For example, in the beginning, the under-trained perception model is highly likely to interpret the images as eq_0= `[1,1,1,1,1]`, which is inconsistent with any binary operations since it has no operator symbol. Observing that, the logic program cannot abduce any consistent hypothesis, therefore ABL will try to substitut the 'possibly incorrect' pseudo-labels in eq_0 with blank Prolog variables, for example, eq_1= `[1,-,1,-,1]`. Then, the logic program can abduce a consistent hypothesis involving the operation rule `op(1,1,[1])` and a list of revised symbols eq_1`'=[1,+,1,=,1]`, where the latter one is used for retrain the perceptual vision model, helping it distinguish images of '+' and '=' from other symbols.

The ABL framework has been tested in two image datasets, which are shown in Figure 10.11. The Digital Binary Additive (DBA) equations were created with images from benchmark handwritten character datasets (LeCun *et al.*, 1998; Thoma, 2017), while the Random Symbol Binary Additive (RBA) equations were constructed from randomly selected characters sets of the Omniglot dataset (Lake *et al.*, 2015) and shared isomorphic structure with the equations in the DBA tasks. In order to evaluate the perceptual generalization ability of the compared methods, the images for generating the training and test equations are disjoint. Each equation is input as a sequence of raw images of digits and operators. Both training and testing data contains equations with lengths 5–26.

The experiments compared ABL with three widely used deep neural nets (DNN) performing end-to-end learning, which do not use any human background knowledge: Differentiable Neural Computer (DNC) (Graves *et al.*, 2016), Transformer

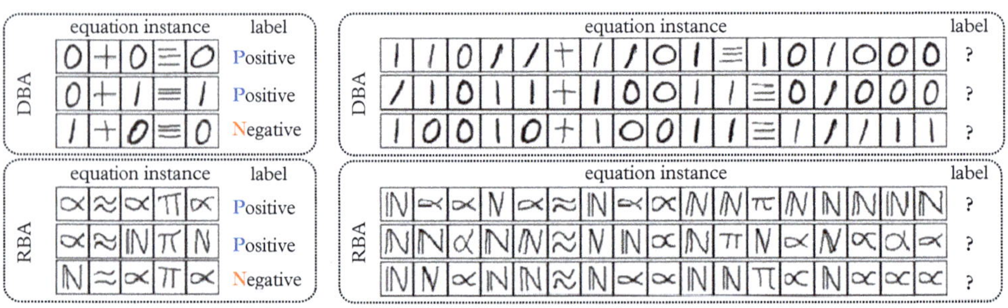

(a) Training examples (b) Test examples

Figure 10.11 *Data examples for the handwritten equations decipherment tasks.*

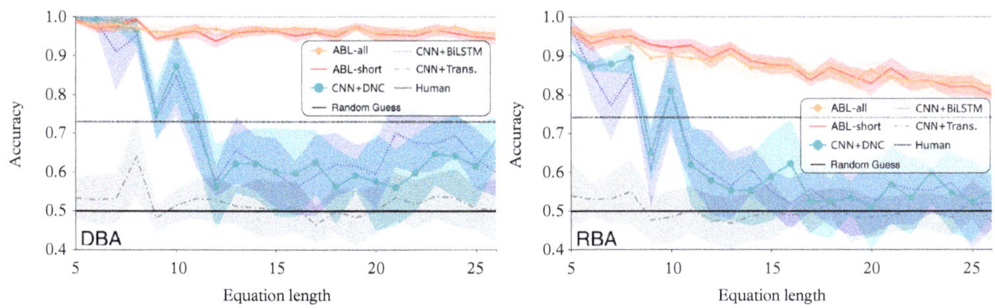

Figure 10.12 *Experimental results of the DBA (left) and RBA (right) tasks.*

networks (Vaswani *et al.*, 2017) and Bidirectional Long Short-Term Memory Network (BiLSTM) (Schuster and Paliwal, 1997). Two different settings have been tried with ABL: the *ABL-all* that uses all training data and the *ABL-short* that only uses training equations of lengths 5–8.

Besides of the machine-learning experiments, a human experiment has also been carried out. Forty volunteers were asked to classify images of equations sampled from the same datasets. Before taking the quiz, the domain knowledge about the bit-wise operation was provided as hints, but specific calculation rules were not available—just like the setting for ABL. Instead of using the precisely same setting as the machine learning experiments, we gave the human volunteers a simplified version, which only contains 5 positive and 5 negative equations with lengths ranging from 5–14.

Figure 10.12 shows that on both tasks, the ABL-based approaches significantly outperform the compared methods, and ABL correctly learned the symbolic rules defining the unknown operations. All the methods performed better on the DBA tasks than RBA because the symbol images in the DBA task are more easily distinguished. The performance of ABL-all and ABL-short have no significant difference, and the performance of the compared approaches degenerates quickly toward the random-guess line as the length of the testing equations grows, while the ABL-based approaches extrapolates better to the unseen data. An interesting result is that the human performance on the two tasks are very close, and both of them are worse than that of ABL. According to the volunteers, they do not suffer from distinguishing different symbols, but machines are better in checking the consistency of logical theories—in which people are prone to make mistakes. Therefore, machine learning systems should make use of their advantages in logical reasoning.

10.5 Conclusion and Future Work

Human beings often learn visual concepts from single or few image presentations (so-called one-shot-learning) (Lake *et al.*, 2011). This phenomenon is hard to explain from a standard machine-learning perspective, given that it is unclear how to estimate any statistical parameter from a single randomly selected instance drawn from an unknown distribution. In this chapter we show that learnable generic logical background knowledge can be used to generate high-accuracy logical hypotheses from few examples. This compares with similar demonstrations concerning one-shot MIL on string transformations (Lin *et al.*, 2014) as well as previous concept learning in artificial images (Dai *et al.*, 2015).

The experiments in section 10.3.2 show that the LV system can accurately identify the position of a light source from a single real image, in a way analogous to scientists such as Galileo observing the moon for the first time through a telescope or Hook observing micro-organisms for the first time through a microscope. We show that logical theories learned by LV from labelled images can also be used to predict concavity and convexity predicated on the assumed position of a light source. The experiments in section 10.4 show that the ABL system is able to learn visual concepts with few training examples even when the low-level perception model is unavailable.

We have studied the failure cases carefully. The main reason causing LV's misclassification is the noise in images. The noise can cause misclassifications of edge_point/1 since it is implemented with statistical models. The mistakes of edge_point detection will further affect the edge detection and shape fitting. As a result, the accuracy of the main object extraction is limited by both the noise level in input images and the power of statistical model of edge_point/1. Therefore, LV will fail too since the wrongly extracted objects are its inputs. However, if we train stronger models for detecting edge_points, the accuracy of LV will not increase either. To solve this problem, ABL proposes to train the statistics-based model with the logic-based model jointly. The experiments show that the abduction-based logical reasoning can automatically revise the incorrectly perceived primitive facts and retrain the statistical model.

In further work we aim to investigate broader sets of visual phenomena which can naturally be treated using background knowledge. For instance, the effects of object obscuration; the interpretation of shadows in an image to infer the existence of out-of-frame objects; the existence of unseen objects reflected in a mirror found within the image. All these phenomena could possibly be considered in a general way from the point of view of a logical theory describing reflection and absorption of light, where each image pixel is used as evidence of photons arriving at the image plane. In this further work we aim to compare our approach once more against a wider variety of competing methods.

In future we hope to demonstrate how performance varies with increasing noise using a synthetic dataset. Moreover, we will systematically investigate the relationship between low-level perception and high-level reasoning in LV and ABL. For example, how does the low-level perceptual noise will affect the time complexity of logical abduction?

The authors believe that LV has long-term potential as an AI technology with the ability to unify the disparate areas of logical-based learning with visual perception.

References

Andrzejewski, D., Zhu, X., Craven, M. *et al.* (2011). A framework for incorporating general domain knowledge into latent dirichlet allocation using first-order logic, in *Proceedings of the 22nd International Joint Conference on Artificial Intelligence*, Barcelona, Spain, 1171–2573.

Borenstein, E. and Ullman, S. (2008). Combined top-down/bottom-up segmentation. *IEEE Transactions on Pattern Analysis and Machine Intelligence*, **30**(12), 2109–25.

Bowditch, C. P. (1910). *The Numeration, Calendar Systems and Astronomical Knowledge of The Mayas*. Cambridge: Cambridge University Press.

Bronkhorst, H., Roorda, G., Suhre, C. *et al.* (2020). Logical reasoning in formal and everyday reasoning tasks. *International Journal of Science and Mathematics Education.* **18** 1673–1694

Cox, D. (2014). Do we understand high-level vision? *Current Opinion in Neurobiology*, **25**, 187–93.

Dai, W.-Z. and Zhou, Z.-H. (2017). Combining logical abduction and statistical induction: Discovering written primitives with human knowledge, in *Proceedings of the 31st AAAI Conference on Artificial Intelligence*, San Francisco, CA, 4392–8. AAAI Press, Palo Alto, California.

Dai, W.-Z., Muggleton, S. H., and Zhou, Z.-H. (2015). Logical Vision: Meta-interpretive learning for simple geometrical concepts, in *Late Breaking Paper Proceedings of the 25th International Conference on Inductive Logic Programming*, 20–22 August 2015. Kyoto, Japan, 1–16. CEUR, Germany.

Dai, W.-Z., Muggleton, S. H., Wen, J. *et al.* (2017). Logic vision: One-shot meta-intepretive learning from real images, in Nicholas Lachiche and Christel Vrain. *Proceedings of the 27th International Conference on Inductive Logic Programming*, 4–6 September, 46–62. Springer-Verlag, Orleans, France.

Dai, W.-Z., Xu, Q., Yu, Y. *et al.* (2019). Bridging machine learning and logical reasoning by abductive learning, in edited by: H. Wallach and H. Larochelle and A. Beygelzimer and F. d'Alché-Buc and E. Fox and R. Garnett. *Advances in Neural Information Processing Systems 32 (NeurIPS 2019)*, 8–14 December 2019 p. in press. Curran Associates, Inc., Red Hook, New York.

Dubba, K. S. R., Cohn, A. G., Hogg, D. C., *et al.* (2015). Learning relational event models from video. *Journal of Artificial Intelligence Research*, **53**, 41–90.

Farid, R. and Sammut, C. (2014). Plane-based object categorization using relational learning. *Machine Learning*, **94**, 3–23.

Felzenszwalb, P. (2011). Object detection grammars, in Dimitris N. Metaxas, Long Quan, Alberto Sanfeliu, Luc Van Gool. *IEEE International Conference on Computer Vision, ICCV 2011*, Barcelona, Spain, November 6–13, 691–691.

Flach, P. A. and Kakas, A. C. (eds) (2000). *Abduction and Induction.* Dordrecht: Springer.

Galilei, Galileo (1610). *The Herald of the Stars.* Edward Stafford trans, Peter Barker ed., London: Byzantium Press, 2004.

Goodfellow, I., Bengio, Y., and Courville, A. (2016). *Deep Learning.* Cambridge, MA: MIT Press.

Graves, A., Wayne, G., Reynolds, M. *et al.* (2016). Hybrid computing using a neural network with dynamic external memory. *Nature*, **538**(7626), 471–6.

Gregory, R. L. (1974). *Concepts and Mechanics of Perception.* London: Duckworth.

Gregory, R. L. (1998). *Eye and Brain: The Psychology of Seeing.* Oxford: Oxford University Press.

Hartz, J. and Neumann, B. (2007). Learning a knowledge base of ontological concepts for high-level scene interpretation, in Arif Wani and Mehmed Kantardzic *Proceedings of the 6th International Conference on Machine Learning and Applications*, 13–15 December 2007, Cincinnati, OH, 436–43. IEEE Press, New York

Heath, D. and Ventura, D. (2016). Before a computer can draw, it must first learn to see, in edited by François Pachet, Amilcar Cardoso, Vincent Corruble, Fiammetta Ghedin *Proceedings of the 7th International Conference on Computational Creativity*, Paris France, 28th June to 1st July 172–9. Sony CSL Paris, France

Horn, B. K. P. (1989). *Obtaining Shape from Shading Information.* Cambridge, MA: MIT Press.
Houston, S. D., Mazariegos, O. C., and Stuart, D. (2001). *The Decipherment of Ancient Maya Writing.* Norman, OK: University of Oklahoma Press.
Jin, Y. and Geman, S. (2006). Context and hierarchy in a probabilistic image model, in Daniel Huttenlocher, David Forsyth, Andrew Fitzgibbon, Camillo Taylor and Yann LeCun *IEEE Computer Society Conference on Computer Vision and Pattern Recognition (CVPR'06)*, 17–22 June 2006, New York, NY, 2145–52.
Kakas, A. C., Kowalski, R. A., and Toni, F. (1992). Abductive logic programming. *Journal of Logic Computation*, **2**(6), 719–70.
Krizhevsky, A., Sutskever, I., and Hinton, G. E. (2012). Imagenet classification with deep convolutional neural networks, in Peter Bartlett *Advances in Neural Information Processing Systems 25: 26th Annual Conference on Neural Information Processing Systems*, 3–6 December 2012, Lake Tahoe, Nevada, USA. 1097–105. Curran Associates, Inc., Red Hook, New York.
Lake, B. M., Salakhutdinov, R., Gross, J. *et al.* (2011). One shot learning of simple visual concepts, in Carlson, L. *33rd Annual Meeting of the Cognitive Science Society 2011 (CogSci 2011)*, 20–23 July 2011, Boston, Massachusetts, USA. 2568–73. Curran Associates, Inc, Red Hook, NY.
Lake, B. M., Salakhutdinov, R., and Tenenbaum, J. B. (2015). Human-level concept learning through probabilistic program induction. *Science*, **350**(6266), 1332–8.
Le, Q. V., Zou, W. Y., Yeung, S. Y. *et al.* (2011). Learning hierarchical invariant spatio-temporal features for action recognition with independent subspace analysis, in *CVPR 2011*, IEEE, 3361–68.
LeCun, Y., Bottou, L., Bengio, Y. *et al.* (1998). Gradient-based learning applied to document recognition. *Proceedings of the IEEE*, **86**(11), 2278–324.
Lin, D., Dechter, E., Ellis, K. *et al.* (2014). Bias reformulation for one-shot function induction, in *Proceedings of the Twenty-first European Conference on Artificial Intelligence*, 525–30.
Liu, J., Huang, Y., Wang, L. *et al.* (2014). Hierarchical feature coding for image classification. *Neurocomputing*, **144**, 509–15.
Maclin, R., Shavlik, J., Walker, T. *et al.* (2005). Knowledge-based support-vector regression for reinforcement learning. In *Reasoning, Representation, and Learning in Computer Games*, 61.
Magnani, L. (2009). *Abductive Cognition: The Epistemological and Eco-Cognitive Dimensions of Hypothetical Reasoning*. Berlin: Springer-Verlag.
Marr, D. (1982). *Vision: A Computational Investigation into the Human Representation and Processing of Visual Information.* New York, NY: Henry Holt & Co., Inc.
Mei, S., Zhu, J., and Zhu, J. (2014). Robust regbayes: Selectively incorporating first-order logic domain knowledge into bayesian models, in *International Conference on Machine Learning*, 253–61.
Muggleton, S. H. (1991). Inductive logic programming. *New Generation Computing*, **8**(4), 295–318.
Muggleton, S. H. (2017). Meta-interpretive learning: achievements and challenges, in *Proceedings of the 11th International Symposium on Rule Technologies, RuleML+RR 2017* (eds R. Kontchakov and F. Sadri), Berlin: Springer-Verlag, 1–7.
Muggleton, S. H., Dai, W.-Z., Sammut, C. *et al.* (2018). Meta-interpretive learning from noisy images. *Machine Learning*, **107**, 1097–118.
Muggleton, S. H., Lin, D., Pahlavi, N. *et al.* (2014). Meta-interpretive learning: application to grammatical inference. *Machine Learning*, **94**, 25–49.
Muggleton, S. H., Lin, D. and Tamaddoni-Nezhad, A. (2015). Meta-interpretive learning of higher-order dyadic datalog: predicate invention revisited. *Machine Learning*, **100**(1), 49–73.

Muggleton, S. H. and Raedt, L. De (1994). Inductive logic programming: theory and methods. *Journal of Logic Programming*, **19,20**, 629–679.

Muggleton, S. H., Raedt, L. De, Poole, D. *et al.* (2011). ILP turns 20: biography and future challenges. *Machine Learning*, **86**(1), 3–23.

Needham, C. J., Santos, P. E., Magee, D. R. *et al.* (2005). Protocols from perceptual observations. *Artificial Intelligence*, **167**, 103–36.

Newell, A. and Simon, H. A. (1976). Computer science as empirical inquiry: symbols and search. *Communications of the ACM*, **19**(3), 113–26.

Ohta, Y., Kanade, T., and Sakai, T. (1978). An analysis system for scenes containing objects with substructures, in *Proceedings of the Fourth International Joint Conference on Pattern Recognitions*, 752–54.

Peirce, S. C. (1955). Abduction and induction, in J Buchler ed., *Philosophical Writings of Peirce*. New York, NY: Dover Publications.

Poppe, R. (2010). A survey on vision-based human action recognition. *Image and Vision Computing*, **28**(6), 976–90.

Porway, J., Wang, Q., and Zhu, S.-C. (2010). A hierarchical and contextual model for aerial image parsing. *International Journal of Computer Vision*, **88**(2), 254–83.

Schuster, M. and Paliwal, K. K. (1997). Bidirectional recurrent neural networks. *IEEE Transactions on Signal Processing*, **45**(11), 2673–81.

Sermanet, P., Eigen, D., Zhang, X. *et al.* (2014). Overfeat: Integrated recognition, localization and detection using convolutional networks, in *2nd International Conference on Learning Representations, ICLR* 2014.

Simon, H. A. and Newell, A. (1971). Human problem solving: the state of the theory in 1970. *American Psychologist*, **26**(2), 145.

Simonyan, K. and Zisserman, A. (2015). Very deep convolutional networks for large-scale image recognition, in *Proceedings of the 3rd International Conference on Learning Representations*, San Diego, CA.

Stuart, David (2006). *Sourcebook for the 30th Maya Meetings Part II: The Malenque Mythology: Inscriptions from the Cross Group at Palenque*. Austin, TX: University of Texas Austin Press.

Tenenbaum, J. B., Kemp, C., Griffiths, T. L. *et al.* (2011). How to grow a mind: Statistics, structure, and abstraction. *Science*, **331**, 1279–85.

Thoma, M. (2017). The HASYv2 dataset. *CoRR*, **abs/1701.08380**, 1–8.

Vaswani, A., Shazeer, N., Parmar, N. *et al.* (2017). Attention is all you need, in Von Luxburg, U. *Advances in Neural Information Processing Systems 30*, 5998–6008. Curran Associates, Inc., Red Hook, New York

von Helmholtz, H. (1962). *Treatise on Physiological Optics Volume 3*. New York, NY: Dover Publications. (Originally published in German in 1825).

Zhang, H., Fritts, J. E., and Goldman, S. A. (2008). Image segmentation evaluation: A survey of unsupervised methods. *Computer Vision and Image Understanding*, **110**(2), 260–80.

Zhang, R., Tai, P. S., Cryer, J. E. *et al.* (1999). Shape-from-shading: a survey. *IEEE Transactions on Pattern Analysis and Machine Intelligence*, **21**(8), 670–706.

Zheng, S., Tu, Z., and Yuille, A. (2007). Detecting object boundaries using low-, mid-, and high-level information. *Computer Vision and Image Understanding*, **114**(10), 1055–67.

Zhou, Z.-H. (2019). Abductive learning: towards bridging machine learning and logical reasoning. *Science China Information Sciences*, **62**(7), 1–3.

11

Apperception

Richard Evans
Imperial College London and DeepMind, UK

11.1 Introduction

'Apperception' (from the French 'apercevoir') is a term introduced by Leibniz in the *New Essays Concerning Human Understanding* (Leibniz, 1765), and is a central term in Kant's *Critique of Pure Reason* (Kant, 1781). Nowadays, the term has two distinct but related meanings: first, to apperceive something is to be self-consciously aware of it; second, to apperceive something is to assimilate it into a larger collection of ideas, to make sense of something by unifying it with one's other thoughts into a coherent whole. In this chapter, I focus on the second sense.

Apperception should be distinguished from mere low-level perception. Consider a neural network trained to classify images into classes. Although the network can perceive the image as a dog, *it has no idea what a dog actually is.* The network may be able to classify the image under the label 'dog', but it does not understand what 'dog' means, since it does not understand the *inferential connections* between the concept 'dog' and other concepts: it does not know that dogs are mammals, that no dog is also a cat, or that dogs are (typically) loyal to their masters. In other words, the neural network may be able to perceive but it is not able to apperceive, since it does not understand the inferential relations between concepts.

To apperceive, then, is to make sense of one's sensory inputs by combining them into a unified whole using concepts and inferential relations between concepts.

In this chapter, I describe a computer model of apperception. This builds on previous work Evans *et al.*, 2021. Our system, the Apperception Engine, takes as input a sensory sequence and constructs an explicit causal theory that both explains the sequence and also satisfies a set of unity conditions designed to ensure that the constituents of the theory—the objects, properties, and propositions—are combined together in a relational structure. In this previous work, we showed, in a range of experiments, how this system is able to outperform recurrent networks and other baselines on a range of tasks, including Hofstadter's *Seek Whence* dataset.

But in our initial implementation, there was one fundamental limitation: we assumed the sensory given had already been discretized. We assumed *some other system had already*

Richard Evans, *Apperception* In: *Human Like Machine Intelligence.* Edited by: Stephen Muggleton and Nick Charter, Oxford University Press.
 DOI: 10.1093/oso/9780198862536.003.0011

parsed the raw sensory input using a set of discrete categories, so that all the Apperception Engine had to do was receive this already-digested discretized input, and make sense of it (Smith, 2019). But what if we don't have access to pre-parsed input? What if our sensory sequence is raw unprocessed information, say, a sequence of noisy pixel arrays from a video camera, for example?

The central contribution of this chapter is a major extension of the Apperception Engine so that it can be applied to raw unprocessed sensory input. This involves two phases. First, we extend our system to receive ambiguous (but still discretized) input: sequences of disjunctions. Second, we use a neural network to map raw sensory input to disjunctive input. Our binary neural network is encoded as a logic program, so the weights of the network and the rules of the theory can be found *jointly* by solving a single SAT problem. This way, we are able to jointly learn how to perceive (mapping raw sensory information to concepts) and apperceive (combining concepts into declarative rules).

We test the Apperception Engine in the Sokoban domain, and show how the system is able to learn the game's dynamics from a sequence of noisy pixel arrays. This system is, to the best of our knowledge, the first system that is able to learn explicit provably correct dynamics of non-trivial games from raw pixel input.

11.2 Method

In describing the Apperception Engine, we shall use three increasingly complex forms of sequential input. We start by assuming that the sensory sequence has already been discretized into ground atoms of first-order logic representing sensor readings. So, for example, the (discrete) fact that light sensor a is red might be represented by the atom $red(a)$. Next, we expand to consider disjunctive input sequences. For example, to represent that sensor a is either red or $orange$ we would represent our uncertainty by $red(a) \vee orange(a)$. Finally, we let go entirely of our simplifying assumption of already-discretised sensory input and consider sequences of *raw unprocessed input*. Consider, for example, a sequence of pixel arrays from a video camera.

11.2.1 Making sense of unambiguous symbolic input

Definition 1: *An* **unambiguous symbolic sensory sequence** *is a sequence of sets of ground atoms. Given a sequence* $S = (S_1, S_2, ...)$, *a* **state** S_t *is a set of ground atoms, representing a* partial *description of the world at a discrete time-step* t.

Example 1: Consider, for example, the following sequence $S_{1:10}$. Here there are two sensors a and b, and each sensor can be on or off.

$$
\begin{array}{ll}
S_1 = \{\} & S_2 = \{off(a), on(b)\} \\
S_3 = \{on(a), off(b)\} & S_4 = \{on(a), on(b)\} \\
S_5 = \{on(b)\} & S_6 = \{on(a), off(b)\} \\
S_7 = \{on(a), on(b)\} & S_8 = \{off(a), on(b)\} \\
S_9 = \{on(a)\} & S_{10} = \{\}
\end{array}
$$

Note that there is no expectation that a sensory sequence contains readings for all sensors at all time steps. Some of the readings may be missing. In state S_5, we are missing a reading for a, while in state S_9, we are missing a reading for b. In states S_1 and S_{10}, we are missing sensor readings for both a and b.

The Apperception Engine makes sense of a sensory sequence by constructing a unified theory that explains that sequence. The key notions, here, are 'theory', 'explains', and 'unified'. We consider each in turn.

Definition 2: *A* **theory** *is a four-tuple* (ϕ, I, R, C) *where:*

- ϕ *is a type signature specifying the types of objects, variables, and arguments of predicates*
- I *is a set of initial conditions (ground atoms that are well-typed according to* ϕ*)*
- R *is a set of rules in Datalog*$^{\ni}$*, an extension of Datalog for representing causal rules*
- C *is a set of constraints*

There are two types of rule in Datalog$^{\ni}$*. A* **static rule** *is a definite clause of the form* $\alpha_1 \wedge ... \wedge \alpha_n \rightarrow \alpha_0$*, where* $n \geq 0$ *and each* α_i *is an* unground *atom consisting of a predicate and a list of variables. This means: if conditions* $\alpha_1, ... \alpha_n$ *hold at the current time-step, then* α_0 *also holds at that time-step. A* **causal rule** *is a clause of the form* $\alpha_1 \wedge ... \wedge \alpha_n \ni \alpha_0$*, where* $n \geq 0$ *and each* α_i *is an unground. A causal rule expresses how facts change over time.* $\alpha_1 \wedge ... \wedge \alpha_n \ni \alpha_0$ *states that if conditions* $\alpha_1, ... \alpha_n$ *hold at the current time-step, then* α_0 *will be true at the* next *time-step. So, for example,* $on(X) \ni off(X)$ *states that if object* X *is currently on, then* X *will become off at the next-time-step.*

There are three types of constraint. A **unary xor constraint** *is an expression of the form* $\forall X{:}t, p_1(X) \oplus ... \oplus p_n(X)$ *where* $n > 1$*. Here,* $\oplus$ *means exclusive-or, so for example* $\forall X{:}t, p_1(X) \oplus p_2(X)$ *means that for every object* X *of type t,* X *either satisfies* p_1 *or* p_2*, but not both. A* **binary xor constraint** *is an expression of the form* $\forall X{:}t_1, \forall Y{:}t_2, r_1(X,Y) \oplus ... \oplus r_n(X,Y)$ *where* $n > 1$*. A* **uniqueness constraint** *is an expression of the form* $\forall X{:}t_1, \exists! Y{:}t_2, r(X,Y)$*, which means that for all objects* X *of type* t_1 *there exists a* unique *object* Y *of type* t_2 *such that* $r(X,Y)$*.*

Example 2: Consider the frame $\phi = (T, O, P, V)$, consisting of types $T = \{s\}$, objects $O = \{a{:}s, b{:}s\}$, predicates $P = \{on(s), off(s), r(s,s), p_1(s), p_2(s), p_3(s)\}$, and variables $V = \{X{:}s, Y{:}s\}$. Consider the theory $\theta = (\phi, I, R, C)$, where:

$$I = \left\{ \begin{array}{l} p_1(b) \\ p_2(a) \\ r(a,b) \\ r(b,a) \end{array} \right\} \quad R = \left\{ \begin{array}{l} p_1(X) \ni p_2(X) \\ p_2(X) \ni p_3(X) \\ p_3(X) \ni p_1(X) \\ p_1(X) \rightarrow on(X) \\ p_2(X) \rightarrow on(X) \\ p_3(X) \rightarrow off(X) \end{array} \right\} \quad C = \left\{ \begin{array}{l} \forall X{:}s,\ on(X) \oplus off(X) \\ \forall X{:}s,\ p_1(X) \oplus p_2(X) \oplus p_3(X) \\ \forall X{:}s,\ \exists! Y{:}s\ r(X,Y) \end{array} \right\}$$

Definition 3: *Every theory* $\theta = (\phi, I, R, C)$ *generates an infinite sequence* $\tau(\theta)$ *of sets of ground atoms, called the* **trace** *of that theory. Here,* $\tau(\theta) = (A_1, A_2, ...)$*, where each* A_t *is the smallest set of atoms satisfying the following conditions:*

- $I \subseteq A_1$
- *If there is a static rule* $\beta_1 \wedge ... \wedge \beta_m \rightarrow \alpha$ *in* R *and a ground substitution* σ *such that* A_t *satisfies* $\beta_i[\sigma]$ *for each antecedent* β_i*, then* $\alpha[\sigma] \in A_t$
- *If there is a causal rule* $\beta_1 \wedge ... \wedge \beta_m \ni\!\!- \alpha$ *in* R *and a ground substitution* σ *such that* A_t *satisfies* $\beta_i[\sigma]$ *for each antecedent* β_i*, then* $\alpha[\sigma] \in A_{t+1}$
- Frame axiom: *if* α *is in* A_{t-1} *and there is no atom in* A_t *that is incompossible with* α *w.r.t constraints* C*, then* $\alpha \in A_t$*. Two ground atoms are* **incompossible** *if they are ruled out by one of the constraints in* C*.*

Note that the state transition function is deterministic: A_{t+1} *is uniquely determined by* A_t*.*

Example 3: The infinite trace $\tau(\theta) = (A_1, A_2, ...)$ for the theory θ of Example 2 begins with:

$$
\begin{aligned}
A_1 &= \{on(a), on(b), p_2(a), p_1(b), r(a,b), r(b,a)\} \\
A_2 &= \{off(a), on(b), p_3(a), p_2(b), r(a,b), r(b,a)\} \\
A_3 &= \{on(a), off(b), p_1(a), p_3(b), r(a,b), r(b,a)\} \\
A_4 &= \{on(a), on(b), p_2(a), p_1(b), r(a,b), r(b,a)\} \\
&\ldots
\end{aligned}
$$

Note that the trace repeats at step 4. It is possible to show in general that the trace always repeats after some finite set of time steps, since the set of ground atoms is finite and the state transition of Definition 3 is Markov.

Now that we have defined what we mean by a 'theory', next we define what it means for a theory to 'explain' a sensory sequence.

Definition 4: *Given two finite sequences* $S = (S_1, ..., S_T)$ *and* $S' = (S'_1, ..., S'_{T'})$*, define* $\sqsubseteq$ *as:*

$$S \sqsubseteq S' \textit{ iff } T \leq T' \textit{ and } \forall 1 \leq i \leq t,\ S_i \subseteq S'_i$$

So far, we assume both sequences are finite. Now we extend this partial order so that S' *may be infinitely long:* $S \sqsubseteq S'$ *if* S' *has a finite subsequence* S'' *such that* $S \sqsubseteq S''$*. If* $S \sqsubseteq S'$*, we say that* S *is* **covered by** S'*, or that* S' *covers* S *(Lee and De Raedt, 2004; Tamaddoni-Nezhad and Muggleton, 2009).*

A theory θ **explains** *a sensory sequence* S *if the trace of* θ *covers* S*, i.e.* $S \sqsubseteq \tau(\theta)$*.*

Example 4: The theory θ of Example 2 explains the sensory sequence S of Example 1, since the trace $\tau(\theta)$ (as shown in Example 3) covers S. Note that $\tau(\theta)$ 'fills in the blanks' in the original sequence S, both predicting final time-step 10 and retrodicting initial time-step 1.

Next, we proceed from explaining a sensory sequence to 'making sense' of that sequence. In providing a theory θ that explains a sensory sequence S, we make S intelligible by placing it within a bigger picture: while S is a scanty and incomplete description of a fragment of the time-series, $\tau(\theta)$ is a complete and determinate description of the whole time-series.

In order for θ to make sense of S, it is *necessary* that $\tau(\theta)$ covers S. But this condition is not, on its own, *sufficient*. The extra condition that is needed for θ to count as 'making sense' of S is for θ to be *unified*. In this paper, we formalize what it means for a theory to be 'unified' using Kant's key notion of the 'synthetic unity of apperception'.[1,2]

Definition 5: *A trace $\tau(\theta)$ is (i) a sequence of (ii) sets of ground atoms composed of (iii) predicates and (iv) objects.*[3] *For the theory θ to be* **unified** *is for unity to be achieved at each of these four levels:*

- *Objects are united in space.*[4] *A theory θ satisfies* **spatial unity** *if for each state A_i in $\tau(\theta) = (A_1, A_2, ...)$, for each pair (x, y) of distinct objects, x and y are connected via a chain of binary atoms $\{r_1(x, z_1), r_2(z_1, z_2), ... r_n(z_{n-1}, z_n), r_{n+1}(z_n, y)\} \subseteq A_i$*
- *Predicates are united via constraints.*[5] *A theory $\theta = (\phi, I, R, C)$ satisfies* **conceptual unity** *if for each unary predicate p in ϕ, there is some xor constraint in C of the form $\forall X{:}t, p(X) \oplus q(X) \oplus ...$ containing p; and, for each binary predicate r in ϕ, there is some xor constraint in C of the form $\forall X{:}t_1, \forall Y{:}t_2, r(X, Y) \oplus s(X, Y) \oplus ...$ or some $\exists!$ constraint in C of the form $\forall X{:}t, \exists! Y{:}t_2, r(X, Y)$.*
- *Ground atoms are united into sets (states) by jointly respecting constraints and static rules*[6]. *A theory $\theta = (\phi, I, R, C)$ satisfies* **static unity** *if every state $(A_1, A_2, ...)$ in $\tau(\theta)$ satisfies all the constraints in C and is closed under the static rules in R.*
- *States (sets of ground atoms) are united into a sequence by causal rules*[7]. *A sequence $(A_1, A_2, ...)$ of states satisfies* **temporal unity** *with respect to a set $R_{\ni}$ of causal rules if, for each $\alpha_1 \wedge ... \wedge \alpha_n \ni \alpha_0$ in $R_{\ni}$, for each ground substitution σ, for each time-step t, if $\{\alpha_1\sigma, ..., \alpha_n\sigma\} \subseteq A_t$ then $\alpha_0\sigma \in A_{t+1}$. Note that, from the definition of the trace in Definition 3, the trace $\tau(\theta)$* automatically *satisfies temporal unity.*

[1] In this chapter, we do not focus on Kant exegesis, but do provide some key references. All references to the *Critique of Pure Reason* use the standard [A/B] reference system, where A refers to the first edition (1781), and B refers to the second edition (1787).

[2] 'The principle of the synthetic unity of apperception is the supreme principle of all use of the understanding' [B136]; it is 'the highest point to which one must affix all use of the understanding, even the whole of logic and, after it, transcendental philosophy' [B134].

[3] Strictly speaking, we mean two *constants* denoting two objects. Throughout, we follow the common practice of treating the Herbrand interpretation as a distinguished interpretation, blurring the difference between a constant and the object denoted by the constant.

[4] See [B203]. See also the *Third Analogy* [A211-15/B256–62].

[5] See [A103-11]. See also: 'What the form of disjunctive judgment may do is contribute to the acts of forming categorical and hypothetical judgments the perspective of their possible *systematic unity*' (Longuenesse, 1998, 105).

[6] See the *schema of community* [A144/B183-4].

[7] See the *schema of causality* [A144/B183].

Example 5: The theory θ of Example 2 satisfies the four unity conditions since:

- For each state A_i in $\tau(\theta)$, a is connected to b via the singleton chain $\{r(a,b)\}$, and b is connected to a via $\{r(b,a)\}$.
- The predicates of θ are $on, off, p_1, p_2, p_3, r$. Here, on and off are involved in the constraint $\forall X{:}s,\ on(X) \oplus off(X)$, while p_1, p_2, p_3 are involved in the constraint $\forall X{:}s,\ p_1(X) \oplus p_2(X) \oplus p_3(X)$, and r is involved in the constraint $\forall X{:}s,\ \exists! Y{:}s\ r(X,Y)$.
- Let $\tau(\theta) = (A_1, A_2, A_3, A_4, ...)$. It is straightforward to check that A_1, A_2, and A_3 satisfy each constraint in C. Observe that A_4 repeats A_1, and the dynamics are Markov, so all subsequent states are copies of A_1, A_2, and A_3. Hence, every state in the infinite sequence $(A_1, A_2, A_3, ...)$ satisfies each constraint.
- Temporal unity is automatically satisfied by the definition of the trace $\tau(\theta)$ in Definition 3.

Now we are ready to define the key notion of this chapter.

Definition 6: *A theory θ* **makes sense** *of a sensory sequence S if θ explains S, i.e. $S \sqsubseteq \tau(\theta)$, and θ satisfies the four conditions of unity of Definition 5.*

Example 6: The theory θ of Example 2 explains the sensory sequence S of Example 1, since $S \sqsubseteq \tau(\theta)$ (Example 4) and θ satisfies the four conditions of unity (Example 5).

Definition 7: *The input to an apperception task is a triple (S, ϕ, C) consisting of a sensory sequence S, a suitable frame ϕ, and a set C of (well-typed) constraints such that (1) each predicate in S appears in some constraint in C and (2) S can be extended to satisfy C: there exists a sequence S' covering S such that S'_i satisfies each constraint in C.*

Given such an input triple (S, ϕ, C), the **simple apperception task** *is to find a lowest cost theory $\theta = (\phi', I, R, C')$ such that ϕ' extends ϕ, $C' \supseteq C$, and θ makes sense of S. Here, $cost(\theta)$ is just the total number of ground atoms in I plus the total number of unground atoms in the rules of R.*

Example 7: Recall that theory θ of Example 2 makes sense of the sequence from Example 1. Now consider an alternative theory based on the type signature $\phi_2 = (T_2, O_2, P_2, V_2)$, where $T_2 = \{s\}$, $O_2 = \{a{:}s, b{:}s, c{:}s\}$, $P_2 = \{on(s), off(s), right(s,s)\}$, and $V_2 = \{X{:}s, Y{:}s\}$. The alternative theory $\theta_2 = (\phi_2, I_2, R_2, C_2)$ where

$$I_2 = \left\{ \begin{array}{lll} on(a) & on(b) & off(c) \\ right(a,b) & right(b,c) & right(c,a) \end{array} \right\}$$

$$R_2 = \left\{ \begin{array}{l} right(X,Y) \wedge off(X) \ni\!\!\!- off(Y) \\ right(X,Y) \wedge on(X) \ni\!\!\!- on(Y) \end{array} \right\}$$

$$C_2 = \left\{ \begin{array}{l} \forall X{:}sensor,\ on(X) \oplus off(X) \\ \forall X{:}sensor,\ \exists! Y{:}sensor,\ right(X,Y) \end{array} \right\}$$

Now θ_2 also makes sense of the sensory sequence S from Example 1. But while θ used three invented predicates (p_1, p_2, and p_3), θ_2 makes use of an invented *object*, c, postulated to explain the sensory sequence more concisely. Note that while θ has cost 16, θ_2 has cost 12.

11.2.2 The Apperception Engine

Next, we describe how our system, the Apperception Engine, solves apperception tasks. Given as input an apperception task (S,ϕ,C), the engine searches for a frame ϕ' and a theory $\theta = (\phi', I, R, C')$ where ϕ' extends ϕ, $C' \supseteq C$ and θ is a unified interpretation of S.

Definition 8: *A* **template** *is a structure for circumscribing a large but finite set of theories. It is a frame together with constants that bound the complexity of the rules in the theory. Formally, a template χ is a tuple $(\phi, N_{\rightarrow}, N_{\ni}, N_B)$ where ϕ is a frame, $N_{\rightarrow}$ is the max number of static rules allowed in R, $N_{\ni}$ is the max number of causal rules allowed in R, and N_B is the max number of atoms allowed in the body of a rule in R.*

Our method, presented in Algorithm 1, is an anytime algorithm that enumerates templates of increasing complexity. For each template χ, it finds the theory θ with lowest cost that satisfies the conditions of unity. If it finds such a θ, it stores it. When it has run out of processing time, it returns the lowest cost θ it has found from all the templates it has considered.

Note that the relationship between the complexity of a template and the cost of a theory satisfying the template is not always simple. Sometimes a theory of lower cost may be found from a template of higher complexity. This is why we cannot terminate as soon as we have found the first theory θ. We must keep going in case we later find a lower cost theory from a more complex template.

Algorithm 1: The Apperception Engine algorithm in outline

input : (S,ϕ,C), an apperception task
output: θ^*, a unified interpretation of S

$(s^*, \theta^*) \leftarrow (max(float), nil)$
foreach *template* χ *extending* ϕ *of increasing complexity* **do**
  $\theta \leftarrow \text{argmin}_\theta\{cost(\theta) \mid \theta \in \Theta_{\chi,C}, S \sqsubseteq \tau(\theta), unity(\theta)\}$
  if $\theta \neq nil$ **then**
    $s \leftarrow cost(\theta)$
    if $s < s^*$ **then**
      $(s^*, \theta^*) \leftarrow (s, \theta)$
    end
  end
  if *exceeded processing time* **then**
    return θ^*
  end
end

The 'heavy lifting' occurs in line 3, where we find, for a given template χ, a lowest cost theory θ satisfying χ that both explains the given sensory sequence S and also satisfies the four unity conditions. In order to jointly abduce a set I (of initial conditions) and induce sets R and C (of rules and constraints), we implement a Datalog$^{\ni}$ interpreter in Answer Set Programming (ASP). This interpreter takes a set I of atoms (represented as a set of ground ASP terms) and sets R and C of rules and constraints (represented again as a set of ground ASP terms), and computes the trace of the theory $\tau(\theta) = (S_1, S_2, ...)$ up to a finite time limit.

Concretely, we implement the Datalog$^{\ni}$ interpreter as an ASP program that computes $\tau(\theta)$ for theory θ. We implement the conditions of unity as ASP constraints, and we implement the cost minimization as an ASP program that counts the number of atoms in each rule and in each initialization atom in I, and uses an ASP weak constraint (Calimeri *et al.*, 2012) to minimize this total. Then we generate ASP programs representing the sequence S, the initial conditions, the rules and constraints. We combine the ASP programs together and ask the ASP solver (`clingo` (Gebser *et al.*, 2014)) to find a lowest cost solution. There may be multiple solutions that have equally lowest cost; the ASP solver chooses one of the optimal answer sets. We extract a readable interpretation θ from the ground atoms of the answer set. We do not go into more detail, for reasons of space. The source code for the Apperception Engine is available at https://github.com/RichardEvans/apperception.

11.2.3 Making sense of disjunctive symbolic input

In section 11.2.1, the sensory sequence $(S_1, ..., S_T)$ was a sequence of sets of unambiguous discrete sensor readings. Each state S_i may be partial and incomplete, but all information that is present is unambiguous. In this section, we extend the Apperception Engine to handle disjunctive sensory input.

Definition 9: *A* **disjunctive input sequence** *is a sequence of sets of disjunctions of ground atoms.*

A disjunctive input sequence generalises the input sequence of section 11.2.1 to handle uncertainty. Now if we are unsure if a sensor a satisfies predicate p or predicate q, we can express our uncertainty as $p(a) \vee q(a)$.

Example 8: Consider, for example, the following sequence $D_{1:10}$. This is a disjunctive variant of the unambiguous sequence from Example 1. Here there are two sensors a and b, and each sensor can be *on* or *off*.

$$\begin{array}{ll} D_1 = \{\} & D_2 = \{\mathit{off}(a), on(b)\} \\ D_3 = \{on(a), \mathit{off}(b)\} & D_4 = \{on(a), on(b)\} \\ D_5 = \{\mathit{off}(a) \vee on(a), on(b)\} & D_6 = \{on(a), \mathit{off}(b)\} \\ D_7 = \{on(a), on(b)\} & D_8 = \{\mathit{off}(a), on(b)\} \\ D_9 = \{\mathit{off}(a) \vee on(a)\} & D_{10} = \{\} \end{array}$$

$D_{1:10}$ contains less information than $S_{1:10}$ from Example 1, since D_9 is unsure whether a is *on* or *off*, while in S_9 a is *on*.

Recall that the $\sqsubseteq$ relation describes when one sequence is explained by another. We extend the $\sqsubseteq$ relation to handle disjunctive input sequences in the first argument.

Definition 10: *Let $D = (D_1, ..., D_T)$ be a disjunctive input sequence and S be an input sequence. $D \sqsubseteq S$ if S contains a finite subsequence $(S_1, ..., S_T)$ such that $S_i \models D_i$ for all $i = 1..T$.*

Example 9: The theory θ of Example 2 explains the disjunctive sensory sequence D of Example 8, since the trace $\tau(\theta)$ (as shown in Example 3) covers D.

The disjunctive apperception task generalizes the simple apperception task of Definition 7 to disjunctive input sequences.

Definition 11: *The input to a disjunctive apperception task is a triple (D, ϕ, C) consisting of a disjunctive input sequence D, a suitable frame ϕ, and a set C of (well-typed) constraints such that (1) for each disjunction featuring predicates $p_1, ..., p_n$ there exists a constraint in C featuring each of $p_1, ..., p_n$. (2) D can be extended to satisfy C.*

Given such an input triple (D, ϕ, C), the **disjunctive apperception task** *is to find a lowest cost theory $\theta = (\phi', I, R, C')$ such that ϕ' extends ϕ, $C' \supseteq C$, $D \sqsubseteq \tau(\theta)$, and θ satisfies the four unity conditions of Definition 5.*

11.2.4 Making sense of raw input

The reason for introducing the disjunctive apperception task is as a stepping stone to the real task of interest: making sense of sequences of *raw uninterpreted sensory input*. Suppose, for example, our input is a sequence of pixel arrays from a video camera.

Definition 12: *Let $\mathcal{R}$ be the set of all possible raw inputs. A* **raw input sequence** *of length T is a sequence in $\mathcal{R}^T$.*

A raw apperception framework uses a neural network $\pi_{\mathbf{w}}$, parameterized by weights $\mathbf{w}$, to map raw subregions of each $\mathbf{r}_i$ into K classes. Then the results of the neural network are transformed into a disjunction of ground atoms. Thus we transform the raw input sequence into a disjunctive input sequence.

Definition 13: *A* **raw apperception framework** *is a tuple $(\pi_{\mathbf{w}}, K, \Delta, \phi, C)$, where:*

- *$\pi_{\mathbf{w}}$ is a perceptual classifier, a multilabel classifier mapping subregions $\mathbf{p}_i^j$ of $\mathbf{r}_i$ to subsets of $\{1, ..., K\}$; π is parameterised by weight vector $\mathbf{w}$*
- *K is the number of classes that the perceptual classifier $\pi_{\mathbf{w}}$ uses*

- *Δ is a "disjunctifier", taking the raw input sequence $(\mathbf{r}_1, ..., \mathbf{r}_T)$ and producing a sequence of sets of disjunctions $(D_1, ..., D_T)$; Δ works by repeatedly applying the perceptual classifier $\pi_{\mathbf{w}}$ to the N subregions $\{\mathbf{p}_i^1, ..., \mathbf{p}_i^N\}$ of $\mathbf{r}_i$, transforming each result (a subset of $\{1, ..., K\}$) into a disjunction of ground atoms*
- *ϕ is a type signature*
- *C is a set of constraints*

The input to a raw apperception task is a raw sequence $r = (r_1, ..., r_T)$ together with a raw apperception framework $(\pi_{\mathbf{w}}, K, \Delta, \phi, C)$. Given sequence $r = (r_1, ..., r_T)$ and framework $(\pi_{\mathbf{w}}, K, \Delta, \phi, C)$, the **raw apperception task** *is to find the lowest cost weights w and theory θ such that θ is a solution to the disjunctive apperception task $((D_1, ..., D_T), \phi, C)$ where $D_i = \Delta(\pi_{\mathbf{w}}(\mathbf{p}_i^1), ..., \pi_{\mathbf{w}}(\mathbf{p}_i^N))$.*

The best $(\theta, \mathbf{w})$ pair is:

$$\underset{\theta,\mathbf{w}}{\text{argmax}} \; \log p(\theta) + \sum_{k=1}^{K} |\{\mathbf{p} \in \mathbf{P} \mid k \in \pi_{\mathbf{w}}(\mathbf{p})\}| \cdot \log \frac{1}{|\{\mathbf{p} \in \mathbf{P} \mid k \in \pi_{\mathbf{w}}(\mathbf{p})\}|}$$

where $\mathbf{P}$ is the set of all subregions $\mathbf{p}_i^j$ in each $\mathbf{r}_i$.

The intuition here is that $p(\theta) = 2^{-len(\theta)}$ represents the prior probability of the theory θ, while the other term penalizes the neural network $\pi_{\mathbf{w}}$ for mapping many subregions to the same class. In other words, it prefers highly selective classes, minimizing the number of subregions that are assigned by $\pi_{\mathbf{w}}$ to the same class. This particular optimization can be justified on Bayesian grounds. We omit the justification for reasons of space.

11.2.5 Applying the Apperception Engine to raw input

In the experiments that follow, we use a binary neural network (Hubara *et al.*, 2016; Kim and Smaragdis, 2016; Rastegari *et al.*, 2016; Cheng *et al.*, 2018; Narodytska *et al.*, 2018) for our parameterized perceptual classifier. Binary neural networks (BNNs) are increasingly popular because they are more efficient (both in memory and processing) than standard artificial neural networks. But our interest in BNNs is not so much in their resource efficiency as in their *discreteness*.

In the BNNs that we use (Cheng *et al.*, 2018), the node activations and weights are all binary values in $\{0, 1\}$. If a node has n binary inputs $x_1, ..., x_n$, with associated binary weights $w_1, ..., w_n$, the node is activated if the total sum of inputs *xnor*-ed with their weights is greater than or equal to half the number of inputs. In other words, the node is activated if

$$\sum_{i=1}^{n} x_i \overline{\oplus} w_i \geq \left\lceil \frac{n}{2} \right\rceil$$

Because the activations and weights are binary, the state of the network can be represented by a set of atoms, and the dynamics of the network can be defined as a logic program. This means we can combine the low-level perception task (of mapping raw data to concepts) and the high-level apperception task (of combining concepts into rules) *into a single logic program* in ASP, and solve both simultaneously using SAT.

11.3 Experiment: Sokoban

In this experiment, we combined the Apperception Engine with a neural network, simultaneously learning the weights of the neural network and also finding an interpretable theory that explains the sensory given. We used 'Sokoban' as our domain. Here, the system is presented with a sequence of noisy pixel images together with associated actions. The system must jointly:

- parse the noisy pixel images into a set of persistent objects, with properties that change over time
- construct a set of rules that explain how the properties change over time as a result of the actions being performed.

We wanted the learned dynamics to be 100% correct. Although next-step prediction models based on neural networks are able, with sufficient data, to achieve accuracy of 99%, this is insufficient for our purposes. If a learned dynamics model is going to be used for long-term planning, 99% is insufficiently accurate as the roll-outs will become increasingly untrustworthy as we progress through time, since 0.99^t quickly approaches 0 as t increases (Buesing *et al.*, 2018).

11.3.1 The data

In this task, the raw input is a sequence of pairs containing a binarized 20×20 image together with a player action from the action space $\mathcal{A} = \{north, east, south, west\}$. In other words, $\mathcal{R} = \mathbb{B}^{20\times 20} \times \mathcal{A}$, and $(r_1, ..., r_T)$ is a sequence of (image, action) pairs from $\mathcal{R}$.

Each array is generated from a 4×4 grid of 5×5 sprites. Each sprite is rendered using a certain amount of noise (random pixel flipping), and so each 20×20 pixel image contains the accumulated noise from the various noisy sprite renderings.

Our trajectory contains a sequence of 17 (image, action) pairs, plus held-out data for evaluation. Because of the noisy sprite rendering process, there are many possible acceptable pixel arrays for time-step 18. These acceptable pixel arrays were generated by taking the true underlying symbolic description of the Sokoban state at time-step 18, and producing many renderings, using the noisy sprite rendering process described above. A set of unacceptable pixel arrays for time-step 18 was generated by considering various symbolic states distinct from the true symbolic state at time-step 18. For each such distinct symbolic state, a pixel rendering was generated. Figure 11.1 shows an example, but here the time-series is shorter to fit on the page.

In our evaluation, a model is considered accurate if it considers every acceptable pixel array at time 18 to be plausible, and considers every unacceptable pixel array at time 18

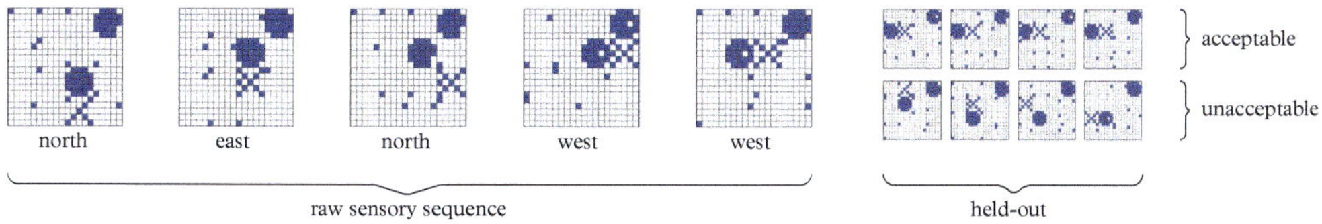

Figure 11.1 *The Sokoban task. The input is a sequence of (image, action) pairs. For the held-out time-step, there is a set of acceptable images, and a set of unacceptable images.*

to be implausible. This is a stringent test. We do not give partial scores for getting some of the predictions correct.

11.3.2 The model

In outline, we convert the the raw input sequence into a disjunctive input sequence by imposing a grid on the pixel array and repeatedly applying a binary neural network to each sprite in the grid. In detail:

1. We impose a grid on the pixel array. We choose a sprite size k, and assume the pixel array can be divided into squares of size $k \times k$. We assume all objects fall exactly within grid cell boundaries. In this experiment, we set $k = 5$.
2. We choose a number m of persistent objects $o_1, ..., o_m$.
3. We choose a number n of distinct types of objects $v_1, ..., v_n$.
4. We add an additional type v_0, where v_0 is a distinguished identifier that will be used to indicate that there is *nothing* at a grid square.
5. We choose a total map $\kappa : \{o_1, ..., o_m\} \rightarrow \{v_1, ..., v_n\}$ from objects to types.
6. We apply a binary neural network (**BNN**) to each $k \times k$ sprite in the grid. The BNN implements a mapping $\mathbb{B}^{k \times k} \rightarrow \{v_0, v_1, ..., v_n\}$. If sprite σ is at (x,y), then $BNN(\sigma) = v_i$ can be interpreted as: it looks as if there is *some object* of type v_i at grid cell (x,y), for $i > 0$. If $BNN(\sigma) = v_0$, it means that there is *nothing* at (x,y). See Figure 11.2.
7. For each time-step, for each grid cell, we convert the output of the BNN into a disjunction of ground atoms: if sprite σ is at (x,y), and $BNN(\sigma) = v_i$, then we create a disjunction featuring each object o of type v_i stating that *any* of them could be at (x,y). See Figure 11.3.
8. We use the Apperception Engine to solve the disjunctive apperception task generated by steps 1–7.

In terms of the formalism of section 11.2.4, the framework (π, ϕ, C) consists of:

- A perceptual classifier π that applies a binary neural network to each 5×5 sprite in the 4×4 grid, and uses the output of the binary network to produce a disjunction

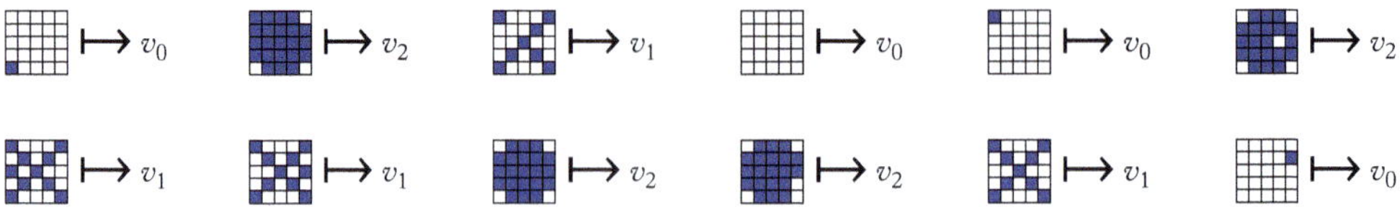

Figure 11.2 *A binary neural network maps sprite pixel arrays to types* $\{v_0, v_1, v_2\}$.

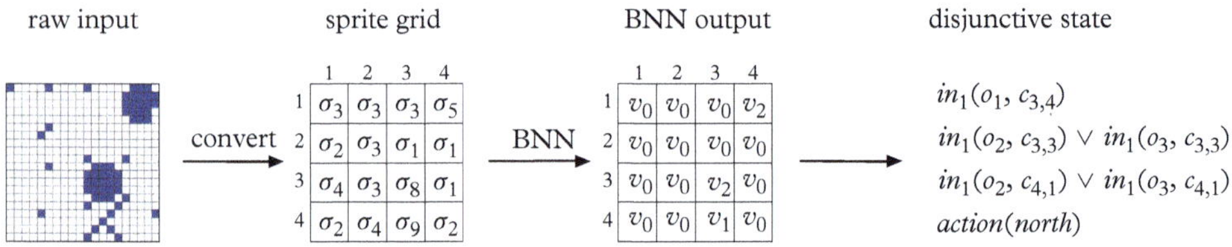

Figure 11.3 *A binary neural network converts the raw pixel input into a set of disjunctions. Here, there is one object* o_1 *of type* v_1 *and two objects* o_2, o_3 *of type* v_2. *If sprite* σ *is at* (x, y), *and* $BNN(\sigma) = v_i$, *then we create a disjunction featuring each object o of type* v_i *stating that any of them could be at* (x, y).

- A frame $\phi = (T, O, P, V)$ consisting of $n+1$ types: $cell, v_1, ..., v_n$, and n predicates $in_i(v_i, cell)$, representing that an object of type v_i is (currently) in a particular cell
- A set of n constraints C insisting that every object of type v_i is always in exactly one cell

The input frame ϕ and initial constraints C are:

$$\phi = \begin{cases} T = \{cell, v_1, ..., v_n, d\} \\ O = \begin{cases} c(x,y) : cell \mid (x,y) \in 4 \times 4 \\ o_1{:}v_1, ..., o_m{:}v_n \\ north{:}d, east{:}d, south{:}d, west{:}d \end{cases} \\ P = \begin{cases} in_i(v_i, cell) \mid i = 1..n \\ action(d) \\ right(cell, cell) \\ below(cell, cell) \end{cases} \\ V = \{C{:}cell, A{:}d\} \cup \{X_i{:}v_i \mid i = 1..n\} \end{cases}$$

$$C = \begin{cases} \forall X_i{:}v_i,\ \exists! C{:}cell,\ in_i(X, C) \mid i = 1..n \\ \exists! A{:}d, action(A) \end{cases}$$

As background knowledge, we provide the spatial arrangement of the grid cells: $right(c_{1,1}, c_{2,1})$, $below(c_{1,1}, c_{1,2})$, etc.

The hyperparameters are n (the number of distinct types of object), m (the number of objects), and k (the sprite-size). We grid-search over n and m, and fix $k = 5$.

Both the Apperception Engine and the binary neural network are implemented as logic programs in ASP. Thus, we can find the binary weights of the neural network and the rules of our theory *simultaneously*, by solving a single SAT problem.

11.3.3 Understanding the interpretations

Figure 11.4 shows the best theory found by the Apperception Engine from one trajectory of 17 time-steps. On the left we show the raw input, the output of the neural network, and the set of atoms that are true at each time-step. On the right, we show the theory that explains the sequence. The top four rules of R describe how the man X moves when actions are performed. The middle four rules define four invented predicates $p_1, ...p_4$. These four unary predicates are used by the theory to describe when a block is being pushed in one of the four cardinal directions. The bottom four rules of R describe what happens when a block is being pushed in one of the four directions.

When neural network next-step predictors are applied to these sequences, the learned dynamics typically fail to generalize correctly to different-sized worlds or worlds with a different number of objects. But the theory leaned by the Apperception Engine applies to all Sokoban worlds, no matter how large, no matter how many objects. Not only is this learned theory correct but it is provably correct.

Figure 11.5 shows the evolving state of the Apperception Engine over time. The grid on the left is the raw perceptual input, a grid of 20×20 pixels. The second element is a 4×4 grid of 5×5 sprites, formed by preprocessing the pixel array. The third element is the output of the binary neural network: a 4×4 grid of predicates v_0, v_1, v_2. If v_i is at (x, y), this means 'it looks as if there is *some* object of type i at (x, y)' (but we don't yet know which particular object). So, for example, the grid in the top row states that there is some object of type 1 at $(3, 4)$, and some object of type 2 at $(4, 1)$. Here, v_0 is a distinguished predicate meaning there is *nothing* at this grid square.

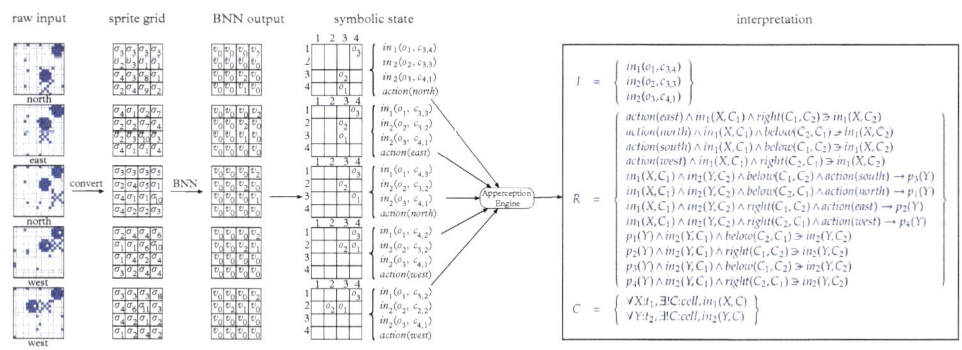

Figure 11.4 *Interpreting Sokoban from raw pixels. Raw input is converted into a sprite grid, which is converted into a grid of types v_0, v_1, v_2. The grid of types is converted into a disjunctive apperception task. The Apperception Engine finds a unified theory explaining the disjunctive input sequence, a theory which explains how objects' positions change over time.*

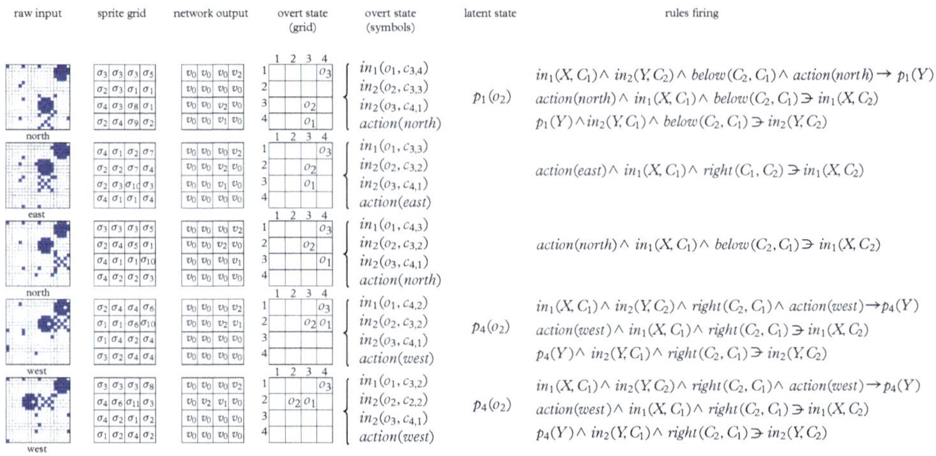

Figure 11.5 *The state evolving over time. Each row shows one time-step. We show the raw pixel input, the sprites, the output of the binary neural network, the set of ground atoms that are currently true, and the rules that fire.*

The fourth element is a 4×4 grid of persistent objects: if o_i is at (x, y) this means: the particular persistent object o_i is at (x, y). The fifth element is a set of ground atoms. This is a re-presentation of the persistent object grid (the fourth element) together with an atom representing the player's action. The sixth element shows the latent state. In Sokoban, the latent state stores information about which objects are being pushed in which directions. Here, in the top row, $p_1(o_2)$ means that the persistent object o_2 is being pushed up. The seventh element shows which rules fire in which situations. In the top row, three rules fire. The first rule describes how the man moves when the *north* action is performed. The second rule concludes that a block is pushed northwards if a man is below the block and the man is moving north. The third rule describes how the block moves when it is pushed northwards.

Looking at how the engine interprets the sensory sequence, it is reasonable—in fact, we claim, inevitable—to attribute *beliefs* to the system. In the top row of Figure 11.5, for example, the engine believes the object at $(3, 3)$ is the same type of thing as the object at $(4, 1)$, but the object at $(3, 4)$ is not the same type of thing as the object at $(4, 1)$. As well as beliefs about a particular moment, the system also believes facts relating two successive moments. For example, the object at $(3, 4)$ at time t_1 is the very same persistent object as the thing that is at $(3, 3)$ at time t_2. As well as beliefs about particular situations, the system also has general beliefs that apply to all situations. For example: whenever the *north* action is performed, and the man is below a block, then the block is pushed upwards. One of the main reasons for using a purely declarative language (such as Datalog$^{\ni}$) as the target language of program synthesis is that an individual clause can be interpreted as a *judgement* with a truth-condition. If the program that generated the trace had been a procedural program, it would have been much less clear what judgement, if any, the procedure represented.

11.3.4 The baseline

Our hypothesis is that the strong inductive bias provided by the Datalog$^{\ni}$ language (Definition 2) and the unity constraints (Definition 5) should allow data-efficient learning of accurate models. To evaluate the hypothesis, we built a neural network baseline to compare data efficiency and accuracy.

The baseline we constructed for the Sokoban task is an auto-regressive model with a continuously-relaxed discrete bottleneck (Figure 11.6). The model applies an array of parameter-sharing multilayer perceptrons (MLPs) to each block of the game state, and concatenates the result with the one-shot representation of the actions before feeding it into an LSTM. The LSTM, combined with a dense layer, produces the parameters of Gumbel-Softmax continuous approximations of the categorical distribution, one per each block of the state. These distributions, when the model is learned well, can encode a close-to-symbolic representation of the current state without direct supervision. The step following is a decoder network consisting of a two-layer perceptron which targets the next raw state of the sequence.

Given that the presented model is a purely generative model over a large state space, in order to compare it to the Apperception Engine, we add a density estimation classifier on its output. The classifier fits a Gaussian per class, trained on log-probabilities of independently sampled acceptable and unacceptable test states calculated over the Bernoulli distribution outputted by the model.

We trained the baseline with the Adam optimizer, varying the learning rate in $[0.05, 0.01, 0.005, 0.001]$, batch size in $[512, 1024]$, and executing each experiment

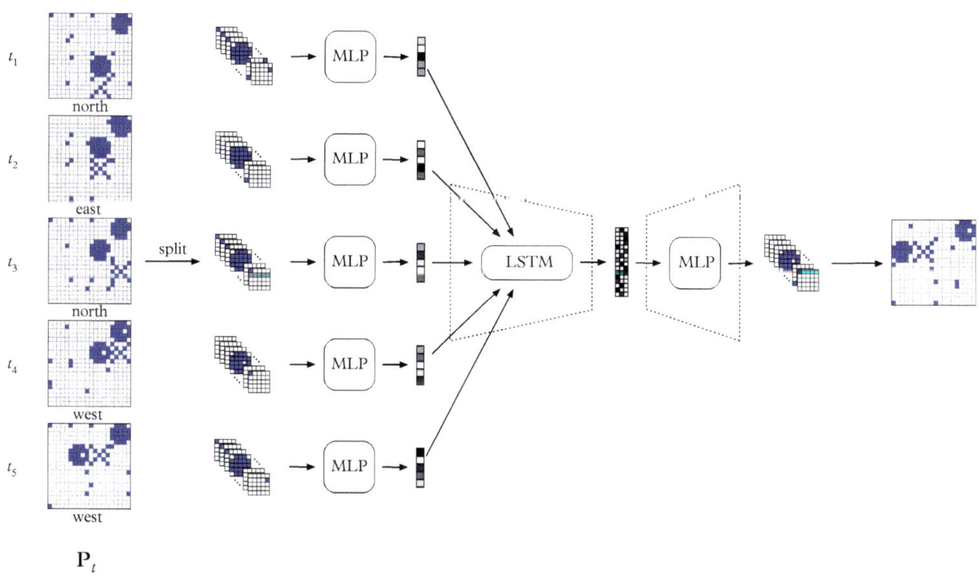

Figure 11.6 *The baseline model for the Sokoban task.*

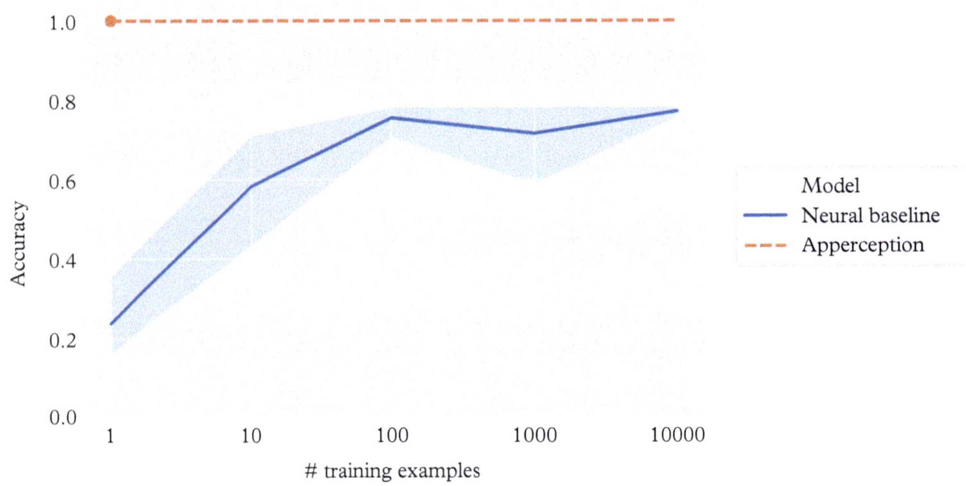

Figure 11.7 *The results on the Sokoban task. Apperception is trained on only a single example and the dashed line represents the apperception results on that single example. The neural baseline is trained on an increasing number of training examples. The shaded area is the 95% confidence interval on 10 runs with different random seeds.*

10 times. We selected the best set of hyperparameters by choosing the parameters with the best development set performance, and averaged the performance across 10 repetitions with different random seeds. During training, we annealed the temperature of the Gumbel-Softmax with an exponential decay, from 2.0 to 0.5 with a per-epoch decay of 0.0009.

As can be seen from Figure 11.7, the neural baseline is not able to correctly distinguish between acceptable and unacceptable next steps, neither from the single example, nor a large number of examples. However, as expected, the accuracy of the baseline increases with increasing size of the training set, though it hits a plateau without reaching the maximum. The Apperception Engine, on the other hand, is able to learn a fully accurate theory from a *single* carefully chosen example.

11.4 Related Work

Unsupervised learning from temporal sequences is central to statistics, engineering, and the sciences, with countless applications. Most sequence models are probabilistic, allowing us to model the uncertainty and ambiguity arising from noisy measurements or partial observability. Traditionally, due to computational limitations, sequence models were informed by strong prior knowledge about particular domains and usually contained few tunable parameters, e.g (Kalman, 1960; Black and Scholes, 1973). The advent of deep learning methods and the availability of large scale data sets has revolutionized sequence modelling (Graves, 2013), by incorporating general learned function approximation to replace strong prior assumptions.

Probabilistic sequence models can broadly be divided into two classes (with some overlap): *auto-regressive* models and *latent variable* models. Auto-regressive models directly capture the temporal dependencies of observed variables by modeling the distribution of each time slice conditioned on the history of observations. This results in conceptually simple and fast model training procedures using maximum-likelihood estimation. At the same time, auto-regressive models for raw sensory data are usually large and therefore costly to evaluate at test time, hindering their application to planning problems, for example, with long horizons where thousands of model evaluations are often necessary. Furthermore, they are usually not humanly interpretable and specifying prior domain knowledge is difficult.

Latent variable models can in principle address these shortcomings (Loehlin, 1987). They aim to capture the statistical dependencies of time series by positing latent variables that underlie the observed data (Ghahramani and Jordan, 1996). This latent structure is usually assumed to be simpler than the observed high-dimensional raw data, and can therefore in principle reduce the computational demands of long-range predictions, therefore facilitating applications to reinforcement learning and planning, for example (Buesing *et al.*, 2018). Furthermore, low-dimensional latent models are often used for finding simple, structured explanations in exploratory data analysis (Byron *et al.*, 2009). Allowing for uncertainty over latent entities in order to capture multiple hypotheses of observed phenomena comes at the price of necessitating sophisticated approaches for fitting models to data. Most approaches either apply explicit probabilistic inference to determine the distributions over latent variables from observations, or they rely on implicit or likelihood-free methods (Li *et al.*, 2017).

Modern latent variable sequence models mostly capture latent structure with continuous variables, as this allows approximate model-fitting methods with gradient descent on parameters as a sub-routine (Kingma and Welling, 2013). However, predictions under these models are often subject to degrading fidelity with increasing prediction horizon. Although often inevitable to some degree, this effect is exacerbated by the inability of latent variables to capture discrete, or categorical structure, resulting in a 'conceptual drift' in domains where the ground truth can be described by discrete concepts. Discrete (or mixed discrete-continuous, Johnson *et al.*, 2016) latent variable models can in principle alleviate this issues (Mnih and Rezende, 2016; van den Oord *et al.*, 2017), however, fitting these efficiently to data remains challenging (Maddison *et al.*, 2016).

11.5 Discussion

In this chapter, we have shown how the Apperception Engine can be combined with a binary neural network in order to learn explicit causal theories from raw unprocessed sensory sequences. We demonstrated, in the *Sokoban* experiments, how this system is able to learn explicit, interpretable models from small amounts of noisy data. This is, to the best of our knowledge, the first system that can learn a provably correct dynamics model of a non-trivial game from raw unprocessed sensory input.

We pause to consider some limitations of our system in its current form. A fundamental limitation of the Apperception Engine is that it assumes that the underlying dynamics can be expressed as rules that operate on *discrete* concepts. While the system is capable of handling raw, noisy, continuous sensory input, it assumes that the *underlying dynamics* of the system can be represented by rules that operate on discrete concepts. There are many domains where the underlying dynamics are discrete while the surface output is noisy and continuous: Raven's progressive matrices, puzzle games, and Atari video games, for example. But our system will struggle in domains where the underlying dynamics are best modelled using continuous values, such as models of fluid dynamics. Here, the best our system could do is find a discrete model that crudely approximates the true continuous dynamics. Extending Datalog$^{\ni}$ to represent continuous change would be a substantial and ambitious project.

Another major limitation of our system is that it assumes that causal rules are strict, universal, and exceptionless. There is no room in the current representation for defeasible causal rules (where normally, all other things being equal, a causes b) or non-deterministic causal rules (where a causes either b or c). In future work, we plan to implement non-deterministic causal rules.

A third limitation of our system is that it suffers from scaling issues, both in terms of memory and processing time. Finding a unified theory that explains the sensory input means searching through the space of logic programs. This is a huge and daunting task. For example, the Apperception Engine takes 5 gigabytes of RAM and 48 hours to make sense of a single Sokoban trajectory consisting of 17 pixel arrays of size 20×20. This is, undeniably, a computationally expensive process.

We would like to scale our approach up so that we can learn the dynamics of Atari games from raw pixels. But this will prove to be challenging, as games such as Pacman are harder than our Sokoban test case in every dimension: it requires us to increase the number of pixels, the number of time-steps, the number of trajectories, the number of objects, and the complexity of the dynamics.

The dominant reason for our system's scaling difficulties is that it uses a maximizing SAT solver to search through the space of logic programs. We encode the program synthesis problem as an ASP program, and find the simplest theory by encoding the theory size as a weak constraint. Finding an optimal solution to an ASP program with weak constraints is in Σ_2^P, but this complexity is a function of the number of ground atoms, and the number of ground atoms is exponential in the size of the Datalog$^{\ni}$ theory.

We are currently evaluating various different ways of improving the performance of our system so we can scale up to harder problems such as Atari, and are excited by the prospect of scaling up the Apperception Engine to the next level, so as to be able to induce robust causal models for Atari. But it will, we believe, require substantial further research.

11.6 Conclusion

When a human opens her eyes, she combines low-level perception (mapping raw unprocessed sensory input into objects and concepts) with high-level apperception (constructing a conceptual theory that makes sense of the stimulus).

Figure 11.8 *Using high-level conceptual information (in this case, the spelling of English words) to disambiguate low-level perceptual information. Here, there is a highly ambiguous symbol (in red) that is used for both the 'H' of 'THE' and for the 'A' of 'CAT'.*

Consider Figure 11.8 (Chalmers *et al.*, 1992). Here, the red letter is ambiguous between an 'A' and an 'H'. When we read the words, our high-level conceptual knowledge of the spelling of English words informs our low-level perceptual processing, so that we can effortlessly disambiguate the images.

We want our machines to do the same, combining low-level perception with high-level apperception. And we want the information to flow in *both directions*: as well as low-level perceptual information informing high-level conceptual theorising, we also want high-level conceptual considerations to inform low-level perceptual processing. The Apperception Engine is a proof of concept that such a system is, indeed, possible.

References

Black, F. and Scholes, M. (1973). The pricing of options and corporate liabilities. *Journal of Political Economy*, **81**(3), 637–54.

Buesing, L., Weber, T., Racaniere, S. *et al.* (2018). Learning and querying fast generative models for reinforcement learning. *arXiv preprint arXiv:1802.03006.*

Byron, M. Y., Cunningham, J. P., Santhanam, G. *et al.* (2009). Gaussian-process factor analysis for low-dimensional single-trial analysis of neural population activity. *Journal of Neurophysiology*, **102**(1), 614–35.

Calimeri, F., Faber, W., Gebser, M. *et al.* (2012). Asp-core-2: Input language format. *ASP Standardization Working Group.*

Chalmers, D. J., French, R. M., and Hofstadter, D. R. (1992). High-level perception, representation, and analogy: A critique of artificial intelligence methodology. *Journal of Experimental & Theoretical Artificial Intelligence*, **4**(3), 185–211.

Cheng, C.-H., Nührenberg, G., Huang, C.-H. *et al.* (2018). Verification of binarized neural networks via inter-neuron factoring, in *Working Conference on Verified Software: Theories, Tools, and Experiments*. Cham: Springer, 279–90.

Evans, R., *et al.* (2021). "Making sense of sensory input." Artificial Intelligence **293** (2021): 103438.

Gebser, M., Kaminski, R., Kaufmann, B. *et al.* (2014). Clingo= asp+ control: Preliminary report. *arXiv preprint arXiv:1405.3694.*

Ghahramani, Z. and Jordan, M. I. (1996). Factorial hidden Markov models. *Machine Learning*, **29**(2), 245–73.

Graves, Alex (2013). Generating sequences with recurrent neural networks. *arXiv preprint arXiv:1308.0850.*

Hubara, I., Courbariaux, M., Soudry, D. *et al.* (2016). Binarized neural networks. *Advances in Neural Information Processing Systems*, **29**, 4107–15.
Johnson, M. J., Duvenaud, D. K., Wiltschko, A. *et al.* (2016). Composing graphical models with neural networks for structured representations and fast inference, in *Advances in Neural Information Processing Systems*, 2946–54.
Kalman, R. E. (1960). A new approach to linear filtering and prediction problems. *Journal of Basic Engineering*, 82(1), 35–45.
Kant, I. (1781). *Critique of Pure Reason*. Cambridge: Cambridge University Press.
Kim, M. and Smaragdis, P. (2016). Bitwise neural networks. *arXiv preprint arXiv:1601.06071.*
Kingma, D. P. and Welling, M. (2013). Auto-encoding variational Bayes, in *Proceedings of 2nd International Conference on Learning Representations 2014*, Ithaca.
Lee, S. D. and De Raedt, L. (2004). Constraint based mining of first order sequences in seqlog, in *Database Support for Data Mining Applications*. Berlin, Heidelberg: Springer, 154–73.
Leibniz, G. W. (1765). *New Essays on Human Understanding*. Cambridge University Press.
Li, J., Monroe, W., Shi, T. *et al.* (2017). Adversarial learning for neural dialogue generation, in *Proceedings of the 2017 Conference on Empirical Methods in Natural Language Processing*, 2157–69.
Loehlin, J. C. (1987). *Latent Variable Models*. New Jersey: Lawrence Erlbaum Publishers, 87–91.
Longuenesse, B. (1998). *Kant and the Capacity to Judge*. Princeton, NJ: Princeton University Press.
Maddison, C. J., Mnih, A., and Teh, Y. W. (2016). The concrete distribution: A continuous relaxation of discrete random variables, in *Proceedings of the International Conference on Learning Representations*.
Mnih, A. and Rezende, D. J. (2016). Variational inference for Monte Carlo objectives, in *Proceedings of the 33rd International Conference on International Conference on Machine Learning*, Vol. 48, 2188–96.
Narodytska, N., Kasiviswanathan, S., Ryzhyk, L. *et al.* (2018). Verifying properties of binarized deep neural networks, in *Proceedings of the AAAI Conference on Artificial Intelligence*, Vol. 32, No. 1.
Rastegari, M., Ordonez, V., Redmon, J. *et al.* (2016). Xnor-net: Imagenet classification using binary convolutional neural networks, in *European Conference on Computer Vision*. Cham: Springer, 525–42.
Smith, B. C. (2019). *The Promise of Artificial Intelligence: Reckoning and Judgment*. Cambridge, MA: MIT Press.
Tamaddoni-Nezhad, A. and Muggleton, S. (2009). The lattice structure and refinement operators for the hypothesis space bounded by a bottom clause. *Machine Learning*, **76**(1), 37–72.
van den Oord, A. *et al.* (2017). Neural discrete representation learning, in *Proceedings of the 31st International Conference on Neural Information Processing Systems*, 6309–18.

12

Human–Machine Perception of Complex Signal Data

Alaa Alahmadi, Alan Davies, Markel Vigo, Katherine Dempsey, and Caroline Jay

University of Manchester, UK

12.1 Introduction

This research explores a new approach to the processing of complex signal data, which exploits an understanding of the human perceptual system to facilitate its interpretation by humans and machines simultaneously. We focus on the interpretation of electrocardiogram (ECG) data, and in particular on the heart condition known as 'long QT syndrome (LQTS)', which is associated with a life-threatening arrhythmia (*Torsades de pointes* (TdP)) that can lead to sudden cardiac death. The syndrome is characterized by a prolongation of the QT-interval on the ECG, which represents the duration of the ventricular depolarization and repolarization cycle, and is measured from the beginning of the Q-wave to the end of the T-wave (Goldenberg and Moss, 2008), as shown in Figure 12.1. Whether or not the interval is considered to be prolonged depends on a number of factors, and particularly heart rate. Given a heart rate of 60 beats per minute (bpm), a normal QT interval would generally be 430 milliseconds (ms) or less (Yap and Camm, 2003; Goldenberg and Moss, 2008).

By manipulating the way the data are represented—in this case, by adding colour to the signal to expose the QT-interval—we produce a simple and accurate detection algorithm, which has the benefit of the human and machine sharing the same representation of the data. We compare the approach with current signal processing techniques and consider its potential usage within clinical practice, from the perspectives of explanation, trust, and transparency. Finally, we discuss future avenues of research, with a particular focus on how using salience within visual images can support other approaches to computer vision.

12.1.1 Interpreting the QT interval on an ECG

The electrocardiogram (ECG), a recording of the complex signal data representing the heart's electrical activity, is widely used in clinical practice for assessing cardiac function

Alaa Alahmadi, Alan Davies, Markel Vigo, Katherine Dempsey, and Caroline Jay, *Human-Machine Perception of Complex Signal Data* In: *Human Like Machine Intelligence.* Edited by: Stephen Muggleton and Nick Charter, Oxford University Press.
 DOI: 10.1093/oso/9780198862536.003.0012

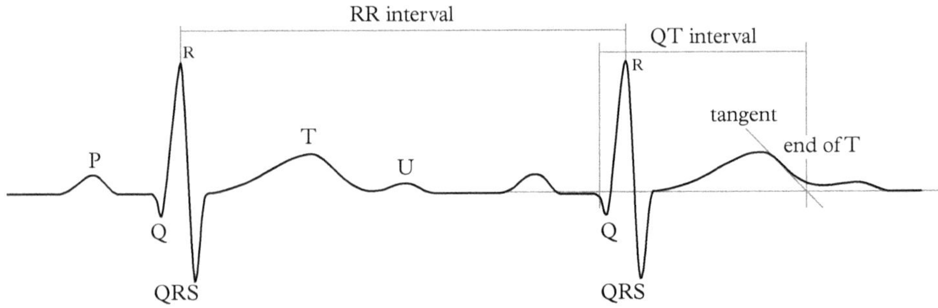

Figure 12.1 *Measurement of the QT-interval on the ECG from the beginning of the Q-wave to the end of the T-wave. Figure taken from (Alahmadi* et al., *2019).*

and detecting pathologies. The standard method for visualizing ECG data is with a two-dimensional line graph showing the amplitude of the recorded electrical signal of the heart on the Y-axis and the time in milliseconds on the X-axis. The ECG 'wave forms' (peaks and troughs) are labelled with letters and represent different stages of the heartbeat, as shown in Figure 12.1. Long QT syndrome is indicated by a prolongation of the QT-interval on the ECG, representing a delay in the ventricular repolarization activity of the heart (Goldenberg and Moss, 2008). Many commonly prescribed medications can prolong the QT-interval, leading to acquired drug-induced long QT syndrome (diLQTS).

Whilst there have been attempts to automate ECG interpretation for several decades, expert human reading of the data presented visually remains the 'gold standard' (Wood *et al.*, 2014; Schläpfer and Wellens, 2017). Nevertheless, ECG interpretation is complex and requires extensive training, with some conditions known to be very challenging to recognize, even for clinicians who routinely read ECGs (Viskin *et al.*, 2005). To date, LQTS has remained difficult to recognize on the ECG, from both a human and a machine perspective (Miller *et al.*, 2001; Viskin *et al.*, 2005; Rautaharju *et al.*, 2009; Tyl *et al.*, 2011; Garg and Lehmann, 2013; Talebi *et al.*, 2015; Kligfield *et al.*, 2018). Here, we demonstrate how a visualization technique that significantly improves human interpretation of ECG data—without the need for prior training—can be used as the basis for an automated human-like algorithm.

12.1.2 Human–machine perception

Many approaches to computer vision, including deep learning, attempt to mimic or improve on human ability, but are often only loosely related to human visual processes (Sinha *et al.*, 2006; Scheirer *et al.*, 2014). A growing area of research is investigating the role that models of human perception may play in improving computer vision. Perception, broadly speaking, is the process of recognising and interpreting sensory information (Hendee and Wells, 1997). The human visual system has a highly developed capability for perceiving different patterns and objects simultaneously, both whole and in parts, without prior training (Sinha *et al.*, 2006; O'Toole *et al.*, 2012; Scheirer *et al.*, 2014). This appears to be due to a primitive perceptual organization process that derives relevant groupings and patterns from an image or scene without prior knowledge of its contents (Lowe, 2012; van der Helm, 2017; Stadler, 2019), a phenomenon first noted in the Gestalt principles of visual perception, which articulate factors that regulate perceptual grouping, including proximity, similarity, closure, continuation and symmetry (Wertheimer, 1923). Today, there is evidence of a multiplicity of perceptual grouping processes that vary in attentional demands (Behrmann and Kimchi, 2003; Kimchi and Razpurker-Apfeld, 2004; Kimchi *et al.*, 2005; Rashal *et al.*, 2017), with many operating before the higher level cognitive system applies top-down knowledge to recognizing a scene (van der Helm, 2017; Stadler, 2019).

Research in computer vision has shown that the capability of a machine to organize and interpret sensory information in a 'human-like' way, termed *machine perception* (Velik, 2010), can dramatically decrease the search space required for object recognition (Lowe, 2012). Combining a perceptual grouping approach with the principle of simplicity, which states that people tend to perceive the simplest possible interpretation of any given visualized information (Hochberg, 1957; Hatfield and Epstein, 1985; Feldman, 2016), has been shown to enhance machine vision further (Darrell *et al.*, 1990; Feldman, 1997; van der Helm, 2015; Feldman, 2016).

Pre-attentive processing theory outlines a set of visual properties known to be detected rapidly and accurately by the human eye, which are important in the perceptual grouping process (Nothdurft, 1993; Theeuwes, 2013; Wolfe and Utochkin, 2019). Examples of pre-attentive properties include colour, shape, and size. Colour, in particular, is known to aid and influence the perceptual organization of the visual scene (Kimchi and Razpurker-Apfeld, 2004). Zavagno and Daneyko (2014) have shown colour to be a relatively strong grouping factor that functions according to the principles of Gestalt theory, and can override other types of pre-attentive property including shape and size. In the technique we describe here, colour serves as the foundation for drawing the human reader's attention to the QT-interval in an ECG image, such that he or she can make a decision about whether it is dangerously prolonged. The salience information used by humans to make this judgement is then mapped to quantitative values, which can be used by an algorithm to automate the detection of prolongation.

12.2 Human–Machine Perception of ECG Data

The majority of machine vision algorithms use machine learning, and in particular neural networks, as the basis for processing image data. Here, we take a different approach, in terms of both the data representation, and the role of human vision in the resulting algorithm. Our starting point is ECG signal data that is presented visually for human interpretation, and is typically analysed computationally using signal processing methods. As a first step, the signal data is visualized with colour, such that a human can easily draw the relevant information from it. Following this, the perceptual process we hypothesize the human is using to interpret the data is modelled, and forms the basis of a simple rule-based algorithm that accurately classifies whether the interval is prolonged.

12.2.1 Using pseudo-colour to support human interpretation

Recognizing QT-interval prolongation on the standard ECG is notoriously difficult, even for trained medical professionals (Viskin *et al.*, 2005). From a perceptual perspective, this is likely to be related to the fact that humans are poor at judging quantity on a horizontal scale (Wittig and Allen, 1984; Liben, 1991; Leroux *et al.*, 2009; Papadopoulos and Koustriava, 2011). Morphological diversity of the wave forms and artifacts in the ECG signal serve to exacerbate this issue (Taggart *et al.*, 2007; Postema *et al.*, 2008; Alahmadi *et al.*, 2018).

Our previous work—motivated by the potential benefits of self-monitoring for diLQTS—considered the problem from the perspective of the lay person with no experience of ECG interpretation. To support this challenging target population we used knowledge of visual perception to enhance the way the ECG is visualized. Colour is a particularly powerful way of attracting visual attention in complex scenes, and aids visual recognition via perceptual grouping (Treisman, 1983; Oliva and Schyns, 2000; Kimchi and Razpurker-Apfeld, 2004). Based on this phenomenon, we produced a visualization technique that highlights the duration of the QT-interval on the ECG using pseudo-colouring, a salient means of representing continuously varying values using a sequence of meaningful colours (Ware, 2012). This shifts the visual encoding process from perceiving a distance between two waves to perceiving colour in terms of hue and intensity. Applying pseudo-colouring to the ECG significantly increased the speed and accuracy of human perception of QT-prolongation, at both a regular 60 bpm heart rate (Alahmadi *et al.*, 2019) and at varying heart rates, with diverse T-wave morphologies (Alahmadi *et al.*, 2020*a*).

Pseudo-colouring method

The first step in the application of pseudo-colouring was to detect the R-peaks on the ECG, which identify heartbeats (see Figure 12.1). Note that it is trivial to detect the R-peaks in the vast majority of ECGs, as they consistently have the greatest amplitude (the amplitude of the other waves varies considerably). Pseudo-colour was then applied as follows:

- **Identifying the risk threshold for QT-prolongation:** The risk threshold for QT-prolongation changes according to heart rate. The QT-nomogram, a clinical assessment method that shows the risk of TdP by considering QT-interval as a function of heart rate, was used to identify the threshold (Chan *et al.*, 2007). Figure 12.2 shows the QT nomogram plot. If the QT/HR pair plots on or above the risk line, the patient is at clinically significant risk of TdP; below the line the patient is not considered at risk. The risk threshold was calculated for each heart rate using the nomogram risk line. The risk threshold for a given heart rate was termed 'QT-value at risk'.

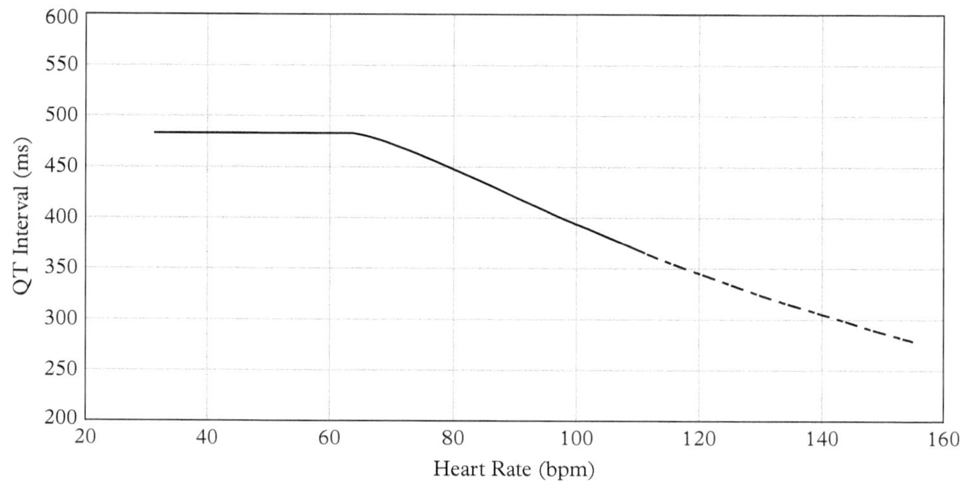

Figure 12.2 *The QT-nomogram (Chan* et al., *2007).*

- **Applying pseudo-colouring to each heartbeat:** In clinical practice, the QT-interval is measured by counting the small squares (each representing 40 ms) on the standard ECG background grid from the beginning of the Q-wave to the end of the T-wave. The time period of interest (i.e., an approximation of the QT-interval) was calculated for each heartbeat from the R-peak minus 20 ms (estimated as the start of the Q-wave) to the maximum potential QT-interval, which was estimated as the QT-value at risk plus two small squares (80 ms). This formed an additional time dimension, to which the pseudo-colour could be mapped. Pseudo-colouring was then applied periodically over each heartbeat to the area between the isoelectric baseline (where amplitude is zero) and the ECG signal, by mapping the relevant area of the heartbeat time to a pseudo-colouring sequence. We used a spectrum-approximation pseudo-colouring sequence, where cool spectral colour codes (purple to blue to green) were used to indicate normal QT-interval ranges, and warm colours (yellow to orange to red) to show abnormal QT-interval ranges. This produced nine indices on the pseudo-colouring scale, where each index was mapped to a colour code and represented a small square on the ECG, starting backwards from the nomogram line showing the QT-value at risk plus two small squares (80 ms) to six small squares (240 ms) below the nomogram line. The QT-value at risk was mapped to dark orange. Forty ms and 80 ms above this were mapped to red and dark red respectively, to indicate higher risk. Within each square, the intensity of the hue changed every millisecond, to show time progression. Figure 12.3 shows an illustration explaining how the pseudo-colouring technique was applied according to the standard ECG background grid. The way in which colours were mapped to the time period of interest according to the nomogram is shown in Table 12.1. Figure 12.4 shows examples of ECGs with normal and very

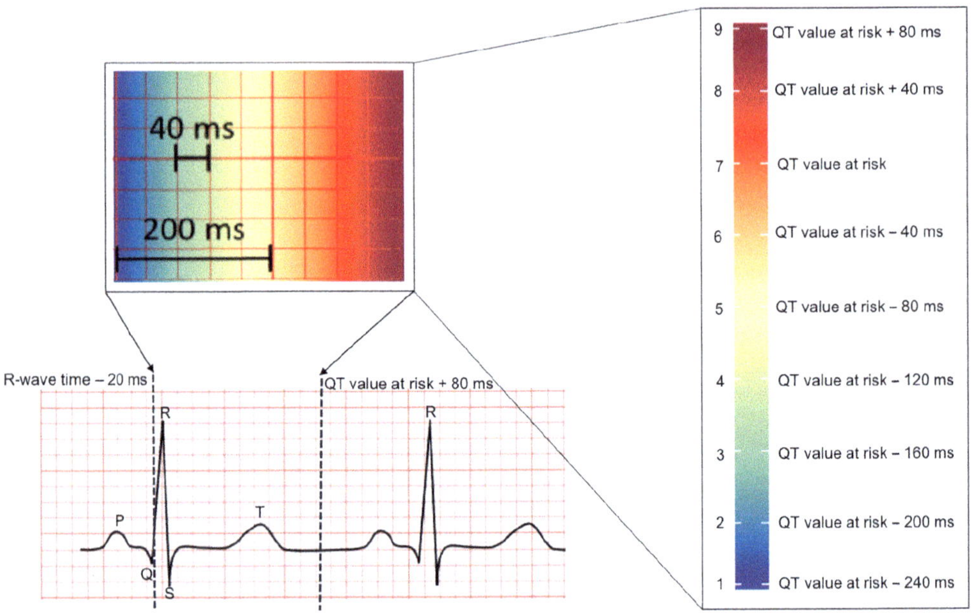

Figure 12.3 *Mapping the pseudo-colouring to the ECG. A small square on the grid is equal to 40 ms.*

Table 12.1 *The nine indices on the pseudo-colouring scale with their corresponding time value in milliseconds (ms) and colour code.*

Index	Corresponding time value (ms)	Color code
1	QT-value at risk − (40 × 6)	Purple
2	QT-value at risk − (40 × 5)	Blue
3	QT-value at risk − (40 × 4)	Green
4	QT-value at risk − (40 × 3)	Lime
5	QT-value at risk − (40 × 2)	Yellow
6	QT-value at risk − (40 × 1)	Orange
7	QT-value at risk	Dark orange
8	QT-value at risk + (40 × 1)	Red
9	QT-value at risk + (40 × 2)	Dark red

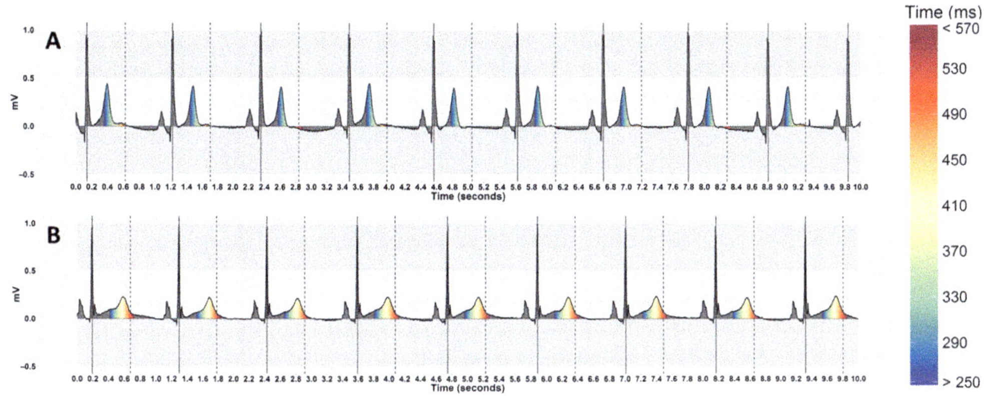

Figure 12.4 *Examples of ECGs with pseudo-colouring, showing (A) a normal QT-interval (HR = 55, QT = 361 ms) and (B) a dangerously prolonged QT-interval (HR = 52, QT = 579 ms).*

prolonged QT-intervals visualized using the pseudo-colouring technique. Figure 12.5 shows how the pseudo-colouring was adjusted according to heart rate (based on the QT-nomogram), where the pseudo-colouring shows the same level of risk, despite heart rate differences. The R script used to implement the visualization technique can be found in (Alahmadi *et al.*, 2020*b*).

The pseudo-colouring technique was shown in an evaluation to be very effective, enabling lay people with no prior training in ECG interpretation to detect QT prolongation with 83% sensitivity, compared with 63% when pseudo-colour was not used. For more information about the visualization technique and its evaluation with humans see (Alahmadi *et al.*, 2019) and (Alahmadi *et al.*, 2020*a*).

12.2.2 Automated human-like QT-prolongation detection

Automated ECG interpretation systems have typically proved poor at detecting LQTS (Miller *et al.*, 2001; Rautaharju *et al.*, 2009; Tyl *et al.*, 2011; Estes III, 2013; Garg and Lehmann, 2013; Talebi *et al.*, 2015; Kligfield *et al.*, 2018). A major challenge for automated QT-detection algorithms is identifying the precise end of the T-wave (the terminal point), particularly when the T-wave's morphology is abnormal (Higham and Campbell, 1994; Morganroth, 2001; Goldenberg *et al.*, 2006; Postema and Wilde, 2014). This is particularly problematic as medications that prolong the QT-interval often change levels of the blood's electrolytes including potassium, calcium, sodium, and magnesium, which can affect T-wave morphology (Vicente *et al.*, 2015). Methadone, a drug that is infamous for prolonging the QT-interval and increasing the risk of TdP, also causes changes in the T-wave that cause the interval to be underestimated (Talebi *et al.*, 2015).

The pseudo-colouring technique was able to communicate QT-prolongation in such a way that humans were able to accurately perceive risk of TdP, as the signal from the colour reduces the need to identify the end of the T-wave. It thus follows that an

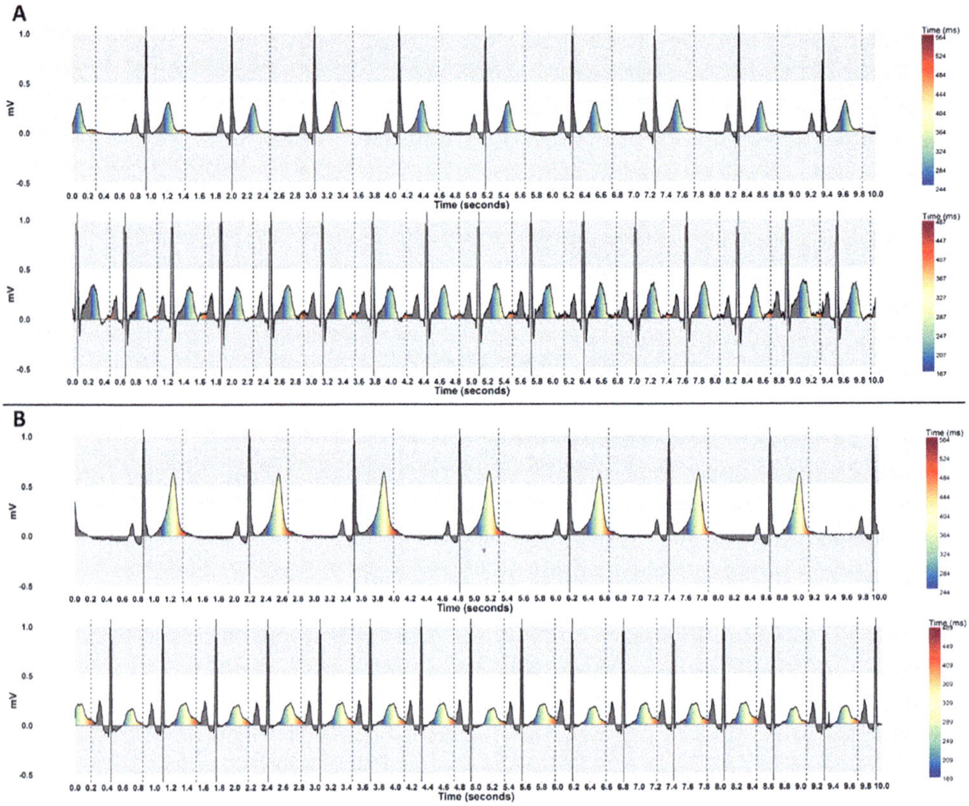

Figure 12.5 *Examples of ECGs with pseudo-colouring that have the same QT-level, but different heart rates. (A) Pseudo-colouring indicates the same normal range of QT-intervals, while (B) shows the same abnormal range of QT-intervals.*

algorithm using the same or an equivalent process to 'perceive' the information encoded in the colour should also be able to perform this task. We hypothesized that quantifying computationally the amount of warm colour displayed in the ECG signal could help a machine to detect LQTS, alleviating the need to measure the QT-interval directly.

The heuristic of quantifying area, rather than identifying the interval *per se*, is effective because the T-wave generally has the largest area under the curve of the ECG signal. The QT-interval is considered normal when the T-wave is located in the cool colour region, and prolonged when in the warm colour region. The pseudo-colour thus highlights the T-wave position in relation to the inter-heartbeat time dimension, without needing to identify either the peak or end of the wave. As such, the first step in the computational human-like algorithm is to calculate the area under the curve of the ECG signal using the trapezoidal rule. In mathematics, the trapezoidal rule is a method commonly used for approximating the definite integral that estimates the area under the curve of a linear function by dividing the area into a number of strips of equal width. That is, given a linear function $f(x)$ of a real variable x and an interval [a, b], the rule estimates the area under the graph of the function $f(x)$ as a trapezoid, calculating its area as follows:

$$\int_a^b f(x)dx \approx \frac{(f(a)+f(b))}{2}(b-a) \tag{12.1}$$

The raw ECG signal has an array of X and Y values, where X represents the time of the ECG signal in milliseconds, and Y represents the amplitude of the ECG signal. We considered [a, b] to be the interval of two successive time stamps in the X array of the ECG signal. If we consider the time interval of x_1 and x_2, the trapezoidal rule was applied by taking the average of amplitude of the ECG signal on the Y-axis of this interval as $f(x_1)$ and $f(x_2)$, and multiplying it by the difference in time between x_1 and x_2. As the time of the ECG is represented by integer numbers ranging from 1 to 10000 milliseconds for a 10-second ECG recording, the difference between any 2 successive times x_1 and x_2 is always equal to 1. Using the finest possible level of granularity maximized the precision of the estimation.

The total area under the curve was calculated for every 40 ms index on the pseudo-colouring scale. Cool spectral colours from indices 1 to 4 represented normal QT-intervals, and warm spectral colours from indices 5 to 9 represented prolonged QT-intervals. The percentage of the area under the curve was then calculated for warm and cool colours respectively. The QT-interval was considered 'prolonged' by the algorithm if the proportion of warm colours was greater than the that of cool colours; otherwise it was considered 'normal'. In summary, the QT-interval was considered 'prolonged' if the warm colours occupied more than 50% of the area under the curve of the ECG signal. The full human-like algorithm is described in the pseudo-code below and the R script used to implement it can be found in (Alahmadi *et al.*, 2020*b*).

Comparison with human interpretation

We evaluated the accuracy of the human-like algorithm by first comparing it with the results of a study conducted with humans (Alahmadi *et al.*, 2020*a*). The ECGs ($n = 40$) were acquired from a clinical study conducted to assess QT-interval prolongation in healthy subjects receiving medication known to cause this issue (Johannesen *et al.*, 2014). As part of the clinical study QT intervals were calculated for all ECGs, and it is these values that were used as ground truth for our subsequent evaluation. ECGs were selected from multiple patients ($n = 17$), and represented different values of the QT-interval and heart rate, with 20 ECGs showing a normal QT-interval, and 20 ECGs showing clinically significant QT-prolongation. The ECGs had different heart rates, with some morphological T-wave changes caused by the QT-prolonging drugs. The ECG datasets can be found in the PhysioNet database (Goldberger *et al.*, 2000), and the clinical trial study is described in (Johannesen *et al.*, 2014).

We measured the sensitivity, specificity, and overall accuracy of the classification. The sensitivity is the ability of the classifier (human/algorithm) to identify correctly those

Algorithm 1 Human-like algorithm

Data: ECG signal with X and Y
Input: X is time list $x[i]$, $i = 1, 2, \cdots, 10000$.
Y is amplitude list $y[i]$, $i = 1, 2, \cdots, 10000$.
Result: Normal or prolonged QT-interval
`// Calculate area under the curve (AUC)`
$i \leftarrow 1$
while $x[i] \leq 10000$ **do**
  $delta\_x \leftarrow x[i+1] - x[i]$
  $AUC[i] \leftarrow ((y[i] + y[i+1])/2)(delta\_x)$
  $i \leftarrow i + 1$
end
`// Acquire the time points of the nine indices of the pseudo-colouring scale`
Pseudo-colouring scale is time list $index[i]$, $i = 1, 2, \cdots, 9$.
for *each index i in the pseudo-colouring scale* **do**
  Calculate the sum of AUCs of the 40 ms interval:
  $Sum\_index[i] \leftarrow \sum_{index[i]}^{index[i+40]} AUC[i]$
end
`// Calculate percentage of AUC for warm colours`
$Sum\_cool\_areas \leftarrow \sum_{index[1]}^{index[4]} AUC[index]$
$Sum\_warm\_areas \leftarrow \sum_{index[5]}^{index[9]} AUC[index]$
$Total \leftarrow Sum\_cool\_areas + Sum\_warm\_areas$
$Warm\_Colours\% \leftarrow (Sum\_warm\_areas/Total) \times 100$

`// Inference rule`
if $Warm\_Colours\% \geq 50$ **then**
  QT is prolonged
else
  QT is normal
end

patients with the disease, and was calculated as the proportion of correctly classified 'prolonged' ECGs. The specificity is the ability of the classifier to identify correctly those patients without the disease, and was calculated as the proportion of correctly classified 'normal' ECGs. The overall accuracy was calculated as the proportion of correct classification of the 40 ECGs, that is the average of the sensitivity and specificity. The results of the human-like algorithm were very similar to those of the human participants, with both showing slightly higher specificity than sensitivity, and the algorithm being slightly more accurate overall (see Table 12.2 and Figure 12.6).

Table 12.2 *The sensitivity, specificity, and overall accuracy of the human-like algorithm and human participants (mean values) when classifying the 40 ECGs.*

	Human-like algorithm	Human	Difference
Sensitivity	0.85	0.83	0.02
Specificity	0.95	0.90	0.05
Overall accuracy	0.90	0.86	0.04

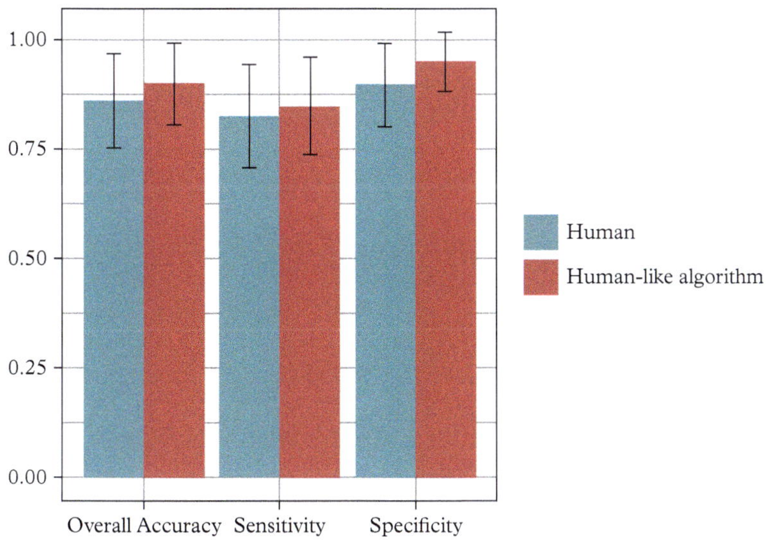

Figure 12.6 *The sensitivity, specificity, and overall accuracy of the human-like algorithm and human participants (mean values) when classifying the 40 ECGs. The error bars represent 95% confidence intervals.*

Comparison with signal processing approaches

The majority of automated QT-interval analysis algorithms are proprietary or unavailable, and as such formally bench-marking the performance of our algorithm is not possible. Here, we compare it with signal processing approaches reported in the literature that have previously been applied to QT-interval measurement. The logic behind the human-like algorithm differs considerably from that used by standard signal processing methods, as it takes a naive perspective, calculating the percentage area of warm colours relative to cool colours, without prior identification and detection of the Q-wave or T-wave. By contrast, traditional signal processing approaches to ECG interpretation are based on the precise determination of the onset and offset of the different waves and complexes (P-wave, QRS complex, T-wave). This process is relatively straightforward if the ECG signal has a normal sinus rhythm, but it quickly becomes challenging in the presence of anomalies, artifacts, or non-standard ECG waves (Schläpfer and Wellens, 2017). ECG wave characteristics are also known to differ substantially across individuals, and are affected by factors including age, race, sex, and health status (Macfarlane *et al.*, 1994; Goldenberg *et al.*, 2006; Hnatkova *et al.*, 2016). At present, there are no standard definitions for the ECG waves (Willems, 1980; Party, 1985; Schläpfer and Wellens, 2017). As a result, differences in signal processing measurements persist. In addition to the challenge of correctly recognizing the different ECG waves, accurate measurements of intervals (PR, QRS, QT) are particularly difficult to make, and thus the methods of determining onset and offset of waves vary among algorithms.

We implemented two common signal processing algorithms that use different methods to identify the end of the T-wave. The first algorithm, proposed by Hermans *et al.* (2017), uses an automated tangent method to identify the point of the maximum T-wave down-slope as the end of the T-wave. The second algorithm uses the 15% threshold method which determines the end of the T-wave as the point in time when the ECG signal crosses the threshold at 15% of the amplitude of the T-wave peak (Hunt, 2005). In both cases, the Q-wave onset was determined using the same method as the visualization technique, which is the time of the R-peak −20 ms.

We compared the sensitivity, specificity, and overall accuracy of the two signal processing methods with the human-like algorithm using the same 40 ECGs tested in the human interpretation study (Alahmadi *et al.*, 2020*a*). The signal processing methods measured the QT-interval of the ECGs as numbers and calculated the heart rate. As the pseudo-colouring technique was adjusted according to the nomogram to detect normal and prolonged QT-intervals across different heart rates, the signal processing methods were implemented to classify the QT-interval as 'normal' if the calculated QT/HR plot was below the nomogram line, and otherwise as 'prolonged'.

Table 12.3 shows the 40 ECGs ordered by QT-level from low-risk (1–3) to high-risk (4–6). The results show that both signal processing methods significantly underestimate the QT-interval, with most ECGs with a prolonged QT-interval being classified as 'normal' based on the nomogram (see Figure 12.7). The mean difference in milliseconds between the actual QT values and the calculated QT values were 61 ms and 62 ms for the automated tangent method and 15% threshold method respectively. The accuracy of the human-like algorithm was considerably better, due to its dramatically higher sensitivity, as shown in Figure 12.7. Table 12.3 illustrates how, as the QT-level relative to the nomogram increased, the percentage of warm colours calculated by the human-like algorithm also increased.

12.3 Human–Machine Perception: Differences, Benefits, and Opportunities

Perception, broadly speaking, involves two forms of processing: bottom-up processing, which is driven by incoming stimuli, and uses perceptual organization to form a representation of an object, and top-down processing, which uses contextual information to aid the perception of patterns (Egeth and Yantis, 1997; Connor *et al.*, 2004;

Table 12.3 *The 40 ECGs with, from left to right, the actual values as measured in the clinical study of the QT interval in milliseconds, the heart rate (HR) in beats per minute, the QT-level at risk, the calculated QT values in milliseconds of the two signal processing methods, and the percentage of warm colours calculated by the human-like algorithm.*

ECG	QT	HR	QT-level	Automated tangent algorithm (QT)	15% threshold algorithm (QT)	Human-like algorithm (warm colours%)
1	370	48	1	324	338	14%
2	361	55	1	319	332	1%
3	350	68	1	437	309	2%
4	343	72	1	417	306	5%
5	329	83	1	361	287	1%
6	335	90	1	332	290	20%
7	401	57	2	344	359	12%
8	389	75	2	306	324	19%
9	339	95	2	317	286	11%
10	419	47	2	354	367	28%
11	396	68	2	330	317	3%
12	355	82	2	364	327	2%
13	445	46	3	384	401	21%
14	441	67	3	378	390	49%
15	431	75	3	360	372	42%
16	417	80	3	343	367	32%
17	371	94	3	320	309	57%
18	444	58	3	390	403	29%
19	424	76	3	325	346	26%
20	363	95	3	315	310	39%
21	487	46	4	454	460	38%
22	468	72	4	417	409	69%
23	451	79	4	358	374	50%
24	445	81	4	367	396	54%
25	486	54	4	419	435	61%
26	485	64	4	405	422	48%

Continued

Table 12.3 *Continued*

ECG	QT	HR	QT-level	Automated tangent algorithm (QT)	15% threshold algorithm (QT)	Human-like algorithm (warm colours%)
27	419	91	4	316	349	71%
28	410	94	4	312	326	74%
29	523	42	5	432	452	57%
30	494	71	5	427	403	69%
31	470	85	5	353	421	63%
32	518	54	5	413	431	48%
33	482	80	5	378	398	55%
34	417	96	5	310	344	51%
35	507	79	6	377	251	62%
36	565	49	6	478	528	65%
37	579	52	6	477	548	65%
38	547	64	6	451	448	68%
39	509	68	6	435	414	79%
40	518	77	6	388	405	87%

Table 12.4 *The sensitivity, specificity, and overall accuracy of the human-like algorithm and the two signal processing algorithms when classifying the 40 ECGs.*

	Human-like algorithm	**Automated tangent algorithm**	**15% threshold algorithm**
Sensitivity	0.85	0.10	0.10
Specificity	0.95	0.100	0.100
Overall accuracy	0.90	0.55	0.55

Fenske *et al.*, 2006). ECG interpretation is thought to be dependent primarily on top-down processing (Wood *et al.*, 2014; Schläpfer and Wellens, 2017). From a human perspective, the visualization technique described here works by harnessing bottom-up processing, drawing visual attention to the critical information contained within the ECG signal. Using a simple model of this process, the human-like algorithm was able not only

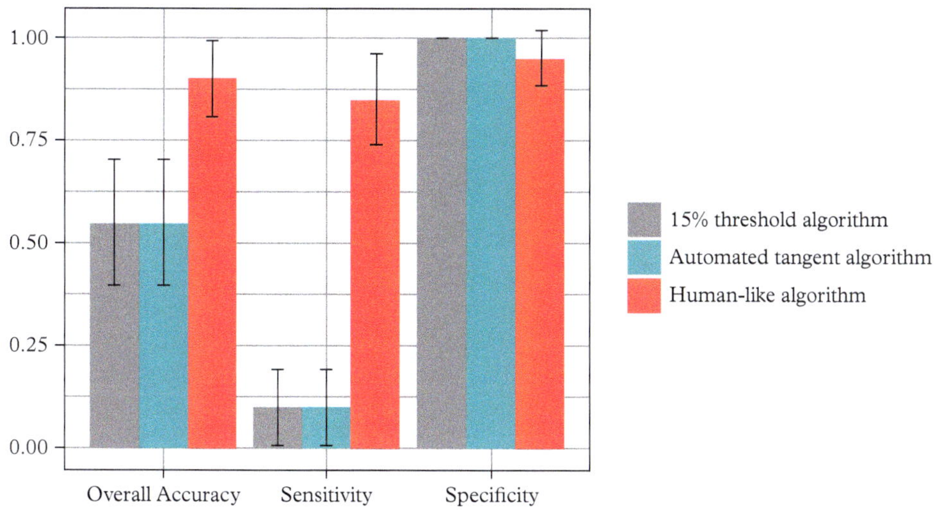

Figure 12.7 *The sensitivity, specificity and overall accuracy of the human-like algorithm and two signal processing algorithms when classifying the 40 ECGs. The error bars represent 95% confidence intervals.*

to match human performance but to exceed it. It is interesting to note that specificity was slightly higher for both the human participants and the algorithm. Whilst we cannot be sure that humans were internally employing a version of the algorithm to make decisions, this indicates that similar processes may be at work.

The shared representation of the data, and the shared model of how to interpret them, are important when considering the real world application of this approach. At present, clinicians do not regard automated ECG interpretation as reliable. In broader terms, the public also have concerns about the use of algorithms for medical decision support. A survey carried out for the Wellcome Trust highlighted transparency as an important factor in automation (Fenech *et al.*, 2018). A human can, in theory, explain an error, and it is therefore possible to establish negligence or malicious intent, an important consideration for the respondents. The majority of those surveyed also stated they would not like machines to suggest treatments or answer medical questions. This sits in contrast to the proposed digitization of healthcare, which includes the potential use of AI and robotics in healthcare delivery (Topol, 2019). To achieve a shift in people's attitudes towards this technology, a number of challenges need to be addressed, including improved data protection and privacy standards, fairness (guarding against bias), and transparency in how technology works and decisions are made (Vayena *et al.*, 2018). While we have not evaluated trust in the algorithm described here, we hypothesize that its inherent transparency will be of benefit in this regard. Its Explainability goes beyond that offered by many rule-based algorithms, which are theoretically explainable, but may nevertheless be extremely complex and difficult for humans to understand. The human-like algorithm uses a representation of the data that matches that used by the person making the decision. We suggest that this operates not just at a conscious level, using deliberate 'system 2' processes, but also at a subconscious, perceptual level, and is thus able to engage fast 'system 1' processes, such that people can understand the data quickly and relatively effortlessly (Kahneman, 2011).

12.3.1 Future work

Biologically inspired algorithms such as deep Convolutional Neural Networks (CNN), are now capable of image classification on a par with adult human capabilities (Zhou and Firestone, 2019). While they have achieved impressive practical successes across a number of application domains including medicine, deep learning models must be trained on large datasets, and the way they represent and use data internally is often unclear (Holzinger *et al.*, 2017). They also exhibit weaknesses, as demonstrated by adversarial learning, a field of study that evaluates the safe use of machine learning techniques in adversarial settings such as spam filtering, cybersecurity, and biometric recognition, by attempting to fool the models through malicious input (Lowd and Meek, 2005). An example of this is provided by Goodfellow *et al.* (2015), who took an image of a panda that was that used in an image classification task and introduced a small perturbation to the image data. This changed the algorithm's classification from 57.7% confidence the image showed a panda, to 99.3% confidence that the image showed a gibbon (Goodfellow *et al.*, 2015). To a human observer, the image still clearly resembles that of a panda, but to a machine, the small change was catastrophic. Further exploration of human information processing may be the key to addressing these challenges.

Human visual perception is described as the construction of efficient representation formalisms of visual information (Cantoni, 2013). The challenges of visual perception have drawn the curiosity of computer scientists for many years, particularly in terms of understanding representation formalisms that may be effective in machine perception and artificial intelligence (O'Toole *et al.*, 2008). Human representation formalisms used in artificial intelligence include *relational* representations based on networks, graphs, or frame schemes, *propositional* representations that use linguistics based on first-order predicate logic, and *procedural* representations, also known as pattern-directed schemes (Cantoni, 2013). A promising direction is using these representations to mimic the sophisticated and flexible perceptual capabilities of human perception to organize information in a preprocessing step. Merging deep learning models with symbolic *procedural* representations based on perceptual schemes has the potential to advance AI systems (Tian *et al.*, 2017). A good example of the potential of this approach is provided by Stettler and Francis (2018), who demonstrated that mimicking the process involved in perceiving visual obstruction can improve the performance of a CNN in a letter recognition task.

The human-like algorithm developed here also has the potential to aid CNN design. For example, using pseudo-colouring to improve information segmentation in a pre-processing step may help to improve a CNN's accuracy in classifying ECGs with LQTS.

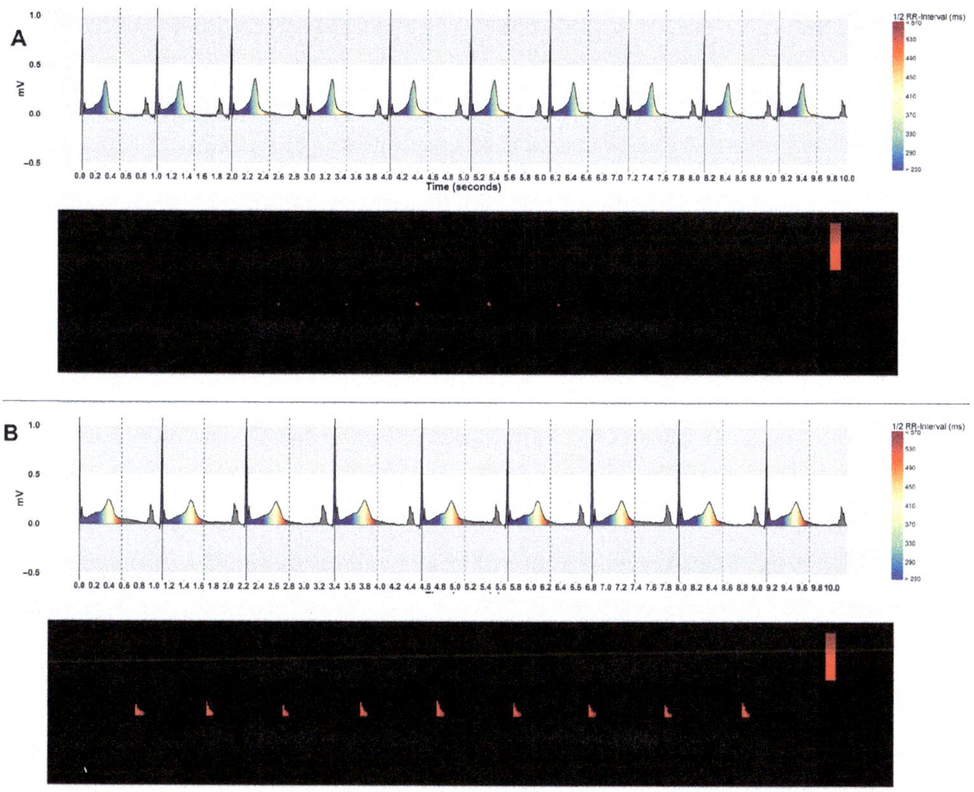

Figure 12.8 *Examples of two ECGs and corresponding images pre-processed to display orange to red pixels, where (A) shows an ECG with a normal QT-interval and (B) shows an ECG with a prolonged QT-interval.*

Figure 12.8 shows two ECGs, one with a normal and one with a prolonged QT-interval. Below each ECG are the results of using an image pre-processing method, available online (Analyst, 2020), for detecting orange to red colours using RGB (red, green, blue) colour space and then masking the ECG image to show only the parts containing the warm colours. The information that is salient to a human can thus be prioritized as input to the CNN, reducing the search space.

The field of human perception has long been a great source of inspiration for developing and improving machine perception. A motivation for our approach was not only improving the accuracy of ECG interpretation but also producing data representations that can be used to provide a transparent, understandable, and explainable interpretation that keeps the human in the loop. Further exploration of the potential that human perceptual processes have for informing machine interpretation is a promising avenue for future research.

References

Alahmadi, A., Davies, A., Royle, J. *et al.* (2019). Evaluating the impact of pseudo-colour and coordinate system on the detection of medication-induced ECG changes, in *Proceedings of the 2019 CHI Conference on Human Factors in Computing Systems*, p. 123. ACM.

Alahmadi, A., Davies, A., Vigo, M. *et al.* (2018). Can lay people identify a drug-induced QT-interval prolongation? A psychophysical and eye-tracking experiment examining the ability of non-experts to interpret an ECG. *Journal of the American Medical Informatics Association*, **26**(5), 404–11. https://doi.org/10.1093/jamia/ocy183

Alahmadi, A., Davies, A., Vigo, M. *et al.* (2020*a*). Pseudo-colouring an ECG enables lay people to detect QT-interval prolongation regardless of heart rate. *PLoS One*, **15**(8). https://doi.org/10.1371/journal.pone.0237854

Alahmadi, A., Vigo, M., and Jay, C. (2020*b*). The QT-interval ECG Visualisation and the Human-like Algorithm. Version 1.0.0. [Computer Software]. https://doi.org/10.5281/zenodo.3866622.

Analyst, Image (Retrieved January 15, 2020). Simplecolordetection(). https://www.mathworks.com/matlabcentral/fileexchange/26420-simplecolordetection, MATLAB Central File Exchange.

Behrmann, M. and Kimchi, R. (2003). What does visual agnosia tell us about perceptual organization and its relationship to object perception? *Journal of Experimental Psychology: Human Perception and Performance*, **29**(1), 19.

Cantoni, Virginio (2013). *Human and Machine Vision: Analogies and Divergencies*. New York: Springer Science & Business Media.

Chan, A., Isbister, G. K., Kirkpatrick, C. M. J., *et al.* (2007). Drug-induced QT prolongation and torsades de pointes: evaluation of a QT nomogram. *QJM: An International Journal of Medicine*, **100**(10), 609–15.

Connor, C. E., Egeth, H. E., and Yantis, S. (2004). Visual attention: bottom-up versus top-down. *Current Biology*, **14**(19), R850–R852.

CSE Working Party (1985). Recommendations for measurement standards in quantitative electrocardiography. *European Heart Journal*, **6**(10), 815–25.

Darrell, T., Sclaroff, S., and Pentland, A. (1990). Segmentation by minimal description, in *[1990] Proceedings Third International Conference on Computer Vision*, Osaka, Japan, 112–16. doi: 10.1109/ICCV.1990.139506

Egeth, H. E. and Yantis, S. (1997). Visual attention: Control, representation, and time course. *Annual Review of Psychology*, **48**(1), 269–97.

Estes III, N. A. M. (2013). Computerized interpretation of ECGs: supplement not a substitute. *Circulation: Arrhythmia and Electrophysiology*, **6**(1), 2–4. doi: 10.1161/CIRCEP.111.000097. PMID: 23424219.

Feldman, J. (1997). Regularity-based perceptual grouping. *Computational Intelligence*, **13**(4), 582–623.

Feldman, J. (2016). The simplicity principle in perception and cognition. *Wiley Interdisciplinary Reviews: Cognitive Science*, 7(5), 330–40.

Fenech, M., Strukelj, N., and Buston, O. (2018). Ethical, social, and political challenges of artificial intelligence in health. Technical report. London: Future Advocacy.

Fenske, M. J., Aminoff, E., Gronau, N. *et al.* (2006). Top-down facilitation of visual object recognition: object-based and context-based contributions. *Progress in Brain Research*, **155**, 3–21.

Garg, A. and Lehmann, M. H. (2013). Prolonged QT interval diagnosis suppression by a widely used computerized ECG analysis system. *Circulation: Arrhythmia and Electrophysiology*, **6**(1), 76–83.

Goldberger, A. L., Amaral, L. A. N., Glass, L. *et al.* (2000). Physiobank, physiotoolkit, and physionet: components of a new research resource for complex physiologic signals. *Circulation*, **101**(23), e215–e220.

Goldenberg, I. and Moss, A. J. (2008). Long QT syndrome. *Journal of the American College of Cardiology*, **51**(24), 2291–300.

Goldenberg, I., Moss, A. J., and Zareba, W. (2006). QT interval: how to measure it and what is 'normal'. *Journal of Cardiovascular Electrophysiology*, **17**(3), 333–6.

Goodfellow, I. J., Shlens, J., and Szegedy, C. (2015). Explaining and harnessing adversarial examples, in *3rd International Conference on Learning Representations (ICLR) 2015, San Diego, CA, May 7–9, 2015, Conference Track Proceedings*, 1–11.

Hatfield, G. and Epstein, W. (1985). The status of the minimum principle in the theoretical analysis of visual perception. *Psychological Bulletin*, **97**(2), 155.

Hendee, W. R. and Wells, P. N. T. (1997). *The Perception of Visual Information*. New York: Springer Science & Business Media.

Hermans, B. J. M., Vink, A. S., Bennis, F. C., *et al.* (2017). The development and validation of an easy to use automatic QT-interval algorithm. *PloS ONE*, **12**(9), e0184352.

Higham, P. D. and Campbell, R. W. (1994). QT dispersion. *British Heart Journal*, **71**(6), 508.

Hnatkova, K., Smetana, P., Toman, O. *et al.* (2016). Sex and race differences in QRS duration. *Ep Europace*, **18**(12), 1842–9.

Hochberg, J. E. (1957). Effects of the gestalt revolution: the Cornell Symposium on perception. *Psychological Review*, **64**(2), 73.

Holzinger, A., Biemann, C., Pattichis, C. S. *et al.* (2017). What do we need to build explainable AI systems for the medical domain? *arXiv preprint arXiv:1712.09923*.

Hunt, A. C. (2005). Accuracy of popular automatic QT interval algorithms assessed by a'gold standard'and comparison with a novel method: computer simulation study. *BMC Cardiovascular Disorders*, **5**(1), 29.

Johannesen, L., Vicente, J., Mason, J. W. *et al.* (2014). Differentiating drug-induced multichannel block on the electrocardiogram: randomized study of dofetilide, quinidine, ranolazine, and verapamil. *Clinical Pharmacology & Therapeutics*, **96**(5), 549–58.

Kahneman, D. (2011). *Thinking, Fast and Slow*. New York, NY: Farrar, Straus and Giroux.

Kimchi, R. and Razpurker-Apfeld, I. (2004). Perceptual grouping and attention: Not all groupings are equal. *Psychonomic Bulletin & Review*, **11**(4), 687–96.

Kimchi, R., Hadad, B., Behrmann, M. *et al.* (2005). Microgenesis and ontogenesis of perceptual organization: Evidence from global and local processing of hierarchical patterns. *Psychological Science*, **16**(4), 282–90.

Kligfield, P., Badilini, F., Denjoy, I. *et al.* (2018). Comparison of automated interval measurements by widely used algorithms in digital electrocardiographs. *American Heart Journal*, **200**, 1–10.

Leroux, G., Spiess, J., Zago, L. *et al.* (2009). Adult brains don't fully overcome biases that lead to incorrect performance during cognitive development: an fMRI study in young adults completing a Piaget-like task. *Developmental Science*, **12**(2), 326–38.

Liben, L. S. (1991). Adults' performance on horizontality tasks: conflicting frames of reference. *Developmental Psychology*, **27**(2), 285.

Lowd, D. and Meek, C. (2005). Adversarial learning, in *Proceedings of the Eleventh ACM SIGKDD International Conference on Knowledge Discovery in Data Mining*, (KDD '05). New York, NY: Association for Computing Machinery, 641–47. https://doi.org/10.1145/1081870.1081950

Lowe, D. (2012). *Perceptual Organization and Visual Recognition.* Vol. 5. New York: Springer Science & Business Media.

Macfarlane, PW, McLaughlin, SC, Devine, B, and Yang, TF (1994). Effects of age, sex, and race on ECG interval measurements. *Journal of Electrocardiology*, 27, 14–19.

Miller, M. D., Coburn, J. P., and Ackerman, M. J. (2001). Diagnostic accuracy of screening electrocardiograms in long QT syndrome I. *Pediatrics*, **108**(1), 8–12.

Morganroth, J. (2001). Focus on issues in measuring and interpreting changes in the QTc interval duration. *European Heart Journal Supplements*, **3**(suppl_K), K105–K111.

Nothdurft, H.-C. (1993). The role of features in preattentive vision: Comparison of orientation, motion and color cues. *Vision research*, **33**(14), 1937–1958.

Oliva, A. and Schyns, P. G. (2000). Diagnostic colors mediate scene recognition. *Cognitive Psychology*, **41**(2), 176–210.

O'Toole, A. J., An, X., Dunlop, J. *et al.* (2012). Comparing face recognition algorithms to humans on challenging tasks. *ACM Transactions on Applied Perception (TAP)*, **9**(4), 16.

O'Toole, A. J., Phillips, P. J., and Narvekar, A. (2008). Humans versus algorithms: comparisons from the face recognition vendor test 2006, in *2008 8th IEEE International Conference on Automatic Face & Gesture Recognition*, Amsterdam, 1–6. doi: 10.1109/AFGR.2008.4813318

Papadopoulos, K. and Koustriava, E. (2011). Piaget's water-level task: The impact of vision on performance. *Research in Developmental Disabilities*, **32**(6), 2889–93.

Postema, Pieter G, De Jong, J. S. S. G, Van der Bilt, I. A. C. *et al.* (2008). Accurate electrocardiographic assessment of the qt interval: teach the tangent. *Heart Rhythm*, **5**(7), 1015–18.

Postema, P. G. and Wilde, A. M. (2014). The measurement of the QT interval. *Current Cardiology Reviews*, **10**(3), 287–94.

Rashal, E., Yeshurun, Y., and Kimchi, R. (2017). Attentional requirements in perceptual grouping depend on the processes involved in the organization. *Attention, Perception, & Psychophysics*, **79**(7), 2073–87.

Rautaharju, P. M., Surawicz, B., and Gettes, L. S. (2009). AHA/ACCF/HRS recommendations for the standardization and interpretation of the electrocardiogram: part IV: the ST segment, T and U waves, and the QT interval a scientific statement from the American Heart Association Electrocardiography and Arrhythmias Committee, Council on Clinical Cardiology; the American College of Cardiology Foundation; and the Heart Rhythm Society Endorsed by the International Society for Computerized Electrocardiology. *Journal of the American College of Cardiology*, **53**(11), 982–91.

Scheirer, W. J., Anthony, S. E., Nakayama, K. *et al.* (2014). Perceptual annotation: Measuring human vision to improve computer vision. *IEEE Transactions on Pattern Analysis and Machine Intelligence*, **36**(8), 1679–86.

Schläpfer, J. and Wellens, H. J. (2017). Computer-interpreted electrocardiograms: benefits and limitations. *Journal of the American College of Cardiology*, **70**(9), 1183–92.

Sinha, P., Balas, B., Ostrovsky, Y. *et al.* (2006). Face recognition by humans: Nineteen results all computer vision researchers should know about. *Proceedings of the IEEE*, **94**(11), 1948–62.

Stadler, M. W. (2019). Thinking, experiencing and rethinking mereological interdependence. *Gestalt Theory*, **41**(1), 31–46.

Stettler, M. and Francis, G. (2018). Using a model of human visual perception to improve deep learning. *Neural Networks*, **104**, 40–9.

Taggart, N. W., Haglund, C. M., Tester, D. J. *et al.* (2007). Diagnostic miscues in congenital long-qt syndrome. *Circulation*, **115**(20), 2613–20.

Talebi, S., Alaleh, A., Sam, Z. *et al.* (2015). Underestimated and unreported prolonged QTc by automated ECG analysis in patients on methadone: can we rely on computer reading? *Acta Cardiologica*, **70**(2), 211–16.

Theeuwes, J. (2013). Feature-based attention: it is all bottom-up priming. *Philosophical Transactions of the Royal Society B: Biological Sciences*, **368**(1628), 20130055.

Tian, Y., Chen, X., Xiong, H. *et al.* (2017). Towards human-like and transhuman perception in AI 2.0: a review. *Frontiers of Information Technology & Electronic Engineering*, **18**(1), 58–67.

Topol, E. (February 2019). The Topol review: preparing the healthcare workforce to deliver the digital future. Technical Report. London: Health Education.

Treisman, A. (1983). The role of attention in object perception, in O. J. Braddick and A. C. Sleigh, eds, *Physical and Biological Processing of Images*, Berlin, Heidelberg: Springer Berlin Heidelberg, 316–25.

Tyl, B., Azzam, S., Blanco, N. *et al.* (2011). Improvement and limitation of the reliability of automated QT measurement by recent algorithms. *Journal of Electrocardiology*, **44**(3), 320–5.

van der Helm, P. (2015). Simplicity in perceptual organization, in J. Wagemans, eds., *The Oxford Handbook of Perceptual Organization*. Oxford University Press.

van der Helm, P. A. (2017). Human visual perceptual organization beats thinking on speed. *Attention, Perception, & Psychophysics*, **79**(4), 1227–38.

Vayena, E., Blasimme, A., and Cohen, I. G. (2018). Machine learning in medicine: Addressing ethical challenges. *PLoS Medicine*, **15**(11), 4–7.

Velik, R. (2010). Towards human-like machine perception 2. 0. *International Review on Computers and Software*, **5**(4), 476–88.

Vicente, J., Johannesen, L., Mason, J. W. *et al.* (2015). Comprehensive T wave morphology assessment in a randomized clinical study of dofetilide, quinidine, ranolazine, and verapamil. *Journal of the American Heart Association*, **4**(4), e001615.

Viskin, S., Rosovski, U., Sands, A. J. *et al.* (2005). Inaccurate electrocardiographic interpretation of long QT: the majority of physicians cannot recognize a long QT when they see one. *Heart Rhythm*, **2**(6), 569–74.

Ware, C. (2012). *Information Visualization: Perception for Design*. Boston: Elsevier.

Wertheimer, Max (1923). Untersuchungen zur Lehre von der Gestalt, II. *Psychologische Forschung*, **4**, 301–50. [Translated extract reprinted as "Laws of organization in perceptual forms," in W. D. Ellis, ed. (1938), *A source book of Gestalt psychology*. London, UK: Routledge & Kegan Paul Ltd., 71–94.]

Willems, J. L. (1980). A plea for common standards in computer aided ECG analysis. *Computers and Biomedical Research*, **13**(2), 120–31.

Wittig, M. A. and Allen, M. J. (1984). Measurement of adult performance on Piaget's water horizontality task. *Intelligence*, **8**(4), 305–13.

Wolfe, J. M. and Utochkin, I. S. (2019). What is a preattentive feature? *Current Opinion in Psychology*, **29**, 19–26.

Wood, G., Batt, J., Appelboam, A. *et al.* (2014). Exploring the impact of expertise, clinical history, and visual search on electrocardiogram interpretation. *Medical Decision Making*, **34**(1), 75–83.

Yap, Y. G. and Camm, A. J. (2003). Drug induced QT prolongation and torsades de pointes. *Heart*, **89**(11), 1363–72.

Zavagno, D. and Daneyko, O. (2014). Perceptual grouping, and color, in R. Luo, ed., *Encyclopedia of Color Science and Technology*. New York: Springer Science+Business Media, 1–5.

Zhou, Z. and Firestone, C. (2019). Humans can decipher adversarial images. *Nature Communications*, **10**(1), 1334.

13

The Shared-Workspace Framework for Dialogue and Other Cooperative Joint Activities

Martin J. Pickering and Simon Garrod
University of Edinburgh and University of Glasgow, UK

13.1 Introduction

As humans, we regularly engage in cooperative joint activities with other humans and such activities involve language and communication. Such an ability is central to human intelligence and clearly sets us apart from other species, but also from most artificial systems. If we want to understand human intelligence and to develop systems that reflect it, and also to develop systems that can interact intelligently with humans, then we need to understand the nature of such interactive intelligence.

In this chapter, we sketch our framework for dialogue and other cooperative joint activities, a framework that models both the individuals engaged in such activities and the system that links them together. Our framework is presented in much greater detail in Pickering and Garrod (2021). It assumes that individuals are not linked directly but rather via a shared workspace. This workspace contains the aspects of reality that are relevant to the joint activity, and the individuals 'post' their contributions to it. After describing the framework, we consider its implications for human-like machine intelligence.

13.2 The Shared Workspace Framework

Imagine two people collaborating to construct a piece of flat-pack furniture. At a particular point, the woman holds two components in place for the man to screw together. The action is successful only because their individual actions mesh together—they take place at the same time and are appropriately located for each other. They share a plan (which might be defined by a set of instructions) and are committed to carrying out the plan.

The two agents are performing what we call a cooperative joint activity (which relates to shared cooperative activity as discussed by Bratman, 1992). They both realize that

Martin J. Pickering and Simon Garrod, *The Shared-Workspace Framework for Dialogue and Other Cooperative Joint Activities* In: *Human Like Machine Intelligence.* Edited by: Stephen Muggleton and Nick Charter, Oxford University Press.
DOI: 10.1093/oso/9780198862536.003.0013

they are both taking part in the activity and they are both committed to its success. Moreover, they both have a joint intention: the woman intends that the woman and the man construct the furniture, and the man intends the same. An important aspect of this intention is that they are mutually responsive – for example, they monitor each other for mistakes (i.e., deviations from the plan) and seek to compensate.

In our example, the plan is of course agreed beforehand, both at the overarching level at which the agents are committed to successful furniture construction, and at the level of subplans that are defined by the set of instructions (e.g., a booklet provided by the manufacturers). In reality, most planning is more partial, with each agent representing the individual aspects of their plan, the shared aspects of their plan, and some (but typically not all) aspects of their partner's plan. Formally representing such collaborative planning is extremely complex (see Grosz and Kraus, 1996). Mutual responsiveness (via communication) is necessary for the online development of such collaborative plans. Our focus is on the role of communication, not primarily with respect to the development of the plan itself but rather to the way in which the plan is carried out.

Our framework accounts for such cooperative joint activities. We assume two agents (*A* and *B*) who both 'post' their individual contributions to the joint action to a shared workspace (see Fig. 13.1). This shared workspace contains the joint action (attaching the components together) and therefore includes the components themselves ('props') as well as the agents' actions. Importantly, the workspace is not a representation (within one or both agents) but is reality; it describes those aspects of the world that are integral to the activity. It contains the shared objects of the agents' internal representations – that is, what they are representing. These objects are defined at the level that is appropriate for the activity (e.g., the props are pieces of furniture rather than lumps of wood or metal

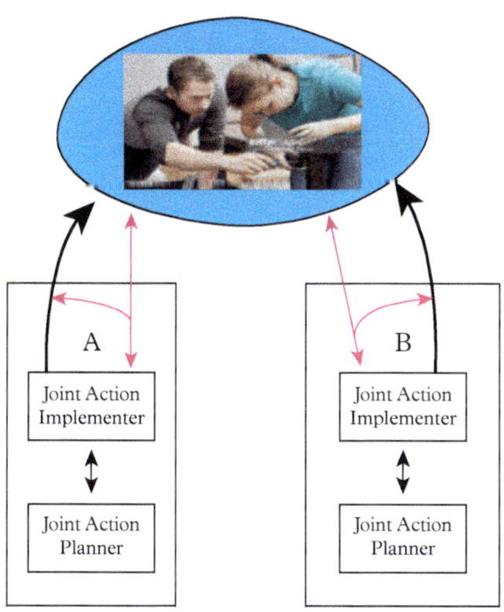

Figure 13.1 *An illustration of the shared workspace framework for cooperative joint action. Each agent 'posts' contributions to the shared workspace using the thick black arrows.*

or arrangements of molecules). The workspace is a *shared* slice of reality because both agents can observe and manipulate it. Furthermore, the contents of the shared workspace are typically manifest to both agents, which means that they are both confident that they can both observe them. In addition, the contents have joint affordances – for example, the components *afford* the joint action of being screwed together by the agents (Gibson, 1979; Norman, 2013).

Each agent has a joint action planner and a joint action implementer. Both agents represent the plan – that is, what they intend to do (build the furniture) and their roles within this plan (e.g., *B* will hold components in a manner that allows *A* to screw them together). They also both represent how they will implement the plan (e.g., *B* will apply force to her arms so that the components are lifted to position X, and *A* will raise the screwdriver to position X and turn it), and their implementations are sensitive to each other's implementations (e.g., if *B* lifts the components higher than expected, then *A* must respond).

This sensitivity is a consequence of the red arrows, which serve to perceive, predict, and monitor the workspace. For example, *A* perceives that *B* is holding the components at a different height from *A*'s screwdriver via the downwards vertical red arrowhead. As a result, *A* detects an error in the joint action, because the position at which *A* perceives *B* to be holding the components does not match *A*'s prediction of this position (carried out via the upwards red arrowhead). *A* monitors this error and can correct by moving the screwdriver (because *A* is committed to being responsive). (The horizontal red arrows are used for internal processing.)

13.3 Applying the Framework to Dialogue

Dialogue is a form of cooperative joint activity. The interlocutors (*A* and *B*) both realize that they are both taking part in the dialogue and are both committed to its success. Moreover, they both have a joint intention; *A* intends that *A* and *B* have a successful dialogue, and *B* intends the same. And they are mutually responsive – for example, *A* queries *B* if *B* appears to make a mistake, or if *A* cannot follow where the conversation is going.

For example, consider this excerpt from a conversation in which two interlocutors seek to navigate themselves around the maze in Figure 13.2 (Garrod and Anderson, 1987).

1. B: 'Tell me where you are?'
2. A: 'Ehm : Oh God!' (laughs)
3. B: (laughs)
4. A: 'Right :two along from the bottom one up.'
5. B: 'Two along from the bottom, which side'?
6. A: 'The left: going from left to right in the second box.'
7. B: 'You're in the second box.'

8. A: 'One up:(1 second) I take it we've got identical mazes?'
9. B: 'Yeah well: right, starting from the left, you're one along.'
10. A: 'Uh-huh.'
11. B: 'and one up?'
12. A: 'Yeah, and I'm trying to get to ...'

Success in dialogue requires that the interlocutors reach the same understanding as each other, that is, they *align* their situation models (Pickering and Garrod, 2004). And the interlocutors are committed to achieving such alignment. In this case, *A* and *B* seeks to make sure that *B* correctly represents *A*'s maze location. And to do this, *A* describes *A*'s position (at 6 and 8) and *B* tries to clarify and perhaps correct *A*'s description (at 9). In 9–12, they play an 'information probing' dialogue game (with *B* as prober in 9 and 11 and *A* as respondent in 10 and 12), a game which they both represent as manifest in their dialogue game models. They represent the same dialogue game (information probing), with the same interlocutors playing the same roles (*B* as prober, *A* as respondent).

Interlocutors have to plan dialogue just as agents have to plan cooperative joint activities. We can apply the shared workspace framework to dialogue by treating the situation model and the dialogue game model as components of the dialogue planner (see Fig. 13.3). In addition, each interlocutor has a dialogue implementer that executes the dialogue. In psycholinguistic terms, it contains the mechanisms of language production and comprehension. The shared workspace contains the utterances by both interlocutors that make up the dialogue, as well as relevant aspects of the world (as 'props') – in this example, the relevant location in the maze. Each interlocutor uses the black arrow to 'post' contributions to the dialogue; for example, *B* produces '*you're one along*' and *A* produces '*uh-huh*'. As in Figure 13.1, the vertical red arrows are used to comprehend, predict, and monitor the workspace (with the horizontal red arrows being used for internal processing). The utterances are meaningful entities; they are elements of reality defined at a level of description appropriate for communication. Just as the workspace contains a piece of furniture rather than an arrangement of molecules, so it contains words rather than a sequence of sounds. It can also contain other signs such as communicative gestures and depictions (e.g., a drawing in the air, or a mocking tone of voice; Clark, 2016).

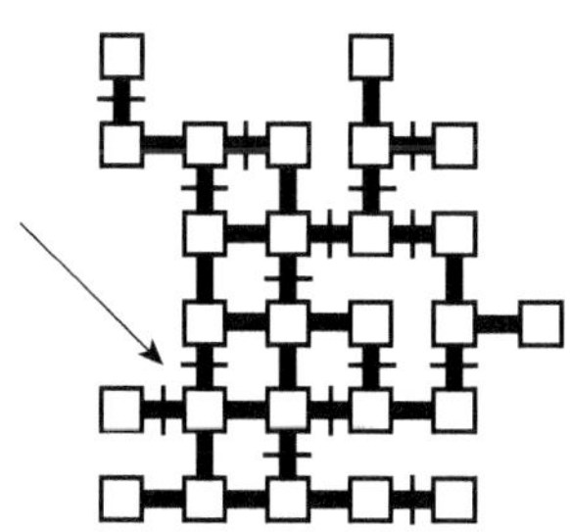

Figure 13.2 *A's location in the maze (indicated by the arrow).*

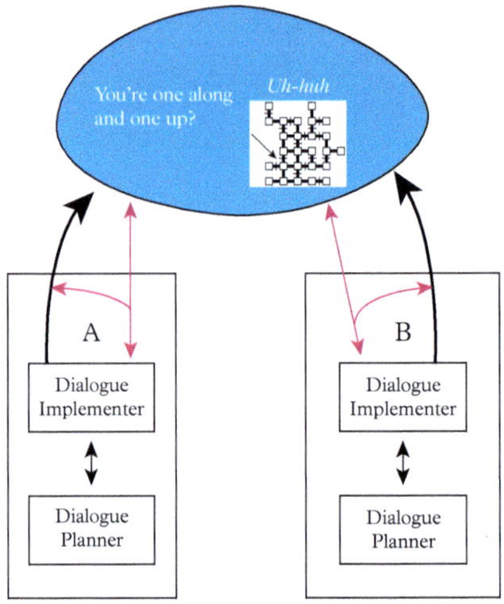

Figure 13.3 *An illustration of the shared workspace framework for dialogue.*

The situation model contains representations of the entities in the workspace (i.e., at the same level of description).

Traditional psycholinguistic accounts of production and comprehension assume that people produce and comprehend isolated utterances. In the shared workspace framework, they of course produce individual contributions (via the black arrows), but they can then act on the contents of the workspace as a whole. So they can comprehend 'you're one along uh-huh and one up' (with *B*'s contribution underlined) as a joint contribution—for example, determining if it makes sense, or drawing inferences from it as a whole. Similarly, they can predict the flow of the dialogue – for example, whether they will successfully play a particular dialogue game. They can perform joint monitoring (e.g., checking if a response is appropriate for an utterance) and can act accordingly, something which is not possible in traditional theories of speech monitoring (e.g., Levelt, 1989).

We have already noted that interlocutors align their situation models in successful dialogue. In our example, such alignment is hard to achieve (and it is perhaps unclear whether the dialogue will end up being successful). If successful, the interlocutors align on a position in the maze and on a description scheme that constitutes a path through the maze (rather than on coordinates such as *B2*). In many dialogues, they align on a description of an event, or information about a character, or on a set of instructions. But in addition, interlocutors align their dialogue game models. Such alignment occurred in our example (they both represented the same game with the same role assignments), though it would not happen if one interlocutor had probed for information and the other did not realize what the first interlocutor was doing. Finally, both situation model and dialogue game alignment can refer to what is happening at the current moment, as in these examples (focal alignment). But it can also refer to the build-up of alignment over an interaction, for example when interlocutors come to understand a description of a complex event or a body of knowledge in the same way as each other (global alignment).

At the same time, interlocutors align their representations of linguistic information. The maze players tend to repeat each other's words and their meanings (Garrod and Anderson, 1987). For example, a pair might align on the use of 'row' or 'level', or they might align on the use of 'row' to mean 'vertical arrangement' (vertical row) as opposed to the more standard 'horizontal' arrangement. Similarly, participants describing cards to each other tend to repeat syntax, for example describing a picture of a red sheep as 'The sheep that's red' after their partner described a different picture as 'The goat that's red' (Cleland and Pickering, 2003). In this case, they are aligning on the use of a relative clause structure.

In Figure 13.4 we illustrate the internal structure of the dialogue planners (containing the situation model and the game model) and the dialogue implementers (containing semantic, syntactic, and phonological representations, as well as a 'binding node' that links them together as an expression such as a word; cf. Jackendoff, 2002). These components are linked across interlocutors via what we call *channels of alignment*. Importantly, these channels are not causal links but rather the consequence of patterns of co-activation. The causal links go through the shared workspace, as a consequence of dialogue itself; there are no telepathic links between minds, or anything corresponding to wireless links across artificial agents (as in shared mental models frameworks; see Scheutz *et al.*, 2017).

Interlocutors distribute control over the progress of the dialogue. Only in special formal circumstances is one interlocutor in 'sole command'. In almost all situations,

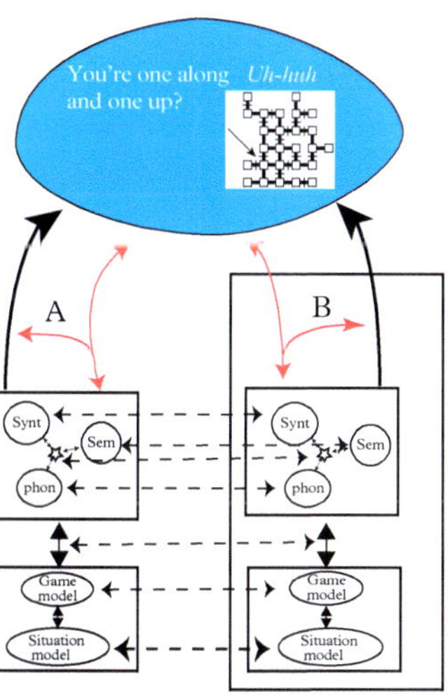

Figure 13.4 *The shared workspace framework including internal structure to the dialogue planners and the dialogue implementers, together with channels of alignment.*

the interlocutors share decisions about the dialogue: who should speak and when, what they should talk about, and what to do if the dialogue appears to be going off-course. Thus, each interlocutor exerts control as necessary. And distributed control can work because interlocutors are typically well-aligned; they activate and use the same linguistic representations, they are engaged in the same dialogue game, and they have similar situation models.

Successful distributed control also depends on prediction and synchronization. Dialogue involves fluent turn-taking with the intervals often being 200 milliseconds or less, even though isolated speakers take much longer to plan utterances (see Levinson, 2016). It appears as though interlocutors predict what their partners will say and when they will say it and prepare their responses accordingly. In our terms, they predict the contents of the workspace and when it will arrive, and they compare what they have predicted with what occurs and use this comparison to develop their dialogue plans and prepare their next contributions. The interlocutors have to make the predictions at the right time; in other words, their predictions must be synchronized with the interlocutor's speech rate. If they converge their speech rates, then they avoid switch costs—that is, they can stick to the same timing when speaking and when predicting each other—and there is some evidence for such convergence (e.g., Cohen Priva *et al.*, 2017).

Control also depends on each interlocutor meta-representing alignment – that is, representing whether or not she is aligned with her interlocutor. In our example, *B* utters 'you're one along' (9) and A responds with 'uh-huh' (10). This is a *positive commentary* that indicates that *A* meta-represents alignment with *B*—specifically, in relation to the horizontal component of the situation model. *B* trusts *A* and therefore also meta-represents alignment, and proceeds accordingly by trying to establish the vertical component with 'and one up?' (11). Similarly, interlocutors meta-represent misalignment, as occurs in the following example (13–16) when *A* utters 'mm'—a *negative commentary*. *B* recognizes the misalignment and attempts to correct it by expanding 'she' to 'Isabelle'. This correction is successful, and so *A* provides a positive commentary to indicate that the misalignment has been resolved. In sum, the positive and negative commentaries are used to steer the dialogue (and are indicative that the interlocutors are committed to the cooperative joint activity).

13. B: 'and um it- you know it's rea- it's it was really good and of course she teaches theology that was another thing.'
14. A: 'mm'
15. B: I- mm- I- Isabelle'
16. A: 'oh that's great.' (Horton and Gerrig, 2005)

The relationship between commentary and alignment is cyclic. In a successful interchange, *A* produces an utterance which leads to *B* aligning on *A*'s linguistic and non-linguistic representations. *B* meta-represents alignment (i.e., realizes that they are aligned) and provides a positive commentary, which leads *A* to also meta-represent alignment. At this point, *A* can therefore instigate another cycle by producing a new contribution. Alternatively, *B* can produce a new contribution for which *A* provides a commentary. If the interchange is not initially successful, the respondent meta-represents misalignment, and provides an appropriate negative commentary that may lead to correction (or cycles of correction). The decision about who should contribute depends on the unfolding dialogue plan, in a manner that respects social conventions about who should speak and when.

13.4 Bringing Together Cooperative Joint Activity and Communication

Our framework applies to both linguistic interaction and other forms of cooperative joint activity, and therefore also applies to 'situated' dialogue. When our actors make furniture (Fig. 13.1), they of course typically use language, and so the activity involves both non-linguistic and linguistic components. To explain such complex activity, we must combine the framework sketched in the two previous sections. In outline, the agents put both actions (such as holding a piece of wood in place) and language (such as words) into the shared workspace. They then construct representations that are both linguistic and non-linguistic in nature, and use those representations in the service of alignment, prediction, control, and so on. Language allows interlocutors to refine the goals underlying joint activity, so that they can plan a series of joint activities (as when determining an itinerary or indeed in jointly constructing furniture).

When we consider the activity without language, we cannot unambiguously 'parse' it into components—the structure of the activity depends on external analysts (as observers). But communicating agents construct meaning by jointly performing the activity, and critically they 'parse' it by the way that they refer to it. Imagine that the woman (*A*) has just pointed to a mark on the wood and said *screw here*. The man (*B*) replies 'OK' and starts screwing. If so, they treat the action as having five components: *A* pointing, *A* speaking, *B* saying *OK*, *A* presenting the wood, and *B* screwing (see Clark and Krych, 2004). So when *A* says 'screw here', the act of screwing plus the screw and its target become components of the joint action. In turn, they become meaningful elements within the shared workspace, and these meaningful elements are interpreted in relation to moves in a dialogue game and the actors' situation models. *A* says 'screw here' and presents the wood as two serial steps, and *B* says 'OK' and screws as two subsequent parallel steps. The five components together constitute an action-seeking game, with *A*'s components (pointing, speaking, and presenting) comprising the initiation move and *B*'s components (the positive commentary *OK* and the act of screwing) comprising the response move. Both agents construct a situation model that represents the action in terms of its components (with 'here' instantiated as the mark on the wood).

Consider another example, in which *A* says 'you now need to screw the components together' and *B* responds by holding up two different screws with a quizzical expression. *A* initiates an action-seeking dialogue game, and *B*'s response is a non-linguistic demonstration serving a negative commentary function (corresponding roughly to the linguistic utterance 'which screw?'). This response therefore initiates an embedded information-seeking dialogue game and leads to *A* meta-representing misalignment. If *A* now points to one of the screws or utters 'the smaller one', then *A* attempts to complete the embedded game and resolve the misalignment. And if *B* nods, then *B* indicates re-alignment, and can complete the original action-seeking game by using that screw.

Now imagine that *A* and *B* are military historians discussing aspects of the Battle of Waterloo over coffee. *A* places her mug on the table and says 'This is Napoleon's cavalry' and points to the sugar bowl, saying 'This is Wellington's infantry'. *A* moves the mug in an arc towards the bowl, thereby representing the indirect troop movement. (The use of a mug to refer to cavalry is symbolic, and the use of an arc to represent indirect movement is iconic.) At this point, *A* takes a sip out of her mug before replacing it in position and says 'Coffee, please', and *B* responds by filling the mug with more coffee. Assuming *B* is following what *A* is doing, the mug is an object in the shared workspace, but an object that has two functions. First, it is interpreted by *A* and *B* as a sign referring to Napoleon's cavalry, and second it is a mug – that is, an object that functions as itself (i.e., as a prop). Importantly, the meaningful object is the same in both cases (i.e., it is a mug). But it is used flexibly—it has two functions at the same time.

If *B* had filled *A*'s mug as part of a simple coffee-drinking activity (involving or not involving language), the mug would be a simple prop – its only role would be as a container of coffee. But our mug is also a sign (standing for Napoleon's cavalry). It takes part in two distinct cooperative joint actions, and has two distinct joint affordances. In the military re-enactment, the mug jointly affords reference to Napoleon's cavalry and jointly affords either or both actors to move other objects in concert with the mug – for example, moving a spoon forwards 'in support' and a sugar bowl backwards 'in retreat'. In the coffee-drinking activity, it jointly affords being a receptacle for coffee, something that can be proffered in order to have coffee poured into it and drunk out of. And for it to play its two roles, both *A* and *B* represent both roles. These roles are manifest to both of them: the coffee-drinking role because *A* proffered the mug to *B* to fill up and *B* drank out of it, and the re-enactment role because *A* labelled the mug as Napoleon's cavalry and *B* accepted the label (as a consequence of *B*'s contribution).

The shared workspace framework is ideally suited for this flexibility. The workspace provides a resource by which the same entities can be non-signs or signs, and can be signs of different types (symbolic or iconic). And it allows the integration of linguistic and non-linguistic contributions to cooperative joint activity. All entities in the shared workspace have a role in the joint plan. Our framework therefore allows us to understand the meshing of communication and cooperative joint activity.

In addition, the framework predicts that people engaged in cooperative joint activities will communicate effectively (i.e., align) on the basis of any meaningful signs placed in the shared workspace. In fact, non-linguistic signs may contribute as much to effective dialogue as linguistic signs, with interlocutors blending them according to circumstances. For example, in a noisy pub a person might request a refill by miming a drinking action rather than asking for the refill. In sum, according to our framework dialogue is not exclusively linguistic.

13.5 Relevance to Human-like Machine Intelligence

The shared workspace framework is developed to explain natural dialogue and associated joint activity but has relevance for technology. One implication is for communicative technologies—how they can be used to facilitate human communication. Another implication is for communication with machines.

13.5.1 Communication via an augmented workspace

Interlocutors can use a physical whiteboard to assist organization or problem solving—a board that is manifest to both of them. If an architect and a builder plan how to tackle the complex task of building a house extension, then they can add information to the board to facilitate their joint planning. This information can be words (e.g., bricks needed, location of door, 600) or sketches (e.g., a floor layout, which might include the location of the door). They can also add information associated with executing the plan, for example ticking off tasks when completed. The board contains meaningful entities but they are not interpreted. Thus, the interlocutors might interpret 'location of door' to refer to different doors but the words themselves are the same. As they perform the joint task, they add new words or sketches and modify or delete existing ones. At the same time, they each monitor what is happening to the board (both as a result of their own and their partner's activities). Therefore, the board and the way it is used are very similar to the shared workspace and the way it is used.

The whiteboard is of course not the shared workspace itself but is used to augment it. It does so in two ways. First, it makes entities salient—for example, the architect can emphasise a candidate location for the door by encircling it in the sketch. In this sense, the board is not primarily used to enrich the workspace, but rather to help both of them navigate it. But in addition, the sketch provides a record, as does a written word. Unlike the spoken word, entities on the board do not disappear. And so the agents can straightforwardly 're-sample' them, by using the downward red arrowheads whenever they need to. Importantly, the board must not become 'cluttered'—each interlocutor must remain aware of the relevant content of the board and be confident that the other interlocutors are aware of it as well, and this is not the case if the board is cluttered. Thus, the content of the board should be manifest, and if so, it will remain in the shared workspace. We propose that many technologies attempt to prevent a cluttered workspace, for example by limiting the amount of information that can be communicated (such as text length, in Twitter; number of pictures, as in Instagram; or persistence of pictures, such as Snapchat). However, such technologies have many characteristics of monologue and so are less relevant for this discussion.

In contrast, some dialogue-like technologies directly incorporate a whiteboard for the purposes of videotelephony and videoconferencing. In Skype, Zoom, and Teams, *A* can project A document onto both *A*'s screen and *B*'s screen. *A* can control editing, or *A* can pass control over to *B*, or (in some systems) both can edit at the same time. *A* and *B* typically assume that they can both see the document—that is, the document is manifest to both of them. (In fact, this can be an illusion and one of them can have replaced the document by email.) If they can both edit, they each have their own thick black arrow (see Fig. 13.1) to insert text or pictures into the document, and a full set of red arrows. If A can edit but not B, then A has a full set of arrows for all aspects of the activity; in contrast, B has a full set of arrows for speech (and gesture, assuming visibility), but only the vertical red arrow for editing.

The technologies possess other interesting properties with respect to the shared workspace framework, though many of these become more important or apparent in multi-party dialogue. For example, interlocutors can type comments as well as speak, something that allows a second means of production (i.e., a different use of the thick black arrow). Another point of interest is that speaking tends to cause the camera to show the speaker's face, which means that the speaker's gestures become apparent and attention is focused on the speaker more generally. Of course, the shared-workspace framework applies to multi-party dialogue. But this more complex situation takes us beyond the scope of this chapter. We consider some of the issues in Pickering and Garrod (2021, Chapter 10), where we also show how our framework can be applied to monologue (and associated technologies).

More generally, Norman (2013) argued that good engineering designers seek to satisfy affordances – that is, they construct artefacts in ways that facilitate their use. For example, they design doorknobs to fit the hand and place them at the best height to make door-opening easy, and they design remote controls for televisions so that they are user-friendly. In other words, they design artefacts so that they have the appropriate affordances for their characteristic uses. Designers of communication technologies (whether Twitter or Skype) make them appropriate for communication – that is, to support alignment. For example, it might be that alignment can be enhanced when interlocutors project a simple diagram during a Zoom meeting, but impaired if the diagram is too complex and the interlocutors focus on different aspects of it. Importantly, alignment is a property of a dyadic system and therefore relates to joint affordances. Thus, a good communication technology has the appropriate joint affordances for interaction.

13.5.2 Making an intelligent artificial interlocutor

So far, we have discussed how technology can affect the medium of communication. Our framework also applies more radically to the replacement of one interlocutor by an intelligent machine. If the machine is concerned with language alone, the intelligent machine is a dialogue system. If it is also concerned with cooperative joint activity more generally, it is a social robot.

We can apply the shared workspace framework to the design of dialogue technologies by designing an artificial interlocutor that shares properties of a real interlocutor (e.g., B in Fig. 13.4). The artificial interlocutor needs an appropriate dialogue implementer, one that includes linguistic representations and processes that allow it to produce and comprehend utterances. But the specific challenge of dialogue technologies primarily relates to the dialogue planner. In everyday conversation, interlocutors play dialogue games such as information probing or information seeking and it should be possible to play such games with artificial interlocutors as well. However, such games are in fact a fairly crude characterisation of communicative activity types (see Levinson, 1979), and we can make distinctions between, for example, a polite request to a shopkeeper, an instruction to a child, and a military command. Such activities constrain participants' contributions. In a professional exchange between shopkeeper and customer (in the United Kingdom), the shopkeeper is conventionally limited to questions such as 'Can I help you?', and the customer is, in turn, conventionally limited to asking whether the shop sells a particular product, where it is on the shelves, or how much it costs; both participants may also make a few generic comments (e.g., about the weather). The shopkeeper then provides little more than answers to these questions or requests for further information (e.g., which brand the customer prefers).

In a similar way, interactions between a person and an artificial agent are constrained, primarily as a consequence of the artificial agent's limitations – a system that makes reservations cannot give financial advice, and the person would not request such advice. The constraints are absolute and not merely normative (in a shop, either person can violate the conventions and will most likely get a response). The dialogue games may well be entirely formalized: for example, the person may be able to query (e.g., 'What is the time of first train from Edinburgh to London?') and therefore initiate an information-seeking game, book (e.g., 'Please book a ticket from Edinburgh to London on the 6 a.m. train') and therefore initiate an action-seeking game, or a handful of alternatives. The artificial agent can respond by completing the game (e.g., answering the query, making the booking) or by using a commentary (e.g., 'which day?') to initiate an information-probing game. To play such games, the interlocutors construct appropriate dialogue game models. The artificial agent has a limited repertoire of such models, and the person realizes that many models are irrelevant for any interaction with the artificial agent.

The artificial agent also has limited situation models. For example, it may represent objects such as origin, destination, type of ticket, and the topology of the railway system. Of course, most of this information is in databases (corresponding to background knowledge in a person), but the relevant information is represented in the situation model. So when a person asks 'How can I go from Edinburgh to London via Leeds?', the system first represents the origin, intermediate stop, and destination in its model, and then accesses the complete route map (and timetable) to incorporate (via reasoning) the appropriate map fragment into the focal situation model. The system then uses this representation to produce a response (e.g., 'Leave Edinburgh 7.30 a.m., arrive Leeds . . .') and the person comprehends this response and develops a situation model that is aligned with the artificial agent's model. Of course, such alignment takes place within a very limited domain.

The system designer has the task of developing a system that supports such minimal alignment as quickly as possible and with no more words than necessary (i.e., 'saying just enough'). Such a system is 'intelligent' in only a very limited sense. A quite different type of dialogue system has at the least the potential to communicate in a much more general way (this is the goal of a Siri or Alexa), and of course makes fundamental use of learning to do so. We propose that a successful system will represent dialogue models and linguistic representations in a functionally equivalent way to the way that people do (i.e., even if they are implemented in different ways), though it may draw on a database that is more extensive in some ways. If these representations are equivalent to people's representations, then the system should have a similar interactive intelligence to people. The shared workspace should be similar to the shared workspace for human interlocutors. Importantly, the artificial agent should also be able to meta-represent alignment (i.e., realize when the interlocutors are aligned or not). It should also use commentaries appropriately, for example negative commentaries to seek clarification and bring the dialogue back on track, and should learn from such clarification. The designer of course needs to incorporate such abilities into this general system, which can then 'take off' on its own.

A social robot that can converse needs to involve an intelligent dialogue system that is situated in the world, for example by perceiving objects and moving them around. To know what it can refer to, it needs to determine which objects are manifest—and it should be able to make them manifest, for example by pointing at them or by having a screen on its 'chest' that shows what it sees or knows. It can use its implementer to produce a combination of utterances and non-linguistic depictions (e.g., 'Look at this', while holding up a piece of furniture), and its dialogue game model can incorporate both types of information. Its situation model can include representations of non-linguistic props, and so it draws on a more elaborate shared workspace than a dialogue system that is not situated in the world. Such robots can deal with limited domains, for example a surgical assistant that uses visual information and limited dialogue to select and present the appropriate size scalpel at just the right time. But a designer of a social robot with more general abilities has a more complex task – that is, to design a robot that can process and learn in a situated manner. Both specialist and general robots should be able to use appropriate linguistic commentary (e.g., 'which screw?') and non-linguistic commentary (e.g., hold up the short screw) to keep the interaction on track.

13.6 Conclusion

We have outlined a shared-workspace framework for dialogue and other cooperative joint activities, described how they can be brought together in situated communication, and then sketched some implications of our account for technologies such as human-like artificial systems. Our framework involves a dyadic system and two individuals-within-the-system, and is incomplete without discussion of both components at the same time. We argue that our framework is opposed to what we term *monadic cognitive science* – that is, the science of individual minds. It is not enough to represent social information in an individual mind; in addition, we must consider how individual minds interact. And this is particularly critical when discussing communication, which requires at least two communicators (even in monologue, when one communicator need not be present).

A related notion is interactive intelligence, which exists in the service of interaction. Thus, people or machines are interactively intelligent if they can solve problems together and communicate appropriately, and it should be possible to study how such interactive intelligence is achieved in people or machines. We hope that our framework provides a way of modelling such interactive intelligence and can contribute to the development of human-like machine intelligence.

References

Bratman, M. E. (1992). Shared cooperative activity. *The Philosophical Review*, **101**(2), 327–41.

Clark, H. H. (2016). Depicting as a method of communication. *Psychological Review*, **123**(3), 324–347.

Clark, H. H., & Krych, M. A. (2004). Speaking while monitoring addressees for understanding. *Journal of Memory and Language*, **50**(1), 62–81.

Cleland, A. A. and Pickering, M. J. (2003). The use of lexical and syntactic information in language production: Evidence from the priming of noun-phrase structure. *Journal of Memory and Language*, **49**(2), 214–30.

Cohen Priva, U., Edelist, L., and Gleason, E. (2017). Converging to the baseline: Corpus evidence for convergence in speech rate to interlocutor's baseline. *The Journal of the Acoustical Society of America*, **141**(5), 2989–96.

Garrod, S. and Anderson, A. (1987). Saying what you mean in dialogue: a study in conceptual and semantic co-ordination. *Cognition*, **27**(2), 181–218.

Gibson, J. J. (1979). *The Ecological Approach to Visual Perception*. Boston, MA: Houghton Mifflin.

Grosz, B. J., and Kraus, S. (1996). Collaborative plans for complex group action. *Artificial Intelligence*, 86(2), 269–357.

Horton, W. S. and Gerrig, R. J. (2005). Conversational Common Ground and Memory Processes in Language Production. *Discourse Processes*, **40**(1), 1–35.

Jackendoff, R. (2002). *Foundations of Language: Brain, Meaning, Grammar, Evolution*. Oxford: Oxford University Press.

Levelt, W. J. (1989). *Speaking: From Intention to Articulation*. Boston, MA: MIT Press.

Levinson, S. C. (2016). Turn-taking in human communication—origins and implications for language processing. *Trends in Cognitive Sciences*, *20*(1), 6–14.

Norman, D. (2013). *The Design of Everyday Things: Revised and Expanded Edition*. New York, NY: Basic Books.

Pickering, M. J. and Garrod, S. (2004). Toward a mechanistic psychology of dialogue. *Behavioral and Brain sciences*, **27**(2), 169–225.

Pickering, M.J. and Garrod, S. (2021). *Understanding Dialogue: Language Use and Social Interaction*. Cambridge: Cambridge University Press.

Scheutz, M., DeLoach, S. A., and Adams, J. A. (2017). A framework for developing and using shared mental models in human-agent teams. *Journal of Cognitive Engineering and Decision Making*, **11**(3), 203–24.

14

Beyond Robotic Speech: Mutual Benefits to Cognitive Psychology and Artificial Intelligence from the Study of Multimodal Communication

Beata Grzyb and Gabriella Vigliocco

University College London, Division of Psychology and Language Sciences, UK

14.1 Introduction

Language has predominantly been studied as a unimodal phenomenon (i.e., speech or text) without much consideration for the physical and social context in which communication takes place. In psychology, the study of language learning, production, and comprehension has largely ignored the face-to-face communicative context in which language is learnt and processed, following a long-standing tradition in linguistics according to which only the linguistic content is 'language proper' (see Vigliocco, 2014 for a discussion). Thus, language has been defined as a rule-governed system which consists of abstract units (phonological, morphological, lexical) that can be separated from other aspects of communicative behaviour, such as gestures, gaze, prosody, and the mouth movements associated with speech. These cues, although omnipresent in face-to-face context, are typically considered as 'non-verbal' or 'non-linguistic' and studied separately as part of non-verbal behaviour. There is, however, now a growing body of literature emphasizing the role of such cues in language learning and processing. For example, gestures such as pointing and iconic gestures (that imagistically evoke aspects of referents) have long been shown to have a powerful effect on language learning, predicting the onset of learning milestones (Rowe and Goldin-Meadow, 2009); they support speakers in organizing their thoughts for speaking (Kita *et al.*, 2017) and help listeners, especially in noisy environments (Drijvers and Özyürek, 2017).

Implementation of language in embodied agents (i.e., artificial agents that interact with their environment through the body including virtual agents and physical robots) has also been reduced to speech only (see Cangelosi and Ogata, 2016 for overview and alternative approaches). This is problematic because in interactions with humans, the

Beata Grzyb and Gabriella Vigliocco, *Beyond Robotic Speech: Mutual Benefits to Cognitive Psychology and Artificial Intelligence from the Study of Multimodal Communication* In: *Human Like Machine Intelligence*. Edited by: Stephen Muggleton and Nick Charter, Oxford University Press. DOI: 10.1093/oso/9780198862536.003.0014

embodied agent is not capable of taking advantage of the wealth of cues produced by the speaker when comprehending and cannot provide these useful cues when producing language. Imagine for example a robot assisting an elderly person. The person asks her to bring him *that* glass (pointing toward the glass he wants). The robot has no way of understanding this indexical expression. As another example, imagine a robot and a human working on a joint task in a factory. The environment is noisy and the person has problems understanding what the robot says, the conversation is constantly interrupted and the person is frustrated. In human-to-human interactions, mouth movements, gesture, and gaze can successfully be used to fill in for missed speech content, but this is not possible with a robot assistant. These examples illustrate why multimodality in human-machine interactions is desirable. As embodied agents, and especially robots, become more common in our living environments (e.g., virtual receptionists and tour guides, lab demonstrators or tutors, assistive robots in home, companion robots for the elderly), they will need to be endowed with face-to-face communication skills, especially in the settings where they have to interact with naïve users. The existing works on effects of robot multimodal behaviour on human-robot interaction have indicated that multimodality does increase the intuitiveness and naturalness (Iio *et al.*, 2011), as well as the effectiveness and robustness of interaction (Breazeal *et al.*, 2005; Riek *et al.*, 2010), and finally, it positively influences the evaluation of agents increasing the chances of these agents to be used and accepted by their users (Krämer *et al.*, 2007; Bergmann *et al.*, 2010; Salem *et al.*, 2013).

Both producing multimodal cues by embodied agents and detecting and understanding these cues in human users are complex problems. Embodied agents have to display cues in appropriate contexts, congruently with speech, and importantly, their responses need to vary. Otherwise their behaviour is unbelievable, and embodied agents rapidly become rigid and unappealing. One limitation of cue generation in physical robots (but not in virtual agents) is the fine control of the movements where physical motors are used to control the appearance and timing of the movements. But more generally, a key limitation is our lack of knowledge of the underlying mechanisms of multimodal language production and comprehension in humans. Indeed, equipping embodied agents with multimodal language requires a detailed understanding of operational models of how the different cues are dynamically assembled and produced in humans.

In this chapter, we advocate that artificial intelligence should take a more comprehensive stance on language to improve the effectiveness, intuitiveness and natural flow of human-machine interactions. In turn, computational methods can support collection and coding of human face-to-face communication data essential for developing artificial systems. Finally, the computational models that produce language in artificial embodied agents can provide a unique research paradigm for investigating the underlying mechanisms that govern language processing and learning in humans. Such a joint approach to the study of multimodal language will be beneficial for psychology/linguistics and artificial intelligence.

Below, we review the literature on the use of multimodal cues in human-to-human communication. We move next to discussing whether—and if so how—humans respond to multimodal cues displayed by embodied agents. Finally, we review how embodied agents can recognize multimodal cues, and how they can produce them. We close discussing the potential benefits to psychologists and computer scientists of joining forces in the study of multimodal communication.

14.2 The Use of Multimodal Cues in Human Face-to-face Communication

In face-to-face communication, the unfolding speech signal, which carries phonological, syntactic, semantic, and discourse information, is accompanied by a multiplex of cues that recruit different articulators (manual, gaze, mouth patterns, and vocal).[1] These cues have all been argued to play an important role in adults' language processing and in children's word learning.

Iconic gestures evoke visual features of the referent (e.g., when describing a room, the speaker can represent the length of a piece of furniture by placing the hands far from each other) as well as properties of actions (e.g., when talking about a recipe, the movement of the hands can act out whisking an omelette). These gestures facilitate comprehension when the message they convey is congruent with speech, however, they can also impede comprehension when the message diverges from speech (Kelly *et al.*, 2010). Furthermore, iconic gestures can help listeners to disambiguate the meaning of words (e.g., the word 'ball' presented with a gesture congruent either with the dance or game meaning) (Holle and Gunter, 2007).

Iconic gestures are important not only for the listener but also for the speaker. Individuals who gesture tend to speak more rapidly (Allen, 2003). In contrast, the prohibition of gestures in narrations makes speech less fluent and causes a higher proportion of filled pauses (Rauscher *et al.*, 1996). Gesture production may facilitate speech formulation processes through decreasing cognitive load (Goldin-Meadow, 2001), priming conceptual information (Krauss, 2000), activating information or maintaining the activation of spatial information (Friedman, 1977; Ruiter, 2000), and finally, preparing and structuring information for processing (Kita, 2000). Furthermore, iconic gestures permit the speaker to express thoughts/concepts that otherwise would be difficult, if not impossible, to be expressed using speech only (e.g., describing the aspects of the coastline) (Goldin-Meadow, 1999). Finally, gestures can also be used to depict more abstract concepts (e.g., mathematical notions, etc. (Alibali *et al.*, 2013)).

Points (i.e., hand actions performed with an extended index finger or with other configurations of the hand) are used to focus their listener's attention on something, for instance, to refer to an object, or to isolate the object from other objects in a complex scene. They are the most common multimodal cues used by caregivers (Iverson, 1999;

[1] The cues we focus on in this chapter do not exhaust the repertoire of multimodal cues available in face-to-face communication as we are not including other aspects such as facial expression and head-body motion which also provide valid cues for learning and processing. The reason for limiting our attention to the cues described above is related to the fact that these have received the greatest attention in psychology.

Ozçalişkan and Goldin-Meadow, 2005) and can facilitate language learning by helping to link a provided label with the referent.

Finally, **object manipulations** (e.g., showing a toy hammer to a child or showing how to use the toy hammer) can be used by speakers in a purely indexical manner (showing the object), or in order to communicate something about the object.

Iconic gestures, points, and object manipulations (i.e., manual cues) have been argued to support language learning by directing attention to specific objects present in the environment, or by bringing to the mind's eye properties of objects absent in the environment (Vigliocco *et al.*, 2019). Iconic gestures and points are present early in in parental input to children (Rowe and Goldin-Meadow, 2009), as well as in the gestural repertoires of children (Acredolo and Goodwyn, 1988). Importantly, it has been shown that parents' gesture use (specifically points) predicts children's gesture use, which in turn predicts later vocabulary development (Rowe *et al.*, 2008), indicating the importance of this cue for overall language development. A link between object manipulation and children's vocabulary learning has also been suggested (Rohlfing, 2011), while iconic gestures may not be present until somewhat later (around 28 months) when they are seen in parental input especially when objects are not present in the environment (Vigliocco *et al.*, 2019).

Gaze can be used by listeners to anticipate, ground, and disambiguate spoken referents (Staudte and Crocker, 2011). Speakers fixate on objects or locations in the immediate surrounding, one second or less before naming them (Griffin and Bock, 2000; Hanna and Brennan, 2007; Yu *et al.*, 2012). Speaker's gaze can be used by listeners to identify the reference and to predict what the speaker is going to say. Indeed, listeners are slower at responding to the speaker's referential utterance when their access to the speaker's gaze is limited (Boucher *et al.*, 2012). Gaze helps listeners to eliminate uncertainty and to disambiguate between referents (Hanna and Brennan, 2007; Staudte and Crocker, 2011). In addition, gaze can signal communicative intent, and hence modulate how listeners process speech and gestures (Holzapfel *et al.*, 2004). Similar to the manual cues, gaze is an important cue from a very young age (Tomasello, 2003; Waxman and Lidz, 2006). Even young children systematically use the speaker's gaze to disambiguate the referent (Baldwin, 1993). More generally a strong link between gaze behaviour (i.e., gaze following, joint attention) and language development has been argued for (Morales *et al.*, 1998; Brooks and Meltzoff, 2005). For instance, gaze-following behaviour at 10–11 months accompanied by vocalizations has been shown to predict language comprehension and gesture production at 18 months (Brooks and Meltzoff, 2005). Gaze is often considered together with pointing as establishing joint attention. However, when in conflict, the two cues dissociate in their developmental trajectory with gaze providing the strongest cue earlier on (about 14 months) and pointing becoming the preferred cue later on (by 24 months) (Paulus and Fikkert, 2014).

Face-to-face communication allows listeners to make use of **mouth movements** that accompany speech. Interestingly, dubbing a voice saying 'b' onto a face articulating 'g' results in hearing 'd'. This is an example of a powerful 'McGurk illusion' (McGurk and Macdonald, 1976), which has been shown both in adults and in infants (Burnham and Dodd, 2004; Kushnerenko *et al.*, 2008). In more naturalistic presentations of the audio-visual speech, it is well established that mouth movements facilitate speech perception by reducing lexical competition, especially in noisy conditions (for a review, see Peelle and Sommers, 2015). They can also speed up speech perception (van Wassenhove, 2005; Castillo *et al.*, 2017) by providing predictive information about forthcoming phonemes. Finally, mouth movements were suggested to influence speech comprehension directly and automatically (Arnold and Hill, 2001).

In addition to visual cues, the **prosodic contour** of the speech provides meaningful information to listeners. While prosody is often considered as part of the linguistic information provided in speech, in most psycholinguistic research, prosodic information is reduced to achieve experimental control. However, prosody is used by the speakers to draw listeners' attention to words that introduce new information (Bolinger, 1972), it can be iconic (e.g., referring to a long trip by saying 'looong'), and it can provide important cues to word boundaries (see review in Cutler *et al.*, 1997). Prosody (especially changes in pitch) has long been recognized as a key property of child-directed speech (Fernald and Simon, 1984) and is helpful in word segmentation (Thiessen *et al.*, 2005), syntactic processing (Hawthorne and Gerken, 2014), and word learning more generally (Grassmann and Tomasello, 2007; Ma *et al.*, 2011; Herold, 2012).

These multimodal cues can be classified in terms of whether they support referent mapping/disambiguation: for example, points, gaze, and object manipulations all direct attention to referents present in the environment (they are indexical) and therefore they help singling the referent out. Iconic gestures and iconic prosody by virtue of being iconic can bring to the mind's eye properties of absent referents. They can also be classified in terms of whether they support semantic processing of the speech, such as iconic gestures, points, gaze, and prosody (iconic prosody and the use of pitch/duration to mark new information), or whether they support sensory processing of the speech (mouth movements). Both classifications emphasize the supportive role of multimodal cues in language learning and use.

14.3 How Humans React to Embodied Agents that Use Multimodal Cues

Human–machine interaction focuses on understanding how the different features of an embodied agent (i.e., what the robot looks like or how it behaves) affect its perception by the human in terms of, for instance, agent's intelligibility and likeability, as well as effectiveness of interaction. Here, studies generally use 'Wizard-of-Oz' techniques (where a person manually chooses the agent's output) or template-based techniques or prerecorded behaviours (i.e., speech, gestures, etc.) without implementing these complex behaviours on an agent.

Because embodied agents, either physical robots or virtual agents, usually have some human-like features (e.g., human-like bodies, or body parts) naïve users expect them to display the whole range of human-like communicative behaviours. Human participants are sensitive to points and gaze displayed by a robot, and their ability to read robot's cues is aligned with their ability to read and interpret the same cues from other humans (Breazeal *et al.*, 2005). Robot gaze can successfully direct attention to object reference

(Admoni *et al.*, 2014), as well as manage conversational turn-taking (Andrist *et al.*, 2014). Similarly, human users can exploit robot's gaze to anticipate and disambiguate spoken references (Staudte and Crocker, 2011), and they adapt their behaviour in response to the robot's gaze (Admoni *et al.*, 2014; Xu *et al.*, 2016). For instance, human participants looked more to a robot's face when the robot produced more looks to ward the person's face, thus creating more opportunities for mutual gaze and eye contact between the two (Xu *et al.*, 2016).

Human–machine interactions are rated as more natural and intuitive when the agent uses a variety of co-speech cues. For instance, Iio *et al.* (2011) showed that human participants rated higher the naturalness and ease of instruction and understanding of a robot that used gaze and pointing in addition to speech as opposed to a robot who used speech and gaze only or speech and pointing only; showing advantages of using multimodal channels for communication. Similarly, embodied agents that produced a range of different gestures were evaluated as more natural than agents that only produced points (McBreen and Jack, 2000). In addition, agents that produce communicative gestures such as movement of the eyebrows, or non-communicative such as body scratching are perceived as more lifelike (Kopp *et al.*, 2008), as well as more expressive, enjoyable, and having positive personality (Buisine and Martin, 2007). Similarly, robot gaze was shown to increase the perception of robot human-likeness, and increase the positive evaluation of the agents (Heylen, 2002; Karreman *et al.*, 2013).

There is evidence that multimodal cues can also improve the effectiveness of human-machine communication. In the study of Breazeal *et al.* (2005), naïve human participants guided a robot to perform a physical task using both speech and gesture. In one case, the robot proactively communicated its internal states both implicitly through gaze cue (e.g., where the robot looks and when the robot makes eye contact with the human participant) and explicitly using other communicative cues (e.g., nods of the head, points, facial expressions). In the other case, the robot communicated only when prompted by the human and using only explicit cues that do not show the robot's internal state (e.g., only looking straight ahead, no expressions of confusions, and only responds with head nods and shakes when prompted by the human). The results of a questionnaire and behavioural analysis of a video revealed that participants could understand the robot's current state and abilities better when the robot was displaying implicit cues. The communication with such a robot was also more efficient and resistant to errors. As in case of gestures, gaze behaviour also improved the agent's evaluations in cooperative tasks (Boucher *et al.*, 2012).

While multimodal cues have a general positive effect in embodied agents, their timing and congruency with the speech is key. Just as in human agents, gestures that are incongruent with the speech have a negative effect on comprehension of embodied agents. For example, in Bergmann et als' study (2010), an embodied agent with no gestures was rated more positively as compared with an agent that randomly produced gestures. Moreover, the ratings of the agent with random gestures were lower for overall comprehension, as well as comprehension of iconic gestures, and the certainty of the agent's communicative intentions; and the incongruent gestures displayed by the robot negatively contributed the participants' task-related performance (Salem *et al.*, 2013).

14.4 Can Embodied Agents Recognize Multimodal Cues Produced by Humans?

Recognition of multimodal cues produced by agents is a prerequisite to using them to improve comprehension by embodied agents especially in situations in which speech recognition is not optimal, like noisy environments (as most of everyday settings). Recognition of multimodal cues is hard for artificial systems. The existing robotic systems can only recognize a few, usually predefined, hand gestures. For instance, a system proposed by Xiao *et al.* (2014) can recognize a set of 12 communicative gestures and actions with objects, while a system based on a deep neural network can robustly recognize 5 gestures performed by different people (Barros *et al.*, 2014). More recently, Castillo *et al.* (2017) proposed a system able to recognize 14 different gestures (i.e., come toward, stop, greeting, pointing left/right). Only a few attempts focused on recognition of iconic gestures (Koons and Sparrell, 1994; Sowa and Wachsmuth, 2002), and these were able to recognize gestures, iconic of a few objects, based on the similarity between the shape of the object and the shape of the hand or its motion. Recognizing pointing is easier for embodied agents—given a well-defined and unique hand posture—and existing systems achieve good accuracy (Nickel and Stiefelhagen, 2003; Noda *et al.*, 2015). For instance, the system devised by Nickel and Stiefelhagen (2003) can detect pointing (with 88% success rate) as well as appropriately indicate the target of pointing (one of eight objects).

A number of studies addressed the problem of detecting gaze direction in human users in order to drive the behaviour of embodied agents. These systems often use the direction of head as an approximation of gaze (Doniec, 2006; Ivaldi *et al.*, 2014). As an alternative, some studies estimate the gaze direction either using head-mounted (Xu *et al.*, 2016) or remote eye-trackers (Palinko *et al.*, 2016). Detection/recognition of objects at which the human user is looking at, however, is difficult, especially in a naturalistic cluttered environment, and the systems usually are limited to a few objects.

Little attention has been paid to prosodic cues in the existing literature. Formolo and Bosse (2016) extract prosodic features with the main purpose of detection of different emotions, Rosis *et al.* (2007) looked at social attitude; Ang *et al.* (2002) at frustration and annoyance.

Why is recognition of multimodal cues so difficult for embodied agents? One reason is the type of sensors that the robots or virtual agents use. For instance, due to low resolution of cameras used in robotics, the systems often use the direction of head as an approximation of gaze (Doniec, 2006; Ivaldi *et al.*, 2014). However, in naturalistic setting people often make short gazes at objects without moving their heads. Systems that only use head direction are not sensitive to detecting such gazes. As an alternative, some robotics systems use head-mounted eye trackers (Xu *et al.*, 2016). However, using external devices prevents the naturalness of the human-machine interaction. In addition, changes in illumination can affect recognition of hand gestures based on visual input, and some studies use wearable gloves, or Microsoft Kinect to capture user's movement.

Another reason is that often the hand gesture recognition problem is limited to a pattern classification problem, in which the given input data (e.g., hand motion trajectory) is classified into one of few predefined categories. Although such an approach works well for recognizing points or predefined gestures (commands that a computer or a robot can understand), it is unsuitable for recognition of other meaningful (iconic) gestures which have no one-to-one mapping between their form and meaning (i.e., many gestures may have the same meaning or alternatively one gesture could have a different meaning, depending on the context).

Crucially, artificial systems usually rely only on a single modality for recognition (e.g., vision or motion) which is error prone, due to background noise and changing light conditions, in real-world environments. Most often, these systems do not take advantage of other available cues that facilitate recognition. It is however the case that some existing studies show that multimodality improves recognition performance. For instance, the ICONIC gesture recognition system (Koons *et al.*, 1993; Koons and Sparrell, 1994) exploits the fact that most gestures co-occur with speech. More specifically, the system scans speech for key words that indicate a possibility of a gesture (e.g,, 'like this') and once such a key word is detected, the system looks for a relevant gesture that accompanies the speech. Recognition of gestures can also be improved with help of prosodic information. Kettebekov *et al.* (2003) investigated how the intonationally prominent parts of speech align with kinematically defined gesture primitives. The results of the analysis were then used to discard the falsely detected points as well as preparation gesture primitives.

Using speech, also improves detection of points and importantly of their referents. For instance, Showers and Si (2018) showed that the accuracy of point and referent detection is higher when a robotic system uses a contextual information such as speech (extracted object names, object properties, and object position) and confidence heuristics (the quality of speech or visual information) as compared to a visual information alone. The same is true for gaze. Morency *et al.* (2006) showed that using three types of contextual features such as lexical features, prosody/punctuation, and timing improves significantly the recognition of gaze.

Although not used explicitly in embodied agents, mouth movements are often used in computational models of speech recognition. For instance, Bregler and Konig (1994) showed that, in the presence of additive noise and crosstalk, such a combined architecture achieves better recognition rates than speech-only architecture. Other systems based on audio-visual (extracted visual information about the lip features) integration show an improvement of speech recognition in normal but also in noisy real-word environments. For instance, Noda *et al.* (2015) proposed a system that combines a deep denoising autoencoder with a visual feature extraction mechanism that predicts the phonemes from the image sequence of the mouth area. They showed that using mouth information further improves the system performance (results in a lower error rate of word recognition).

Some studies attempted to improve the recognition of naturalistic conversational speech by utilizing information from deictic cues such as points and gaze. The object reference ambiguities in the system of Holzapfel *et al.* (2004) are resolved using either speech or through the interpretation of pointing. In addition, gaze can improve automatic speech recognition systems by highlighting the subset of vocabulary that relates to a person's visual attention. For instance, Cooke *et al.* (2014) proposed a model, which selectively uses the information from gaze (i.e., depending on the detected gaze role), and showed that under varying noise conditions, such a system achieves lower word error recognition rates. Finally, there have been some attempts to use prosody to aid automatic speech recognition. Here Fu *et al.* (2015) show a decrease in word error rate when recognition network uses not only previous words but also prosodic features as an input.

14.5 Can Embodied Agents Produce Multimodal Cues?

Humans produce a wide range of multimodal cues that are tightly linked to and time-aligned with the accompanying speech. These cues have been shown both to facilitate language processing in listeners and to facilitate language production in speakers. We already know that human users are sensitive to the cues produced by embodied agents, but can embodied agents produce them?

Few studies have proposed computational systems that provide embodied agents with the ability to produce speech and co-speech gestures. Kopp and Wachsmuth (2004) proposed a system that produces gestures from (manually) annotated descriptions which comprise speech transcript as well as gesture types (hand shape, palm orientation, hand location, etc.). Contrary to this model, the BEAT system (2001) was able to infer from input text what gestures to produce and when. The system was provided with a 'knowledge base' including information about objects, actions as well as gestures. In order to decide what gestures to use for a given text, the system first performs semantic analysis of the text marking novel words, and then it applies a set of rules derived from studies of human communication. For instance, it produces iconic gestures when during the explanation part of the utterance an object name is uttered and when this object has (as per knowledge base) some unusual features. Similarly, iconic gestures are produced for any action for which knowledge based contained a gesture description. Ng-hing *et al.* (2010) proposed instead a system that analysed input text in order to produce co-speech gestures. It used an automatic tagger to determine for each word of input text the part of speech of the word and a text-to-speech engine to extract word timing information. Then, input text together with any automatically extracted tags was processed by several grammars, one for each gesture type, and several candidate gestures for each word of the text were proposed.

One main drawback of these gesture-production systems is that gestures are highly deterministic (Cassell *et al.*, 2001; Kopp and Wachsmuth, 2004) meaning that the exact same gestures are produced for the same input text. One way to solve this problem is to select gestures probabilistically, from the range of possible candidates, as it has been done by Ng-hing *et al.* (2010) or to learn the probabilistic distributions of speech and gestures from a multimodal data collected from human participants (Kipp *et al.*, 2007; Bergmann and Kopp, 2009). Such an approach was followed by Kipp *et al.* (2007), who derived a gesture lexicon (i.e., all the gestures that are shared between speakers) as well as speaker style (i.e., a statistical model which captured the individual differences in the use and frequency of each gestures) from a manually annotated video corpus. Their system was then able to produce gestures for any arbitrary text input in the style of a particular speaker, though the text still had to be annotated (i.e., provide additional information on word boundaries, utterances, and information structure and focus of each utterance). Similarly, the model proposed by Bergmann and Kopp (2009) could learn idiosyncratic features (i.e., features that are characteristic of an individual speaker) from human communicative data. In addition, they also modelled more universal gesture features (such as shape/gesture form) defined by a set of rules.

Often gesture production systems in embodied agents use gesture templates which lead to almost identical gestures. For instance, Ng-hing *et al.* (2010) created a gesture lexicon based on the analysis of video corpora where gesture phases (start, stroke, retraction) were annotated manually. However, human gestures are also highly variable in terms of their trajectory (i.e., the shape of gesture) and such a template-based approach again is highly deterministic. Other systems have attempted to produce gestures on the fly which (in principle) may add variability. For instance, in the Kopp, Bergmann, Wachsmuth system (Kopp *et al.*, 2008) gesture form was not fixed, but produced from multimodal representations of objects, locations, and their spatial relations. More recently, Ferstl and McDonnell (2018) used a recurrent neural network to produce gesture motion directly from prosodic speech features. Interestingly, the network was first pre-trained with a motion modelling task before training the final speech-to-gesture model. A similar approach was taken in Ferstl, Neff, McDonnell (2019), where generative adversarial networks were used for producing gestures from input speech.

Finally, to be effective and believable, iconic gestures and points must be tightly linked to and perfectly time-aligned with the accompanied speech. The BEAT system of Cassell *et al.* (2001) uses word and phoneme timings to construct a multimodal animation schedule. Whereas the system of Kopp and Wachsmuth (2004) coordinates gestures and speech based on the assumption that both speech and gesture could be divided into successive, single units. More specifically, gesture motion is divided into gesture units (e.g., preparation, stroke, retraction, and hold) while speech into intonation units (e.g., intonation phases with exactly one primary pitch accent, i.e., nucleus). Gesture units then are aligned with speech units in a way to allow the gesture stroke to start before the affiliated word (or nucleus) and lasts throughout. To maintain the synchrony between the gesture and speech, the gesture is adapted to the structure and timing of accompanied speech.

Gaze has also been implemented on virtual agents and robots. Some models used heuristics derived from the analysis of human dyadic conversations. For instance, (Cassell *et al.*, 2001) analysed the distributions of gaze and head movements. They found that speakers look away from the listener at the beginning of a theme (what is talked about, the topic, with .70 probability) and look toward the listener at the beginning of a rheme (what it is said about the topic, with .73 probability). Mutlu *et al.* (2006) extended this empirical model with parameters derived from the analysis of data collected from a professional storyteller. More specifically, they first determined main gaze locations. Then, they derived probabilities of looking towards and looking away from the listener from the analysis of the frequencies of the storyteller's gaze at each location. The gaze durations followed a normal distribution with the mean and standard deviation values of the storyteller's gaze. Similar approach was used by Mutlu *et al.* (2012), where the analysis of empirical studies directly informed the gaze mechanisms of a humanoid robot. In addition to previous works, this model also implemented topic-signalling mechanism: for each new topic, the robot produced a series of gaze shifts based on the gaze patterns identified for each conversational scenario.

Several models produce head and gaze movements based on prosodic features extracted from speech signal. For instance, in the model of Albrecht *et al.* (2002), the head and eyebrow raise depending on the value of pitch, while gaze directed at an immobile, fixed location during pauses for thinking and word search, and random movement are produced during normal speech. Le *et al.* (2012) also proposed a system that produces head motion, gaze, and eyelid motion simultaneously based on speech input. Separate statistical model for each behaviour were learned from data collected during dyadic face-to-face conversation and single subject's speaking scenario. More believable performance of a virtual agent can be achieved by producing behaviours based on semantics in addition to prosody. For instance, Marsella *et al.* (2013) system produced head movements and gaze (as well as other multimodal cues) from the prosodic and semantic analysis of the speech. The system is based on a set of rules derived from psychological literature, but also based on data collected from human face-to-face interactions which were annotated and analysed to extract the dynamics of multimodal cues.

Another important cue, mouth movement, makes the communication with an embodied agent more natural and believable. The existing systems produce mouth animations from either natural speech or a text-to-speech engine. Often mouth animation is done by first creating a canonical set of mouth shapes (i.e., visemes) that map one or more phoneme to a corresponding viseme, and later on, given the sequence of visemes, by interpolating the visemes to animate the mouth. However, such an approach leads to poor results due to discontinuities in the sequences of mouth movements. Better results are obtained taking into account co-articulation, where the shape of the current viseme is affected by adjacent phonemes. For instance, Cohen and Massaro (1993) modeled co-articulation using dominance functions, which for each speech unit described the influence of its preceding and following viseme. Some approaches use statistical models to learn the co-articulation. For instance, Ding and Pelachaud's system (2015) takes as an input a spoken text decomposed into phonemes and their durations. Gaussian Mixture Models (GMM) are built to infer the shape of mouth for each phoneme. For producing smooth mouth animation, the HMM interpolations are used as interpolation function. Similarly, Luo *et al.* (2014) used GMM for producing mouth movements from speech. They incorporated previous visual features into the model which helped to eliminate the discontinuities in the mouth movement. Interestingly, Xu *et al.* (2013) showed that lip animation can be achieved using only a canonical set of phoneme pairs. More recent approaches use machine learning for lip movement production. For instance, Aneja and Li (2019) proposed a deep learning system to automatically produce from input audio mouth movements (i.e., viseme sequences) for two-dimensional virtual characters.

Embodied agents use text-to-speech (TTS) systems to produce speech. For the speech to sound more natural (be more variable yet consistent), TTS systems incorporate prosody models to produce prosodic properties of speech (i.e., pitch, intonation, duration, stress, and style). Traditionally, these systems focused on predicting phoneme duration, and later on predicting fundamental frequency (F0) (Taylor *et al.*, 1998; Stanton *et al.*, 2018). Recent TTS systems, represent prosody implicitly within a neural network (Ping *et al.*, 2018; Wang *et al.*, 2018). Although these speech synthesis models produce natural prosody and expression, it is unclear how they can produce more variable, hence more human-like, prosodic features. Here several studies focused on modelling emotions in speech (Wang *et al.*, 2018; Wan *et al.*, 2019) as well as the different ways in which the same text can be said as well as modelling. For instance, Wang's and colleagues (2018) goal was to provide the systems with the capability to choose a speaking style appropriate for the given context, moving beyond modelling the naturalness of speech. Similarly, Wan *et al.* (2019) presented a new system able to produce more variations in produced synthetic speech.

It is important to note that, again, the focus is predominantly on single (e.g., iconic gestures or gaze), rather than on combined cues. Only a few attempts considered the generation of more than one multimodal cue. For instance, Salem *et al.* (2011) looked at both gaze and gestures that accompany speech. The complexity of robot's gaze behaviour, however, was reduced to directional looks (i.e., looking right when pointing right). A more complex multimodal interaction was attempted by Huang and Mutlu (2013) who applied a learning-based approach to model how humans coordinate speech, gaze, and gestures during storytelling. The alignment among these cues was modelled using a dynamic Bayesian network that learned the distribution and alignment parameters automatically from annotated data from a human multimodal corpus.

14.6 Summary and Way Forward: Mutual Benefits from Studies on Multimodal Communication

From a cognitive neuroscience/psychology perspective, a main goal is to understand communication and its neural circuitry in humans. From an AI perspective, a main goal is to develop embodied agents that can communicate with humans in the most effective manner. In both camps, the investigation of multimodal communication is a very recent development but as it is clear from the previous sections, a promising one. A growing literature is emerging showing that looking at human communication from a multimodal perspective is providing novel insights into our understanding of the psychological and neural mechanisms (Vigliocco, 2014; Holler and Levinson, 2019) and that agents that use multimodal cues have a number of benefits over agents that only use speech such as being perceived as better communication partners and improving the effectiveness of the communication.

However, there are a number of general issues that at the moment preclude faster development of this area. The most important and general one is perhaps that in both camps much of the existing work has considered only how to integrate one of the multimodal cues (e.g., either gesture or gaze) at a time with speech, rather than considering the joint contribution of the different multimodal cues to production or comprehension. This is limiting the potential understanding of the phenomenon as human communication is characterized by speech accompanied by all the multimodal cues we have discussed here (see e.g., Zhang et al., in prep). Considering a single cue at a time, may also render the problem of recognition and production by embodied agents harder. This is because the different multimodal cues are correlated among one another and correlated with speech, hence they can provide additional constraints (e.g., as discussed above, the referent of a point can be better disambiguated using speech combined with gaze (Holzapfel *et al.*, 2004)). Thus, approaches that attempt to understand and use the different multimodal cues at the same time may be the most promising way forward.

14.6.1 Development and coding of shared corpora

As already mentioned, a major bottleneck is the lack of data on (1) when and which of the multimodal cues are used across topics and across speakers; (2) how the cues are coordinated—how they are correlated—in face-to-face communication. For example, words accompanied by gestures tend to be marked by prosodic stress (Holler and Levinson, 2019), and increased prosodic contour is correlated with larger mouth movements. In order to answer these questions, corpora in which each cue is annotated and time-aligned with the speech are needed. This is really hard. From naturalistic face-to-face interactions, the different cues need to be extracted and then coded. The current practice is to hand-code a number of these cues (e.g., type of gestures) and then to time-lock the coded cues along with segmented speech. Even for those cues that can be automatically extracted from video (e.g., gaze) or audio (e.g., prosodic contour), manual intervention is still often necessary (e.g., for gaze, the objects being looked at often need to be hand-coded), or desirable (e.g., for prosody where automatic methods still lack high levels of precision). For mouth movements as well as gestures, although automatic systems, in principle, can extract motion patterns in a quite reliable manner (especially if speakers wear motion markers), these need then to be interpreted in order to be usable. Thus, coded and time-aligned data is still very expensive.

Collaborations among psychologists and computer scientists bringing together expertise in designing behavioural experiments with state-of-the-art sensing and ML technology in the development of corpora of face-to-face communication can offer a way out of the empasse created by manual coding.

14.6.2 Toward a mechanistic understanding of multimodal communication

Building embodied agents that use multimodal cues in a human-like manner calls for a detailed understanding of how these work in humans. On the other hand, developing computational models (which can be implemented in embodied agents) of human-like multimodal language processing and production provides cognitive scientists with explicit mechanistic accounts of the processes that can be tested. There is a long-standing tradition in cognitive science and neuroscience in using computational models to provide explicit descriptions of cognitive processes (e.g., (Seidenberg and McClelland, 1989; Munakata *et al.*, 1997; Rabovsky *et al.*, 2018)). Especially for complex problems, like multimodal communication, such an approach promises to provide novel and important insights.

It has been argued that in multimodal communication, signals are processed at different levels in a predictive manner (Holler and Levinson, 2019). For example, seeing speaker's lips shaped to produce a 'w' sound may restrict the search space for predictions about upcoming words to a phonetically congruent set of candidates. This candidate set may be pruned further by other co-occurring signals both in the visual modality, such as raised eyebrows and a lifted palm-up open hand, and in the auditory modality, such as raising pitch. Such co-occurring multimodal combinations of signals would then trigger the prediction of a question being produced. This prediction, in turn, could then feed downwards, increasing the expectation for a 'wh' word being uttered, plus a question-typical syntactical structure, and so forth. Implementing such prediction-based mechanisms in computational models (or embodied agents) provides the most powerful way to assess their plausibility as an account of human multimodal communication (see Rabovsky *et al.*, 2018, for an example).

14.6.3 Studying human communication with embodied agents

Language has been studied predominantly in its unimodal instantiation as speech or text. This has been done primarily for experimental control reasons. Recent developments in virtual technology and humanoid robotics, provide a way of studying human language processing in a controlled, yet, naturalistic face-to-face setting. Importantly, both VR and humanoid robots provide a good balance between experimental control and ecological validity (Wykowska, 2016; Pan and Hamilton, 2018; Peeters, 2019). Using robots and virtual avatars as experimental stimuli (i.e., conversational partners) ensures consistency and replicability of their behaviours between participants, and hence, the reproducibility of the results. It also allows (to a certain degree) to study communication in a social interaction setting, instead of often used 'observational' paradigms. Moreover, the behaviour of robots and virtual agents, can be controlled with relatively high precision, allowing to investigate questions that would not be possible using more traditional methods. One such problem could be how the timing of different multimodal cues affects language processing. Using, for instance, an avatar we can manipulate separately the different cues (gestures, gaze, prosody, etc.) and in a controlled way change their presence/absence as well as their relative timings. This would not be possible without virtual agents or humanoid robots.

In addition, embodied agents can be used as a test bed for testing computational models (e.g., (Morse *et al.*, 2015; Di Nuovo and J. L. McClelland, 2019; Nagai, 2019)). This is particularly important when we consider the multimodal nature of language, that utilizes a wide range of different cues, and hence, different articulators. Computational models can be designed following the insights from empirical studies in humans, and subsequently validated on an agent that, similar to humans, interacts with its physical and social environment through its body. The results from such computational studies can, in turn, lead to revision of the models but also, more generally, to a better understanding of underpinning mechanisms of human communication.

Acknowledgements

When writing this chapter, the authors were supported by European Research Council Advanced Grant (ECOLANG, 743035) to GV. GV was further supported by a Royal Society Wolfson Research Merit Award (WRM370016).

References

Acredolo, L. P. and Goodwyn, S. (1988). Symbolic gesturing in normal infants. *Child Development*, **59**(2), 450–66.

Admoni, H., Datsikas, C., and Scassellati, B. (2014). Speech and gaze conflicts in collaborative human-robot interactions, in *Proceedings of the Annual Meeting of the Cognitive Science Society (CogSci)*. Quebec City, Canada, 104–109.

Albrecht, I., Haber, J., and Seidel, H. (2002). Automatic generation of non-verbal facial expressions from speech, in J. Vince and R. Earnshaw, eds, *Advances in Modelling, Animation and Rendering*. London: Springer, 283–93.

Alibali, M. W., Nathan, M. J., Church, R. B., *et al.* (2013). Teachers' gestures and speech in mathematics lessons: forging common ground by resolving trouble spots. *ZDM: Mathematics Education*, **45**, 425–40.

Allen, G. L. (2003). Gestures accompanying verbal route directions: do they point to a new avenue for examining spatial representations? *Spatial Cognition & Computation*, **3**(4), 259–68.

Andrist, S., Tan, X. Z., Gleicher, M., and Mutlu, B. (2014). Conversational gaze aversion for humanlike robots, in *Proceedings of the 2014 ACM/IEEE International Conference on Human–Robot Interaction*. Bielefeld, Germany, 25–32.

Aneja, D., and W. Li, W. (2019). Real-Time Lip Sync for Live 2D Animation (2019). url: arXiv:1910.08685.

Ang, J., Dhillon, R., Krupski, A., Shriberg, E., and Stolcke, A. (2002). Prosody-based automatic detection of annoyance and frustration in human-computer dialog, in *Proceedings of the 7th International Conference on Spoken Language Processing (ICSLP 2002)*. Denver, Colorado, 2037–40.

Arnold, P. and F. Hill, F. (2001). Bisensory augmentation: a speechreading advantage when speech is clearly audible and intact. *British Journal of Psychology*, **92**(2), 339–55.

Baldwin, D. A. (1993). Early referential understanding: Infants' ability to recognize referential acts for what they are. *Developmental Psychology*, 29(5), 832–43.

Barros, P., Parisi, G. I., Jirak, D., and Wermter, S. (2014). Real-time gesture recognition using a humanoid robot with a deep neural architecture, in *2014 IEEE-RAS International Conference on Humanoid Robots*. Madrid, Spain, 83–8.

Bergmann, K. and S. Kopp, S. (2009). GNetIc: using Bayesian decision networks for iconic gesture generation, in Z. Ruttkay, M. Kipp, A. Nijholt, and H. H. Vilhjálmsson, eds, *Intelligent Virtual Agents*. Berlin: Springer, 76–89.

Bergmann, K., Kopp, S., and Eyssel, F. (2010). Individualized gesturing outperforms average gesturing: evaluating gesture production in virtual humans, in J. Allbeck, N. Badler, T. Bickmore, C. Pelachaud, and A. Safonova, eds, *Intelligent Virtual Agents*. Berlin: Springer, 104–17.

Bolinger, D. Le M. (1972). *Degree Words*. Paris: Mouton.

Boucher, J.-D., Pattacini, U., Lelong, A., *et al.* (2012). I reach faster when I see you look: gaze effects in human–human and human–robot face-to-face cooperation. *Frontiers in Neurorobotics*, 6, 3.

Breazeal, C., Kidd, C. D., Tomaz, A. L., Hoffman, G., and Berlin, M. (2005). Effects of nonverbal communication on efficiency and robust- ness in human-robot teamwork, in *2005 IEEE/RSJ International Conference on Intelligent Robots and Systems*. Edmonton, Alta, Canada, 708–13.

Bregler, C. and Konig, Y. (1994, April). Eigenlips for robust speech recognition, in *Proceedings of ICASSP '94. IEEE International Conference on Acoustics, Speech and Signal Processing*. Adelaide, SA, Australia, 2, II/669–II/672.

Brooks, R. and Meltzoff. A. N. (2005). The development of gaze following and its relation to language. *Developmental Science*, **8**(6), 535–43.

Buisine, S. and Martin, J.-C. (2007). The effects of speech–gesture cooperation in animated agents' behavior in multimedia presentations. *Interacting with Computers*, **19**(4), 484–93.

Burnham, D. and Dodd, B. (2004). Auditory-visual speech integration by prelinguistic infants: perception of an emergent consonant in the McGurk effect. *Developmental Psychobiology*, **45**(4), 204–20.

Cangelosi, A. and Ogata, T. (2016). Speech and language in humanoid robots, in A. Goswami and P. Vadakkepat, eds, *Humanoid Robotics: A Reference*. Amsterdam: Springer Netherlands, 1–32. doi: 10.1007/978-94-007-7194-9_135-1.

Cassell, J., Pelachaud, C, Badler, N., *et al.* (1994). Animated conversation: rule-based generation of facial expression, gesture & spoken intonation for multiple conversational agents, in *Proceedings of the 21st Annual Conference on Computer Graphics and Interactive Techniques*, 413–20.

Cassell, J., Bickmore, T., Campbell, L., Vilhjálmsson, H., and Yan, H. (2001). More than just a pretty face: conversational protocols and the affordances of embodiment. *Knowledge-Based Systems*, **14**(1), 55–64.

Castillo, J. C., Cáceres-Domínguez, D., Alonso-Martín, F., Castro-González, Á., and Salichs, M. Á. (2017). Dynamic gesture recognition for social robots, in A. Kheddar, E. Yoshida, Ge, S. S., *et al.*, eds, *Social Robotics. ICSR 2017. Lecture Notes in Computer Science*, Vol. 10652. Switzerland: Springer, Cham, 495–505.

Chandrasekaran, C., Trubanova, A., Stillittano, S., Caplier, A., and Ghazanfar, A. A. (2009). The natural statistics of audiovisual speech. *PLoS Computational Biology*, 5(7), e1000436.

Cohen, M. N. and Massaro, D. W. (1993). Modeling coarticulation in synthetic visual speech, in N. M. Thalmann and D. Thalmann, eds, *Models and Techniques in Computer Animation*. Tokyo: Springer Japan, 139–56.

Cooke, N., Shen, A., and Russell, M. (2014). Exploiting a 'gaze-Lombard effect' to improve ASR performance in acoustically noisy settings, in *2014 IEEE International Conference on Acoustics, Speech and Signal Processing (ICASSP)*. Florence, Italy, 1754–8.

Cutler, A., Dahan, D., and van Donselaar, W. (1997). Prosody in the comprehension of spoken language: a literature review. *Language and Speech*, **40**(2), 141–201.

Di Nuovo, A. and McClelland, J. L. (2019). Developing the knowledge of number digits in a child-like robot. *Nature Machine Intelligence*, **1**(12), 594–605.

Ding, Y. and Pelachaud, C. (2015). Lip animation synthesis: a unified framework for speaking and laughing virtual agent, in *FAAVSP – The 1st Joint Conference on Facial Analysis, Animation and Auditory-Visual Speech Processing*. Vienna, Austria, 78–83.

Doniec, M. W., Sun, G., and Scassellati, B. (2006). Active learning of joint attention, in *2006 6th IEEE-RAS International Conference on Humanoid Robots*. Genova, Italy.

Drijvers, L. and Özyürek, A. (2017). Visual context enhanced: the joint contribution of iconic gestures and visible speech to degraded speech comprehension. *Journal of Speech, Language, and Hearing Research*, **60**(1), 212–22.

Droeschel, D., Stückler, J., and Behnke, S. (2011). Learning to interpret pointing gestures with a time-of-flight camera, in *2011 6th ACM/IEEE International Conference on Human–Robot Interaction (HRI)*. Lausanne, Switzerland.

Fernald, A., and Simon, T. (1984). Expanded intonation contours in mothers' speech to newborns. *Developmental Psychology*, **20**(1), 104–13.

Ferstl, Y., and McDonnell, R. (2018). Investigating the use of recurrent motion modelling for speech gesture generation, in *Proceedings of the 18th International Conference on Intelligent Virtual Agents*. Sydney, Australia, 93–8.

Ferstl, Y., Neff, M., and McDonnell, R. (2019). Multi-objective adversarial gesture generation, in *Motion, Interaction and Games (MIG '19)*. Association for Computing Machinery, New York, USA, Article 3, 1–10.

Formolo, D. and Bosse, T. (2016, June). A Conversational Agent that Reacts to Vocal Signals, in R. Poppe, J.-J. Meyer, R. Veltkamp, and M. Dastani, eds, *Intelligent Technologies for Interactive Entertainment. 8th International Conference, Revised Selected Papers, Vol. 178*. Utrecht, The Netherlands: Springer Verlag, 285–91.

Friedman, L. A. (1977). *On the Other Hand: New Perspectives on American Sign Language*. London, United Kingdom: Academic Press.

Fu, T., Han, Y., Li, X., Liu, Y., and Wu, X. (2015). Integrating prosodic information into recurrent neural network language model for speech recognition, in *2015 Asia-Pacific Signal and Information Processing Association Annual Summit and Conference (APSIPA)*. Hong Kong, 1194–7.

Goldin-Meadow, S. (1999). The role of gesture in communication and thinking. *Trends in Cognitive Sciences*, **3**(11), 419–29.

Goldin-Meadow, S., Nusbaum, H., Kelly, S. D., and Wagner, S. (2001). Explaining math: gesturing lightens the load. *Psychological Science*, **12**(6), 516–22.

Grassmann, S. and Tomasello, M. (2007). Two-year-olds use primary sentence accent to learn new words. *Journal of Child Language*, **34**(3), 677–87.

Griffin, Z. M. and Bock, K. (2000). What the eyes say about speaking. *Psychological Science*, **11**(4), 274–9.

Hanna, J. E. and Brennan, S. E. (2007). Speakers' eye gaze disambiguates referring expressions early during face-to-face conversation. *Journal of Memory and Language*, **57**(4), 596–615.

Hawthorne, K. and Gerken, L. A. (2014). From pauses to clauses: prosody facilitates learning of syntactic constituency. *Cognition*, **133**(2), 420–8.

Herold, D. S., Nygaard, L. C., and Namy, L. L. (2012). Say it like you mean it: mothers' use of prosody to convey word meaning. *Language and Speech*, **55**(3), 423–36.

Heylen, D. K. J., van Es, I., Nijholt, A., and van Dijk, E. M. A. G. (2002). Experimenting with the gaze of a conversational agent, in J. van Kuppevelt, L. Dybkjaer, N.O. Bernsen, eds, *Proceedings International CLASS Workshop on Natural, Intelligent and Effective Interaction in Multimodal Dialogue Systems*. Copenhagen: NIS Lab, 93–100.

Holle, H. and Gunter, T. C. (2007). The role of iconic gestures in speech disambiguation: ERP evidence. *Journal of Cognitive Neuroscience*, **19**(7), 1175–92.

Holler, J. and Levinson, S. C. (2019). Multimodal language processing in human communication. *Trends in Cognitive Sciences*, **23**(8), 639–52.

Holler J., Schubotz L., Kelly S., *et al.* (2014). Social eye gaze modulates processing of speech and co-speech gesture. *Cognition*, **133**(3), 692–7.

Holzapfel, H., Nickel, K., and Stiefelhagen, R. (2004). Implementation and evaluation of a constraint-based multimodal fusion system for speech and 3D pointing gestures, in *Proceedings of the 6th International Conference on Multimodal Interfaces (ICMI '04)*. Association for Computing Machinery, New York, NY, USA, 175–82.

Huang, C.-M. and Mutlu (2013, June). Modeling and evaluating narrative gestures for humanlike robots, in P. Newman, D. Fox, and D. Hsu, eds, *Proceedings of Robotics: Science and Systems*. Berlin, Germany.

Iio, T., Shiomi, M., Shinozawa, K., *et al.* (2011). Investigating entrainment of people's pointing gestures by robot's gestures using a WOZ method. *International Journal of Social Robotics*, **3**(4), 405–14.

Ivaldi, S., Anzalone, S. M., Rousseau, W., *et al.* (2014). Robot initiative in a team learning task increases the rhythm of interaction but not the perceived engagement. *Frontiers in Neurorobotics*, **8**(5), 1–16.

Iverson, J. M., Capirci, O., Longobardi, E., and Caselli, M. C. (1999). Gesturing in mother-child interactions. *Cognitive Development*, **14**(1), 57–75.

Karreman, D. E., Bradford, G. S., van Dijk, B., Lohse, M., and Evers, V. (2013). What happens when a robot favors someone? How a tour guide robot uses gaze behavior to address multiple persons while storytelling about art, in *Proceedings of the 8th ACM/IEEE International Conference on Human–Robot Interaction*. Tokyo, Japan, 157–8.

Kelly, S. D., Ozyürek, A., and Maris, E. (2010). Two sides of the same coin: speech and gesture mutually interact to enhance comprehension. *Psychological Science*, **21**(2), 260–7.

Kettebekov, S., Yeasin, M., and Sharma, R. (2003). Improving continuous gesture recognition with spoken prosody, in *2003 IEEE Computer Society Conference on Computer Vision and Pattern Recognition*. Madison, WI, USA.

Kipp, M., Neff, M., Kipp, K. H., and Albrecht, I. (2007). Towards natural gesture synthesis: evaluating gesture units, in C. Pelachaud, J-C. Martin, André, E., *et al.*, eds, *A Data-Driven Approach to Gesture Synthesis*. Berlin, Heidelberg: Springer, 15–28.

Kita, S. (2000). How representational gestures help speaking, in D. McNeill ed., *Language and Gesture*. Cambridge: Cambridge University Press, 162–85.

Kita, S., Alibali, M. W., and Chu, M. (2017). How do gestures influence thinking and speaking? The gesture-for-conceptualization hypothesis. *Psychological Review*, **124**(3), 245–66.

Koons, D. B., and Sparrell, C. J. (1994). Iconic: speech and depictive gestures at the human-machine interface, in *Conference Companion on Human Factors in Computing Systems*. Association for Computing Machinery, New York, NY, 453–4.

Koons, D. B., Sparrell, C. J., and Thórisson, K. R. (1993). Integrating simultaneous input from speech, gaze, and hand gestures, in M. Mayberry, ed., *Intelligent Multimedia Interfaces*. Cambridge, MA: AAAI Press/MIT Press, 257–76.

Kopp, S., Bergmann, K., and Wachsmuth, I. (2008). Multimodal communication from multimodal thinking: towards an integrated model of speech and gesture production. *International Journal of Semantic Computing*, **2**, 115–36.

Kopp, S. and Wachsmuth, I. (2004). Synthesizing multimodal utterances for conversational agents. *Computer Animation and Virtual Worlds*, **15**(1), 39–52.

Krämer, N. C., Simons, N., and Kopp, S. (2007). The effects of an embodied conversational agent's nonverbal behavior on user's evaluation and behavioral mimicry, in C. Pelachaud, J-C. Martin, E. André, G. Chollet, K. Karpouzis, and D. Pelé, eds, *Proceedings of Intelligent Virtual Agents (IVA 2007). LNAI, 4722*. Berlin, Heidelberg: Springer, 238–51.

Krauss, R. M., Chen, Y., and Gottesman, R. F. (2000). Lexical gestures and lexical access: a process model, in D. McNeill, ed., *Language and Gesture*. Cambridge: Cambridge University Press, 261–83.

Kushnerenko, E., Teinonen, T., Volein, A., and Csibra, G. (2008). Electrophysiological evidence of illusory audiovisual speech percept in human infants. *Proceedings of the National Academy of Sciences of the United States of America*, **105**(32), 11442–5.

Le, B. H., Ma, X., and Deng, Z. (2012). Live speech driven head-and-eye motion generators. *IEEE Transactions on Visualization and Computer Graphics*, **18**(11), 1902–14.

Luo, C., Yu, J., and Wang, Z. (2014, May). Synthesizing real-time speech-driven facial an- imation, in *2014 IEEE International Conference on Acoustics, Speech and Signal Processing (ICASSP)*. Florence, Italy, 4568–72.

Ma, W., Golinkoff, R. M., Houston, D., and Hirsh-Pasek, K. (2011). Word learning in infant- and adult-directed speech. *Language Learning and Development*, 7(3), 185–201.

Marsella, S., Xu, Y., Lhommet, M., *et al.* Virtual character performance from speech, in *Proceedings of the 12th ACM SIGGRAPH/Eurographics Symposium on Computer Animation*. Association for Computing Machinery, New York, NY, USA, 25–35.

McBreen, H., and Jack, M. A. (2000). *Empirical Evaluation of Animated Agents in a Multi-Modal E-Retail Application*. Technical report FS-00-04. The AAAI Press, Menlo Park, California, 122–7.

Mcgurk, H., and Macdonald, J. (1976). Hearing lips and seeing voices *Nature*, 264.5588, 746–8.

Morales, M., Mundy, P., and Rojas, J. (1998). Following the direction of gaze and language development in 6-month-olds. *Infant Behavior & Development*, **21**(2), 373–7.

Morency, L.-P., Christoudias, C. M., and Darrell, T. (2006). Recognizing gaze aversion gestures in embodied conversational discourse, in *Proceedings of the 8th International Conference on Multimodal Interfaces*. Association for Computing Machinery, New York, NY, USA, 287–94.

Morse, A. F., Benitez, V. L., Belpaeme, T., Cangelosi, A., and Smith, L. B. (2015). Posture affects how robots and infants map words to objects. *PLoS ONE*, **10**(3), e0116012.

Munakata, Y., McClelland, J. L., Johnson, M. H., and Siegler, R. S. (1997). Rethinking infant knowledge: toward an adaptive process account of successes and failures in object permanence tasks. *Psychological Review*, **104**(4), 686–713.

Mutlu, B., Forlizzi, J., and Hodgins, J. (2006). A storytelling robot: modeling and evaluation of human-like gaze behavior, in *Proceedings of 2006 6th IEEE-RAS International Conference on Humanoid Robots*. Genova, Italy, 518–23.

Mutlu, B., Kanda, T., Forilizzi, J., *et al.* (2012). Conversational gaze mechanisms for humanlike robots. *ACM Transactions on Interactive Intelligent Systems (TiiS)*, **1**(2), 12:1–12:33.

Nagai, Y. (2019). Predictive learning: its key role in early cognitive development. *Philosophical Transactions of the Royal Society B: Biological Sciences*, **374**(1771), 20180030.

Nickel, K. and Stiefelhagen, R. (2003). Real-time recognition of 3D-pointing gestures for human-machine-interaction, in B. Michaelis and G. Krell, eds, *Pattern Recognition*. Berlin: Springer, 2003, 557–65.

Noda, K., Yamaguchi, Y., Nakadai, K., Okuno, H. G., and Ogata, T. (2015). Audio-visual speech recognition using deep learning *Applied Intelligence*, **42**(4), 722–37.

Ozçalişkan, S. and Goldin-Meadow, S. (2005). Gesture is at the cutting edge of early language development. *Cognition*, 96(3), B101–113.

Palinko, O., Rea, F., Sandini, G., and Sciutti, A. (2016). Robot reading human gaze: Why eye tracking is better than head tracking for human-robot collaboration, in *2016 IEEE/RSJ International Conference on Intelligent Robots and Systems (IROS)*. Daejeon, South Korea, 5048–54.

Pan, X. and Hamilton, A. F. de C. (2018). Why and how to use virtual reality to study human social interaction: the challenges of exploring a new research landscape. *British Journal of Psychology*, **109**(3), 395–417.

Paulus, M. and Fikkert, P. (2014). Conflicting social cues: fourteen-and 24-month-old infants' reliance on gaze and pointing cues in word learning. *Journal of Cognition and Development*, **15**(1), 43–59.

Peelle, J. E. and Sommers, M. S. (2015). Prediction and constraint in audiovisual speech perception. *Cortex*, **68**, 169–81.

Peeters, D. (2019). Virtual reality: a game-changing method for the language sciences. *Psychonomic Bulletin & Review*, **26**(3), 894–900.

Ping, W., Peng, K., Gibiansky, A., *et al.* (2018). Deep Voice 3: scaling text-to-speech with convolutional sequence learning. url: arXiv:1710.07654.

Rabovsky, M., Hansen, S. S., and McClelland, J. L. (2018). Modelling the N400 brain potential as change in a probabilistic representation of meaning. *Nature Human Behaviour*, 2, 693–705.

Rauscher, F. H., Krauss, R. M., and Chen, Y. (1996). Gesture, speech, and lexical access: the role of lexical movements in speech production. *Psychological Science*, 7(4), 226–31.

Riek, L. D., Rabinowitch, T-C., Bremner, P., *et al.* (2010). Cooperative gestures: Effective signaling for humanoid robots, in *2010 5th ACM/IEEE International Conference on Human–Robot Interaction (HRI)*. Osaka, Japan, 61–8.

Rohlfing, K. J. (2011). Meaning in the objects, in J. Meibauer and M. Steinbach, eds. *Experimental Pragmatics/Semantics*, **175**, 151–76.

de Rosis, F., Batliner, A., Novielli, N., and Steidl, S. (2007). 'You are Sooo Cool, Valentina!' recognizing social attitude in speech-based dialogues with an ECA, in A. C. R. Paiva, R. Prada, and R. W. Picard, eds, *Affective Computing and Intelligent Interaction*. Berlin: Springer, 179–90.

Rowe, M. L. and Goldin-Meadow, S. (2009). Early gesture selectively predicts later language learning. *Developmental Science*, **12**(1), 182–7.

Rowe, M. L., Özçalişkan, Ş., and Goldin-Meadow, S. (2008). Learning words by hand: gesture's role in predicting vocabulary development. *First Language*, **28**(2), 182–99.

de Ruiter, J. (2000). The production of gesture and speech, in D. McNeill ed., *Language and Gesture*. Cambridge: Cambridge University Press, 248–311.

Salem, M., Rohlfing, K., Kopp, S., and Joublin, F. (2011). A friendly gesture: investigating the effect of multimodal robot behavior in human-robot interaction. In *2011 RO-MAN*. Atlanta, GA, USA, 247–52.

Salem, M., Eyssel, F., Rohlfing, K., Kopp, S., and Joublin, F. (2013). To err is human(-like): effects of robot gesture on perceived anthropomorphism and likability. *International Journal of Social Robotics*, **5**(3), 313–23.

Seidenberg, M. S. and McClelland, J. L. (1989). A distributed, developmental model of word recognition and naming. *Psychological Review*, 96(4), 523–68.

Showers A. and Si M. (2018). Pointing estimation for human-robot interaction using hand pose, verbal cues, and confidence heuristics, in G. Meiselwitz ed., *Social Computing and Social Media. Technologies and Analytics*. Berlin: Springer International Publishing, 403–12.

Sowa, T. and Wachsmuth, I. (2002). Interpretation of shape-related iconic gestures in virtual environments, in I. Wachsmuth and T. Sowa eds, *GW 2001: Gesture and Sign Language in Human-Computer Interaction: International Gesture Workshop Revised Papers*, Vol. 2298. Berlin: Springer, 21–33.

Stanton, D., Wang, Y., and Skerry-Ryan, R. J. (2018). Predicting expressive speaking style from text in end-to-end speech synthesis, in *2018 IEEE Spoken Language Technology Workshop (SLT)*. Athens, Greece, 595–602.

Staudte, M. and Crocker, M. W. (2011). Investigating joint attention mechanisms through spoken human–robot interaction. *Cognition*, **120**(2), 268– 91.

Taylor, P., Black, A. W., and Caley, R. (1998). The architecture of the Festival speech synthesis system, in *Proceedings 3rd ESCA Workshop on Speech Synthesis*. Jenolan Caves, Australia, 147–51.

Thiessen, E. D., Hill, E. A., and Saffran, J. R. (2005). Infant-directed speech facilitates word segmentation. *Infancy*, 7(1), 53–71.

Ng-Thow-Hing, V., Pengcheng, L., and Okita, S. Y. (2010). Synchronized gesture and speech production for humanoid robots, in *2010 IEEE/RSJ International Conference on Intelligent Robots and Systems*, 4617 – 4624.

Tomasello, M.. *Constructing a Language:A Usage-Based Theory of Language Acquisition*. Cambridge, MA: Harvard University Press, 2003.

Vigliocco, G., Perniss, P., and Vinson, D. (2014). Language as a multimodal phenomenon: implications for language learning, processing and evolution. *Philosophical Transactions of the Royal Society B: Biological Sciences*, **369**(1651), 20130292.

Vigliocco, G., Murgiano, M., Motamedi, Y., *et al.* (2019). Onomatopoeia, gestures, actions and words: how do care- givers use multimodal cues in their communication to children? in *Proceedings of the 41th Annual Meeting of the Cognitive Science Society*, 1171–7.

Wan, V., Chun-an, C., Kenter, T., Vit, J., and Clark, R. (2019). CHiVE: varying prosody in speech synthesis with a linguistically driven dynamic hierarchical conditional variational network. url: arXiv:1905.07195.

Wang, Y., Stanton, D., Zhang, Y., *et al.* (2018). Style tokens: unsupervised style modeling, control and trans- fer in end-to-end speech synthesis. url: arXiv:1803.09017.

Vvan Wassenhove, V., Grant, K. W., and Poeppel, D. (2005). Visual speech speeds up the neural processing of auditory speech. *Proceedings of the National Academy of Sciences of the United States of America*, **102**(4), 1181–6.

Waxman, S. R. and Lidz, J. L. (2006). Early Word Learning *Handbook of Child Psychology: Cognition, Perception, and Language*, Vol. 2, 6th edn. Hoboken, NJ: John Wiley & Sons Inc, 2006, 299–335.

Wykowska, A., Chaminade, T., and Cheng, G. (2016). Embodied artificial agents for understanding human social cognition. *Philosophical Transactions of the Royal Society of London. Series B, Biological Sciences*, **371**(1693), 20150375.

Xiao, Y., Zhang Z., and Beck A. (2014). Human-robot interaction by understanding upper body gestures. *Presence: Teleoperators and Virtual Environments*, **23**(2), 133–54.

Xu, T. L., Zhang, H., and Yu, C. (2016). See you see me: the role of eye contact in multimodal human-robot interaction. *ACM Transactions on Interactive Intelligent Systems*, **6**(1), 2.

Yu, C., Schermerhorn, P, and Scheutz, M. (2012). Adaptive eye gaze patterns in interactions with human and artificial agents, in *ACM Transactions on Interactive Intelligent Systems (TiiS)*, **1**(2), 13:1–13:25.

Xu, Y., Feng, A. W., Marsella, S., and Shapiro, A. (2013). A practical and configurable lip sync method for games, in *Proceedings of Motion on Games*. Association for Computing Machinery, New York, USA, 131–140.

Zhang, Y., Frassinelli, D, Toumainen, J., Skipper, J. & Vigliocco, G. (2020). Word predictability, prosody, gesture and mouth movements in face-to-face language comprehension. **doi:** https://doi.org/10.1101/2020.01.08.896712

Part 4

Human-like Representation and Learning

15

Human–Machine Scientific Discovery

Alireza Tamaddoni-Nezhad[1,2], David Bohan[3], Ghazal Afroozi Milani[2], Alan Raybould[4], and Stephen Muggleton[2]

[1] *University of Surrey,* [2] *Imperial College London, UK,* [3] *INRA, France, and* [4] *University of Edinburgh, UK*

15.1 Introduction

Humanity is facing existential, societal challenges related to the well-being and sustaining a growing population of 7.7 billion people, and issues such as food security, the use of biotechnology in agriculture and medicine, antimicrobial resistance (AMR), and the emergence of new pathogens and pandemic diseases are on the international agenda.

Scientists are today equipped with an ever-growing volume of human knowledge and empirical data in addition to advanced technologies such as Artificial intelligence (AI). AI and machine learning are already playing an important role in tackling these new scientific challenges. For example, AI in the form of deep learning has recently been used in the discovery of a new candidate antibiotic which has been successfully tested against a range of antibiotic-resistant strains of bacteria (Stokes et. al., 2020).

Despite great potential for new scientific discoveries, most current AI approaches, including deep learning, are limited when it comes to 'knowledge transfer' with humans. It is difficult to incorporate existing human knowledge and the output knowledge is not human comprehensible. Knowledge transfer is, however, a critically important part of human–machine discovery which is necessary for collaboration between humans and AI.

Human–machine knowledge transfer is the subject of Human-Like Computing, also known as the Third Wave of AI. Human-Like Computing (HLC) research aims to endow machines with human-like perception, reasoning, and learning abilities which support collaboration and communication with human beings. Such abilities should support computers in interpreting the aims and intentions of humans based on learning and accumulated background knowledge.

Figure 15.1 shows the change in perspective which HLC represents in AI research, in particular with regards to knowledge transfer with humans. The idea of incorporating human knowledge in AI is not new and it was the basis of Expert Systems in 1980s where machines were dependent on being fed explicit knowledge from human experts

Alireza Tamaddoni-Nezhad, David Bohan, Ghazal Afroozi Milani, Alan Raybould, and Stephen Muggleton, *Human–Machine Scientific Discovery* In: *Human Like Machine Intelligence.* Edited by: Stephen Muggleton and Nick Charter, Oxford University Press.
 DOI: 10.1093/oso/9780198862536.003.0015

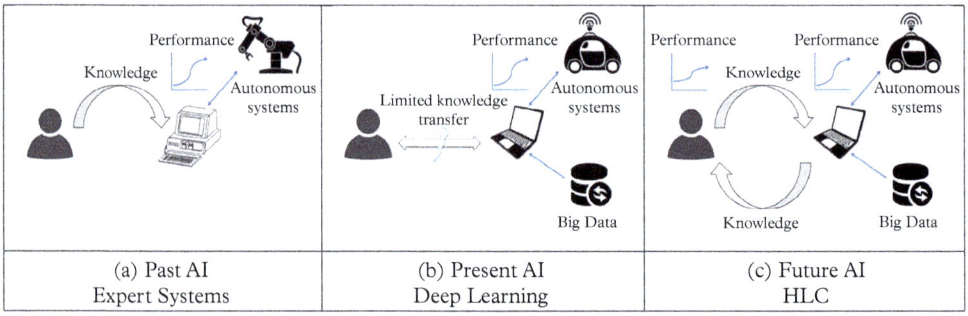

Figure 15.1 *Perspective of human-machine knowledge transfer as variants of AI research. a) Expert Systems (1980s) with a dependence on manual encoding of human knowledge, b) Deep Learning and Big Data in which humans are excluded from the encoded knowledge and c) Human-Like Computing (HLC) in which Humans and Computers jointly develop and share knowledge.*

(Fig. 15.1a). However, incorporating existing knowledge and knowledge transfer are limited in the present black-box forms of AI where computers learn from Big Data, while humans are excluded from both the knowledge development cycle and the understanding of output knowledge (Fig. 15.1b). In HLC, by contrast, humans and machines are viewed as co-developers of knowledge (Fig. 15.1c). In the HLC world, we envisage a symmetric form of learning in which humans derive explicit knowledge from machines, and machines learn from humans and other data sources.

This form of two-way human–machine learning is also related to *ultra-strong* machine learning as defined by Michie (1988). Michie's aim was to provide operational criteria for various qualities of machine learning which include not only predictive performance but also comprehensibility of learned knowledge. His *weak* criterion identifies the case in which the machine learner produces improved predictive performance with increasing amounts of data. The *strong* criterion additionally requires the learning system to provide its hypotheses in symbolic form. Lastly, the *ultra-strong* criterion extends the strong criterion by requiring the learner to teach the hypothesis to a human, whose performance is consequently increased to a level beyond that of the human studying the training data alone.

In this chapter, we demonstrate how a logic-based machine learning approach could meet the *ultra-strong* criterion and how a combination of this machine learning approach, text mining, and domain knowledge could enhance human–machine collaboration for the purpose of automated scientific discovery where humans and computers jointly develop and evaluate scientific theories.

As a case study, we describe a combination of the logic-based machine learning (which included human-encoded ecological background knowledge) and text mining from scientific publications (to evaluate machine-learned hypotheses and also to identify potential novel hypotheses) for the purpose of automated discovery of ecological interaction networks (food-webs) from a large-scale agricultural dataset. Many of the learned trophic links were corroborated by the literature, in particular, links ascribed with high probability by machine learning corresponded with those having multiple references in the literature. In some cases, previously unobserved but high probability links were suggested and subsequently confirmed by experimental studies.

These machine-learned food-webs were also the basis of a recent study (Ma *et. al.*, 2019) revealing resilience of agro-ecosystems to changes in farming management using genetically modified herbicide-tolerant (GMHT) crops.

This chapter is organized as follows. Section 15.2 describes the scientific problem and dataset. The knowledge gap for modelling agro-ecosystems is discussed in Section 15.3. Section 15.4 describes a machine learning approach for automated discovery of ecological networks. The ecological evaluation of the results and subsequent discoveries are discussed in Section 15.5. Section 15.6 concludes the chapter.

15.2 Scientific Problem and Dataset: Farm Scale Evaluations (FSEs) of GMHT Crops

Humanity is facing great challenges to feed the growing population of 7.7 billion people, and sustainable management of ecosystems and growth in agricultural productivity is at the heart of the United Nations' Sustainable Development Goals for 2030. Innovative agricultural management will be required to minimize greenhouse gas emissions and enrich biodiversity, provide sufficient nutritious food, and maintain farmers' livelihoods and thriving rural economies. Predicting system-level effects will be crucial to introducing management that optimises delivery of many potentially conflicting objectives of agricultural, environmental, and social policy.

Replacing existing conventional weed management with GMHT crops, for example, might reduce herbicide applications and increase crop yields. However, this requires an evaluation of the risks and opportunities owing to concerns about potential adverse impacts of GMHT crop management on biodiversity and the functioning of the agro-ecosystems.

The Farm Scale Evaluations (FSE) was a three-year study to test the effects of GMHT crop management on farmland biodiversity across the United Kingdom, and the details of farmland selection and crop field design are described in Champion *et al.*, 2003 and Bohan *et al.*, 2005. To summarize, a split-field design was used in 64 beet, 57 maize, 65 spring-sown oilseed rape and 65 winter-sown oilseed rape sites in the United Kingdom (see Fig. 15.2). Each crop field was split approximately in half, and a conventional and GMHT variety of one of the crops assigned randomly to each half. Plant and invertebrate species were sampled using a variety of standard ecological protocols. Taxa identity and abundance information were recorded within the field across all the sites. Approximately 60,000 field visits were made, sampling some 930,000 plants and 650,000 seeds that were identified to species. In excess of 2 million invertebrates were sampled, and 24,000 bees and 18,000 butterflies counted on the transect walks.

The overarching null hypothesis for the FSEs was that 'there was no effect of the herbicide management of GMHT crops on biodiversity', but with the expectation that effects on biodiversity would be mediated by a combination of the direct effects of herbicides killing weed plants and indirect effects on wider biodiversity through the loss of refuge and food resources provided by these weeds. The FSE scientists and steering

Figure 15.2 *Map of study fields in the FSEs. The circles show the locations of the field sites of spring-sown beet, maize, and oilseed rape, and winter-sown oilseed rape overlain across the United Kingdom.*

committee agreed that a biologically significant effect on any taxon was a change in amount (count, density, biomass) of 50%, either up or down.

The sample data were analysed on a taxon-by-taxon basis using statistical approaches such as ANOVA (Perry *et al.*, 2003). The null hypothesis was tested with a paired randomization test using the treatment effect, d (computed as $d = \log_{10}(\mathrm{GM}+1) - \log_{10}(\mathrm{C}+1)$), for the difference in count for a taxon due to management in the GM and conventional half-fields. The results of the analyses demonstrated that there were significant changes to the amounts of some taxa of weeds, surface dwelling invertebrates, and bees and butterflies in the different crops, with some going up and others down in the GMHT.

Assessment of the probable changes to biodiversity from adopting GMHT crops was used to inform decision-making by regulatory authorities and companies. For a variety of environmental policy and commercial reasons, none of the crops were commercialised in the United Kingdom. Nevertheless, the FSE dataset is the largest agro-ecological census dataset collected to date and it provided the agricultural Big Data used in the human–machine discovery of agro-ecological networks described in this chapter. Network reconstruction was done from invertebrate abundances sampled in the Vortis suction sampling and Pitfall trapping protocols. A new study using these machine-learned food-webs has also revealed that network-level responses in GMHT crop fields are remarkably similar in their composition, network properties, and responses to simulated trajectories of species removals, to their conventional counterparts, suggesting the resilience of agro-ecosystems to changes in farming management using GMHT crops (see section 15.5).

15.3 The Knowledge Gap for Modelling Agro-ecosystems: Ecological Networks

The agro-ecological mechanistic underpinning of ecosystem services, their response to change, and how they interact is still poorly understood, as exemplified by the so-called optimist's scenario (Pocock *et al.*, 2012), which may be summarized as 'the management of one ecosystem service, for improved outcomes, benefits the outcomes of all ecosystem services'. The specific dependencies of one service on any other are only poorly understood and the validity of this scenario at system-relevant scales can only be guessed.

Since ecosystems are structured by flows of energy (biomass) between primary producer plants (autotrophs) and consumers (heterotrophs), such as invertebrates, mammals, and birds (Lindeman, 1942; Dickinson and Murphy, 1998), food-webs are key explanations of ecosystem structure and dynamics that could be used to understand and predict responses to environmental change (Odum, 1974; Caron-Lormier *et al.*, 2009; Cohen *et al.*, 2009, Woodward *et al.*, 2012).

Still relatively few ecosystems have been described and detailed using food-webs because establishing interactions, such as predation, between the many hundreds of species in an ecosystem is resource-intensive, requiring considerable investment in field observation and laboratory experimentation (Ings *et al.*, 2009). Across such large datasets, it is often difficult to relate observational data sampled in protocols that have different basic metrics such as density or activity density or absolute abundance. Increasing the efficiency of testing for trophic links by filtering out unlikely interactions is typically not possible because of uncertainty about basic background knowledge of the network, such as whether any two species are likely even to come into contact and then interact (Ings *et al.*, 2009). In addition, it may require considerable analysis and interpretation to translate from the ecological 'language' of sample data (count, abundance, density, etc.) to the network language of nodes and links within a trophic network. Consequently, of those ecosystems that have been studied using trophic network approaches, component communities that provide known, valuable ecosystem services or those that are experimentally tractable or under threat have most often been evaluated (Ings *et al.*, 2009).

To make good decisions about ecosystem management, e.g. the management of agricultural land for the optimal delivery of ecosystem services, it is necessary to have theories that predict the effects of perturbation on ecosystems. Network ecology, and in particular food-webs, hold great promise as an approach to modeling and predicting the effects of perturbation on ecosystems. Networks of trophic links (i.e., food-webs) that describe the flow of energy/biomass between species are important for making predictions about ecosystem structure and dynamics. However, relatively few ecosystems have been studied through detailed food-webs because establishing predation relationships between the many hundreds of species in an ecosystem is expensive and in many cases impractical. This is mainly because establishing predation relationships between the many hundreds of species in an ecosystem requires specialist expertise in species identification and considerable investment in field observation and laboratory experimentation.

The difficulties in deriving ecological networks therefore severely limit our ability to model and predict responses to changes in ecosystem management and any technique which can automate the discovery of plausible trophic links from ecological data is highly desirable.

15.4 Automated Discovery of Ecological Networks from FSE Data and Ecological Background Knowledge

Many forms of machine learning, such as neural nets (NNs) and support vector machines (SVMs), cannot make use of domain knowledge (i.e., ecological knowledge in this study). By contrast, Inductive Logic Programming (ILP) techniques (Muggleton, 1991; Muggleton and De Raedt, 1994) support the inclusion of such background knowledge and allow the construction of hypotheses that describe structure and relationships between sub-parts. ILP systems use given example observations E and background knowledge B to construct a hypothesis H that explains E relative to B. The components E, B, and H are each represented as logic programs. Since logic programs can be used to encode arbitrary computer programs, ILP is arguably the most flexible form of machine learning, which has allowed it to be successfully applied in complex problems (Tsunoyama *et al.*, 2008; Bohan *et al.*, 2011; Santos *et al.* 2012).

In this section, we describe an abductive ILP approach which has been used to automatically generate plausible and testable food-web theories from ecological census data and existing ecological background knowledge. The main role of abductive reasoning in machine learning of scientific theories is to provide hypothetical explanations of empirical observations (Flach and Kakas, 2000). Then, based on these explanations, we try to inject back into the scientific theory new information that helps complete the theory. This process of generating abductive explanations and updating theory can be repeated as new observational data become available. The process of abductive learning can be described as follows. Given a theory, T, that describes our incomplete knowledge of the scientific domain and a set of observations, O, we can use abduction to extend the current theory according to the new information contained in O. The abduction generates hypotheses that entail a set of experimental observations subject to the extended theory being self-consistent. Here, entailment and consistency refer to the corresponding notions in formal logic. Abductive Logic Programming (Kakas *et al.*, 1993) is typically applied to problems that can be separated into two disjoint sets of predicates: the observable predicates and the abducible predicates. In practice, observable predicates describe the empirical observations of the domain that we are trying to model. The abducible predicates describe underlying relations in our model that are not observable directly but can, through the theory T, bring about observable information. Hence, the hypothesis language (i.e. abducibles) can be disjoint from the observation language. We may also have background predicates (prior knowledge), which are auxiliary relations that help us link observable and abducible information.

In many implementations of abductive reasoning, such as that of Progol 5.0 (Muggleton and Bryant, 2000), as used in this chapter, the approach taken is to choose the explanation that 'best' generalizes under some form of inductive reasoning. This link to induction then strengthens the role of abduction to machine learning and the development of scientific theories. We refer to this approach as Abductive ILP (A/ILP).

A/ILP has been used in a series of studies involving the inference of biological network models from example data. In Tamaddoni-Nezhad *et al.* (2006) encoding and revising logical models of biochemical networks was done using A/ILP to provide causal explanations of rat liver cell responses to toxins. The observational data consisted of up and down regulation patterns found in high throughput metabonomic data. This approach was further extended by Sternberg *et al.* (2013), where a mixture of linked metabonomic and gene expression data was used to identify biosynthetic pathways for capsular polysaccharides in Campylobacter jejuni. In this case, ILP was shown to provide a robust strategy to integrate results from different experimental approaches.

A/ILP was also used in Tamaddoni-Nezhad *et al.* (2012) to infer probabilistic ecological networks from the FSE data described in section 15.2. The Vortis and Pitfall datasets used for the machine learning were year total data, produced by summing the counts from each sample date, for each taxon in each half-field. This raw data was used to measure a treatment effect ratio: counts from each conventional and GMHT half-field pair were converted into a geometric treatment ratio, as used in Haughton *et al.* (2003). Counts were log-transformed, using formula $L_{ij} = \log_{10}(C_{ij} + 1)$, where C_{ij} is the count for a species or taxon in treatment i at site j. Sites where $(C_{1j} + C_{2j}) \leq 1$ were removed from the learning dataset (as in Haughton *et al.*, 2003). The treatment ratio, R, was then calculated as $R = 10^{d}$ where $d = (L_{2j} - L_{1j})$. Following the rationale in Squire *et al.* (2003), important differences in the count between the two treatments were considered to be greater than 50%. Thus, treatment ratio values of $R < 0.67$ and $R > 1.5$ were regarded as important differences in count with direction of *down* (decreased) and *up* (increased) in the GMHT treatment, respectively. This information on *up* and *down* abundances is considered as our observational data for the learning and can be represented by predicate *abundance(X, S, up)* (or *abundance(X, S, down)*) stating the fact that the abundance of species X at site S is *up* (or *down*).

The knowledge gap that we initially aimed to fill was a predation relationship between species. Thus, we declare abducible predicate *eats(X, Y)* capturing the hypothesis that species X eats species Y. It is clear that this problem has properties that require an abductive learning approach such as A/ILP: firstly, the theory describing the problem is incomplete, and secondly, the problem requires learning in the circumstance in which the hypothesis language is disjoint from the observation language.

In order to use abduction, we also need to provide the rules which describe the observable predicate (*abundance*) in terms of the abducible predicate (*eats*):

```
abundance(X, S, Dir):-
        predator(X),
        bigger_than(X, Y),
        abundance(Y, S, Dir)
        eats(X, Y)
```

where *Dir* can be either *up* or *down*. This Prolog rule expresses the inference that following a perturbation in the ecosystem (caused by the management), the increased

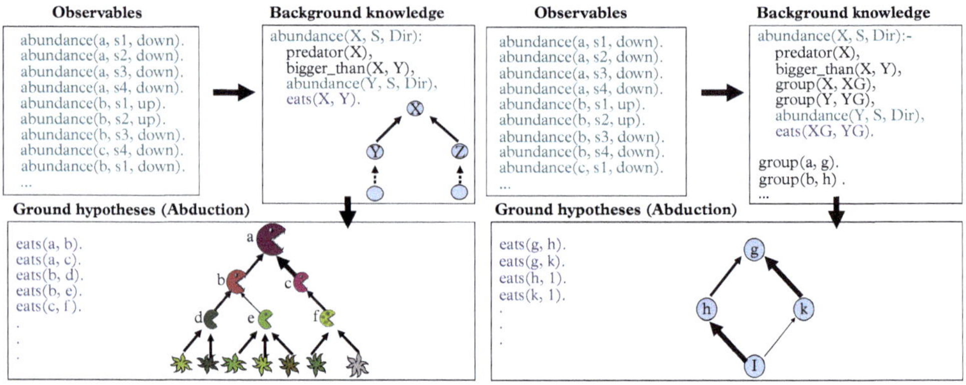

Figure 15.3 *Machine learning of species (left) and functional (right) food-webs from ecological data using Abductive ILP.*

(or decreased) abundance of species X at site S can be explained by X eating species Y and the abundance of species Y is increased (or decreased). This rule also includes additional conditions to constraint the search for abducible predicate *eats(X, Y)*. These constraints are 1) X should be a predator and 2) X should be bigger than Y. Predicates *predator(X)* and *bigger_than(X, Y)* are provided as part of the background knowledge. The 'ecological' background knowledge that a predator should be bigger than a prey was provided by the domain expert.

Given this model and the observable data, the Abductive ILP system Progol 5.0 (Muggleton and Bryant, 2000) was used to generate a set of ground abductive hypotheses in the form of 'eats' relations between species as shown in Figure 15.3. These abductive hypotheses are generated by matching observable input against the background knowledge (which includes the rule describing the observable predicate in terms of abducible predicate). In general, many choices for matching could be made, leading to a variety of alternative hypotheses and a preference is imposed by Progol 5 using an information-theoretic criterion known as compression (Muggleton and Bryant, 2000). Here, compression can be defined as $p - n - h$, where p is the number of observations correctly explained by the hypothesis, n is the number incorrectly explained and h is the length of the hypothesis (e.g., 1 for a single fact such as a trophic link). The set of ground hypotheses can be visualized as a network of trophic links (a food-web) as shown in Figure 15.4. In this network, a ground fact *eats(a, b)* is represented by a directed trophic link from species b to species a.

A Probabilistic ILP (PILP) approach, called Hypothesis Frequency Estimation (HFE) (Tamaddoni-Nezhad *et al.*, 2012), was used for estimating the probabilities of hypothetical trophic links based on their frequency of occurrence when randomly sampling the hypothesis space. HFE is based on direct sampling from the hypothesis space. In some ILP systems, including Progol 5.0, training examples act as seeds to define the hypothesis space (e.g. a most specific clause is built from the next positive example). Hence, permutation of the training examples leads to sampling from different

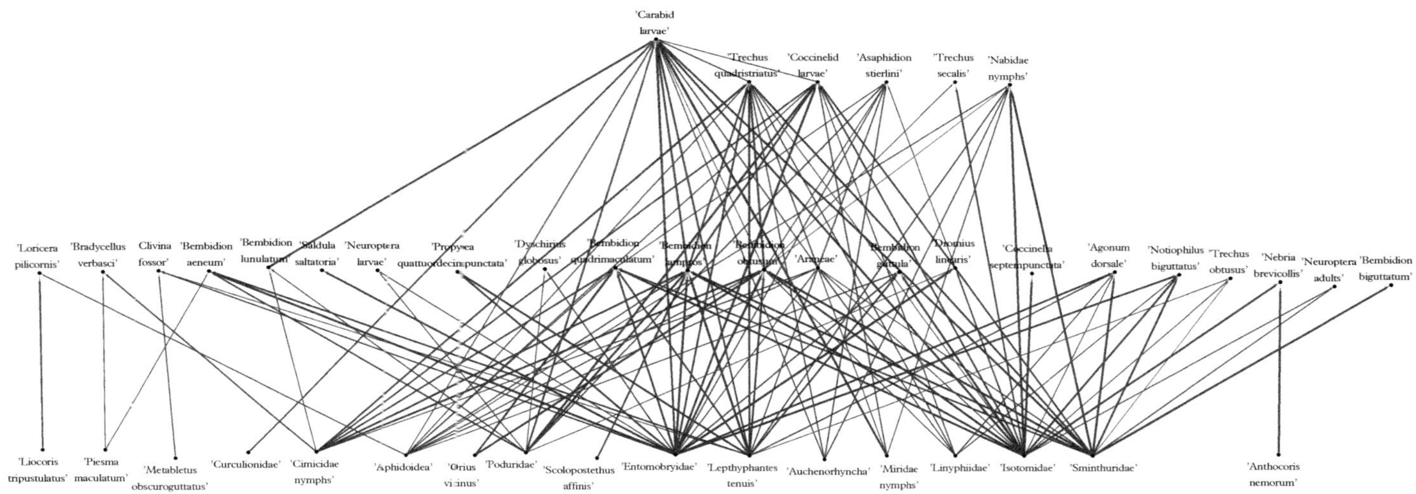

Figure 15.4 *Species food-web learned from farm-scale evaluations (FSE) of GMHT crops data collected using Vortis sampling method from 257 fields across the United Kingdom. Thickness of trophic links represents probabilities which are estimated using Hypothesis Frequency Estimation (HFE).*

parts of the hypothesis space. Using this technique, the thickness of trophic links in Figure 15.4 (and Figure 15.5) represent probabilities which are estimated based on the frequency of occurrence from 10 random permutations (a user-selected parameter) of the training data (and hence different seeds for defining the hypothesis space).

A probabilistic trophic network can be also represented using standard PILP representations such as SLPs (Muggleton, 1996) or ProbLog (De Raedt *et al.*, 2007). For this we can use relative frequencies in the same way probabilities are used in PILP. We can then use the probabilistic inferences based on these representations to estimate probabilities. For example, the probability *p(abundance(a, s, up))* can be estimated by relative frequency of hypotheses that imply *a* at site *s* is *up*. Similarly, *p(abundance(a, s, down))* can be estimated and by comparing these probabilities we can decide to predict whether the abundance is *up* or *down*.

Species food-web (Figure 15.4) can be used to explain the structure and dynamics of a particular ecosystem. However, functional food-webs which represent trophic interactions between functional groups of species might be more important for predicting changes in agro-ecosystem diversity and productivity (Caron-Lormier *et al.*, 2009). Species in FSE data can be classified into 'trophic-functional types' using general traits that reflect their functional type, primarily resource acquisition, and attributes (Caron-Lormier *et al.*, 2009).

By assuming that the background knowledge includes information on the functional group of each species, trophic networks for functional groups can be also learned from ecological data using the machine learning approach described above (See Figure 15.3). Here we need a rule which describes the observable predicate in terms of *eats* relation between functional groups:

abundance(X,S,Dir):-
    predator(X),
    bigger_than(X, Y),
    group(X, XG),
    group(Y, YG),
    abundance(Y, S, Dir),
    eats(XG, YG)

Given this new model and background information, i.e. functional group of species in the form of *group(X, XG)*, trophic networks can be constructed for functional groups in a learning setting similar to the one described above for individual species.

Figure 15.5 shows a functional food-web learned from the FSE data (Vortis). This food-web is constructed by learning trophic interactions between functional groups rather than individual species. Each functional group is represented by a species which can be viewed as an archetype for the functional group.

Evaluating food-webs learned from a set of crops on unseen data from a different crop was done by repeatedly constructing food-webs from all crops data, excluding test data from a particular crop, and measuring the predictive accuracy on this test data. Figure 15.6 shows predictive accuracies of Vortis species-based and functional food-webs on different crops. The average predictive accuracies (the proportions of correctly

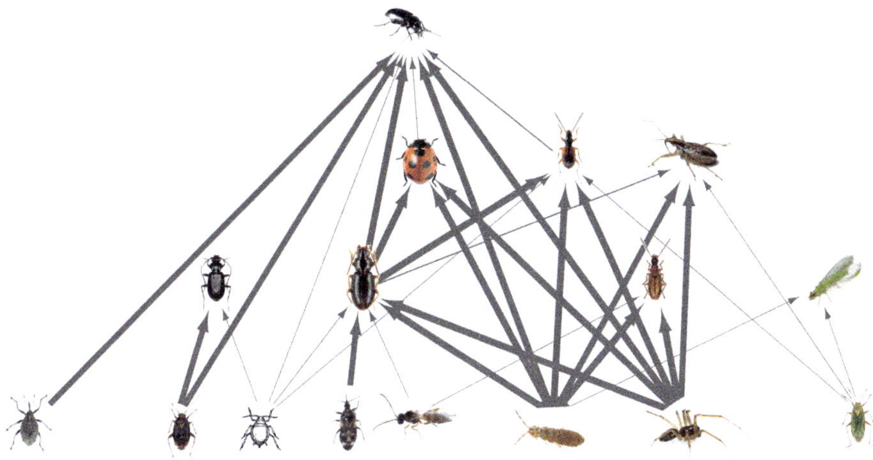

Figure 15.5 *Functional food-web learned from FSE data (Vortis). Each group in the functional food-web is represented by a species which can be viewed as an archetype for that functional group.*

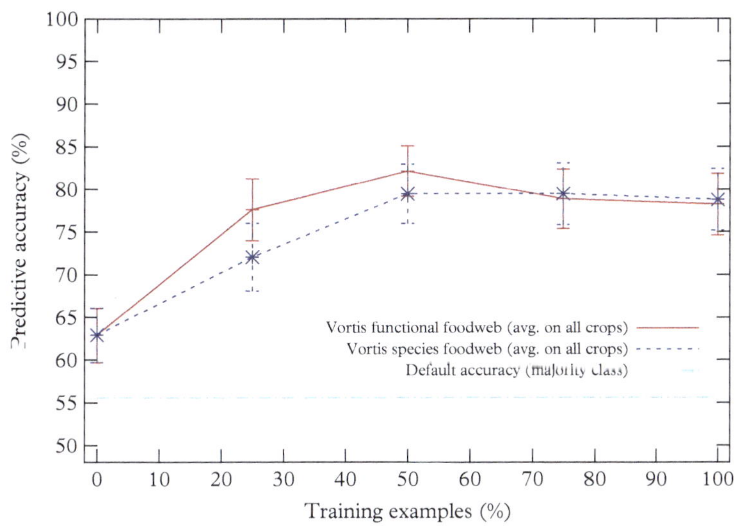

Figure 15.6 *Predictive accuracies of functional food-web versus species food-web from cross-validation tests on different crops.*

predicted left-out test examples) are reported with standard errors associated with each point where 0% to 100% of the training examples are provided.

In these experiments, Hypothesis Frequency Estimation (HFE) (Tamaddoni-Nezhad *et al.*, 2012), was used for estimating probabilities of hypothetical trophic links based on the frequency of occurrence from 10 random permutations of the training data.

The HFE method was also used in the leave-one-out cross-validation to compare the predictive accuracies of species food-web versus functional food-webs (food-webs shown in Figures 15.4 and 15.5). The experimental materials and methods are described in Tamaddoni-Nezhad *et al.*, 2013.

According to this figure, the predictive accuracies of the learned food-webs were significantly higher than the default accuracy of the majority class (around 55%). Predictive accuracies for the functional food-webs were the same or higher than their species-based counterpart, particularly at low to medium percentages of training examples. This suggests that the functional food-webs are at least as accurate as their species-based counterpart, but are much more compact (parsimonious). We also expect the higher predictive accuracy of the functional food-web to be more evident if the food-webs are evaluated on a different agricultural system where different species (not present in the training of species food-webs) may exist.

15.5 Evaluation of the Results and Subsequent Discoveries

The initial species food-webs discovered by machine learning, were examined in Bohan *et al.* (2011) by domain experts from Rothamsted Research UK and it was found that many of the learned trophic links, in particular those ascribed with high probability by machine learning are corroborated by the literature. In some cases, novel and high probability links were suggested, and some of these were tested and corroborated by subsequent empirical studies (Davey *et al.*, 2013).

Manual examination of the food-webs was used to corroborate some of known trophic links and also to identify potential novel hypotheses as shown in Figure 15.7. However, manual corroboration of hypothetical trophic links is difficult and requires significant amounts of time and is error prone. Hence, a text-mining technique was adopted (Tamaddoni-Nezhad *et al.*, 2013) for automatic corroboration of hypothetical trophic links from ecological publications. This was particularly useful for larger foodwebs from merged Vortis and pitfall data.

Figure 15.8 illustrates how a literature network can be generated based on the co-occurrences of predators/prey species in the relevant context, directly from the literature. The pairs of species (from a given food-web) and the interaction lexicons (from a dictionary file) are used to generate queries. Then the text-mining module searches through the text of available publications to match each query. The publications can be in a local database or accessed via a search engine (e.g., Google Scholar). The output of the text-mining for each query is the number of publications that matched that query (number of hits). The output for a whole food-web can be represented by a literature network in which the number associated with each edge is related to the number of papers where the co-occurrences of the predator / prey species have been found with at least one trophic interaction lexicon (eat, feed, prey, or consume). We have shown that the frequencies of trophic links (using HFE) are significantly correlated with the total number of hits for these links in the literature networks (Tamaddoni-Nezhad *et al.*, 2013). Moreover, the proposed approach was used to identify hypothetical trophic relations for which there are little or no information in the literature (potential novel hypotheses).

	Anthocoris nemorum	Bembidion lampros	Bembidion lunulatum	Bembidion obtusum	Cimicidae nymphs	Curculionidae	Entomobryidae	Isotomidae	Lepthyphantes tenuis	Liocoris tripustulatus	Miridae nymphs	Orius vicinus	Poduridae	Scolopostethus affinis	Sminthuridae
Agonum dorsale							9 [13]	4	5 [6]						10 [13]
Bembidion aeneum							10 [11]		9 [6]				9 [11]		
Bembidion biguttatum															10 [11]
Bembidion guttula							7 [11]	10 [11]	9 [6]						
Bembidion lampros					9		10 [11]	10 [11]	10 [6]				10 [11]	9	10 [11]
Bembidion obtusum					9		10 [11]	10 [11]	10 [6]			10			10 [11]
Bembidion quadrimaculatum					9		10 [11]	10 [11]	9 [6]			10			9 [11]
Bradycellus verbasci					8										
Carabid larvae		9	10	3		9	10 [2]	10 [2]	9		10		9 [2]		10 [2]
Clivina fossor							7 [12,5]		2						
Coccinelid larvae		9			9		10 [14,16]	9 [14,16]			1		10 [14.16]		9 [14,16]
Coccinella septempunctata								10 [14,16]							
Dromius linearis							10 [1]	7					3		
Loricera pilicornis										9					
Nabidae nymphs				3				10	7		2 [10,7]				10 [9]
Nebria brevicollis	10							10 [15]							
Notiophilus biguttatus							10 [4,11,3]	10 [4,11,3]							10 [4,11,3]
Propylea quattuordecimpunctata							10 [14,16]		10 [16]						
Saldula saltatoria							10 [8]								
Trechus quadristriatus		9		9	9		9 [15,3]	4 [15,3]	9 [6]		2				10 [15,3]
Trechus secalis				2				8 [15,3]							

Figure 15.7 *Manual corroboration of trophic links for some prey (columns) and predator (rows) species combination from Figure 15.4. Each pairwise hypothesised link has a strength (i.e., frequency between 1 to 10) followed by references (in square brackets) in the literature (see Appendix 1 in Tamaddoni-Nezhad et al., 2012) supporting the link. Multiple references are indicated by yellow and green ellipses and potential novel hypotheses by dashed red ellipses.*

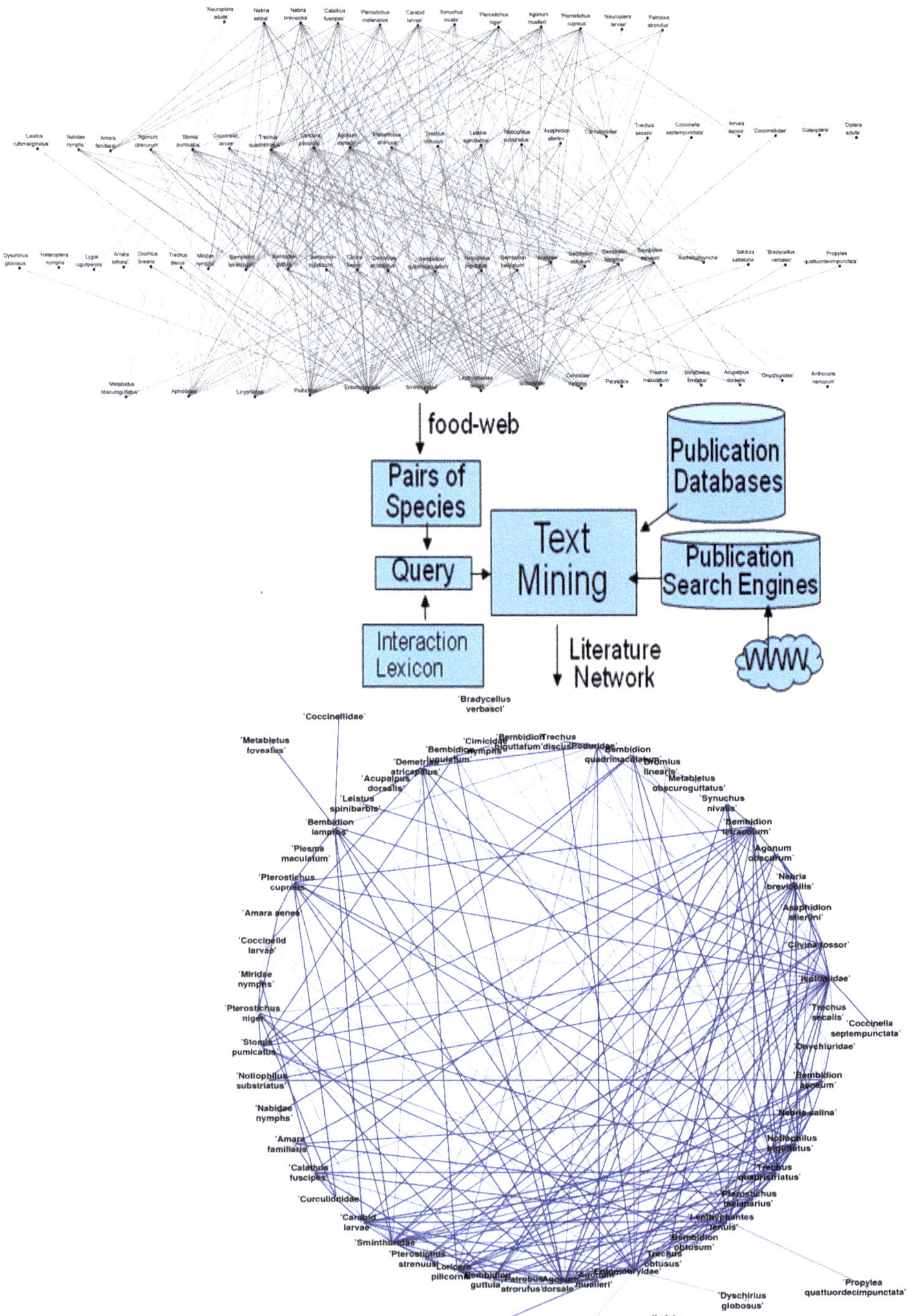

Figure 15.8 *Automatic corroboration of the merged Vortis and pitfall food-web. A literature network is automatically generated from a food-web using text mining of pairs of species from publications. Thickness of the links in a literature network is related to the number of papers with the co-occurrences of the pairs of species (number of hits).*

The manual corroboration table (Figure 15.7) represents prey (columns) and predator (rows) species combination from the Vortis food-web. Each pairwise hypothesised link has a strength (i.e., frequency between 1 to 10, from the HFE method) followed by references (in square brackets) in the literature (see Appendix 1 in Tamaddoni-Nezhad *et al.*, 2012) supporting the link. This table shows that many of the links, suggested by the model, are corroborated by the literature. In particular, links in the model ascribed with high frequency correspond well with those having multiple references in the literature. For example, there are 15 links with more than two references and 8 of these are with frequency 10, and from these all the 3 links with 3 references (marked by green ellipses) have frequency 10. In addition, there are also highly frequent links with no references in the literature, and these could potentially be novel hypotheses for future testing with targeted empirical data. For example, one surprising result was the importance of carabid larvae as predators of a variety of prey and in some cases with no reference in the literature (see Figure 15.7). As another example, some species of spiders appeared as prey for other predators; a result that was unexpected because spiders are obligate predators. This hypothesis was tested in a subsequent study using molecular analysis of predator gut contents and it was found that this hypothesised position in an animal–animal network is correct (Davey *et al.*, 2013), and spiders do appear to play an important role as prey at least for part of the agricultural season. Thus, even though some of the hypothesized links were unexpected, these were in fact confirmed later and this provided an extremely stringent test for this human–machine scientific discovery approach.

The food-webs constructed and validated using this human–machine discovery approach were also the basis of a recent study revealing resilience of agro-ecosystems to changes in farming management using GMHT crops. Ma et al. (2019) constructed replicated food-webs using the merged Vortis and pitfall food-webs, populated on the basis of the sampled taxonomic and abundance information of each half of the split-field in FSE and obtained a total of 502 food-webs (251 conventional and 251 GMHT). A network analysis approach was used to characterize the structural properties of all the individual food-webs. The network analysis metrics include: C, connectance; ϕ, core link density; core size; R_R, robustness via random removal; R_T, robustness via targeted removal of highest degree nodes, as defined in Ma et al. (2019). Each metric is averaged across all webs of a given variety and normalized by its overall range. The effects of crop type can be visualized by comparing results from conventional crops horizontally as shown in Figure 15.9. As shown in this figure, food-web properties varied significantly between crop types. However, this figure suggests that the food-web properties remain unaltered between conventional and GMHT food-webs. The network analysis approach by Ma et. al. (2019) also revealed that network-level responses of GMHT crops are remarkably similar in their composition and responses to simulated trajectories of species removals, to their conventional counterparts. These results suggest that crop type was by far the dominant driver of differences in web structure and robustness, across several organizational levels, ranging from sub-structural to whole-network attributes; inter-annual variation is probably greater than differences between conventional and GMHT.

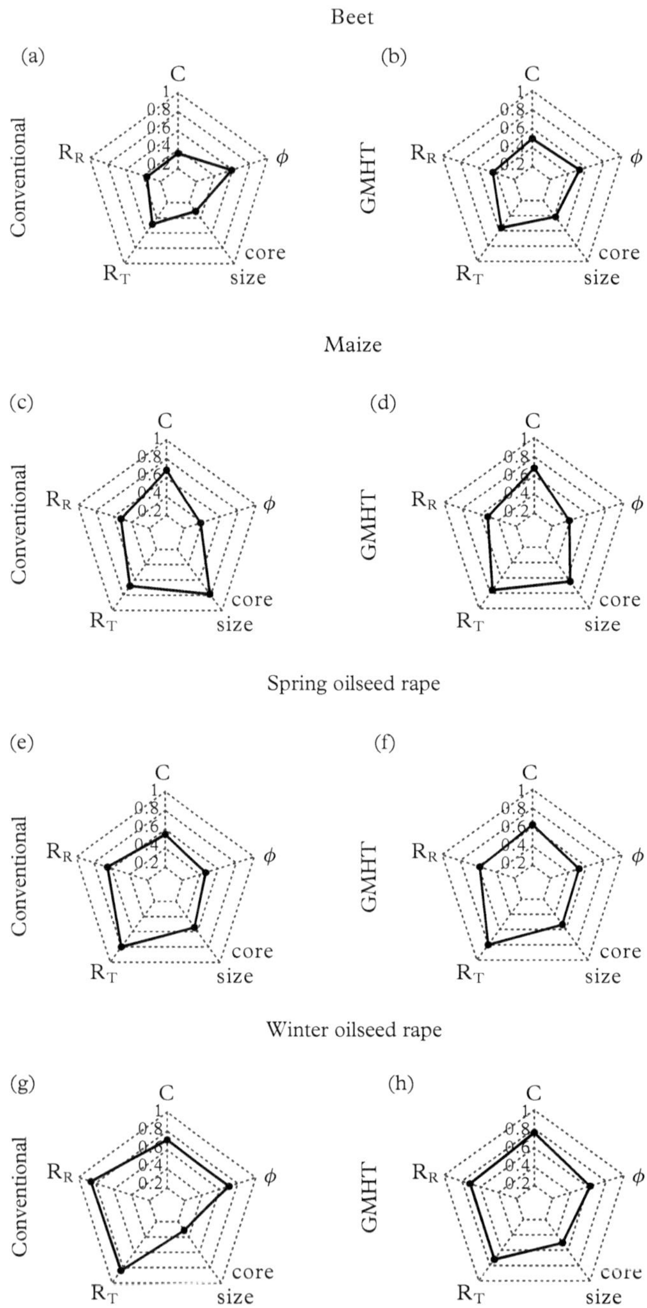

Figure 15.9 *Pairwise comparisons of structural properties of individual crop food-webs between conventional and GMHT managements (**a**, **b**, beet; **c**, **d** maize; **e**, **f**, spring oilseed rape; **g**, **h**, winter oilseed rape). C represents network connectance; ϕ, core link density; R_R, robustness via random removal; R_T, robustness via targeted removal of highest degree nodes as described in Ma et al. (2019).*

15.6 Conclusions

In this chapter, we have demonstrated how a combination of comprehensible machine learning, text mining, and expert knowledge was used to generate plausible and testable food-web hypotheses automatically from ecological census data. The logic-based machine learning included human-encoded ecological background knowledge, e.g. size relationship between predator and prey and taxonomical functional types. Text mining from scientific publications was initially used to verify machine-learned hypotheses, but it was also useful for identifying potential novel hypotheses, i.e. high probability hypotheses suggested by machine learning with no references in the literature. The results included novel food-web hypotheses, some confirmed by subsequent experimental studies (e.g. DNA analysis of gut contents) and published in scientific journals. This case study shows the potentials of human–machine collaboration/communication for the purpose of hypothesis generation in scientific discovery.

Figure 15.10 shows the cycle of hypothesis generation and experimentation in (biological) scientific discovery. In this cycle, machine learning is usually used for 'Model Construction' from 'New Data'. However, the purpose of human-machine discovery is to also automate other steps of this cycle by combining machine learning, text mining, and domain knowledge, as in the case study described in this chapter.

We argue that with ever-growing amount of human knowledge and empirical data as well as advances in AI, human–machine discovery where humans and computers jointly develop and evaluate scientific theories will be important for the advancement of the science in future.

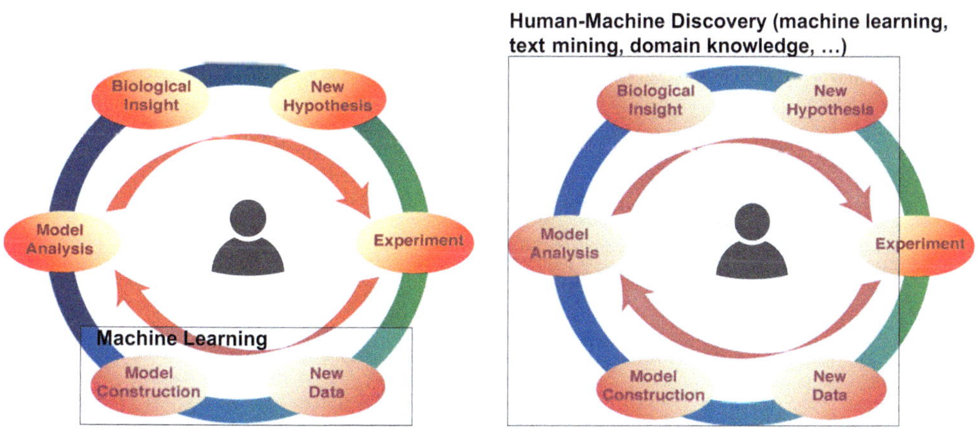

Figure 15.10 *Machine learning vs human-machine discovery.*

References

Bohan, D. A., Boffey, C. W. H., Brooks, D. R. *et al.* (2005). Effects on weed and invertebrate abundance and diversity of herbicide management in genetically modified herbicide-tolerant winter-sown oilseed rape. *Proceedings of the Royal Society B: Biological Sciences, 272*, 463–74.

Bohan, D. A., Caron-Lormier, G., Muggleton, S. *et al.* (2011). Automated discovery of food webs from ecological data using logic-based machine learning, *PloS ONE*, 6, e29028.

Caron-Lormier, G., Bohan D. A, Hawes, C. *et al.* (2009). How might we model an ecosystem? *Ecol Model* 220, 1935–49.

Champion, G. T., May, M. J., Bennett, S. *et al.* (2003). Crop management and agronomic context of the farm scale evaluations of genetically modified herbicide-tolerant crops. *Philosophical Transactions of The Royal Society of London, Series B*, **358**, 1801–18.

Cohen, J. E., Schittler, D. N., Raffaelli, D. G. *et al.* (2009). Food webs are more than the sum of their tritrophic parts. *Proceedings of the National Academy of Sciences USA*, **106**, 22335–40.

Davey, J. S., Vaughan, I. P., Andrew King, R., Bell, J. R. *et al.* (2013). Intraguild predation in winter wheat: prey choice by a common epigeal carabid consuming spiders. *Journal of Applied Ecology*, **50**(1), 271–9.

De Raedt, L., Kimmig, A., and Toivonen, H. (2007). Problog: A probabilistic prolog and its applications in link discovery, in R. Lopez de Mantaras and M. M. Veloso, eds., *Proceedings of the 20th International Joint Conference on Artificial Intelligence*, 804–9 (Vol. 7, pp. 2462–7), Melbourne. San Francisco: Morgan Kaufmann.

Dickinson, G. and Murphy, K. (1998). *Ecosystems: A Functional Approach.* London: Routledge.

Flach, P. and Kakas, A. C. (2000). Abductive and inductive reasoning: Background and issues, in P. A. Flach and A. C. Kakas, eds, *Abductive and Inductive Reasoning, Pure and Applied Logic.* Alphen aan den Rijn, Netherlands: Kluwer.

Haughton, A., Champion, G., Hawes, C. *et al.* (2003). Invertebrate responses to the management of genetically modified herbicide–tolerant and conventional spring crops. II. Within-field epigeal and aerial arthropods, *Philosophical Transactions of the Royal Society of London Series B: Biological Sciences*, **358**, 1863–77.

Ings, T. C., Montoya, J. M., Bascompte, J. *et al.* (2009). Ecological networks: beyond food webs. *Journal of Animal Ecology* **78**, 253–69.

Kakas, A. C., Kowalski, R. A. and Toni, F. (1993). Abductive logic programming. *Journal of Logic and Computation*, **2**(6), 719–70.

Kakas, A. C. and Riguzzi, F. (2000). Abductive concept learning. *New Generation Computing*, **18**, 243–94.

Ma, A., Lu, X., Gray, C. *et al.* (2019). Ecological networks reveal resilience of agro-ecosystems to changes in farming management. *Nature Ecology & Evolution*, **3**(2), 260–4.

Michie, D. (1988). Machine learning in the next five years, in *Proceedings of the third European working session on learning.* London: Pitman, 107–22.

Muggleton, S. H. (1991). Inductive logic programming. *New Generation Computing*, **8**(4), 295–318.

Muggleton, S. H. (1996). Stochastic logic programs, in L. de Raedt, ed, *Advances in Inductive Logic Programming.* Amsterdam: IOS Press, 254–64.

Muggleton, S. and Bryant, C. (2000). Theory completion using inverse entailment, in Proceedings of the 10th International Conference on Inductive Logic Programming, 130–46, Springer, Berlin, Heidelberg.

Muggleton, S.H. and De Raedt, L. (1994). Inductive logic programming: theory and methods. *Journal of Logic Programming*, **19**(20), 629–79.

Odum, E. (1971). *Fundamentals of Ecology*, 3rd edn. New York, NY: Saunders Press.

Perry, J. N., Rothery, P., Clark, S. J., *et al.* (2003). Design, analysis and statistical power of the Farm Scale Evaluations of genetically modified herbicide-tolerant crops. *Journal of Applied Ecology* 40, 17–31.

Pocock, M.J.O., Evans, D.M., and Memmott, J. (2012). The robustness and restoration of a network of ecological networks. Science, **335**, 973–7.

Reuman, D. C., Mulder, C., Banašek-Richter, C. *et al.* (2009). Allometry of body size and abundance in 166 food webs. *Advances in Ecological Research*, **41**, 1–44.

Santos, J. C. A., Nassif, H., and Muggleton, S. H. *et al.* (2012). Automated identification of protein-ligand interaction features using inductive logic programming: A hexose binding case study. *BMC Bioinformatics*, **13**(162), pp. 1–11.

Squire, G. R., Brooks, D. R., Bohan, D. A. *et al.* (2003). On the rationale and interpretation of the farm-scale evaluations of genetically-modified herbicide-tolerant crops. *Philosophical Transactions of the Royal Society, Series B*, **358**, 1779–1800.

Sternberg, M. J. E., Tamaddoni-Nezhad, A., Lesk, V.I. *et al.* (2013). Gene function hypotheses for the campylobacter jejuni glycome generated by a logic-based approach. *Journal of Moleular Biology*, **425**(1), 186–97.

Stokes, J. M., Yang, K., Swanson, K. *et al.* (2020). A deep learning approach to antibiotic discovery. *Cell*, **180**(4), 688–702.

Tamaddoni-Nezhad, A., Bohan, D., Raybould, A. *et al.* (2012). Machine learning a probabilistic network of ecological interactions, in Proceedings of the 21st International Conference on Inductive Logic Programming, LNAI 7207, 332–46, Springer, Berlin, Heidelberg.

Tamaddoni-Nezhad, A., Chaleil, R., Kakas, A. *et al.* (2006). Application of abductive ILP to learning metabolic network inhibition from temporal data. *Machine Learning*, **64**, 209–30.

Tamaddoni-Nezhad, A., Chaleil, R., Kakas, A., *et al.* (2007). Modeling the effects of toxins in metabolic networks. *IEEE Engineering in Medicine and Biology*, **26**, 37–46

Tamaddoni-Nezhad, A., Milani, G., Raybould, A., *et al.* (2013). Construction and validation of food-webs using logic-based machine learning and text-mining. *Advances in Ecological Research*, **49**, 225–89

Tsunoyama, K., Amini, A., Sternberg, M. J. E. *et al.* (2008). Scaffold hopping in drug discovery using inductive logic programming. *Journal of Chemical Information and Modelling*, **48**(5), 949–57.

Woodward, G., Brown, L. E., Edwards, F. K. *et al.* (2012). Climate change impacts in multispecies systems: drought alters food web size structure in a field experiment. *Philosophical Transactions of the Royal Society, B: Biological Sciences*, **367**(1605), 2990–7.

16

Fast and Slow Learning in Human-Like Intelligence

Denis Mareschal and Sam Blakeman

Birkbeck College, University of London, UK

16.1 Do Humans Learn Quickly and Is This Uniquely Human?

It is a commonly held belief in the world of artificial intelligence (AI) and machine learning that humans can learn general relations rapidly from a single or at least a small set of exemplars whereas non-human animals find this very difficult if not impossible. Such impressive and rapid learning has been called by different names, including "fast mapping" and "one-shot" learning. It is often pointed to by protagonists of the uniqueness of human cognition as something that characterizes human intelligence and not that of other species for which learning is driven by Thorndike's laws of incremental associative learning (Thorndike, 1911). Rapid learning is also appealed to by opponents of neural network or statistical learning accounts of human cognition as evidence that such systems cannot account for human cognition because (contra to humans) they require very substantial numbers of examples to learn anything and do not generalize well (Marcus, 1998; Lake *et al.*, 2018).

But how well founded are these beliefs? For many years, psycholinguists, behaviour analysts, and animal cognition researchers have investigated how individuals learn new arbitrary relations without explicit training (Wilkinson *et al.*, 1998). On the one hand, psycholinguists have studied young children's 'fast mapping' of new words—a phenomenon believed to underly the explosion of word learning that occurs in the early preschool years (e.g., Carey and Bartlett, 1978; Golinkoff *et al.*, 1994). On the other hand, behaviour analysts have studied so-called exclusion performances, in which participants immediately display arbitrary matching relations without explicit differential reinforcement (Dixon, 1977; McIlvane *et al.*, 1992). Animal cognition researchers have used similar methods to examine the behaviour of higher primates, dolphins, and sea lions (e.g., Premack, 1976; Schusterman and Kreiger, 1984). Despite well-known differences in philosophy and terminology between behaviour analysists and

Denis Mareschal and Sam Blakeman, *Denis Mareschal and Sam Blakeman* In: *Human Like Machine Intelligence.* Edited by: Stephen Muggleton and Nick Charter, Oxford University Press.
DOI: 10.1093/oso/9780198862536.003.0016

psycholingusts, these two research communities have developed similar procedures and reported similar findings (Huntley and Ghezzi, 1993; Wilkinson *et al.*, 1996), suggesting that the communities are studying the same phenomenon.

Exclusion or fast mapping typically occurs in the context of a well learned matching-to-sample task. For example, an experimenter might show an array of familiar items and speak the name of one of these items. The participant then has to select the item that was named. With appropriate training procedures, most humans and many nonhuman animals demonstrate good performance on this task (e.g., Carter and Werner, 1978). Next, the experimenter, without any special instruction, shows the participant an unfamiliar item in the array and gives an unfamiliar name. Here, the participants might be asked something like 'Which one is the blinket?' or 'Which one is the zipo?'. Almost all participants will spontaneously select the unfamiliar item in this situation. This behaviour is referred to as "exclusion" by behaviour analysts and non-human animal cognition/language researchers (Dixon, 1977; Schusterman *et al.*, 1993) and, the "disambiguation effect" or "mutual exclusivity" in psycholinguistics (Merriman and Bowman, 1989). Such behaviours are very widely spread, having been frequently documented in young children (Markman, 1989; Merriman & Bowman, 1989; Golinkoff *et al.*, 1992), children with specific language impairments (Oollaghan, 1987), people with severe mental retardation (Mcllvane *et al.*, 1992; Mcllvane and Stoddard, 1981; McIlvane and Stoddard, 1985), as well as marine mammals (Herman *et al.*, 1984; Schusterman and Kreiger, 1984; Schusterman *et al.*, 1993).

This chapter will explore what might underlie such fast learning. To this end, we will explore the evidence that there is indeed one-shot learning in humans. As suggested above, much of the evidence in support of rapid learning comes from the word learning literature and is hypothesized to explain the rapid way in which young children acquire a large vocabulary so quickly. This raises two further questions: (1) assuming the phenomenon does exist reliably, then is it limited to world learning only, and (2) does it require specialist or dedicated learning mechanisms, or conversely, is it the consequence of domain general learning mechanisms? We will also examine the extent to which fast learning is present only in children or whether it is also manifested in human adults. In other words, is this something that bootstraps learning early in development, only to fade away as the child grows older, or is it present throughout our lifetime? If only present in children, then a complete account of cognition should explain why and how it disappears.

A second often tacit assumption surrounding fast learning is that the ability to learn quickly is unique to humans (Lake *et al.*, 2018). We will survey the evidence demonstrating one-shot or fast mapping in non-human animals, and again ask whether these are limited to "word learning" type paradigms, or whether in non-human animals other contexts lead to rapid, one-shot learning. Finally, we will explore whether fast or slow learning are engaged depends on the kind information being acquired (semantics, episodic, sensory-perceptual, sensory motor). Indeed, another tacit assumption of the fast-learning brigade is that there is a unique type of learning, or at least that the most interesting component of learning is that in which fast learning is evidenced.

We will conclude that while there is evidence of rapid learning under specials circumstance throughout the life span, it is also found in a broad range of non-human species and does not necessitate special purpose learning mechanisms. Moreover, fast learning is almost always accompanied by slow learning to consolidate acquisition over longer periods of time. Prior knowledge is the key element that gives the illusion of rapid learning in contexts that resemble inference more than learning per se. We will describe a simple neural network model of task learning that uses previously stored episodes to help learn new tasks rapidly and avoid the need to large number of training examples. This model suggests that reward prediction error modulates the extent to which fast or slow learning is engaged.

16.1.1 Evidence of rapid learning in infants, children, and adults

There is a veritable cornucopia of data, drawing on the experimental design described above, to suggest that children are able to learn new words from very sparse data (Carey and Bartlett, 1978; Landau *et al.*, 1988; Markman, 1989; Smith *et al.*, 2002; Xu and Tenenbaum, 2007). They may only need to see a few examples of the concepts before they largely 'get it', grasping the boundary of the set that defines each concept from the set of all possible objects. This raises the question of what underlies such rapid learning. Are there specialized induction mechanisms adapted for language acquisition? Or, perhaps children possess strong inductive biases (such a bias to generalize on the basis of shape) that reduce the scope of possible inferences?

One can equally ask if the rapid learning found in young children and infants is only present in word learning contexts, or if it is a more general phenomenon? To address this question, Casler and Kelemen (2005) investigated the use of novel tools by young. Indeed, tool use is another example of a skill that is held up as marking the distinctiveness of human intelligence (e.g., McGrew, 1996; Byrne, 1997; Povinelli, 2000; Jalles-Filho *et al.*, 2001). They found that young children do not treat all objects with appropriate properties as equally good means to a currently desired end. Instead, after just one observation of an adult intentionally using a novel tool, young children will rapidly construe the artifact as 'for' that privileged purpose, consistently returning to the object to perform that function over time. Thus, instrumental observational learning can also show this "one-shot" learning profile characteristic of word learning in the right context.

Perhaps children are just more practiced than adults at learning new things (learning roughly 9 or 10 new words each day after beginning to speak through the end of high school; Carey, 1978). However, the ability for rapid one-shot learning does not disappear in adulthood. An adult may only need to see a single image or movie of a novel two-wheeled vehicle to infer the boundary between this concept and others, allowing him or her to discriminate new examples of that concept from similar looking objects of a different level. Indeed, adults learning new categories can often generalize successfully from just a single example (Lake *et al.*, 2015).

Coutanche and Thompson-Schill (2015) have further suggested that while rapid word learning may help build vocabulary during childhood, fast mapping might continue into adulthood as a means of rapidly integrating new information into the memory networks of the adult neocortex. When we are explicitly taught information, the new material is successfully learned must eventually be consolidated and integrated with existing knowledge. This happens through interactions between the hippocampus and regions of neocortex. At first, incoming information is rapidly encoded by the hippocampus, and following time and often sleep, it is gradually consolidated into long-term memory in neocortex (McClelland *et al.*, 1995). Models of memory predict that, through consolidation, learned material becomes connected with existing knowledge. In a recent study, Coutanche and Thompson-Schill, (2014) found that names for animals learned through fast mapping (but not explicit encoding) resulted in reaction-time behavioural markers characteristic of successful integration into lexical and semantic memory networks in the cortex. These response-time markers typically emerged days after learning, yet after fast mapping they emerged after just 10 minutes of training. The day following training, fast-mapped (but not explicitly encoded) words began to affect responses to semantically related items, suggesting integration into memory systems. This accelerated influence of newly learned words on existing long-term knowledge supports the idea that fast mapping in adults accelerates the incorporation of new items into long-term semantic memory.

So, while rapid learning may be the hallmark of early child development, it also continues well into adulthood, albeit as part of a different learning process. In young children, fast mapping is associated with the rapid but short-lived learning of a label, whereas in adults fast mapping is associated with quicker assimilation of new knowledge into existing semantic knowledge base.

16.1.2 Does fast learning require a specific mechanism?

Fast mapping may initially appear to be specialized because it seems to diverge from common incremental learning trajectories (Thorndike, 1911). However, this assumption is questionable (McMurray, 2007). Models of incremental learning would predict that learning times should be normally distributed in a sample of to-be-learned associations. That is, children will learn most associations after an intermediate number of exposures, but a small proportion will be learned after one or two exposures (i.e., fast mapping) and a few will be learned very slowly. In support of this, Deák and Wagner (2003) found that four- and five-year-olds learned some words quite slowly, requiring many exposures. The question then becomes when and why are some words apparently learned so rapidly?

To test more directly whether label learning involved specific or general mechanisms, Deak and Toney (2013) explored how four- and five-year-olds learned three kinds of abstract associates for novel objects: words, facts, and pictograms. To test fast mapping (i.e., one-trial learning) and subsequent learning, comprehension was tested after each of four exposures. Production was also tested, as was children's tendency to generalize the learned items to new objects in the same taxonomic category. To test for a bias toward mutually exclusive associations, children learned either one-to-one or many-to-many mappings. Deak and Toeny found that children learned words more slowly than facts and pictograms. In all cases, pictograms and facts were generalized more systematically than words. Children also learned one-to-one mappings faster than many-to-many mappings, but only when cognitive load was high. Finally, although children learned facts faster than words, they remembered all items equally well a week later. The results suggest that word learning follows non-specialized memory and associative learning processes. Similarly, Markson and Bloom (1997) also found that when three- and four-year-olds and adults were taught a novel name and a novel fact about an object, fast mapping was not limited to word learning, suggesting that the capacity to learn and retain new words is the results of learning and memory abilities not specific to language.

In summary, "fast mapping" is not specific to words and it does not necessitate any kind of special purpose mechanism. Instead, domain general associative mechanisms can give rise to apparent fast mapping when the match between the new material to be learned and previous knowledge is highly consistent.

16.1.3 Slow learning in infants, children, and adults

Learning words or labels is not the same as learning the concepts that underlie those words. In many ways, labels are far easier to learn, whereas concepts are hard and slow to learn (Snedeker and Glietman, 2004). In fact, in many species real conceptual knowledge requires many many trials to be acquired (Mareschal, 2013). Research on children's and adults' concepts embody very different assumptions of how concepts are structured, as reflected in their experimental designs. If, in fact, concepts are more like those tested in adult than child experiments, research on word learning may be misleading (Murphy, 2002). Indeed, the standard matching (or non-matching) to sample task used in almost all fast-learning studies (e.g., Brown, 1957; Katz *et al.*, 1974; Casey and Bartlett, 1978) is very artificial because the possible referents are very clearly constrained. Carey and Bartlett (1978) introduced the term 'fast mapping' that has become central to developmental psychology's narrative about how words are learned. Yet in Carey and Bartlett's famous "chromium" study, fast mapping was not so successful. Fewer than one in ten of the three-year-olds appeared to have linked the word to its intended meaning (olive green). The children who had been exposed to the word in the study's naturalistic teaching context ('bring me the chromium one; not the red one, the chromium one') were just as likely as controls to pick out the correct referent from an array of colour patches upon hearing the word (Swingley, 2010).

To capture this distinction, Kucker *et al.*, (2015) suggest that word learning in children can be described as a sequence of events: (1) an initial fast-mapping process in which children form preliminary links between words and referents, followed by (2) a slow-mapping process that builds on these episodic memories to integrate the new words within their existing lexicon or semantic memory. Dynamic real-time control processes like novelty seeking and attention as well as ecological factors like the properties of the body and communicative context operate in concert to enable children to use words intelligently in a rapidly evolving world, (McGregor, 2014). Gradual learning links referents to word forms and refines these links via statistical learning and the slow accumulation of small bits of knowledge. We do not need to posit specialized learning mechanisms to account for children's smart, in-the-moment behaviours. Instead situation-time dynamics supports partial knowledge to yield behaviour that, though occasionally fragile, looks accurate.

What evidence is there for this proposal? When hearing a novel name, children tend to select a novel object rather than a familiar one. Using online processing measures with 18-, 24-, and 30-month-olds, Biron *et al.*, (2013) investigated how the development of this selection bias relates to word learning. As expected, children's proportion of looking time to a novel object after hearing a novel name related to their success in retention of the novel word, and also to their vocabulary size. However, their skill in disambiguation and retention of novel words developed gradually: 18-month-olds did not show a reliable preference for the novel object after labeling; 24-month-olds reliably looked at a novel object on disambiguation trials but showed no evidence of retention; and 30-month-olds succeeded on disambiguation trials and showed only fragile evidence of retention. Biron and colleagues concluded that the ability to find the referent of a novel word in ambiguous contexts is a skill that improves from 18 to 30 months of age, and that word learning is really an incremental process. Further support of this view can be found in the recent report that mutual exclusivity does not hold in bilingual toddlers from 24 months onwards (Kalashnikova, 2017). While disambiguation is initially applied by younger bilingual infants, the bilingual toddlers soon come to realize that this bias does not hold in their environment (in which an object can have more than one name) and they therefore cease to rely on it.

A further case in point is the learning of colour words. According to early reports, children at the turn of the twentieth century did not acquire the meanings of colour words until as late as eight years of age. Recent reports suggest that children now acquire colour words earlier, around three or four years of age, but nevertheless struggle to learn them (e.g., Backscheider and Shatz, 1993; Sandhofer and Smith, 1999). These are concrete nouns, and yet, they have a very prolonged acquisitions period. Most current accounts of colour word acquisition propose that the delay between children's first production of colour words and adult-like understanding is due to problems abstracting colour as a domain of meaning. However, Wagner *et al.*, (2013) suggest that the delay between production and adult-like understanding of colour words is not due to difficulties abstracting colour but rather is largely attributable to the problem of determining the colour boundaries marked by specific languages. The same is true in the more abstract domains of number (Wynn, 1990) and time (Shatz *et al.*, 2010). When children learn words that describe number, time, space, and colour, they typically produce the words, recognize them as belonging to distinct lexical classes, and even use them in response to questions like 'What colour is this?', well before they acquire their adult-like meanings (Shatz *et al.*, 2010).

Even when fast mapping of labels occurs, children often fail to retain the new words over intervals as short as five minutes. Monruo *et al.*, (2012) asked whether the memory process of encoding or consolidation is the bottleneck for retention of fast mapped words. In this study, 49 2- to 3-year-olds, were exposed to 8 2- or 3-syllable nonsense neighbours of words in their existing lexicons. Explicit training consisted of six exposures to each word in the context of its referent, an unfamiliar toy. Productions were elicited four times: immediately following the examiner's model, and at 1-minute, 5-min-, and multiday retention intervals. At the final two intervals, the examiner said the first syllable and provided a beat gesture highlighting target word length in syllables as a cue following any erroneous production. The children were highly accurate at immediate post-test but accuracy fell sharply over the 1-min retention interval and again after an additional 5 min. Performance then stabilized such that the 5-min and multiday post-tests yielded comparable performance. Given this time course, the authors conclude that it was not the post-encoding process of consolidation but the process of encoding itself that presented the primary bottleneck for retention. Patterns of errors and responses to cueing upon error suggested that word forms were particularly vulnerable to partial decay during the time course of encoding.

In summary, even word learning has a slow and a fast component. Fast- and short-lived label learning can happen on the basis of very few examples, but deeper word learning that links to the underlying concept, semantic memory, or lexicon takes much more time to be consolidated. Some common words and concepts such as colour can take years to be learned. Labels that are not assimilated within a existing rich semantic network are quickly forgotten.

16.1.4 Beyond word and concept learning

Of course, there are many other forms of learning beyond word or concept learning. For example, perceptual learning represents one kind of skill learning whereby relatively permanent and consistent changes in perception take place with repeated practice or experience. Perceptual learning involves consolidation of implicit memories formed by training (Squire, 2004), and its time course has drawn a great deal of attention (Karni & Sagi, 1993; Tremblay *et al.*, 1998; Atienza *et al.*, 2002; Mednick *et al.*, 2005; Alain *et al.*, 2007; Yotsumoto *et al.*, 2008). It has been shown that a naive subject's performance on a simple discrimination task can be significantly improved with only a few trials (e.g., Poggio *et al.*, 1992). This fast, within-the-first-session learning is followed by relatively slow learning that accumulates across many training sessions and training days. One important finding about slow learning is that perceptual learning may even occur between sessions when no actual training is conducted (Karni and Sagi, 1993)!

Perceptual learning occurs not only within the first training session but also between sessions. Once acquired, the learning effects can last for a long time. By examining the time course of learning-associated Event Related Potential (ERP) changes, Qu (2010) explored whether fast and slow visual perceptual learning contribute to long-term preservation. Subjects first participated in a visual task for three training sessions, and were then given one test session six months later. ERP results showed that fast learning effects, as reflected by the decrement of posterior N1 and increment of posterior P2 components within Session 1, were preserved in Session 3 but not in the test session. However, slow-learning effects, as reflected by the increment of posterior N1 and decrement of frontal P170 component between Sessions 1 and 3, were retained completely in the test session. This study indicates that perceptual learning induces different changes in the human adult brain, with only the delayed changes of brain activity can be preserved for at least six months.

This is just one example of skill learning. In our everyday life, a wide range of motor, perceptual, and cognitive abilities are gradually and implicitly acquired through our continuous interaction with the environment. Converging data indicate that skill learning is a multiple step process that cannot be reduced to the acquisition episode only (Maquet *et al.*, 2020). Initially, while somebody is practising a task, their performance improves asymptotically with continued practice. This corresponds to a process coined as fast learning by Karni and co-workers (Karni and Sagi, 1993; Karni *et al.*, 1995 although note that multiple trials are still required here). Remarkably, however, the initially formed memory trace apparently continues to be reprocessed after the training has ended. Consequently, when tested at a later date, up to several days to weeks later, performance on the task is markedly improved without any intervening training sessions. This so-called slow component of learning has been observed in humans for both perceptual and motor skill learning (Karni and Sagi, 1993; Karni *et al.*, 1995), and seems to depend critically on sleep rather than simply on time or initial practice (Maquet, 2001; Peigneux *et al.*, 2001).

The exact influence of sleep on the slow component of skill learning is still unclear but what is clear is the need for consolidation time for durable learning to occur. The search for the neural substrates mediating the incremental acquisition of skilled motor behaviours has been the focus of a large body of animal and human studies. Much less is known, however, with regard to the dynamic neural changes that occur in the motor system during the different phases of learning. Ungerleider *et al.*, 2002) reviewed recent fMRI findings, which suggest that: (1) the learning of sequential movements produces a slowly evolving reorganization within primary motor cortex over the course of weeks, and (2) this change follows more dynamic, rapid changes in the cerebellum, striatum, and other motor-related cortical areas over the course of days.

Lest we think that such slow prolonged progress is only the domain of motor and perceptual skills alone, a quick look at the expertise literature suggests that this is absolutely not the case. Expertise in all kinds of domains, from sports to music but also history, mathematics, chess, business negotiations, acting, and teaching to name but a few examples, is slowly acquired and requires an estimated to take 10,000 hours of practice. (Erikson *et al.*, 1993). Slow and gradual practice effects are the hallmark of human expertise, not rapid learning. Those who excel in specific domains are simply those who have put in the time (Howe *et al.*, 1998)

In sum, while there is evidence that fast learning occurs in infants, children, and adults, this is not unique to word or concept learning and may reflect the character and context of the tasks used to assess participants. In addition, even word learning typically has a fast labelling phase followed by a slow assimilation phase in which the new word is assimilated within the existing knowledge base. Finally, slow learning is the defining trait of many other forms of learning such as motor and perceptual skill learning, as well as more complex expertise typical of intelligent human behaviour.

16.1.5 Evidence of rapid learning in non-human animals

Is rapid learning unique to humans? To answer this question, we could start by examining word or label learning since it has been so much linked to fast mapping and one-shot learning in humans. Does word learning exist in non-human species. The answer to this is... yes, it does. Kaminski *et al.*, (2004) found that even domestic dogs can show one-shot word (label) learning. Using a paradigm similar to that used to test word learning in children, these authors provided evidence that a border collie, Rico, was able to fast map. Rico knew the labels of over 200 different items. He inferred the names of novel items by exclusion learning and correctly retrieved those items right away as well as four weeks after the initial exposure. Once again, fast mapping appears to be mediated by general learning and memory mechanisms also found in other animals and not by a language acquisition device that is special to humans.

One-shot learning that does not involve auditory labels is also found in many non-human animals. In fact, many animals have very poor auditory discrimination and prefer to rely on different sensory modalities, such as olfaction to recognize objects and situations (see Lea (2013), for a discussion of how focusing on human friendly sensory cues underestimates the cognitive capacity of other species that do not naturally rely on these sensory cues; indeed imagine yourself trying to learn several hundred novel odour-based discriminations as a proxy for what it must be like for some animals to have to learn auditory words). For example, in rats, odour recognition can be very fast, often performed within a single sniff (Uchida *et al.*, 2003). This leads to one-shot food aversion learning when a noxious substance is paired with a previously palatable food time. A single negative experience is sufficient for the rat to avoid the substance in the future (Welzl *et al.*, 2001).

Even insects show one-shot odour recognition (Nowotny *et al.*, 2005). In insects, the information collected by receptor cells in the antenna is projected to glomeruli in the antennal lobe (AL). Olfactory information is encoded as combinatorial activation patterns which result in a discrete spatiotemporal snapshot of activity. These snapshots are analogous to sniffing behaviours in mammals, albeit on a different time scale. On the single snapshot level, the system needs to perform a one-shot pattern recognition task with noisy patterns. The pattern recognition task addressed can be interpreted as the rapid recognition of an initial activity pattern in the AL in response to an odour.

More impressive is the finding that all kinds of species (including fish) learn rapidly from social punishment (Raihani, *et al.*, 2012). Punishment and altruism are behaviours found in many species in which the collective good (e.g., the survival of the school of fish) are of survival benefit to the individuals (Clutton-Brock and Parker, 1995). Punishments occur to extinguish behaviours dangerous to the collective, such as overfeeding, or to establish breeding hierarchies. Individuals need to learn these restrictions rapidly for their own survival as they risk direct harm or exclusion from the collective. Rapid, one-shot learning is therefore the rule in this situations

In summary, rapid learning can also occur in non-human animals, sharing many of the characteristics of human rapid learning, namely, labelling events (whether through smell or audition) and recording events with high rewards value or punishment, such as social ostracisms.

16.2 What Makes for Rapid Learning?

What mechanisms might underlie fast mapping? One clue comes from another memory phenomenon that allows rapid learning: the effect of having a structure of existing related knowledge (Van Kersteren *et al.*, 2012; McClelland *et al.*, 2013). This existing structure allows new material to be learnt after very few presentations. The possible role of existing knowledge in fast mapping (through ruling out a known item during the inference task) may draw on similar mechanisms. Of course, this can also slow down learning if the prior knowledge is inconsistent with the new learning (Mareschal, 2016).

Coutanche and Thompson-Schill (2015) have suggested that accessing the memory representation for the known item during the fast-mapping task may activate the neuronal population underlying the new item. Models of semantic memory suggest that similar items in knowledge are represented by partially shared 'units'. The activation of neurons underlying the newly learned item (via the known item) may support rapid learning. A further possibility is that the inference task during fast mapping leads to deeper (more elaborative) encoding. One challenge to this view is that adults rarely show superior declarative memory from fast mapping (without extensive training), and that presenting participants with the perceptual question (without a known item) does not produce certain fast mapping effects. This latter finding underscores the fact that much of the heavy work that underlies fast mapping is in fact inference rather than learning per se (Coutanche and Thompson-Schill, 2014).

So, there appear to be two distinct learning strategies: (1) incremental learning, in which we gradually acquire knowledge through trial and error, and (2) one-shot learning, in which we rapidly learn from only one or a few items. Greve and colleagues (2017) suggest that the amount of uncertainty about the relationship between the items mediates the transition between incremental and one-shot learning in the brain. Specifically, the more uncertainty there is about a relationship, the higher the learning rate that is assigned to that stimulus pair. By imaging the brain while participants were performing a learning task, they found that uncertainty about the association is encoded in the ventrolateral prefrontal cortex (VLPC) and that the degree of coupling between this region and the hippocampus increases during one-shot learning. Thus, they suggest that the VLPC may act as a switch, turning on and off one-shot learning as required.

To look more closely at what functional neural systems underpin the rapid and slow acquisition of new semantic information, Holdstock and colleagues (2002) studied the learning abilities of two patients with differing brain pathologies. They found a partial double dissociation between the patterns of new learning shown by these two patients. Rapid acquisition was impaired in a patient (YR) who had relatively selective hippocampal damage, but it was unimpaired in another patient (JL) who, according to structural magnetic resonance imaging (MRI), had an intact hippocampus but damage to anterolateral temporal cortex accompanied by epileptic seizures. Slow acquisition was impaired in both patients, but was impaired to a much greater extent in JL. The dissociation suggests that the mechanisms underlying rapid and slow acquisition of new semantic information are at least partially separable. The findings indicate that rapid acquisition of semantic as well as episodic information is critically dependent on the hippocampus. However, they suggest that hippocampal processing is less important for the gradual acquisition of semantic information through repeated exposure, although it is probably necessary for normal levels of such learning to be achieved

Curiously, though, the hippocampus is not yet fully developed at the age when children can learn words rapidly, and children with abnormal hippocampal development appear capable of learning new words at a typical rate (Vargha-Khadem *et al.*, 1999). Results such as these raise the idea that fast mapping during childhood may not depend on the hippocampus (or at least a fully functional hippocampus). Recent findings have suggested the intriguing possibility that the fast mapping procedure may evoke a unique set of neural processes even in adulthood. Holdstock and his colleagues therefore asked (1) whether learning through fast mapping operates independently of the hippocampus, and (2) whether it accelerates the integration of new information into existing memory networks. First, can fast mapping enable learning that operates independently of the hippocampus? This possibility was first addressed in a study of amnesic patients with hippocampal damage. Despite their very impaired explicit learning, patients were able to learn names for unfamiliar fruits, vegetables, flowers, and animals through fast mapping (Sahron *et al.*, 2011), as shown in a surprise recognition test 10 minutes, and then 1 week, after learning. Although prior studies have reported some semantic learning in amnesic patients, this usually requires extensive training, in contrast to the two trials of fast mapping that were sufficient here.

This surprising result generated a flurry of interest in fast mapping in adulthood. However, the matter of whether fast mapping can 'bypass' the hippocampus is far from settled. A direct replication attempt with another group of hippocampal patients with more severe amnesia has failed to find learning after fast mapping (Smith *et al.*, 2014). One additional study deviated from the firsts fast-mapping paradigm in several ways (such as using a less elaborative task (clicking the new item) and introducing the study as a word-learning investigation), which may explain the atypical pattern of recognition memory performance found, making the results difficult to interpret (Warren and Duff, 2014). An additional finding does, however, challenge the hippocampal-independence theory, as hippocampal volume in young and older adults predicted recognition performance after fast mapping (Greve *et al.*, 2014). Direct comparisons of subjects with differing hippocampal integrities is not the only means to address the question of hippocampal involvement (or lack thereof) in fast mapping. Even in subjects with normal hippocampal functioning, a prediction that arises from computational models of memory is that hippocampally bypassed information should not benefit from hippocampally mediated resistance to interference.

Finally, given that there are different modes of learning relying on different functional neural systems, what controls which form of learning is preferentially engaged? In other words, how does the brain know when to deploy the episodic memory system as opposed to relying on incremental learning? Almost nothing is known about how the brain is capable of switching between different types of learning strategy. However, one suggestion that is gaining traction in both behavioural and modelling circles is the idea that prediction (or reward prediction) errors are the trigger for switching between gradual and one-shot learning in the brain (Lee *et al.*, 2015; Greve *et al.*, 2017).

In summary, fast and slow learning occur in tandem. This is underpinned by interactions between cortical and hippocampal systems. Although normal learning involves the interactions of these two systems, each one can still impact on learning individually.

One suggestion is that prediction errors control the relative balance of fast and slow learning in the brain. To illustrate this, in the next section, we describe a complementary learning systems model that uses reward prediction error to balance between fast and slow learning.

16.3 Reward Prediction Error as the Gateway to Fast and Slow Learning

Complementary Learning Systems (CLS) theory posits that the hippocampus and neocortex have opposing properties that allow them to complement each other during the learning process (McClelland *et al.*, 1995). On the one hand, the neocortex slowly learns overlapping representations that generalize across multiple experiences, whereas the hippocampus rapidly learns pattern-separated representations of individual experiences. These properties suggest that in order to achieve fast mapping, the fast learning of the hippocampus would need to be combined with the generalization properties of the neocortex. In particular, Reinforcement Learning (RL) represents a promising framework for combining the properties of neocortical and hippocampal learning systems. Central to RL is the striatum which is thought to be responsible for evaluating states and actions for decision-making (Schultz *et al.*, 1992; Houk *et al.*, 1995; Schultz, 1998; Setlow *et al.*, 2003; Roesch *et al.*, 2009). These evaluations are modified by Reward Prediction Errors (RPEs) that represent the difference between predicted and actual reward, and are encoded by phasic dopaminergic midbrain neurons (Schultz *et al.*, 1997; Schultz, 2016). Importantly, both the neocortex and the hippocampus project to the striatum (Groenewegen *et al.*, 1987; Thierry *et al.*, 2000), suggesting that both may contribute information for evaluation and that striatal-based reward learning could be a locus in the brain for combining the properties of a neocortical and hippocampal learning system. While RPEs appear to be responsible for modifying predictions made by either system, they may also provide a mechanism of communication between them. It has been shown that RPE-encoding dopaminergic midbrain neurons project to the hippocampus (Lisman and Grace, 2005; Lemon and Manahan-Vaughan, 2006; Rosen *et al.*, 2015) and it has been proposed that this may help to bias the encoding of episodic memories (Shohamy and Adcock, 2010; Jang *et al.*, 2019; Rouhani *et al.*, 2018).

Within the field of AI and machine learning, RL has received a lot of attention in recent years. In particular, there has been a surge in the use of deep neural networks to approximate key functions important for RL (e.g., value functions or parameterized policies; Francois-Lavet *et al.*, 2018). Conceptually, Deep Neural Networks (DNNs) share several key properties with a neocortical learning system in that they slowly learn overlapping representations across multiple experiences. These properties are useful in Deep RL because they allow for generalized predictions in environments with many dimensions. However, one of the major criticisms of these new Deep RL approaches is that they lack flexibility and are sample inefficient (Lake *et al.*, 2017). With this in mind, CLS theory suggests that the addition of a hippocampal learning system that quickly learns pattern-separated representations of individual experiences may help to alleviate some of these limitations (Gershman and Daw, 2017). In particular, the fast learning of individual pattern-separated experiences provided by a hippocampal learning system would allow for fast learning and reduced interference or 'state aliasing'. A purely hippocampal system however would struggle without the generalization properties of a neocortical system because it would require a lot of experience to sufficiently sample the different states of the environment and a lot of computational resources to store their associated reward predictions. It therefore follows that an architecture that utilizes both systems would be beneficial, with a neocortical system that generalizes over the state space and a hippocampal learning system that quickly learns violations of these generalizations to help guide decision-making.

In recent work (Blakeman and Mareschal, 2020) we explore a Deep RL model that combines the properties of a neocortical and a hippocampal learning system to improve both efficiency and flexibility. The model, termed Complementary Temporal Difference Learning (CTDL) has several key properties: (1) both learning systems make reward predictions, (2) the two systems learn in parallel, and (3) the systems communicate via RPEs. The general architecture of CTDL can be seen in Figure 16.1. More specifically, a DNN is used to represent the neocortical learning system and a Self-Organizing Map (SOM) is used to represent the hippocampal learning system. RPEs created by the DNN are used to train the SOM so that the SOM stores environmental states and outcomes

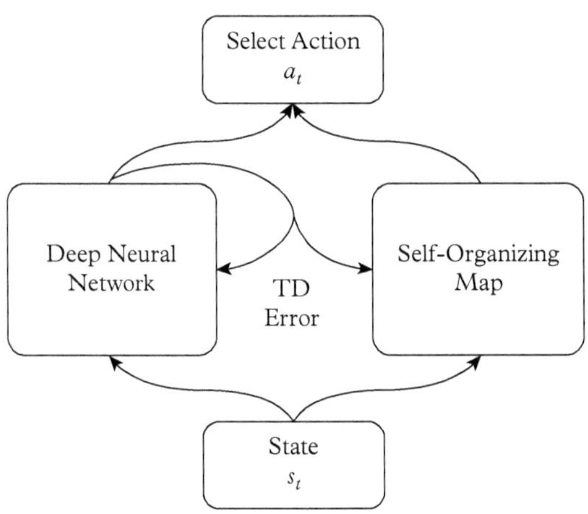

Figure 16.1 *Schema of the CTDL agent architecture. The agent observes the state of the environment (st) and is tasked with selecting an action (at) that will maximize future reward. The agent uses a Deep Neural Network (DNN) and a Self-Organizing Map (SOM) to evaluate the different options and select the best action. The DNN shares similarities with a neocortical system in that it slowly learns overlapping representations over several experiences. The SOM shares similarities with a hippocampal learning system in that it quickly learns pattern-separated representations of individual experiences. Importantly, the Reward Prediction Error (RPE) / Temporal Difference (TD) error from the DNN is used to train both the DNN and the SOM so that the SOM learns to predict states that the DNN is poor at evaluating.*

that the DNN is poor at predicting. In this way CTDL can use semantic knowledge to generalize over the state space and use episodic knowledge to encode violations of these generalizations for fast and flexible learning.

CTDL was compared to classic Deep RL algorithms that do not use an explicit hippocampal learning system (DQN Mnih *et al.*, 2015, and A2C Mnih *et al.*, 2016) on three common RL tasks: Grid World—the agent has to move around a randomly generated grid and reach a goal location while avoiding hazards; Cart-Pole—the agent has to move a cart left and right in order to keep an attached pole upright, Continuous Mountain Car tasks—the agent has to move a car left and right in order to gain momentum and traverse a valley. The three tasks increase in complexity based on their associated state and action spaces. Grid World uses a discrete state and action space, Cart-Pole uses a continuous state space but discrete action space, and Continuous Mountain Car uses both a continuous state and action space.

In the case of the Grid World task, CTDL was better at reaching the goal location in the majority of the randomly created Grid Worlds. To examine the importance of the RPEs that allow the neocortical system to communicate with the hippocampal system we compared CTDL to a version of CTDL that did not use RPEs but just randomly selected states to store in episodic memory. For the majority of Grid Worlds, CTDL with RPEs outperformed CTDL without RPEs. This demonstrates that RPEs are crucial to the performance of CTDL and highlights how structured communication between the two learning systems is a key property of CTDL. To further probe the behaviour of CTDL we ran the model on a series of three structured Grid Worlds as opposed to randomly generated ones. In the first Grid World the agent simply had to travel upwards to reach the goal. This represents a very simple generalization; an increase in the y-axis corresponds to an increase in value (predicted reward). Conversely the other two Grid Worlds were the same but included obstacles that violated this simple generalization. When comparing CTDL to a Deep RL model that didn't include a hippocampal learning system we found that CTDL performed worse on the first Grid World (no obstacles) but better on the others. We also investigated the content of the SOM and found that the states stored in the SOM corresponded to positions just before the obstacles. This suggests that a hippocampal system is particularly valuable when it is required to encode violations of the generalizations made by a neocortical system and that this interaction can greatly speed up learning. As a final experiment on the Grid Worlds task, we ran CTDL on the first Grid World followed immediately by the third Grid World. Typically Deep RL systems experience catastrophic failure when faced with task changes because the deep neural network has entered a region of the parameter space that it cannot recover from. Interestingly, CTDL was much better equipped to deal with the change in task as evidenced by a smaller drop and quicker recovery in performance. It therefore seems that the addition of a hippocampal learning system is not only valuable for speeding up learning but also for increasing flexibility in the face of a changing environment.

In addition to Grid Worlds, CTDL was also ran on the Cart-Pole and Continuous Mountain Car tasks. On the Cart-Pole task CTDL demonstrated marginally slower learning than a Deep RL model lacking a hippocampal learning system but the learning was noticeably more stable and robust. In comparison on the Continuous Mountain

Car task CTDL performed significantly better and was more stable. These results are important because it shows that CTDL can confer advantages even on problems where the state space is continuous and a purely hippocampal learning system is not viable. The Continuous Mountain Car task also requires the selection of continuous actions and different solution methods to the other tasks. We therefore demonstrate that CTDL is a general framework that is independent of the exact solution method as long as it involves some form of value calculation that generates RPEs.

The aforementioned work demonstrates how the combination of a neocortical and hippocampal learning system can promote more efficient and flexible learning. In particular, the RPEs generated by a neocortical learning system can be used to guide the learning of episodic memories in a hippocampal learning system so that they encode violations of the generalizations made by the neocortical learning system. This architecture is parsimonious with a wealth of animal literature exploring the involvement of the hippocampus in different behavioural paradigms. For example, findings in the rodent literature indicate that CA3 neurons in the hippocampus appear to encode decision points in T mazes that are different from the rodent's current position (Johnson and Redish, 2007). CTDL predicts that these decision points should be encoded by the hippocampus because they represent deviations in the animal's general direction, as was seen in the Grid World experiments. In addition, CTDL highlighted the importance of a hippocampal learning system for dealing with changes in the environment. This finding is consistent with studies that implicate activity in the hippocampus with reversal learning (Dong *et al.*, 2013; Vila-Ballo *et al.*, 2017). Finally, it is interesting that CTDL appeared to be more beneficial in the case of the Continuous Mountain Car task as opposed to the Cart-Pole task. One interpretation of this finding is that both the Grid World and Continuous Mountain Car tasks involve important discrete events that are highly informative, that is reaching a goal location. In comparison the Cart-Pole task does not involve a single informative event and instead involves a smooth function over states and actions. It therefore follows that having a hippocampal learning system would be useful for the Grid World and Continuous Mountain Car tasks because it can quickly remember the informative event and use it to guide decision-making. Indeed, this may explain why the hippocampus is often heavily involved in spatial navigation tasks where remembering key locations is pivotal to performance (Burgess *et al.*, 2002).

16.4 Conclusion

One-shot learning is far from ubiquitous in humans. Yes, humans of all ages show evidence of rapid one-shot learning or fast mapping. However, this is by no means the dominant mode of learning. Fast mapping is almost always accompanied by slower, more gradual learning. In fact, when it is not accompanied by such learning new knowledge (such as new label values) is quickly forgotten. This is because the slow learning component is necessary to assimilate the newly acquired information into a complex pre-existing knowledge base. Fast mapping that is not accompanied by slow learning is more akin to on-the-fly inference resulting from the application of strong prior knowledge or beliefs to the current context, rather than learning in the sense of impacting on future behaviour.

One-shot learning is not unique to humans. Many other species including other land and sea mammals, fish and even insects can show one-shot learning. This usually occurs in the context of highly charged events (such as social punishment or consumption of noxious food) in which there is a large prediction reward error. These episodes are then stored in what must be equivalent to an episodic memory and can be retrieved for use when a similar context recurs.

One-shot learning is of questionable value UNLESS the event is of high reward relevance. The rareness or temporary nature of one-shot learning makes perfect sense in a changing environment. High salience (through prediction error) is a marker of what is not currently assimilated within one's knowledge base and therefore something that needs to be attend to for possible future learning. However, it is never clear if a new experience is an exception or the rule in the changing environment. Thus, the extinction of rapidly learned salient events that do not recur avoids the unnecessary overlearning of immediate events that may not be useful for future action. Indeed, one-shot learning is only useful in order to hold an item in memory long enough to determine whether it has sufficient value to be stored in long-term memory.

Learning is the product of multiple interacting systems. Even just a cursory overview of modern cognitive neuroscience makes it clear that there are multiple learning systems that sub-serve similar goals and functions in different ways (see Shallice and Cooper, 2012). One simple classification would include: (1) skill (motor) learning which involves learning to control the motor system in order to carry out some skills; (2) perceptual learning which involves tuning the perceptual system (often with a top-down goal in mind) in order to identify clear features on which to base response decisions, (3) semantic knowledge which captures long-term regularities of the world, most closely identified with knowledge of the world; and (4) episodic memory which comprise records of individual experiences the agent may have encountered. All of these functional systems are simultaneously involved in human cognition. They all have different roles, representational formats, learning mechanisms, and inductive inferences guiding learning. Importantly, they never function in isolation (Mareschal *et al.*, 2007). Each system only ever functions in the context of the others and therefore the computations that they carry out are co-dependent on those of the other functional systems. The solutions that the brain/cognitive system comes up with are based on the involvement of all of these components. Focusing on how only one of these systems operating in isolation, or attributing more importance to one over the other fundamentally misunderstands how human cognition works.

In conclusion, intelligent behaviour is the product of multiple interacting systems. This is true not just for understanding learning but also for understanding human intelligence. The key to human intelligence (and to intelligence in general) is the parallel functioning of multiple partially redundant systems. The control of behaviour (which is what we observe and chose to describe as intelligent or not) is the result of both cooperation and competition between these functional systems. Understanding human-like intelligence therefore consists of: first, mapping what functional systems exist and what their computational properties are, and secondly, understanding how the brain controls and selects the output from these systems. Developing human-like artificial intelligence will require implementing multiple learning and inference systems governed by a flexible control system with equal capacity to that of the human control systems. In reality, focusing on a single aspect of human learning such as fast mapping or one-shot learning is unlikely to be fruitful in and of itself.

Acknowledgements

The writing of this chapter was partly funded by a Biotechnology and Biological Sciences Research Council (BBSRC) LiDO studentship awarded to Sam Blakeman.

References

Alain, C., Snyder, J. S., He, Y. *et al.* (2007). Changes in auditory cortex parallel rapid perceptual learning. *Cerebral Cortex*, **17**, 1074–84.

Atienza, M., Cantero, J. L., and Dominguez-Marin, E. (2002). The time course of neural changes underlying auditory perceptual learning. *Learning & Memory*, **9**, 138–50.

Backscheider, A. G. and Shatz, M. (1993). Children's acquisition of the lexical domain of color, in K. Beals *et al.*, eds, What We Think, What We Mean, And How We Say It, Papers From The Parasession On The Correspondence Of Conceptual, Semantic And Grammatical Representations, CLS 29, Vol. 2, pp. 11–21. Chicago, IL: The Chicago Linguistic Society,

Blakeman, S. and Mareschal, D. (2020). A complementary learning systems approach to temporal difference learning. *Neural Networks*, **122**, 218–30.

Bion, R. A., Borovsky, A., and Fernald, A. (2013). Fast mapping, slow learning: disambiguation of novel word-object mappings in relation to vocabulary learning at 18, 24, and 30 months. *Cognition*, **126**, 39–53.

Botvinick, M., Ritter, S., Wang, J. X. *et al.* (2020). Reinforcement learning, fast and slow. *Trends in Cognitive Sciences*, **23**, (5), 408–422

Brown, R. (1957). Linguistic determinism and the part of speech. *Journal of Abnormal and Social Psychology*, **55**, 1–5.

Burgess, N., Maguire, E. A., and Keefe, J. O. (2002). The human hippocampus and spatial and episodic memory. *Neuron*, **35**, 625–641.

Byrne, R. W. (1997). The technical intelligence hypothesis: an additional evolutionary stimulus to intelligence, in A. Whiten and R. W. Byrne, eds, Machiavellian Intelligence II: Extensions and Evaluations, Cambridge: Cambridge University Press, 289–311).

Carey, S., and Bartlett, E. (1978). Acquiring a single new word. in Proceedings of the Stanford Child Language, Conference, Stanford University **15**, 17–29.

Carter, D. E. and Werner, T. J. (1978). Complex learning and information processing by pigeons: A critical analysis. *Journal of the Experimental Analysis of Behavior*, **29**, 565–601.

Casler, K. and Kelemen, D. (2005). Young children's rapid learning about artifacts. *Developmental Science*, **8**, 472–80.

Clutton-Brock, T. H. and Parker, G. A. (1995). Punishment in animal societies. *Nature*, **373**, 209–15.

Coutanche, M. N. and Thompson-Schill, S. L. (2014). Fast mapping rapidly integrates information into existing memory networks. *Journal of Experimental Psychology: General*, **143**, 2296–303.
Coutanche, M. N. and Thompson, S. L. (2015). Rapid consolidation of new knowledge in adulthood via fast mapping. *Trends in Cognitive Sciences*, **19**, 486–88.
Deak, G. O. and Toney, A. J. (2013). Young children's fast mapping and generalization of words, facts, and pictograms. *Journal of Experimental Child Psychology*, **115**, 273–96.
Dixon, L. (1977). The nature of control by spoken words over visual stimulus selection. *Journal of the Experimental Analysis of Behavior*, 27, 433–42.
Dong, Z., Bai, Y., Wu, X. *et al.* (2013). Hippocampal long-term depression mediates spatial reversal learning in the Morris water maze. *Neuropharmacology*, **64**, 65–73.
Ericsson, K. A., Krampe, R. T., and Tesch-Romer, C. (1993). The role of deliberate practice in the acquisition of expert performance. *Psychological Review*, **100**, 363–406.
Vargha-Khadem, F., Gadian, D. G., Watkins, K. E. *et al.* (1999). Differential effects of early hippocampal pathology on episodic and semantic memory. *Science*, **277**, 376–80.
Francois-lavet, V., Henderson, P., Islam, R. *et al.* (2018). An introduction to deep reinforcement learning. arXiv doi:10.1561/2200000071.Vincent, arXiv:arXiv:1811.12560v2.
Gershman, S. J. and Daw, N. D. (2017). Reinforcement learning and episodic memory in humans and animals: an integrative framework. *Annual Review of Psychology*, **68**, 101–28.
Golinkoff, R. M., Mervis, C. B., and Hirsh-Pasek, K. (1994). Early object labels: The case for a developmental lexical principles framework. *Journal of Child Language*, **21**, 125–55.
Groenewegen, H., Vermeulen-Van der Zee, E., Te Kortschot, A. *et al.* (1987). Organization of the projections from the subiculum to the ventral striatum in the rat. A study using anterograde transport of Phaseolus vulgaris leucoagglutinin. *Neuroscience*, **23**, 103–20.
Greve A., Cooper, E., and Henson, R.N. (2014). No evidence that "fast-mapping" benefits novel learning in healthy Older adults. *Neuropsychologia*, **60**, 52–9.
Greve, A., Cooper, E., Kaula, A. *et al.* (2017). Does prediction error drive one-shot learning? *Journal of Memory and Language*, **94**, 149–65.
Herman, L. M., Richards, D. G., and Wolz J. P. (1984). Comprehension of sentences by bottlenosed dolphins. *Cognition*, **16**, 129–219
Holdstock, J. S., Mayes, A. R., Isaac, C. I. *et al.* (2002) Differential involvement of the hippocampus and temporal lobe cortices in rapid and slow learning of new semantic information. *Neuropsychologia*, **40**, 748 68.
Houk, J. C., Adams, J. L., and Barto, A. G. (1995). A model of how the basal ganglia Generate and use neural signals that predict reinforcement. Computational neuroscience, in J. C. Houk, J. L. Davis, and D. G. Beiser, eds, *Models of Information Processing in the Basal Ganglia*, Cambridge, MA: MIT Press, pp. 249–270.
Howe, M. J. A., Davidson, J. W. and Sloboda, J. A. (1998). Innate talents: reality or myth? *Behavioral and Brain Sciences*, **21**, 99–407.
Huntley, K. R. and Ghezzi, P. M. (1993). Mutual exclusivity and exclusion: converging evidence from two contrasting traditions. *The Analysis of Verbal Behavior*, **11**, 63–76.
Jalles-Filho E., Teixeira Da Cunha, R. G., and Salm R. A. (2001). Transport of tools and mental representation: is capuchin monkey tool behaviour a useful model of Plio-Pleistocene hominid technology? *Journal of Human Evolution*, **40**, 365–77.
Jang, A. I., Nassar, M. R., Dillon, D. G. and Frank, M. J. (2019). Positive reward prediction errors strengthen incidental memory encoding. Nature Human Behavior, **3**, 719–732.
Johnson, A. and Redish, D.A. (2007). Neural ensembles in CA3 transiently encode paths forward of the animal at a decision point. *Journal of Neuroscience*, **27**, 12176–89.

Kalashnikova, M., Escudero, P. and Kidd, E. (2018). The development of fast-mapping and novel word retention strategies in monolingual and bilingual infants. *Developmental Science*, **21**, e12674.

Kaminski, J., Call, J. and Fischer, J. (2004). Word learning in a domestic dog: evidence for "Fast Mapping". *Science*, 304, 1682–3.

Karni, A. and Sagi, D. (1993). The time course of learning a visual skill. *Nature*, **365**, 250–2.

Karni, A., Meyer, G., Jezzard, P. *et al.* (1995). Functional MRI evidence for adult motor cortex plasticity during motor skill learning. *Nature*, **377**, 155–158.

Katz, N., Baker, E., and Macnamara, J. (1974). What's in a name? A study of how children learn common and proper names. *Child Development*, **45**, 469–73.

Kucker, S. C., McMurray, B., and Samuelson, L. K. (2015). Slowing down fast mapping: redefining the dynamics of word learning. *Child Development Perspectives*, 9, 74–8.

Lake, B., Salakhutdinov, R. and Tenenbaum, J. B. (2015). Human-level concept learning through probabilistic program induction. *Science*, **350**, 1332–8.

Lake, B. M., Ullman, T. D., Tenenbaum, J. B. *et al.* (2017). Building machines that learn and think like people. *Behavioral and Brain Sciences*, **40**, e253.

Lea, S. E. G. (2013). Concept learning in nonprimate mammals: in search of evidence, in D. Mareschal., P. C. Quinn, and S. E. G Lea, eds, The Making of Human Concepts, Oxford: Oxford University Press, 173–200.

Lee, S. W., O'Dogherty, J. P., and Shimojo, S. (2015). Neural computations mediating one-shot learning in the human brain, *PLoS Biology*, **13**, e1002137.

Lemon, N. and Manahan-Vaughan, D. (2006). Dopamine D 1 / D 5 receptors gate the acquisition of novel information through hippocampal long-term potentiation and long-term depression. *Journal of Neuroscience*, **26**, 7723–29.

Lisman, J. E. and Grace, A. A. (2005). The Hippocampal-VTA Loop : Controlling the Entry of Information into Long-Term Memory. *Neuron*, **46**, 703–13.

Maquet, P., Laureys, S., Perrin, F. *et al.* (2020). Festina lente: evidences for fast and slow learning processes and a role for sleep in human motor skill learning. *Learning & Memory*, **10**, 237–9.

Marcus, G. F. (1998). Rethinking eliminative connectionism. *Cognitive Psychology*, **37**, 243–82.

Mareschal, D., Johnson, M. H., Sirois, S. *et al.* (2007). *Neuroconstructivism, Vol. I: How the Brain Constructs Cognition.* Oxford: Oxford University Press.

Mareschal, D., Quinn, P. C. and Lea, S. E. G. (2010). *The Making of Human Concepts.* Oxford: Oxford University Press.

Mareschal, D. (2016). The neuroscience of conceptual learning in science and mathematics. *Current Opinion in Behavioral Sciences*, **10**, 14–18.

Markman, E. M. and Wachtel, G. F. (1988). Children's use of mutual exclusivity to constrain the meanings of words. *Cognitive Psychology*, **20**, 121–57.

Markson, L. and Bloom, P. (1997). Evidence against a dedicated system for word learning in children. *Nature*, **385**, 813–15.

McGregor, K. K. (2014). What a difference a day makes: Change in memory for newly learned word forms over twenty-four hours. *Journal of Speech, Language, and Hearing Research*, **57**, 1842–50.

McGrew, W. (1996). *Chimpanzee Material Culture: Implications for Human Evolution.* New York, NY: Cambridge University Press.

Mcilvane, W. J. and Stoddard, L. T. (1981). Acquisition of matching-to-sample performances in severe retardation: learning by exclusion. *Journal of Mental Deficiency Research*, **25**, 33–48.

Mcilvane, W. J., Kledaras, J. B., Lowry, M. J. *et al.* (1992). Studies of exclusion in individuals with severe mental retardation. *Research in Developmental Disabilities*, **13**, 509–32.

McClelland, J. L. (2013). Incorporating rapid neocortical learning of new schema-consistent information into complementary learning systems theory. *Journal of Experimental Psychology: General*, **142**, 1190–210.

McClelland, J. L., McNaughton, B. L., and O'Reilly, R. C. (1995). Why there are complementary learning systems in the hippocampus and neocortex: insights from the successes and failures of connectionist models of learning and memory. *Psychological Review*, **102**, 419–57.

McGregor, K. K. (2014). What a difference a day makes: Change in memory for newly learned word forms over twenty-four hours. *Journal of Speech, Language, and Hearing Research*, **57**, 1842–50.

Mednick, S. C., Arman, A. C., and Boynton, G. M. (2005). The time course and specificity of perceptual deterioration. *Proceedings of the National Academy of Sciences of the United States of America*, 102, 3881–3885.

Merhav, M., Karni, A., and Gilboa, A. (2014). Neocortical catastrophic interference in healthy and amnesic adults: A paradoxical matter of time. *Hippocampus*, **24**, 1653–62.

Merriman, W. E. and Bowman, L. (1989). The mutual exclusivity bias in children's word learning. Monographs of the Society for Research in Child Development, 54, pp. 1–129.

Mnih, V., Kavukcuoglu, K., Silver, D. *et al.* (2015). Human-level control through deep reinforcement learning. *Nature*, **518**, 529–33.

Munro, N., Baler, E., McGrgor, K. *et al.* (2012). Why word learning is not fast. *Frontiers in Psychology*, **3**,

Murphy, G. (2002) Fast-mapping children vs. slow-mapping adults: Assumptions about words and concepts in two literatures. *Behavioral Brain Sciences*, **24**, 1112–13.

Povinelli, D. (2000). *Folk Physics for Apes: The Chimpanzee's Theory of How the World Works.* Oxford: Oxford University Press.

Premack, D. (1976). *Intelligence in Ape and Man.* Hillsdale, NJ: Erlbaum.

Nowotny, T., Huerta, R., Abarbanel, H. D. I. *et al.* (2005). Selforganization in the olfactory system: one shot odor recognition in insects. *Biological Cybernetics*, **93**(6), 436–46.

Qu, Z., Song, Y., and Ding, Y. (2010). ERP evidence for distinct mechanisms of fast and slow visual perceptual learning. *Neuropsychologia*, **48**, 869–1874.

Raihani, N. J., Thornton, A., and Bshary, R (2012). Punishment and cooperation in nature. *Trends and in Ecology and Evolution*, **27**, 288–95.

Roesch, M. R., Singh, T., Brown, P. L. *et al.* (2009). Ventral striatal neurons encode the value of the chosen action rats deciding between differently delayed or sized rewards. *Journal of Neuroscience*, **29**, 13365–76.

Rosen, Z.B., Cheung, S., and Siegelbaum, S.A. (2015). Midbrain dopamine neurons bidirectionally regulate CA3-CA1 synaptic drive. *Nature Neuroscience*, **18**, 1763–71.

Rouhani, N., Norman, K.A., and Niv, Y. (2018). Dissociable effects of surprising re- wards on learning and memory. *Journal of Experimental Psychology: Learning, Memory and Cognition*, 44, 1430–43.

Sandhofer, C. and Smith, L. B. (1999). Learning color words involves learning a system of mappings. *Developmental Psychology*, **35**, 668–79.

Schultz, W., Apicella, P., Scarnati, E. *et al.* (1992). Neuronal Activity in monkey ventral striatum related to the expectation of reward. *Journal of Neuroscience*, **12**, 4595–610.

Schultz, W., Dayan, P., and Montague, P. R. (1997). A Neural Substrate of Prediction and Reward. *Science*, **275**, 1593–9.

Schultz, W. (1998). Predictive reward signal of dopamine neurons. *Journal of Neurophysiology*, **80**, 1–27.

Schultz, W. (2016). Dopamine reward prediction error coding. *Dialogues in Clinical Neuroscience*, **18**, 23–32.

Schusterman, R. J. and Kreiger, K. (1984). California sea lions are capable of semantic comprehension. *The Psychological Record*, **34**, 3-23.

Schusterman, R. J., Gisiner, R., Grimm, B. K. *et al.* (1993). Behavior control by exclusion and attempts at establishing semanticity in marine mammals using match-to-sample paradigms, in H. Roitblat, L. Herman, and P. Nachtigal, eds, *Language and Communication: Comparative Perspectives*, Hillsdale, NJ: Erlbaum. pp. 249–74).

Setlow, B., Schoenbaum, G., and Gallagher, M. (2003). Neural encoding in ventral striatum during olfactory discrimination learning. *Neuron*, **38**, 625–36.

Shallice, T. and Cooper, R. P. (2011). *The Organisation of Mind.* Oxford: Oxford University Press.

Sharon, T., Moscovitch, M., Gilboa A. (2011). Rapid neocortical acquisition of long-term arbitrary associations independent of the hippocampus. *Proceedings of the National Academy of Science*, **108**, 1146–51.

Shatz, M., Tare, M., Nguyen, S. P. *et al.* (2010). Acquiring non-object terms: The case for time words. *Journal of Cognition and Development*, **11**, 16–36.

Shohamy, D. and Adcock, R. A. (2010). Dopamine and adaptive memory. *Trends in Cognitive Sciences*, 14, 464–72.

Smith, C. N., Urgolites, Z. J., Hopkins, R. O. *et al.* (2014). Comparison of explicit and incidental learning strategies in memory-impaired patients. *Proceedings of the National Academy of Science*, **111**, 475–9.

Smith, J. D., Beran, M. J., Crossley, M. J. *et al.* (2010). Implicit and explicit category learning by macque (Macaca mulatta) and Humans (Homo sapiens). *Journal of Experimental Psychology: Animal Behavioral Processes*, **36**, 54–65.

Smith, L. B., Jones, S. S., Landau, B. *et al.* (2002). Object name learning provides on-the-job training for attention. *Psychological Science*, **13**, 13–19.

Snedeker, J. and Gleitman, L. (2004). Why it is hard to label our concepts? in D. G. Hall and L. Gleitman, eds, *Weaving a Lexicon*, Cambridge, MA: MIT Press, 257–93).

Spiegel, C. and Halberda, J. (2011). Rapid fast-mapping abilities in 2-year-olds. *Journal of Experimental Child Psychology*, **109**, 132–40.

Squire, L. R. (2004). Memory systems of the brain: A brief history and current perspective. *Neurobiology of Learning and Memory*, **82**, 171–7.

Swingley, D. (2010). Fast mapping and slow mapping in children's word learning. *Language Learning and Development*, **6**, 179–83.

Thierry, A. M., Gioanni, Y., Degenetais, E. *et al.* (2000). Hippocampo-prefrontal cortex pathway: anatomical and electrophysiological characteristics. *Hippocampus*, **10**, 411–19.

Thorndike, E. L. (1911). *Animal Intelligence.* New York, NY: The McMillian Company.

Tremblay, K., Kraus, N., and McGee, T. (1998). The time course of auditory perceptual learning: Neurophysiological changes during speech-sound training. *Neuroreport*, **9**, 3557–60.

Trueswell, J. C., Medina, T. N., Hafri, A. *et al.* (2013). Propose but verify: Fats mapping meetings cross-situational word learning. *Cognitive Psychology*, **66**, 126–56.

Uchida, N. and Mainen, Z. F. (2003). Speed and accuracy of olfactory discrimination in the rat. *Nature Neuroscience*, **6**, 1224–9.

Ungerleider, L. G., Doyon, J. & Karni, A. (2002). Imaging brain plasticity during motor skill learning. *Neurobiology of Learning and Memory*, **78**, 553–64.

Van Kesteren M. T. R., Ruiter, D. J., Fernandez, G. *et al.* (2012) How schema and novelty augment memory formation. *Trends in Neuroscience*, **35**, 211–219.

Wagner, K., Dobkins, K., and Barner, D. (2013). Slow mapping: color word learning as a gradual inductive process. *Cognition*, **127**, 307–17.

Warren, D. E. and Duff, M. C. (2014). Not so fast: hippocampal amnesia slows word learning despite successful fast mapping. *Hippocampus*, **24**, 920–33.

Welzl, H., D'Adamo, P. and Lipp, H. P. (2001). Conditioned taste aversion as a learning and memory paradigm. *Behavioural Brain Research*, **125**, 205–13.

Wilkinson, K. M., Dube, W. V., and MCilvane, W. J. (1998). Fast mapping and exclusion (emergent matching) in developmental language, behavior analysis, and animal cognition research. *Psychological Research*, **48**, 407–22.

Wilkinson, K. M., Dube, W. V., and Mcilvane, W. J. (1996). A crossdisciplinary perspective on studies of rapid word mapping in psycholinguistics and behavior analysis. *Developmental Review*, **16**, 125–48.

Wynn, K. (1990). Children's understanding of counting. *Cognition*, 36, 155–193.

Xu, F., and Tenenbaum, J. B. (2007). Word learning as Bayesian inference. *Psychological Review*, **114**, 245–72.

Yotsumoto, Y., Watanabe, T., and Sasaki, Y. (2008). Different dynamics of performance and brain activation in the time course of perceptual learning. *Neuron*, 57, 827–33.

17

Interactive Learning with Mutual Explanations in Relational Domains

Ute Schmid

Cognitive Systems Group, University of Bamberg, Germany

> '*We [AI researchers] never asked ourselves—what if it really works?*'
> Stuart Russell, Cumberland Lodge, July 2019

17.1 Introduction

Machine learning went through several changes of research perspective since its beginnings more than 50 years ago. Initially, machine learning algorithms were inspired by human learning (Michalski *et al.*, 1983). Early algorithms addressed policies for game playing (Samuel, 1963; Michie and Chambers, 1968) as well as concept learning (Hunt *et al.*, 1966; Michalski, 1987). Inductive Logic Programming (ILP) (Muggleton, 1991) and explanation-based generalization (Mitchell *et al.*, 1986) were introduced as integrated approaches which combine reasoning in first-order logic and inductive learning.

ILP addresses learning of complex rules involving relations and recursion (Gulwani *et al.*, 2015). It compares to human learning of explicit rules—that is, those that are capable of being inspected and verbalized—to classify entities in complex domains and to generate complex action sequences. Examples are classification of chemical structures (Srinivasan *et al.*, 1994), learning recursive concepts such as sorted list (Muggleton and Feng, 1990; Hofmann *et al.*, 2009) or ancestor (Schmid and Kitzelmann, 2011; Muggleton *et al.*, 2018), learning the transitivity rule (Schmid and Kitzelmann, 2011), the solution procedure for the Tower of Hanoi (Schmid and Kitzelmann, 2011), or strategies for robots (Wernsdorfer and Schmid, 2013; Cropper and Muggleton, 2015).

With the rise of statistical approaches to machine learning, focus shifted from human-like learning to optimizing machine-learning approaches for high predictive accuracy. Approaches such as multi-layer perceptrons, support-vector machines, and the recent deep learning architectures (Goodfellow *et al.*, 2016) resulted in data-intensive, black-box approaches. However, since machine learning increasingly moves from the lab to

Ute Schmid, *Interactive Learning with Mutual Explanations in Relational Domains* In: *Human Like Machine Intelligence*. Edited by: Stephen Muggleton and Nick Charter, Oxford University Press.
DOI: 10.1093/oso/9780198862536.003.0017

the real world, researchers and practitioners alike realize that interpretable, human-like approaches to machine learning are necessary for several reasons (see also Marcus, 2018):

- Some domains only have **small datasets**, for instance specific research domains such as (stress) phenotyping of plants (Stocker *et al.*, 2013; Singh *et al.*, 2016), or chemical processes;
- Datasets often are highly **imbalanced**, especially in medical diagnosis, such as cancer screening where many more examples are available for healthy probes than for cancer (Mazurowski *et al.*, 2008; Mennicke *et al.*, 2009; Schmid and Finzel, 2020);
- **Sampling biases** may be unavoidable, especially in complex domains where the underlying probability distributions are unclear;
- **Ground truth labelling** is either not existent or expensive in many real world applications. While in some domains, such as image net, assigning classes to images is rather straightforward, albeit involving high effort, in other domains, the true class of an instance can be only assessed in retrospect. This is true in many domains of medical diagnosis, but also in industrial quality control, or in assessment of the value of an employee.

Due to these factors, the initial appeal of purely data-driven, highly data-intensive end-to-end learning approaches, such as convolutional neural networks, is waning. It might even be possible that—analogous to the knowledge engineering bottleneck of expert system research (Buchanan, 2005)—the next AI winter will be caused by the *data engineering bottleneck*.

In the following, I will first argue for interpretable and interactive machine-learning as human-like approaches which can counteract the problems inherent in data-intensive, black-box machine learning. Afterwards, different types of explanations are introduced and exemplified for relational domains. How interactive learning can be realised with ILP is presented next. Finally, an application of interactive learning with ILP is presented which supports humans to identify and get rid of irrelevant digital content.

17.2 The Case for Interpretable and Interactive Learning

One possible approach to address the above-named restrictions is to use interpretable, white-box approaches to machine learning, such as decision tree learners or random forests. Indeed, logistic regression (also known as perceptrons) and random forests are the machine-learning approaches which are most successfully applied in practice. These approaches are competitive or even superior to multi-layer perceptrons on many benchmark problems (Fernández-Delgado *et al.*, 2014; Molnar *et al.*, 2020; Rudin, 2019).

Recently, it has been argued that the often used proposition that in machine learning there necessarily is a trade-off between accuracy and interpretability does not hold for many application domains (Rudin, 2019). If instances can be represented in terms of naturally meaningful features, there is no advantage of end-to-end learning. It even has been proposed that for computer vision domains which typically have highly nonlinear concept boundaries and complex interactions of features interpretable approaches can be applied (Dai *et al.*, 2019; Rabold *et al.*, 2019; Rudin, 2019). This observation relates well to models of human memory where different formats of representation have been proposed for explicit declarative and implicit perceptual and procedural memory (Atkinson and Shiffrin, 1968). This distinction has also been covered in computational models of cognition such as ACT-R (Anderson, 2000) or Clarion (Sun *et al.*, 2005) and is currently discussed as fast (system 1) and slow (system 2) thinking (Kahneman, 2011).

Taking into account theories and findings from cognitive science, we can compare the implicit end-to-end black-box machine learning with domains, which are inherently non-declarative for humans, for example, sensory-motor coordination tasks. In many other domains, classes can be at least partially described declarative. For perceptual categories, it typically is at least possible to verbalize reasons why some entity does *not* belong to a class (Markman and Gentner, 1996). For instance, one can reject the classification of an animal as a cat if it has neither whiskers nor claws. Likewise, one can reject the classification of a person as some relative or some celebrity due to some relevant characteristic which is missing or wrong (Ellis *et al.*, 1979). This proposition also is true for procedural knowledge: in many games, a player can give arguments for preferring one move over another in a given situation. This is also true for cognitive puzzles such as the Tower of Hanoi where it has been empirically shown that humans are able to generalize the recursive solution rule from self-play episodes (Welsh, 1991).

For domains where semantic features are grounded in perception, be it visual, auditory or another channel, hybrid approaches combining black-box and white-box learning seem to be promising. Currently, two strategies are being researched: (1) introducing semantics into deep networks which is, for example, the case for capsule networks (Sabour *et al.*, 2017) or embeddings of knowledge graphs (Ji *et al.*, 2015), and (2) learning interpretable surrogate models where features are extracted from the black-box models (Dai *et al.*, 2019; Rabold *et al.*, 2019; Schmid and Finzel, 2020).

Interpretable approaches to machine learning typically have only a small number of hyper-parameters to be adjusted during learning. Therefore, in contrast to deep neural networks, learning is not dependent on the availability of huge amounts of labelled data. Furthermore, predictive accuracy estimation can be based on multiple learning episodes, making use of k-fold cross validation (Flach, 2012) which gives a more reliable estimate than a loss-function evaluation based on only one separation of the data in a training and a test set (Goodfellow *et al.*, 2016). One might speculate that the high performance of large deep learning networks is due to over-fitting rather than predictive generalization.

To overcome the above-named restrictions of purely data-driven machine-learning approaches, human knowledge can be exploited to make learning more efficient. If knowledge is available, less training instances are needed and the search for a model can be guided by constraints derived from a knowledge-base. Knowledge can be made available either in advance of learning or during learning. Furthermore, knowledge can be gained after learning.

Knowledge provision in advance has been proposed in the context of rule learning as explanation-based learning (Mitchell *et al.*, 1986). This approach allows incorporation of a background theory which can be used to draw inferences and thereby enrich the training data. Knowledge can also be introduced in learning deep neural networks, for instance as an explanatory graph, which represents the knowledge hierarchy hidden in the convolution layers of a CNN (Zhang *et al.*, 2018). Making use of knowledge during the learning process is realized in interactive learning approaches where users are allowed to correct system decisions (Fails and Olsen Jr., 2003; Teso and Kersting, 2019).

Providing background theories depends on the availability of explicit knowledge. Such knowledge exists in many domains. For instance, physical laws can be exploited when learning which objects are saved to stack on each other (Mitchell *et al.*, 1986). Interactive learning, on the other hand, has the advantage that humans can introduce their knowledge implicitly into the learning process—by accepting or rejecting system decisions. Extracting rules after training might provide a possibility to combine knowledge-based and machine-learning approaches into a reasoning-learning loop for incremental knowledge gain (Telle *et al.*, 2019; Schmid, 2003).

Combining the strengths of humans with the strength of machine learning can result in efficient human–AI partnerships as envisioned by Michie (1988), who proposed that a learned model should be productive for humans in such a way that it enables humans to solve problems which they could not have solved without the machine-learning model (Muggleton *et al.*, 2018). Joint decision-making profits from the ability of machine learning to detect complex patterns and relations in data, and from the experience-based and context-sensitive decision heuristics of humans. Such systems have been proposed as beneficial artificial intelligence by Russell *et al.* (2015).

Despite these promises, there are also some downsides to joint human–machine decision-making. How admissible it is to allow a human to shape and correct a learned model is strongly dependent on the domain. For example, in a personal assistant, the end-user is expert on his or her individual preferences and therefore, corrections are an obvious way to adapt a system (Holzinger, 2014; Kulesza *et al.*, 2015). On the other hand, there are domains, such as medical diagnosis, where only a human expert has the competence to make appropriate corrections (Holzinger, 2016; Schmid and Finzel, 2020). Furthermore, new methods to evaluate predictive accuracy become necessary when labels in supervised learning are considered no longer as absolute, that is, ground truth.

17.3 Types of Explanations—There is No One-Size Fits All

In human–human cooperation, asking for and receiving an explanation is a natural way to make sense of decisions and actions of others (Miller, 2019). During the last years, explainability has been promoted for machine learning to make system decisions transparent and comprehensible for humans and thereby inspire trust in AI systems (Ribeiro *et al.*, 2016; Gunning and Aha, 2019; Adadi and Berrada, 2018). Starting with LIME (Ribeiro *et al.*, 2016), over the last few years many approaches to explainability

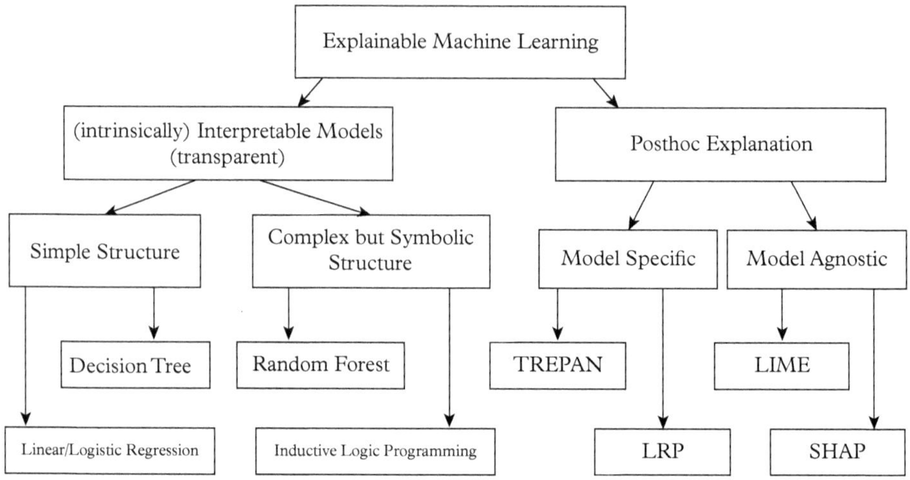

Figure 17.1 *Intrinsic versus post-hoc explanations.*

have been proposed (see Fig.17.1). In general, one can discriminate between approaches which are intrinsically transparent—often named interpretable machine learning—and approaches which can provide post-hoc explanations. Interpretable approaches are comparable to explicit programs. While short and simple programs can be easily understood and debugged, this is not true for complex software. Likewise, learned white-box models can be simply structured or complex. In the second case, humans need assistance to comprehend a learned model. Post-hoc explanations can be either model-specific or model-agnostic. Well-known model specific approaches are the classic TREPAN which extracts trees from neural nets (Craven and Shavlik, 1996) as well as LRP (layer-wise relevance propagation) which allows to identify which pixels in an input image contribute most strongly to a class decision (Samek *et al.*, 2017). LIME (Ribeiro *et al.*, 2016) and SHAP (Lundberg and Lee, 2017) are model-agnostic approaches. To explain why a specific instance is classified in a specific way, LIME generates perturbed instances of the original instance by selecting subsets of words or sub-parts of images. The perturbations of the original instance are submitted to the black-box classifier. Sub-parts which when omitted change the original class decision for the instance are identified as relevant and used as an explanation.

Explanations for image classifications given as highlighting of relevant pixels (as LRP does) or showing relevant areas (super-pixels) in the image (as LIME does) have gained the most attention in the context of explainable AI research. However, such **visual explanations** have strong restrictions with respect to what information they can convey: highlighting is helpful to identify over-fitting (Lapuschkin *et al.*, 2019) and thereby these approaches are important tools for developers. However, to explain complex decisions to end-users, a mere conjunction of relevant areas is not enough (Schmid, 2018):

- **Feature values:** highlighting the area of the eye in an image is not helpful to understand that it is important for the class decision that the lids are tightened (indicating pain) in contrast to eyes which are wide open (indicating startle: Weitz *et al.*, 2019);
- **Quantification:** highlighting all blowholes on the supporting parts of a rim does not make clear that the rim is not a reject because *all* blowholes are smaller than 0.5 millimetres;
- **Negation:** highlighting the flower in the hand of a person does not transport the information that this person is *not* a terrorist because he or she does *not* hold a weapon;
- **Relations:** highlighting all windows in a building cannot help to discriminate between a tower, where windows are *above* each other and a bungalow, where windows are *beside* each other (Rabold *et al.*, 2019);
- **Recursion:** highlighting all stones within a circle of stones cannot transport the information that there must be a sequence of an arbitrary number of stones with increasing size (Rabold *et al.*, 2018).

While decision trees and variants do capture the relation of feature values and class labels (Fürnkranz and Kliegr, 2015; Lakkaraju *et al.*, 2016), they are not expressive enough to capture the types of information enumerated above. General approaches to inductive programming such as inductive functional programming and ILP, however, can teach models which correspond to the expressive power of programs (Gulwani *et al.*, 2015). Prolog programs induced with an ILP approach are intrinsically comprehensible but they presuppose some background in computer science and are often too complex to be helpful to end-users. Here, trace-based explanation generation as it has been proposed for expert systems (Clancey, 1983) can be applied and the reasoning traces why a specific class decision has been derived can be made transparent with **verbal explanations**. How a simple, template approach can be used to translate a trace into a natural language explanation is shown at the end of this chapter in Figure 17.4.

To explain image classifications, a combination of visual and verbal explanations can be very helpful. For instance, sub-concepts occurring in the verbal explanation can be illustrated by reference to the image. This is, for example, realised in the TraMeExCo system which explains tumour classifications from tissue samples (Schmid and Finzel, 2020) or in the PainComprehender project where explanations are given based on the identification of action units (Gromowski *et al.*, 2020).

Besides intrinsic and post-hoc explanations which can be given visual or verbal, explanations can be local, explaining the current classification, or global, explaining the model. In general, it is easier to explain why a specific instance is classified in a certain way, such as why some tissue sample is classified as a specific tumour class, than to explain the full model which, for the tissue classification would be a comprehensive characterization of this type of tumour. In medical textbooks, images which are prototypical for a specific disease are used. In general, prototypes are an efficient way to make the general characteristics of a class comprehensible (Bien *et al.*,

2011). If class decisions depend on subtle differences between feature values or relational configurations, contrasting examples can be helpful. Contrasting examples come in two varieties: as counter-examples in feature-based domains (Wachter *et al.*, 2017; Poyiadzi *et al.*, 2020) and as near misses in relational domains (Winston, 1975).

Counter-examples are an intuitive way to highlight what elements are crucial for a class decision by showing which minimal changes in feature values would have resulted in a different decision. For instance if the classification of a mushroom as poisonous depends on the difference between a yellowish or greyish cap, an image showing both species next to each other is very helpful. Verbal counter examples are especially important in the context of insurance and banking, such as; You were denied a loan because your annual income was 30,000. If your income had been 45,000, you would have been offered a loan' (Wachter *et al.*, 2017). In philosophy, counter examples have been characterized by the concept of a 'closest possible world', that is, the smallest change required to obtain a different (and more desirable) outcome (Pollock, 1976).

In the context of explainable artificial intelligence, there are only few approaches which consider contrastive examples (Adadi and Berrada, 2018) and these are focusing on feature value explanations. However, counter examples are also helpful in relational domains. In early AI research, Winston has introduced the concept of near misses to provide minimal sets of examples for learning relational concepts such as an arc (Winston, 1975). In cognitive science research it has been shown that alignment of structured representations helps humans to understand and explain concepts (Gentner and Markman, 1994). Gentner and Markman found that it is easier for humans to find the differences between pairs of similar items than to find the differences between pairs of dissimilar items. For example, it is easier to explain the concept of a light bulb by contrasting it with a candle than by contrasting it with a cat (Gentner and Markman, 1994).

The different types of explanations discussed in this section are summarised in Table 17.1. Different types of explanations have different functionalities. Consequently, explanation generation to support human understanding of AI decisions needs to take into account the specific demands a specific person has in a specific situation, that is, explanations cannot be one-size fits all (Gromowski *et al.*, 2020; Niessen *et al.*, 2020). A future challenge for explainable AI research will be to provide the best explanation for a given context. To achieve this, either the personality of the user has to be assessed

Table 17.1 *Types of explanations.*

Model	intrinsic	post-hoc
Mode	verbal	visual
Generality	global	local
Reference	current instance	example
		contrasting / prototypical

or current intentions have to be identified. Since the latter typically involves monitoring which invades privacy, when ever possible, the user should be involved by giving him or her a choice between different types of explanation.

17.4 Interactive Learning with ILP

The main motivation of human-in-the-loop interactive machine learning is to build systems that improve their learning outcome with the help of a human expert who interacts with them. Interactive machine learning has been proposed in the context of human–computer interaction research as a means to adaptive personal assistance (Fails and Olsen Jr., 2003). It has also been proposed in the context of medical decision making (Holzinger, 2016; Schmid and Finzel, 2020). In medicine and other highly specialized domains, interactive methods are mainly introduced to provide valid class labels for supervised learning (Kabra *et al.*, 2013). Active learning approaches (Wiener *et al.*, 2015) can be applied to select examples guided by preference order (Kabra *et al.*, 2013; Teso and Kersting, 2019).

However, humans can give more information besides class labelling and class corrections: explanations might not be considered as a one-way street from an AI system to the human but as a mutual exchange where the system makes the underlying reasons for a class decision transparent to the human and the human can accept or correct this explanation. Such a correction can be performed when the class decision is wrong to constrain model adaption. However, it can also be the case that the system classifies an instance correctly but based on a wrong—for example, over-fitted—model (Siebers and Schmid, 2019; Teso and Kersting, 2019; Gromowski, Siebers and Schmid, 2020). If not only the class label but also the explanation is correctable, model adaption can be constrained such that specific information is explicitly excluded or included. It can be assumed that this stronger involvement of a human also helps to counteract negative effects of human–AI interaction such as losing trust due to false alarms or becoming overly trustful in system decisions (Lee and See, 2004).

It has been shown for logic-based reasoning that explanations can be used to revise current models (Falappa *et al.*, 2002). Similar mechanisms can be applied when learning logical models with ILP. A model of such a mutual explanation system is given in Figure 17.2: the core of the system is the application of inductive logic programming (ILP) to generate expressive interpretable models. Starting with an initial ILP model, a new example e is classified. The class decision is presented to the human who can accept it or ask for an explanation. ILP learned models are given as sets of logical rules and classification of a new example is realised as logical inference. Consequently, verbal explanations can be generated from the reasoning trace in a straightforward manner. The explanation might be convincing and the human accepts the class decision. If the explanation is not acceptable, the class decision can be corrected. In addition, the user can correct the explanation by introducing or removing predicates. The correction together with the new class label are used to adapt the model.

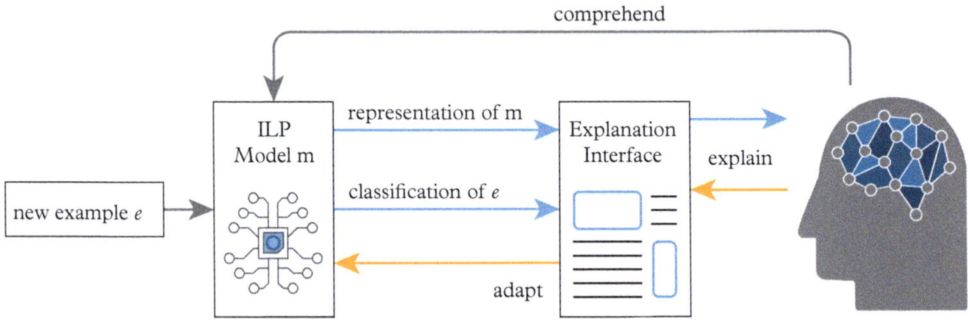

Figure 17.2 *A framework for interactive learning with mutual explanations (Fig. 1 from Schmid & Finzel, 2020).*

Typically, ILP approaches such as Aleph (Srinivasan, 2001) or Metagol (Muggleton *et al.*, 2015), are designed as batch learners, making use of the set of all training data for model induction. For interactive learning, an incremental variant of ILP would be most plausible. An incremental variant of Metagol has been proposed by Siebers and Schmid (2018). However, it only allows specialisation not generalisation of models. For example, an over-general rule to classify a year as leapyear—`leapyear(A) :- divisible(A,4)`—might be specialized to by including the predicate `divisible(A,100)` in the body of the rule. Thereby false alarms will be reduced. However, the proposed incremental algorithm does not allow to generalise a rule by removing a predicate from the body. That is, the number of misses cannot be reduced by this technique. One possible way to overcome this restriction is to introduce a new positive example for which the current, over-specific model does not hold. This kind of interactive learning strategy has been proposed in the context of end-user programming (Gulwani, 2011).

The mutual explanation framework is described in detail in Gromowski, Siebers and Schmid (2020). It has been applied to classification of pain from facial expressions (Schmid, 2018), to classification of tissue samples (Schmid and Finzel, 2020), and to identifying irrelevant digital objects which will be described in the following.

17.5 Learning to Delete with Mutual Explanations

Digitalisation in many domains—be it industrial production, smart homes, administration, or personal assistance—results in an ever-growing volume of digital data. Digital hording occurs in private as well as working contexts resulting in increasing energy needs and costs for storage. Supporting humans to identify irrelevant digital objects is used as a test bed to explore the mutual explanation framework introduced above. The demonstrator system Dare2Del is an ILP system which interactively learns to classify files as being irrelevant or not (Siebers and Schmid, 2019). Learned rules might be:

```
irrelevant(F) :- notAccessedSince(F, oneyear).
irrelevant(F) :- inSameDirectory(F, F2), newer(F2, F).
irrelevant(F) :- inDirectory(F, D), name(D, "tmp").
```

Such rules have the expressiveness of Horn clauses, allowing to represent rich information of the kind described in section 17.3: The classification of a file as irrelevant might depend on the specific value of an attribute, such as that it has been not accessed since a specific time, as in `notAccessedSince(F, oneyear)`. During learning such an attribute value might be made more restrictive (e.g., `onemonth`), resulting in a more specific rule, or more relaxed (e.g., `fiveyears`), resulting in a more general rule. Predicates can specify relations between different files, such as `newer(F2, F)` or between files and directories, such as `inDirectory(F, D)`. Recursion can be used to express complex relations over an arbitratry number of objects. For instance, a predicate can express a the transitivity of the subdirectory relation.

```
subdirectory(D1,D2) :- issubdirectory(D1,D2).
subdirectory(D1,D2) :- issubdirectory(D1,D), subdirectory(D,D2)
```

The Dare2Del domain has many demanding requirements given to the sensitive nature of files and other types of data in the digital world of work (Siebers *et al.*, 2017):

- Transparency and comprehensibility: deletion or archiving of files and other digital objects is a highly sensitive task. Therefore, the system must be able to explain its decisions by making the criteria that led to its decision explicit.
- Adaptation to needs and preferences of specific users: interactive learning is a plausible strategy to adapt to personal preferences where there is no objective criterion to label data as relevant or irrelevant.
- Obeying laws and regulations: in many working contexts, storage of digital data has to be compliant with laws and regulations. Furthermore, users might have explicit preferences with respect to what data should or should not deleted. Therefore, the system must allow to take into account predefined rules.
- Context awareness to users and situations: the interaction process and the offered explanations should agree with the specific demands of a user and situation.

ILP allows to combine logic and learning in a natural way. General rules can be provided as background theories. For instance, a rule characterizing that one file is newer than another one can be formulated with respect to the generation time of files:

```
newer(F1,F2) :- genTime(F1,T1), genTime(F2,T2), T1 < T2.
```

Likewise, general regulations, such as that invoices must be kept for a specific amount of time, can be formulated as rules which can be taken into account.

```
irrelevant(F) :- not invoiceToKeep(F).
invoicetoKeep(F) :- isInvoice(F), timeSpan(F,T),
retentionReg(R), T < R.
```

Starting with an initial, preliminary set of rules to characterize the target predicare `irrelenant(F)`, learning is realized in episodes of presenting a selection of files and their current classification. The user can accept or reject the current irrelevancy decision and, in addition, make corrections in the given explanation. Learning, in general, can be realised by changing argument values, introducing or deleting predicates, and introducing or deleting rules. Currently, the ILP system Aleph is used where learning is based on step-wise specialization of rules (Srinivasan, 2001).

A screenshot of the demonstrator system Dare2Del is given in Figure 17.3. The relevant part of the file system is shown to the left. Deletion suggestions are shown on the top right. An explanation why the selected file may be deleted is given below. Highlightings are used to relate the explanation to the domain of discourse, that is the actual files. The files presented to the user are drawn from the sample of all files in a specific context. For instance, the directory where the last change occurred is used as starting point and sampling is done up to a certain distance in the directory graph. The sampled files are tested against the current set of rules to classify a file as irrelevant or not irrelevant.

The verbal explanations of why a file is considered irrelevant are generated based on simple templates to rewrite a reasoning trace when classifying a given file as irrelevant into natural language. An excerpt of the procedure is given in Figure 17.4.

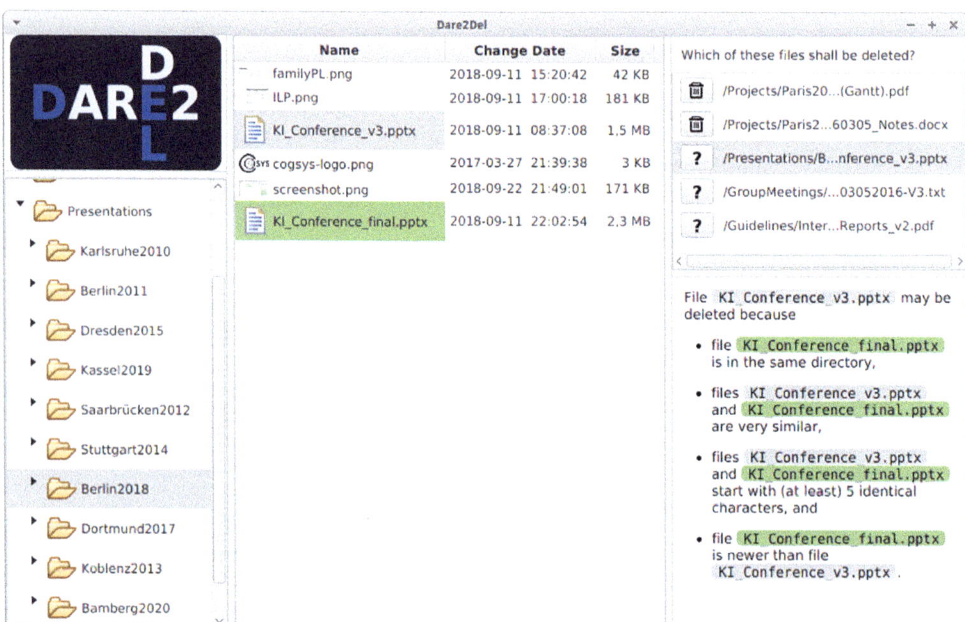

Figure 17.3 *Screenshot of the Dare2Del system (Fig. 4 from Siebers and Schmid, 2018).*

$$\mathbf{T}_{\texttt{newer}}(s_1, s_2) = s_1 \cdot \text{' is newer than'} \cdot s_2$$
$$\mathbf{T}_{\texttt{notnewer}}(s_1, s_2) = s_1 \cdot \text{' is as old as or older than'} \cdot s_2$$
$$\mathbf{T}^1_{\texttt{newer}}(t) = \text{'file'} \cdot n \quad \text{iff } B_D \cup B_T \cup \mathbf{I} \models \texttt{name}(t, n)$$
$$\mathbf{T}^2_{\texttt{newer}}(t) = \mathbf{T}^1_{\texttt{newer}}(t)$$

$$\mathbf{T}_{\texttt{not\_accessed\_since}}(s_1, s_2) = s_1 \cdot \text{' was not accessed within the last'} \cdot s_2$$
$$\mathbf{T}_{\texttt{notnot\_accessed\_since}}(s_1, s_2) = s_1 \cdot \text{' was accessed within the last'} \cdot s_2$$
$$\mathbf{T}^1_{\texttt{not\_accessed\_since}}(t) = \mathbf{T}^1_{\texttt{newer}}(t)$$
$$\mathbf{T}^2_{\texttt{not\_accessed\_since}}(t) = \begin{cases} t/86400 \cdot \text{' days'} & \text{if } t \bmod 86400 = 0 \\ t/3600 \cdot \text{' hours'} & \text{otherwise,} \\ & \text{if } t \bmod 3600 = 0 \\ t \cdot \text{' seconds'} & \text{otherwise} \end{cases}$$

$$\begin{aligned} \mathscr{T}_L(\texttt{newer}(item\_1, item\_2)) &= \mathbf{T}_{\texttt{newer}}(\mathbf{T}^1_{\texttt{newer}}(item\_1), \mathbf{T}^2_{\texttt{newer}}(item\_2)) \\ &= \mathbf{T}_{\texttt{newer}}(\text{'file c.doc', 'file b.doc'}) \\ &= \text{'file c.doc'} \cdot \text{' is newer than '} \cdot \text{'file b.doc'} \\ &= \text{'file c.doc is newer than file b.doc'} \end{aligned}$$

Figure 17.4 *Extract of template based generation of verbal explanations from logical rules (Fig. 3 from Siebers and Schmid, 2018).*

Mutual explanation involves specialization by introducing additional literals or restricting the range of variables, or generalization by deleting literals from an irrelevancy rule or relaxing the range of variables. For instance, the third condition given in the explanation in Figure 17.3 might be relaxed such that only the first three letters of the prefix of two files need to be identical.

Currently, the ILP system Aleph is used for Dare2Del. In Aleph, user-defined constraints can be applied to guide the generation of alternative clauses (Srinivasan, 2001). For example, if a clause is required to contain some predicate p with arguments X and a, where X is a variable and a is a constant, a typical constraint is represented as follows:

```
false :- hypothesis(Head,Body,_), not(in(Body,p(X,a))).
```

In the current system, adaptation to specific demands of the user or the situation, is not performed automatically. Instead, the user decides when Dare2Del should be active and how many files are offered to check for irrelevancy.

17.6 Conclusions and Future Work

Given that machine learning is applied in an increasing number of real world domains, it becomes highly important that humans and AI systems can efficiently interact. Explanations are considered to be a crucial ingredient for transparent, comprehensible, and trustworthy AI. There is a growing number of approaches to generate explanation (Molnar *et al.*, 2020). In summer 2019, IBM launched the AI Explainability 360 Open Source Toolkit (Arya *et al.*, 2019). However, the majority of work is addressing visual explanation generation for black-box classifiers.

In this chapter, I have presented arguments for interpretable approaches, especially for ILP, which allows to combine logic and learning and to induce highly expressive models. Verbal explanations can be generated from such model which can express complex decision criteria involving feature attributes, relations, and recursion. Furthermore, I have argued for interactive learning to allow to exploit human expertise in machine learning. Currently, only a proof-of-concept realization of such a framework for learning with mutual explanations exists. Future challenges include extending existing techniques of ILP such that incremental learning can be efficiently realized, designing hybrid approaches combining end-to-end deep learning with ILP, providing methods to guide corrections of explanations for end-users, and coming up with criteria how to evaluate the quality of an interactively learned model to decide when the human interaction might have negative effects on the predictive accuracy of the model.

Hopefully, the third wave of AI with its focus on explanations will provide solutions to these challenges such that the transition of AI from the lab to the real world will be beneficial for humanity.

Acknowledgements

This research is supported by the German Research Foundation (DFG, grant 318286042), project Dare2Del which is part of the priority program Intentional Forgetting (SPP 1921). Thanks to Michael Siebers and to Bettina Finzel who are working on mutual explanations in their doctoral research.

References

Adadi, A. and Berrada, M. (2018). Peeking inside the black-box: a survey on explainable artificial intelligence (XAI). *IEEE Access*, **6**, 52138–60.

Anderson, J. R. (2000). *Learning and Memory: An Integrated Approach*. New Jersey: John Wiley & Sons Inc.

Arya, V., Bellamy, R. K. E., Chen, P.-Y. *et al.* (2019). One explanation does not fit all: a toolkit and taxonomy of ai explainability techniques. arXiv preprint arXiv:1909.03012.

Atkinson, R. C. and Shiffrin, R. M. (1968). *Human Memory: A Proposed System and its Control Processes*. New York, NY: Academic Press.

Bien, J., and Tibshirani, R. (2011). Prototype selection for interpretable classification. *Annals of Applied Statistics*, **5**(4), 2403–24.

Buchanan, B. G. (2005). A (very) brief history of artificial intelligence. *AI Magazine*, **26**(4), 53.

Clancey, W. J. (1983). The epistemology of a rule-based expert system: a framework for explanation. *Artificial Intelligence*, **20**(3), 215–51.

Craven, M. and Shavlik, J. W (1996). Extracting tree-structured representations of trained networks, in M. C. Mozer, M. I. Jordan, and T. Petsche, eds, *Advances in Neural Information Processing Systems 9*. Boston: MIT Press, 24–30.

Cropper, A. and Muggleton, S. H. (2015). Learning efficient logical robot strategies involving composable objects, in *International Joint Conferences on Artificial Intelligence*, AAAI Press.

Dai, W.-Z., Xu, Q., Yu, Y. *et al.* (2019). Bridging machine learning and logical reasoning by abductive learning, in H. Wallach, H. Larochelle, A. Beygelzimer, *et al.*, eds, *Advances in Neural Information Processing Systems* 32, 2815–26.

Ellis, H. D., Shepherd, J. W., and Davies, G. M. (1979). Identification of familiar and unfamiliar faces from internal and external features: some implications for theories of face recognition. *Perception*, **8**(4), 431–39.

Fails, J. A. and Olsen Jr., D. R. (2003). Interactive machine learning, in *Proceedings of the 8th International Conference on Intelligent User Interfaces*. Boston: AAAI Press, 39–45.

Falappa, M. A., Kern-Isberner, G., and Simari, G. R. (2002). Explanations, belief revision and defeasible reasoning. *Artificial Intelligence*, **141**(1–2), 1–28.

Fernández-Delgado, M., Cernadas, E., Barro, S. *et al.* (2014). Do we need hundreds of classifiers to solve real world classification problems? *The Journal of Machine Learning Research*, **15**(1), 3133–81.

Flach, P. (2012). *Machine Learning: The Art and Science of Algorithms that Make Sense of Data*. Cambridge: Cambridge University Press.

Fürnkranz, J. and Kliegr, T. (2015). A brief overview of rule learning, in *International Symposium on Rules and Rule Markup Languages for the Semantic Web*. Berlin: Springer, 54–69.

Gentner, D. and Markman, A. B. (1994). Structural alignment in comparison: No difference without similarity. *Psychological Science*, **5**(3), 152–8.

Goodfellow, I., Bengio, Y., and Courville, A. (2016). *Deep Learning*. Cambridge, MA: MIT Press.

Gromowski, M., Siebers, M., and Schmid, U. (2020). A process framework for inducing and explaining datalog models. *Advances in Data Analysis and Classification*, 14, 821–35.

Gulwani, S. (2011). Automating string processing in spreadsheets using input-output examples. *ACM Sigplan Notices*, **46**(1), 317–30.

Gulwani, S., Hernández-Orallo, J., Kitzelmann, E. *et al.* (2015). Inductive programming meets the real world. *Communications of the ACM*, **58**(11), 90–9.

Gunning, D., & Aha, D. (2019). DARPA's Explainable Artificial Intelligence (XAI) Program. *AI Magazine*, **40**(2), 44–58.

Hofmann, M., Kitzelmann, E., and Schmid, U. (2009). A unifying framework for analysis and evaluation of inductive programming systems. B. Goerzel, P. Hitzler, and M. Hutter, eds, *Proceedings of the Second Conference on Artificial General Intelligence (AGI-09, Arlington, Virginia, 6–9 March, 2009)*, Amsterdam: Atlantis Press, 55–60.

Holzinger, A. (2014). Trends in interactive knowledge discovery for personalized medicine: cognitive science meets machine learning. *The IEEE Intelligent Informatics Bulletin*, **15**(1), 6–14.

Holzinger, A. (2016). Interactive machine learning for health informatics: when do we need the human-in-the-loop? *Brain Informatics*, **3**(2), 119–31.

Hunt, E. B., Marin, J., and Stone, P. J. (1966). *Experiments in Induction.* New York, NY: Academic Press.

Ji, G., He, S., Xu, L. *et al.* (2015). Knowledge graph embedding via dynamic mapping matrix, in *Proceedings of the 53rd Annual Meeting of the Association for Computational Linguistics and the 7th International Joint Conference on Natural Language Processing (Vol. 1: Long Papers).* Boston: MIT Press, 687–96.

Kabra, M., Robie, A. A., Rivera-Alba, M. *et al.* (2013). Jaaba: interactive machine learning for automatic annotation of animal behavior. *Nature Methods,* **10**(1), 64.

Kahneman, Daniel (2011). *Thinking, Fast and Slow.* New York, NY: Farrar, Straus and Giroux.

Kulesza, T., Burnett, M., Wong, W.-K. *et al.* (2015). Principles of explanatory debugging to personalize interactive machine learning, in *Proceedings of the 20th International Conference on Intelligent User Interfaces,* Atlanta. New York: ACM, 126–37.

Lakkaraju, H., Bach, S. H., and Leskovec, J. (2016). Interpretable decision sets: a joint framework for description and prediction, in *Proceedings of the 22nd ACM SIGKDD International Conference on Knowledge Discovery and Data Mining,* San Francisco. New York: ACM, 1675–84.

Lapuschkin, S., Wäldchen, S., Binder, A. *et al.* (2019). Unmasking clever hans predictors and assessing what machines really learn. *Nature Communications,* **10**(1), 1–8.

Lee, J. D. and See, K. A. (2004). Trust in automation: designing for appropriate reliance. *Human Factors,* **46**(1), 50–80.

Lundberg, S. M. and Lee, S.-I. (2017). A unified approach to interpreting model predictions, in U. V. Luxberg and I. Guyon, eds, *Advances in Neural Information Processing Systems 30.* Long Beach; New York: Curran Associates, 4765–74.

Marcus, G. (2018). Deep learning: A critical appraisal. *CoRR,* **abs/1801.00631**.

Markman, A. B. and Gentner, D. (1996). Commonalities and differences in similarity comparisons. *Memory & Cognition,* **24**(2), 235–49.

Mazurowski, M. A., Habas, P. A., Zurada, J. M. *et al.* (2008). Training neural network classifiers for medical decision making: the effects of imbalanced datasets on classification performance. *Neural Networks,* **21**(2-3), 427–36.

Mennicke, J., Münzenmayer, C., Wittenberg, T. *et al.*(2009). An optimization framework for classifier learning from image data for computer-assisted diagnosis, in *Proceedings 4th European Conference of the International Federation for Medical and Biological Engineering,* Antwerp. Berlin, Heidelberg: Springer, 629–32.

Michalski, Ryszard S. (1987). Learning strategies and automated knowledge acquisition. In *Computational Models of Learning* (ed. L. Bolc), pp. 1–19. Springer.

Michalski, R. S., Carbonell, J. G., and Mitchell, T. M. (eds) (1983). *Machine Learning—An Artificial Intelligence Approach.* Wellsboro, Pennsylvania: Tioga.

Michie, Donald (1988). Machine learning in the next five years. In *Proceedings of the Third European Working Session on Learning, EWSL 1988, Turing Institute, Glasgow, UK, October 3-5, 1988* (ed. D. H. Sleeman), pp. 107–22. Pitman Publishing.

Michie, D. and Chambers, R. A. (1968). Boxes: an experiment in adaptive control, in E. Dale and D, Michie, eds, *Machine Intelligence,* Vol. 2. Edinburgh: Oliver and Boyd, 137–52.

Miller, T. (2019). Explanation in artificial intelligence: Insights from the social sciences. *Artificial Intelligence,* **267**, 1–38.

Mitchell, T. M., Keller, R. M., and Kedar-Cabelli, S. T. (1986). Explanation-based generalization: a unifying view. *Machine learning,* **1**(1), 47–80.

Molnar, C., Casalicchio, G., and Bischl, B. (2020). *Interpretable Machine Learning – A Brief History, State-of-the-Art and Challenges.* arXiv preprint arXiv:2010.09337.

Muggleton, S. (1991). Inductive logic programming. *New generation Computing*, **8**(4), 295–318.

Muggleton, S.H., and Feng, C. (1990). Efficient induction of logic programs, in *Proceedings of the First Conference on Algorithmic Learning Theory*. Tokyo: Ohmsha, 368–81.

Muggleton, S., Schmid, U., Zeller, C. *et al.* (2018). Ultra-strong machine learning: comprehensibility of programs learned with ilp. *Machine Learning*, **107**(7), 1119–40.

Muggleton, S. H., Lin, D., and Tamaddoni-Nezhad, A. (2015). Meta-interpretive learning of higher-order dyadic datalog: predicate invention revisited. *Machine Learning*, **100**(1), 49–73.

Niessen, C., Göbel, K., Siebers, M. *et al.* (2020). Time to forget: a review and conceptual framework of intentional forgetting in the digital world of work. *Zeitschrift für Arbeits- und Organisationspsychologie*, **64**(1), 30–45.

Pollock, J. L. (1976). The possible worlds analysis of counterfactuals. *Philosophical Studies*, **29**(6), 469–476.

Poyiadzi, R., Sokol, K., Santos-Rodriguez, R. *et al.* (2020). Face: feasible and actionable counterfactual explanations, in *Proceedings of the AAAI/ACM Conference on AI, Ethics, and Society*. New York: ACM, 344–50.

Rabold, J., Deininger, H., Siebers, M. *et al.* (2019). Enriching visual with verbal explanations for relational concepts: combining LIME with Aleph, in *Joint European Conference on Machine Learning and Knowledge Discovery in Databases*. Cham: Springer, 180–92.

Rabold, Johannes, Siebers, Michael, and Schmid, Ute (2018). Explaining black-box classifiers with ILP - empowering LIME with aleph to approximate non-linear decisions with relational rules. In *Inductive Logic Programming - 28th International Conference, ILP 2018, Ferrara, Italy, September 2-4, 2018, Proceedings* (ed. F. Riguzzi, E. Bellodi, and R. Zese), Volume 11105, Lecture Notes in Computer Science, pp. 105–17. Springer.

Ribeiro, M. T., Singh, S., and Guestrin, C. (2016). Why should I trust you?: explaining the predictions of any classifier, in *Proceedings of the 22nd ACM SIGKDD International Conference on Knowledge Discovery and Data Mining*. New York: ACM, 1135–44.

Rudin, C. (2019). Stop explaining black box machine learning models for high stakes decisions and use interpretable models instead. *Nature Machine Intelligence*, **1**(5), 206–15.

Russell, S., Dewey, D., and Tegmark, M. (2015). Research priorities for robust and beneficial artificial intelligence. *AI Magazine*, **36**(4), 105–14.

Sabour, S., Frosst, N., and Hinton, G. E. (2017). Dynamic routing between capsules, in *Advances in Neural Information Processing Systems 30*. New York: Curran Associates, 3856–66.

Samek, W., Wiegand, T., and Müller, K.-R. (2017). Explainable artificial intelligence: Understanding, visualizing and interpreting deep learning models. arXiv preprint arXiv:1708.08296.

Samuel, A. L. (1963). Some studies in machine learning using the game of checkers, in E. Feigenbaum and J. Feldman, eds, *Computers and Thought*, New York, NY: McGraw Hill, 71–105.

Schmid, U. (2003). *Inductive Synthesis of Functional Programs: Universal Planning, Folding of Finite Programs, and Schema Abstraction by Analogical Reasoning (Lecture Notes in Computer Science Vol. 2654)*. Springer Science & Business Media.

Schmid, U. (2018). Inductive programming as approach to comprehensible machine learning. In *Proceedings of the 7th Workshop on Dynamics of Knowledge and Belief (DKB-2018) and the 6th Workshop KI & Kognition (KIK-2018) co-located with 41st German Conference on Artificial Intelligence (KI 2018), Berlin, Germany, September 25, 2018* (ed. C. Beierle, G. Kern-Isberner, M. Ragni, F. Stolzenburg, and M. Thimm), Volume 2194, CEUR Workshop Proceedings, pp. 4–12. CEUR-WS.org.

Schmid, U. and Finzel, B. (2020). Mutual explanations for cooperative decision making in medicine. *Künstliche Intelligenz, Special Issue Challenges in Interactive Machine Learning*, **34**.

Schmid, U. and Kitzelmann, E. (2011). Inductive rule learning on the knowledge level. *Cognitive Systems Research*, **12**(3), 237–48.

Siebers, M., Gobel, K., Niessen, C. *et al.* (2017). Requirements for a companion system to support identifying irrelevancy, in *Proceedings of 2017 International Conference on Companion Technology (ICCT 2017)*, Ulm. New York: IEEE, 1–2.

Siebers, M. and Schmid, U. (2018). Was the year 2000 a leap year? Step-wise narrowing theories with metagol, in *International Conference on Inductive Logic Programming*, Ferrara. Cham: Springer, 141–56.

Siebers, M. and Schmid, U. (2019). Please delete that! Why should I?: Explaining learned irrelevance classifications of digital objects. *Künstliche Intelligenz*, **33**(1), 35–44.

Singh, A., Ganapathysubramanian, B., Singh, A. K. *et al.* (2016). Machine learning for high-throughput stress phenotyping in plants. *Trends in Plant Science*, **21**(2), 110–24.

Srinivasan, A. (2001). *The Aleph manual.* http://www.di.ubi.pt/~jpaulo/competence/tutorials/aleph.pdf

Srinivasan, A., Muggleton, S., King, R. D. *et al.* (1994). Mutagenesis: Ilp experiments in a non-determinate biological domain, in *Proceedings of the 4th International Workshop on Inductive Logic Pxogramming*. Gesellschaft fur Mathematik und Datenverarbeitung MBH, GMD-Studien Nr 237.

Stocker, C., Uhrmann, F., Scholz, O. *et al.* (2013). A machine learning approach to drought stress level classification of tobacco plants, in *Proceedings of Workshop on Learning, Knowledge and Adaptivity*, Bamberg. University of Bamberg, 163–7.

Sun, R., Slusarz, P., and Terry, C. (2005). The interaction of the explicit and the implicit in skill learning: a dual-process approach. *Psychological Review*, **112**(1), 159.

Telle, J. A., Hernández-Orallo, J., and Ferri, C. (2019). The teaching size: computable teachers and learners for universal languages. *Machine Learning*, **108**(8-9), 1653–75.

Teso, S. and Kersting, K. (2019). Explanatory interactive machine learning, in *Proceedings of the 2019 AAAI/ACM Conference on AI, Ethics, and Society*, Honolulu. New York: ACM, 239–45.

Wachter, S., Mittelstadt, B., and Russell, C. (2017). Counterfactual explanations without opening the black box: automated decisions and the GDPR. *Harvard Journal of Law & Technology*, **31**(2), 2018.

Weitz, K., Hassan, T., Schmid, U. *et al.* (2019). Deep-learned faces of pain and emotions: Elucidating the differences of facial expressions with the help of explainable ai methods. *Technisches Messen*, **86**(7-8), 404–12.

Welsh, M. C. (1991). Rule-guided behavior and self-monitoring on the tower of hanoi disk-transfer task. *Cognitive Development*, **6**(1), 59–76.

Wernsdorfer, M. and Schmid, U. (2013). From streams of observations to knowledge-level productive predictions, in *Human Behavior Recognition Technologies: Intelligent Applications for Monitoring and Security*. Hershey: IGI Global, 268–81.

Wiener, Y., Hanneke, S., and El-Yaniv, R. (2015). A compression technique for analyzing disagreement-based active learning. *The Journal of Machine Learning Research*, **16**(1), 713–45.

Winston, P. H. (1975). Learning structural descriptions from examples, in P. Wilson, ed., *The Psychology of Computer Vision*, New York, NY: McGraw Hill, 157–210.

Zhang, Q., Cao, R., Shi, F. *et al.* (2018). Interpreting CNN knowledge via an explanatory graph, in *Proceedings of the AAAI Conference on Artificial Intelligence (Vol. 32, No. 1)*, New Orleans. Boston: AAAI.

18

Endowing machines with the expert human ability to select representations: why and how

Mateja Jamnik[1] and Peter Cheng[2]

[1] *University of Cambridge and* [2] *University of Sussex, UK*

18.1 Introduction

To achieve efficient human computer collaboration, computers need to be able to represent information in ways that humans can understand. Picking a good representation is critical for effective communication and human learning, especially on technical topics. To select representations appropriately, AI systems must have some understanding of how humans reason and comprehend the nature of representations. In this interdisciplinary research, we are developing the foundations for the analysis of representations for reasoning. Ultimately, our goal is to build AI systems that select representations intelligently, taking users' preferences and abilities into account.

Alternative representations commonly used in problem solving include formal mathematical notations (logics), graphs and other diagrams (data plots, networks), charts, tables and, of course, natural language. Particular knowledge domains typically have specialised variants of these representations. So, creating an AI system that is capable of selecting an effective representation for a given individual working on a particular class of problems is an ambitious goal. First, individuals vary greatly in their knowledge of the problem domain, and they will have differing degrees of familiarity in alternative representations. How can an automated system take these factors into account? Second, the process of picking a good representation, even for a *typical* problem solver, is a substantial challenge in itself. Problem solvers clearly have the meta-cognitive ability to recognise when they are struggling to find a solution and do things like trying an alternative strategy when they reach an impasse. Nevertheless, deliberately attempting to switch to an alternative representation is difficult for typical problem solvers; it seems this is an expert skill. Ordinary problem solvers often need instructors to tell them when to change representations and what representation to change to.

Mateja Jamnik and Peter Cheng, *Endowing machines with the expert human ability to select representations: why and how* In: *Human Like Machine Intelligence.* Edited by: Stephen Muggleton and Nick Charter, Oxford University Press.
DOI: 10.1093/oso/9780198862536.003.0018

Classic research in cognitive science highlights the challenge of changing representations. For example, in an experiment on analogy by Gick and Holyoak (Gick, 1989), participants were given a *source* story about how an army must storm a fortress but all the roads converging on the fortress are mined to stop the whole army passing along any one of them. To solve the problem, the general split up the army and sent each platoon along a separate road knowing that small groups would not trigger the mines. Shortly after the story, the participants in the experiment were given an isomorphic *target* problem about directing a strong x-ray beam through a patient to kill the tumour and asked to solve the problem of how to avoid damaging the healthy tissue around the tumour. The results of the experiment are normally interpreted as showing the power of analogy as 75% of participants given the source story successfully solved the target problem, but only 10% of the control group without the source story found the solution. However, in terms of representation switching, that is, the ability to extrapolate the isomorphic story to an analogous target problem, only 40% from the successful cohort (of 75%) were successful without additional hints, even though they had recently been told the story by the experimenter. The remaining 35% from the successful cohort (of 75%) were successful when told to try to apply the story (i.e., to switch representation). The implication of many such studies is that changing representation is hard.

The change of representation demanded in the analogy is relatively small, but for real problems finding an effective representation typically involves switching format, say from algebra to a diagram, or a table to a network. Such switches are far more demanding, both for problem solvers to adapt to the new representation, but most importantly, for our present goal to select the alternative representation in the first place. Cognitive problem solving theory (Newell and Simon, 1972) suggests that switching representation may itself be interpreted as a form of problem solving that involves changing the initial state, satisfying goal states, problem state expressions, finding operators to transform problem states, or a mixture of them all. This is why selecting alternative representations has, so far, been a task for instructors and domain experts.

In this chapter we explore how to give computers the ability to select effective representations for humans. One obvious benefit is that giving users representations that are better suited to their knowledge and levels of experience should enhance their ability to comprehend, solve problems and to learn about the target domain. There is another *human-like* computing benefit. Computer systems typically use some logical symbolic language, which may be efficient for the computer but is often inaccessible and thus a barrier for their interaction with humans. Thus, endowing a system with the ability to select a representation that is suited to its current human user may improve the communication between the system and the human. The system might translate its own inference steps and outputs from its logical symbolic language into the representation preferable for the user, and hence be able to provide explanations of its reasoning in a form that can be comprehended by the human.

The general hypothesis of our work is that *if we use foundational analysis of the user's expertise and the cognitive and formal properties of problems and representations in an AI system, then we can improve human interaction with it and their success at solving a task at hand.* Testing this hypothesis is ongoing work in the *rep2rep* project,[1] and here we describe some of its current contributions:[2]

- representation selection theory consisting of the language and measures to analyse representations;
- theory of cognitive properties for assessing the efficacy and suitability of a representation for particular users;
- computational algorithms and implementations for carrying out the analysis and assessing the suitability of a representation for a particular task and user.

In order to test that our theories of representations and computational models based on them are indeed more human-like and lead to humans being more successful in solving a task, we carried out preliminary empirical studies with encouraging results, but much more still needs to be done for evaluation. A practical application of this work in the domain of education and AI tutors is planned for future work.

The structure of this chapter is as follows. We start by exemplifying in Section 18.2 what changing a representation means. In Section 18.3 we discuss the benefits of changing representations that have been recognised in empirical studies. Then in Section 18.4 we explore more deeply the wide range of impacts that switching representations may have on human problem solving, and hence the complexity and challenge of selecting an effective representation. We explore in Section 18.5 how representations can be analysed computationally in terms of formal and cognitive properties, and how attributes of representations may relate across representations. Then, in Section 18.6 we demonstrate how this analysis can be automated in an intelligent system. Finally, in Section 18.7 we discuss how the automated representation choice based on the user model and the cognitive and formal properties could be applied in AI tutoring systems in order to personalise interaction and to improve users' abilities to solve problems.

18.2 Example of selecting a representation

To illustrate the approach to choosing representations in our framework, consider this *Birds* problem in probability:

> *One quarter of all animals are birds. Two thirds of all birds can fly. Half of all flying animals are birds. Birds have feathers. If X is an animal, what is the probability that it's not a bird and it cannot fly?*

[1] http://www.cl.cam.ac.uk/research/rep2rep

[2] Some of the work reported in this chapter has previously been published in (Raggi, Stapleton, Stockdill, Jamnik, Garcia Garcia and Cheng, 2020; Stockdill, Raggi, Jamnik, Garcia Garcia, Sutherland, Cheng and Sarkar, 2020; Raggi, Stockdill, Jamnik, Garcia Garcia, Sutherland and Cheng, 2020).

Here are three different ways one can go about solving this (see Figure 18.1):

(a) You could divide areas of a rectangle to represent parts of the animal population that can fly and parts that are birds.

(b) You could use contingency tables to enumerate in its cells all possible divisions of animals with relation to being birds or being able to fly.

(c) You could use formal Bayesian notation about conditional probability.

Which of these are effective representations for the problem? It depends; the first is probably best for school children; the last for more advanced mathematicians. Can this choice of appropriate representation be mechanised, and how? In our work we lay the foundations for new cognitive theories that would allow us to understand the relative benefits of different representations of problems and their solutions, including taking into account individual differences. We automate an appropriate choice of problem representation for both humans and machines to improve human-machine communication (see Sections 18.5 and 18.6).[3] But first, let us examine what the benefits of switching representations are and why doing so is hard.

(a)
Geometric representation—the solution is the area of the solid shaded region $\frac{3}{4} - \left(\frac{2}{3}\right)\left(\frac{1}{4}\right) = \frac{7}{12}$:

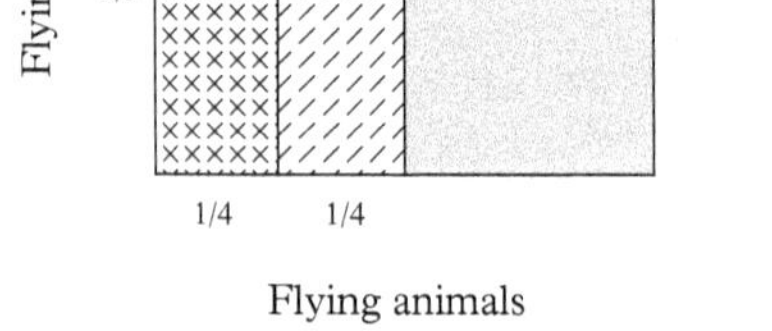

(b)
Contingency table representation – the solution is in the shaded cell:

	birds	non-birds	total
flying	(2/3)(1/4)	(2/3)(1/4)	
non-flying		3/4−(2/3)(1/4)	
total	1/4	3/4	1

(c) Bayesian representation:

$$\begin{aligned}\Pr(\bar{b} \cap \bar{f}) &= \Pr(\bar{b}) - \Pr(\bar{b} \cap f) \\ &= \Pr(\bar{b}) - \Pr(b \cap f) \\ &= (1 - \Pr(b)) - \Pr(f \mid b)\Pr(b) \\ &= \tfrac{3}{4} - \left(\tfrac{2}{3}\right)\left(\tfrac{1}{4}\right) = \tfrac{7}{12}\end{aligned}$$

Figure 18.1 *The* Birds *example.*

[3] It is worth pointing out that we are not devising new representations. Instead, we are introducing a new language and framework that enable us to describe and compare different representations with respect to the task and the user.

18.3 Benefits of switching representations

The three examples in the previous section illustrate the potential for alternative representation to provide varying benefits to different users. But do the benefits justify the effort needed for the development of automated systems to select representations for humans? We review the wide range of *epistemic* and *cognitive* factors from the cognitive and computational literature to show that switching from a poor representation to a good one does provide substantial benefits in reasoning and problem solving.

18.3.1 Epistemic benefits of switching representations

At the most fundamental level, alternative representations can differentially support problem solving through the particular information they encode and thus make available. This is not merely a matter of *what* information about the target domain is provided by a representation's expressions, but concerns *how* information is encoded. For example, here is a transitive inference problem: "A is to the left of B, A is to the left of C; what is the relation between B and C?". In capturing the premises in a sentential notation, such as "left-of(A,B)" and "left-of(A,C)", nothing is said about the relation between B and C. However, in this diagram "A C B" we easily see that C is to the left of B. The diagram naturally has greater specificity: we are forced to draw B somewhere relative to C. Using the diagram we can successfully answer the question, but whether this specificity is a benefit or not depends on whether the extra information in the diagram is appropriate for the task in the first place, and whether the extra information is uniquely determined or it needs to be split into cases. The correct answer could have been "not known". Nevertheless, the general point still holds that alternative representations may usefully encode more, or disadvantageously less, relevant domain information, or vice versa. The idea has been extensively discussed in the literature; for example, by Stenning and Oberlander (1995) and Shimojima (2015).

The information that can be directly encoded by a representation may improve problem solving performance of users. For example, for propositional calculus, diagrammatic representations are an alternative to the traditional formulae and truth table representations, which are commonly used for teaching. A review of Frege, Wittegnstein, Pierce and Gardner's representations by Cheng (2020) shows how their different formats substantially impact the accessibility of information: not just the content of propositional relations but also the form of inference rules and the strategic information needed to manage proof making. Cheng (2020) designed a novel diagrammatic representation for propositional calculus that makes directly accessible these multiple levels of information that are not easily available in the sentential and other representations (Figure 18.2a). Furthermore, Cheng (2011*a*) designed a novel representation for probability theory that simultaneously encodes information that would normally be distributed over algebraic expressions, set notations, contingency tables and tree diagrams (Figure 18.2b). Experimental and classroom studies have shown the beneficial impact on user performance of representations that directly express the full range of the relevant domain information,

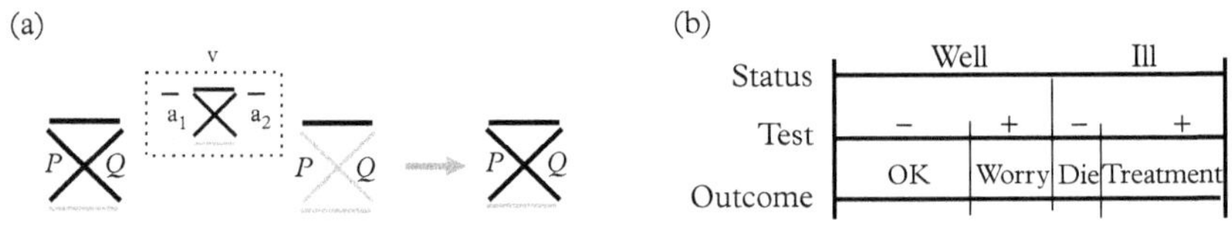

A Truth Diagram for the inference (P ∨ Q) ∨ (P ∧ Q) → P ∨ Q.

A Probability Space Diagram for disease diagnosis.

Figure 18.2 *Novel accessible diagrammatic representations.*

in particular, that these novel representations substantially improve students' problem solving and learning (e.g., Cheng (2002), Cheng and Shipstone (2003)).

The key point here is that different representations can substantially impact how effectively users solve problems by providing them with the information that is needed in a readily accessible form. In this regard, a poor representation will force the problem solver to spawn sub-problems in order to obtain the missing information, either by pursuing extra chains of reasoning or by bearing the cost of switching among multiple representations, when domain knowledge is dispersed.

18.3.2 Cognitive benefits of switching representations

The impact of alternative representations is attributable to cognitive processing (rather than informational differences) when the representations being compared are informationally equivalent: specifically, when all the information held by one representation can also be inferred from the information in the other representation (Larkin and Simon, 1987). Even when two representations are informationally equivalent, the user's ease of problem solving can dramatically vary. For example, in isomorphic, informationally equivalent, versions of the Tower of Hanoi puzzle, the difficulty of reaching the goal can vary by a factor of 18 (Kotovsky, Hayes and Simon, 1985) when operators that move disks between pegs are replaced by operators that transform the size of objects. The seminal work of Larkin and Simon (1987) showed (using computational models of a pulley system problem) that diagrams are (sometimes) superior to sentential representations. Namely, they permit the simple perceptual matching of information to inference rules and the efficient use of locational indexing of information that substantially reduces the need to search deliberately for relevant information. In a study using the same pulley system representations, Cheng (2004) found that solutions using diagrams were arrived at six times faster than with sentential representations.

Another aspect differentiating good from poor representations concerns how coherently elements of the representations encode domain concepts. Many have recognised the importance of isomorphic mappings between domain concepts and the tokens standing for them (Gurr, 1998; Moody, 2009; Barwise and Etchemendy, 1995). One-to-one mappings are easier for users than one-to-many mappings. Cheng (2011*b*) goes further contending that effective representations for rich domains should maintain such isomorphic mappings at successively higher conceptual and symbolic levels.

Cognitive differences may not only impact immediate problem solving but might determine the long term survival and propagation of representations. Numeration systems are informationally equivalent, but Zhang and Norman (1995) demonstrate that different ways of encoding number information can influence how simply numbers are expressed and also how much they can cause unnecessary work in computations. Arguably, the Hindu-Arabic number system's particular properties, its medium size base and separation of number and power information into different perceptual dimensions, are responsible for its world wide adoption.

At the opposite end of the spectrum, significant cognitive differences can occur with relatively small variants in representations. For instance, Peebles and Cheng (2003) investigated how reading Cartesian graphs is affected when data for two dependent measures varying with an independent measure is plotted in different ways (e.g., oil and coal prices over time): either as a conventional "function" graph with the two dependent variables on the y-axis, or as an unfamiliar "parametric" graph with both dependent variables plotted on each of the two primary axes and the independent variable plotted as data points along the curve in the graph. Despite participants' lack of familiarity with the parametric graphs, they responded significantly more quickly than with the conventional graph, without sacrificing accuracy. Thus, it is quite feasible to improve task performance not just by changing the format of the representations, but by changing the particular way in which different types of variables are plotted.

This is but a small sampling of a diverse literature (for more see the review by Hegarty (2011)). The key point here is that different representations can and do substantially impact on how effectively users solve problems in cognitive terms, potentially by over an order of magnitude. Myriad of factors contribute to explanations of why alternative representations may enhance or hinder problem solving, which presents a major challenge to our goal of building an automated system for representation selection. We next explore the reasons why selecting a good representation is challenging.

18.4 Why selecting a good representation is hard

The examples in Section 18.2 showed that alternative representations require different knowledge about the formats (structures) of the representations, and thus support distinct problem solution strategies. In Section 18.3 we saw the large array of benefits that can flow from switching representations. But selecting representations is intrinsically challenging: simply, the myriad of ways and the extent to which representations directly impact problem solving means that there is no small set of core factors that determine the relative efficacy of representations. However, we contend that it may nevertheless be feasible to systematically define a space that organises the wealth of cognitive factors to serve as a cognitive analysis framework.

18.4.1 Representational and cognitive complexity

Why is using a cognitive theory to selected an effective representation so challenging? In general terms, the number of different aspects of cognition that conceivably have an impact are manifold and each of these aspects are themselves complex (Markman, 1999; Anderson, 2000; Stillings, Weisler, Chase, Feinstein, Garfield and Rissland, 1995). Let us consider two examples.

Despite the growing renaissance in cognitive neuroscience, how the brain physically implements mental representations remains largely unknown. So, cognitive scientists still find it essential to formulate accounts in terms of information structures and processes such as: declarative propositions stored as semantic networks with spreading activation (Markman, 1999; Anderson, 2000); mental imagery that exploits much of the functionality of the visual perceptual system but in the mind's eye (Finke, 1989); procedural information encoded as condition-action rules (Klahr, Langley and Neches, 1987; Anderson, 2007); hierarchies of concepts stored in discrimination networks to aggregate related concepts and separate dissimilar ones (Gobet, Lane, Croker, Cheng, Jones, Oliver and Pine, 2001); and, even heterogenous structures that tightly coordinate declarative, procedural and diagrammatic information (Koedinger and Anderson, 1990). Accounts of how problem solvers work with a representation must invoke several of these internal mental representations; for example, the table in Figure 18.1b includes declarative propositions, rules for computing cell values, and a hierarchical conceptual scheme to coordinate the rows and columns. Each representation imposes different cognitive demands, which interact, so it is by no means simple to assess the overall cognitive cost of using each representation.

The idea that the demands placed on working memory (WM) could be a key to assess task difficulty is wide spread and it is tempting to apply it to evaluating alternative representations. For instance, how much does the use of a certain representation tend to breach the user's WM capacity? However, the notation of WM capacity is tricky. Although the famous magic number of 7 ± 2 items (Miller, 1956) is widely known, the modern estimate of approximately four chunks (Cowan, 2001) is a more appropriate general capacity limit for typical tasks. Nevertheless, finer grained models of the sub-processes of the cognitive architecture suggest that WM capacity should not be considered as one fixed-limit general store, but as a collection of lower level mechanisms each possessing their own small capacity buffer (Anderson, 2007, Anderson, 2000). Thus, the prospects of developing a generic account based on WM loading is remote.

For other aspects, a similar tale of complex interactions between representations and cognitive structures and processes can be told. Novices and experts mentally represent information in very different ways (Koedinger and Anderson, 1990). As real world cognition involves continuing cycles of perception, internal cognition, and motor output, we can use the external environment to off-load information from WM or to replace laborious mental reasoning with perceptual inferences (Larkin and Simon, 1987). We can use different strategies on the same problem, which may be knowledge-rich or knowledge-lean (Newell and Simon, 1972). Hierarchies of goals are used to solve problems but the organisation of goals can vary from person to person and from representation to representation (Stillings, Weisler, Chase, Feinstein, Garfield and Rissland, 1995). In sum, there is no doubt that developing a method to predict the efficacy of alternative representation is a substantial endeavour.

18.4.2 Cognitive framework

Although the difficulty of selecting effective representations is revealed by studies in cognitive science, the discipline has matured to the point where it now provides a reasonable map covering the terrain for comparing representations that is to be explored (albeit incomplete and with connections between areas still sketchy). This map combines two fundamental dimensions considered in cognition.

The first is a dimension ranging across the size of cognitive objects that encode meaning, or in other words, a granularity scale of representations. One can think of it as a decision tree, where the scale spans a hierarchy of levels from symbols at the leaves to whole systems of representations at the trunk. The branches in between are constituted by compound cognitive forms such as expressions, chunks and schemas (Stillings, Weisler, Chase, Feinstein, Garfield and Rissland, 1995; Gobet, Lane, Croker, Cheng, Jones, Oliver and Pine, 2001; Schank, 1982).

The second dimension is time. Newell (1990) and Anderson (2002) both identify multiple temporal levels at which cognitive processes operate, ranging from 100 milliseconds to years, for instance, from the time to retrieve a fact from memory to the time required to acquire expertise. Both authors recognise the relatively strong interactions between processes at a particular characteristic time scale, and relatively weak interactions between different time scales. Thus, cognitive processes with durations differing by an order of magnitude may be treated as nearly independent for the sake of analysis, although short processes will cumulatively impact long processes.

Our novel framework, composed of the representational granularity dimension and a temporal dimension, will be elaborated in the following section.

18.5 Describing representations: *rep2rep*

To endow a machine with the ability to select representations, it must first be able to describe them to then analyse them for ranking and selection. The analysis explores if a representation is informationally adequate to express a particular problem, before we even consider whether it will be effective for human users.

We devised a framework, within which different representations can be described and analysed for their suitability.[4] This analysis is based on two main measures: *informational suitability* and *cognitive costs* (Raggi, Stapleton, Stockdill, Jamnik, Garcia Garcia and Cheng, 2020). Informational suitability is described in terms of *formal properties* of a representation, whereas cognitive cost is described in terms of *cognitive properties* (based on our two dimensional—spatial and temporal—cognitive map) that may be

[4] The *rep2rep* framework is a developing project, so formalisations, mechanisms and implementations are evolving. Current *rep2rep* implementation can be found here: https://github.com/rep2rep/robin.

related to formal properties and are crucially dependent also on the user profile. We formalise and describe the implementation of informational suitability and cognitive cost measures in Section 18.6. But first, we introduce the framework components for describing representations.

We distinguish between *cognitive* and *formal* properties of a representation, in an approach that radically, but systematically, reconfigures previously descriptive accounts of the nature of representations and notations (Moody, 2009; Hegarty, 2011; Engelhardt and Richards, 2018). We use this to devise methods for measuring competency in alternative representation use, and also to engineer a system to automatically select representations. *Cognitive properties* characterise cognitive processes demanded of a particular representation (e.g., problem state space characteristics; applicable state space search methods; attention demands of recognition; inference operator complexity (Cheng, Lowe and Scaife, 2001; Cheng, 2016)). *Formal properties* characterise the nature of the content of the representation domain (e.g., operation types like associative or commutative, symmetries, coordinate systems, quantity or measurement scales) (Raggi, Stapleton, Stockdill, Jamnik, Garcia Garcia and Cheng, 2020).

18.5.1 A description language for representations

The language for describing representations in terms of their properties is general as it must be able to deal with very diverse objects of representations. For example, in the above *Birds* problem, the candidate representations include elements like natural language, formal notation, a geometric figure, and a table. Each has its own symbols, grammar and manipulation rules. Their differences yield a different cognitive cost, that is, the effort that is demanded of the user when working with a particular representation. For example, the simpler and fewer inferences in the Geometric representation will result in a less costly solution than in the say Bayesian representation.

Our language describes representations in terms of tokens, expressions, types, laws, tactics and patterns. Each of these can have attributes which specify records associated with them. The attributes encode the informational content of the problem and a representation. Tokens are the atomic symbols from which expressions are built. Types classify expressions and tokens. Tactics specify how to manipulate the representation, and laws determine rules around how tactics can work. Patterns specify the structure of expressions using types, tokens, and attribute holes. For example, if a representation uses a token + four times, we can describe this as:

$$\begin{aligned}\text{token} + : \{&\text{type} := \texttt{real} \times \texttt{real} \to \texttt{real},\\ &\text{occurrences} := 4\}\end{aligned}$$

Moreover, the pattern for expressing conditional probability $\Pr(\_|\_) = \_$ can be described as:

$$\begin{aligned}\text{pattern } \texttt{CP} : \{&\text{type} := \texttt{formula},\\ &\text{holes} := [\texttt{event}^2, \texttt{real}],\\ &\text{tokens} := [\Pr, |, =, (,)]\}.\end{aligned}$$

18.5.2 Importance

Some properties of a representation are more important than others, and some may even be irrelevant (noise) and need to be ignored in the analysis of suitability of representation for solving a problem.[5] We use the notion of *importance* to express this. Clearly, the importance is strictly relevant to the task, so we express it only when describing a problem (like the *Birds* problem above) in a particular representation (e.g., in Bayesian representation). Importance is defined as a function from the properties to the interval ranging between 0 and 1, where 0 is noise and 1 denotes a maximally informationally relevant property. For example, the token *Pr* for the Bayesian representation of the *Birds* problem is important. Assigning importance is like finding good heuristics – in our framework, the *domain expert* who is setting up our framework for deployment assigns these values. In the future, we will explore if there is a principled approach to assigning these values, and if these importance parameters can be generated automatically by analysing a sufficiently large set of problems. We propose to use the theory of observational advantages of representations that we developed elsewhere (Stapleton, Jamnik and Shimojima, 2017; Stapleton, Shimojima and Jamnik, 2018) as a starting point.

18.5.3 Correspondences

We use the notion of *correspondences* to encode informational links between different representations. For example, in the *Birds* problem above, there is a correspondence between the areas used in the Geometric representation and the probability used in the Bayesian representation. These links can be, for example, analogies between representations or structures that are preserved through transformations. We formalise correspondences probabilistically which also allows us to inherit some provable consequences such as reversability of a property (if property a is related to property b, then b is related to a), or composability (if property a is related to property b, and b is related to c, then a is related to c). Similarly to the importance parameter of a property, some correspondences are stronger than others, for example, a token "intersection" in the Natural language representation strongly corresponds to $\cap$ in the Bayesian representation. We use the notion of *strength* to capture this, so a correspondence is a triple $\langle a, b, s\rangle$ where a and b are properties and s is the strength of their correspondence. Similarly to importance, we leave it with the domain expert to assign its value. In addition, we have devised algorithms that automatically generate some correspondences and compute their strengths probabilistically, which eases the load on the domain expert (Stockdill, Raggi, Jamnik, Garcia Garcia, Sutherland, Cheng and Sarkar, 2020).

[5] One could imagine that a property of a representation may be detrimental if the representation is used for a particular problem solving task (e.g., loss of accuracy, correctness), so its importance may need to be negatively accounted for in the suitability analysis. Our framework does not currently provide this feature—it is left for future work.

To illustrate, let us formalise the correspondence between areas in the Geometric representation and events in the Bayesian representation. Specifically, there is a strong correspondence between the type 'region' and the type 'event':

$$\langle \text{type region, type event}, 1.0 \rangle.$$

Immediately we know the reverse is also a correspondence, although less strong:[6]

$$\langle \text{type event, type region}, 0.8 \rangle.$$

Similarly, we might consider how intersection is represented in a Geometric representation – we intersect areas by *overlapping* them:

$$\langle \text{token } \cap, \text{pattern overlap}, 0.8 \rangle.$$

Consider again the statement of the *Birds* problem in the Natural language. We see tokens such as 'all', and 'and'. These have clear analogues in the Bayesian representation with which we can make correspondences (where Ω is the probability space, i.e., the universe).

$$\langle \text{token all, token } \Omega, 0.9 \rangle$$
$$\langle \text{token and, token } \cap, 1.0 \rangle$$

The more strong correspondences occur between representations, the better a potential representation is as a candidate for re-representation.

Note that we have a correspondence between the Natural language and the Geometric representation by composition: we know that 'and' corresponds to $\cap$, and that $\cap$ corresponds to overlapping, and thus we can derive the correspondence from 'and' to overlapping:

$$\frac{\langle \text{token and, token } \cap, 1.0 \rangle \quad \langle \text{token } \cap, \text{pattern overlap}, 0.8 \rangle}{\langle \text{token and, pattern overlap}, 0.8 \rangle}$$

The value for the strength of the resulting correspondence can be computed, here by multiplication of the originating correspondences (or probabilistically from a dataset of co-occurring properties, if such a dataset is available).

Further details of the formalisation and implementation of correspondences, their strength, automatic generation and use in the analysis of the suitability of a representation can be found in (Stockdill, Raggi, Jamnik, Garcia Garcia, Sutherland, Cheng and Sarkar, 2020).

[6] The strength is reduced because there are many ways to encode events in the Geometric representation. An event might become a line segment, or a point in space.

18.5.4 Formal properties for assessing informational suitability

We can now give an example of how some of the representations from our *Birds* problem can be described. Notice that we need to be able to describe *representational systems* in general. In addition, we must be able to express the problem (i.e., the *question*), for example, our *Birds* problem, in these representational systems. For the problem, the notion of importance is used too.

We catalogue formal properties using templates of attributes that (currently) the domain expert who wants to use our framework assigns values to. Table 18.1 gives snippets from a formal property catalogue for the *Birds* problem stated in the Natural language representation. The colours code the importance of the property relative to the information content (top to bottom in decreasing importance). Table 18.2 gives snippets of the catalogue of formal properties for the Bayesian representational system (used in the solution in Figure 18.1c).

It is important to note that our representation language does not provide a complete formal description of a representation as, for example, formal logics do. For many representations, this is not possible. Instead, our language describes representations with sufficient specificity to be able to analyse them, draw analogies between them, and to assess their informational suitability and cognitive cost (see Section 18.6).

Table 18.1 *Formal properties of the* Birds *problem in the Natural language representation (note colour for* importance *parameter).*

Kind	Value
error allowed	0
answer type	ratio
tokens	probability : {occurrences := 1}, and :{occurrences := 1}, not : {occurrences := 1}
types	ratio, class
patterns	Class-ratio : { holes := [class ⇒ 2, ratio ⇒ 1], tokens := [of, are], occurrences := 3, token-registration := 1}, Class-probability : { holes := [class ⇒ 2], tokens := [probability, is], occurrences := 1, token-registration := 1}
laws	Bayes' theorem, law of total probability, unit measure, additive inverse, ...
tactics	re-represent : {occurrences := 1,inference-type := transformation}
tokens	one : {occurrences := 1}, quarter :{occurrences := 1}, all : {occurrences := 3}, animals :{occurrences := 2}, birds : {occurrences := 4}, ...
mode	sentential
tokens	feathers : {occurrences := 1}

Table 18.2 *A section of formal properties for Bayesian representational system.*

Kind	Value
mode	sentential
rigorous	TRUE
types	real, event
tokens	Ω : {type := event }, $\emptyset$: {type := event }, 0 : {type := real }, 1 : {type := real }, = : {type := $\alpha \times \alpha \rightarrow$ formula}, + : {type := real $\times$ real $\rightarrow$ real}, $-$: {type := real $\times$ real $\rightarrow$ real}, $\times$: {type := real $\times$ real $\rightarrow$ real}, $\div$: {type := real $\times$ real $\rightarrow$ real}, $\cup$: {type := event $\times$ event $\rightarrow$ event}, $\cap$: {type := event $\times$ event $\rightarrow$ event}, $\setminus$: {type := event $\times$ event $\rightarrow$ event}, $^{-}$: {type := event $\rightarrow$ event}, Pr : {type := event $\times$ event $\rightarrow$ real}, $\vert$: {type := delimiter }
patterns	Conditional-probability : { holes := [event $\Rightarrow$ 2, real $\Rightarrow$ 1], tokens := [Pr, $\vert$, =] }, Simple-probability : { holes := [event $\Rightarrow$ 1, real $\Rightarrow$ 1], tokens := [Pr, =] }, Joint-probability : { holes := [event $\Rightarrow$ 2, real $\Rightarrow$ 1], tokens := [$\cap$, Pr, =] }, Equality-chain : { holes := [real $\Rightarrow$ $O(n)$], tokens := [=] }
laws	Bayes' theorem, law of total probability, non-negative probability, unit-measure, sigma-additivity, commutativity, ...
tactics	rewrite, apply lemma, arithmetic calculation

18.5.5 Cognitive properties for assessing cognitive cost

Every human user of a system has a different background, expertise and preferences in terms of their ability to solve a problem. In our approach we account for these cognitive aspects through assessing representations in terms of their cognitive properties. Since these are not general characteristics of a representational system but are specific to the problem that the user wants to solve, we attribute cognitive properties to specific representations of specific problems. We then assess the cost of these cognitive properties with respect to particular users. This enables personalisation in terms of adapting representations to users.

Picking up the cognitive map sketched in Section 18.4, we propose 9 key cognitive properties to populate the space defined by the dimensions of *notation granularity* and *temporal scale* (see Table 18.3).

We now give a brief explanation of these 9 key cognitive properties. We computationally modelled these properties and implemented algorithms for calculating the *cognitive cost* that they entail (see Section 18.6).

Registration: refers to the process of identifying some object in a representation as a token (or expression), and acknowledging their existence and location. The registration of tokens depends on users ability to observe their role in the pattern. Analogously, the registration of patterns depends on the *mode* (which describes a higher level of notational granularity) of the representation. Patterns are given an attribute *token registration*,

Table 18.3 *Cognitive properties organised according to spatial (columns) and temporal (rows) aspects of cognitive processing.*

	token	expression	whole
registration	registration		subRS variety
semantic encoding	number of types		
	concept mapping		
inference	quantity scale	expression complexity	
		inference type	branching factor
solution			solution depth

which assigns *icon*, *notation index* or *search* to the tokens used by the pattern, with their corresponding individual costs in the increasing order as implied by (Larkin and Simon, 1987).

Number of types: refers to processing semantics, that is, identifying the types of tokens and expressions in a representation. A larger variety in the *types* of tokens and expressions means a higher semantic processing cost for the user.

Concept-mapping: refers to the mapping of tokens and expressions to their corresponding concepts in the user's internal representation (Zhang, 1997). Its cost is associated with the accumulated effort of processing various defects of representations: specifically, *excess* (symbol that does not match to an important concept), *redundancy* (two symbols for the same concept), *deficit* (a concept with no symbol to represent it), or *overload* (one symbol for multiple concepts). These incur cognitive cost increasing in the order implied by (Gurr, 1998; Moody, 2009). The total cost is a weighted sum of these individual defect costs.

Quantity scale: refers to a well documented scale hierarchy, specifically, *nominal*, *ordinal*, *interval* or *ratio*, all of which effect cognitive costs (Zhang and Norman, 1995). These are associated with arithmetic operations, so we use the correspondences to the Arithmetic representation to estimate the cost.

Expression complexity: encodes the assumption in cognitive science that more complex expressions demand greater processing resources, and that complexity rises with increasing the breadth and depth of expressions (all else being equal). Our algorithm takes each pattern and instantiates its holes recursively with other patterns or tokens of the appropriate type until no holes remain uninstantiated. This results in an encoding of *parse trees* for expressions. Thus, we can generate, for every pattern, a sample of possible expression trees that satisfy it. The average number of nodes in each tree gives us a measure of the complexity of such a pattern.

Inference type: refers to the difficulty intrinsic to applying tactics. We assume an attribute *inference type* for each tactic, valued as *assign*, *match*, *substitute*, *calculate*, or *transform*. These classes are associated with cost that is increasing in the order listed here. Note that we assume that we have a specific solution for a problem in the relevant representation.

subRS variety: considers heterogeneity of a representation, that is, if it consists in part from a sub-representation that could be considered an independent representation in its own right (e.g., a table where the cell values are arithmetic formulae has two subRSs). This heterogeneity incurs a heavy cognitive cost (Van Someren, Reimann, Boshuizen *et al.*, 1998): we estimate it from the number of modes.

Branching factor: refers to the breadth of possible manipulations (estimated from tactics and their attributes, like branching factor).

Solution depth: is simply the total number of tactic uses (from the tactic attribute *uses*). Note that we assume that we have a specific solution for a problem in the relevant representation.

The challenge of assessing the cognitive cost of a representation is greater than just taking cognitive properties into account, because individuals vary in their degree of familiarity and hence proficiency in using particular representations. To adjust cognitive costs from a typical user to individual's abilities, we are devising a small but diverse set of user profiling tests—this is currently under development. The measures extracted from these profiles should enable us to scale the level of contributions of each cognitive property to the overall cost of a representational system for an individual. This can give us a basis for automating the representation selection that is sensitive to individual's cognitive differences, which we address next.

18.6 Automated analysis and ranking of representations

Within the *rep2rep* framework we describe the representations and problems with the language of formal and cognitive properties outlined in Section 18.5. In addition, we built algorithms that automatically analyse these encodings for a given problem (like the one in Table 18.1) with respect to candidate representational systems (like the one in Table 18.2) in order to rank the representations, and ultimately suggest the most appropriate one. This analysis is based on the evaluation of the informational suitability and cognitive cost.

Informational suitability. We define *informational suitability* in terms of correspondences between the formal properties of the problem q in the given representation (e.g., the *Birds* problem in the Natural language representation) and the formal properties of a candidate alternative representation r (e.g., the Bayesian representation). This is modulated by the *importance* score of the property and the *strength* s of the correspondence. If there is no correspondence between the properties of the original representation of q and the candidate representation r, this means that the candidate representation r cannot convey the information carried in the original representation of the problem q. Thus, the algorithm for computing informational suitability first takes the original representation of the problem q, identifies the corresponding properties of the alternative candidate representation r, and then sums the strengths of all corresponding properties multiplied by their importance.

The set C of identified correspondences is computed based on minimally redundant and maximally covering properties. Minimally redundant properties means that for any two pairs of corresponding properties, they must be independent. This ensures that redundancy of informationally similar properties is avoided. Maximally covering properties in the set C are those important properties that are the most information carrying to express the problem q.

More formally, given a minimally redundant and maximally covering set of corresponding properties C for a problem q and a candidate representation r, the informational suitability IS can be computed as:

$$IS(q,r) = \sum_{\langle p_1,p_2,s\rangle \in C} s \cdot \text{importance}_q(p_1) \tag{18.1}$$

where p_1 is a property of q, p_2 is a corresponding property of r, s is a strength of that correspondence $\langle p_1,p_2,s\rangle$, and $\text{importance}_q(p_1)$ is the importance of property p_1 for q.

In the *Birds* example given in the Natural language representation, we computed the IS for Bayesian representation, Contingency tables representation, Geometric representation and Natural language representation. We compared these automatically produced rankings with the rankings that human experts would give for a very similar probability example (the *Medical* problem) in an online survey, and the results are comparable (higher value is better) – see Table 18.4.

Cognitive cost. Informational suitability only takes into account how well a representation can express a problem in terms of information theory. It is the *cognitive cost* that takes a particular user into account in terms of cognitive properties described in Section 18.5. The total cognitive cost Cost is defined as:

$$\text{Cost}(q,u) = \sum_{p} c_p(u) \cdot \text{norm}_p(\text{cost}_p(q,u)) \tag{18.2}$$

where:

- q is the problem;
- u is the user expertise parameter where we assign each user a value $0<u<1$ where $0<u<1/3$ represents a *novice*, $1/3<u<2/3$ an *average* user, and $2/3<u<1$ an *expert*;
- p is a particular cognitive property listed in Section 18.5;
- $\text{cost}_p(q,u)$ is a cost of an individual cognitive property p; it encodes values for the attributes that determine the cost of that property ordered as described above; for example, for registration, according to the literature and as explained above, we need to give an increasing cost to icons, notation index and search, correspondingly,

Table 18.4 *Informational suitability was computed for the* Medical *and* Birds *example (including Probability trees and Set algebra representations). The original statement of both problems was given in Bayesian representation (bay) and for the* Birds *problem also in Natural language (nl) representation. Human experts were surveyed about the use of representations for the Medical problem. Here is one-tailed Pearson's correlation* $r = 0.89$ $(p = 0.053)$. *The r-value indicates that there is a strong positive correlation between the scores the algorithm assigns, and the scores the experts assign. The p-value gives a good indication that this is of statistical significance.*

Informational Suitability				
	Medical		Birds	
	survey	computed (bay)	computed (nl)	computed (bay)
Bayesian	6	17.4	12.6	18.9
Geometric	4.8	11.4	12.8	12.2
Contingency	4.9	8.38	8.5	9.4
NatLang	3.5	6.9	11.9	9.0
Pr-trees	–	9.04	5.7	9.5
Set Algebra	–	10.3	4.4	12.8

but what the exact value should be is unclear; in the future this value could be empirically informed; for now we set them based on the cognitive science literature;

- $c_p(u)$ is a moderating factor for cost of a property according to the expertise of the user: higher-granularity property costs are inflated for novices and deflated for experts;
- norm_p is a function that normalises and makes the scales of each property comparable; this too should be empirically informed, but for now, we pick these provisional values.

Notice that, in principle, to personalise recommendations to individuals, the user can be profiled, and all of the parameter values above can be adjusted using this profile. Based on the literature and expertise, we used provisional values for our prototype *rep2rep* framework and based them for a typical average user ($u = 0.5$). Currently, we devised a profiling test specifically for the quantity scale property. In future work, we plan to design profiling methods for the other properties to inform the values of these parameters. Table 18.5 gives the results of computing the total cognitive cost of a representation, and similarly to the IS score, these are in line with what the experts reported in our survey (lower is better).

Table 18.5 *Computed cognitive costs for* $u = 0.5$ *(average user). For this u, we make* $c_p(u) = 1$ *for all properties: tr = token registration, er = expression registration, tt = number of token types, et = number of expression types, cm = concept-mapping, qs = quantity scale, ec = expression complexity, it = inference type, sr = subRS variety, bf = branching factor, sd = solution depth. Also,* η_p *normalises the value to a number between 0 and 100* ($\eta_p(x) = 100(x - \min_p)/(\max_p - \min_p)$)*, while a constant scales it according to the p's proposed total effect on cognitive cost. The rest of the columns are representational systems.*

	$\text{norm}_p(x)$	nl	bay	geo	cont	tree	s-alg	eul
tr	$0.5 \cdot \eta_{tr}(x)$	6.6	0	20.4	29.8	50	14.2	14.1
er	$0.5 \cdot \eta_{er}(x)$	0.8	0	0	2.2	50	1.3	0.2
tt	$1 \cdot \eta_{tt}(x)$	27.8	88.9	27.8	0	88.9	11.1	100
et	$1 \cdot \eta_{et}(x)$	4.2	12.5	45.8	58.3	100	0	75
qs	$1 \cdot \eta_{qs}(x)$	10.7	48.9	100	10.3	16.1	0	4
cm	$2 \cdot \eta_{cm}(x)$	108	0	200	198.4	181.7	124.7	143.8
ec	$2 \cdot \eta_{ec}(x)$	22.3	0	46.8	71.1	200	9.6	20.1
it	$2 \cdot \eta_{it}(x)$	200	25.5	17	0	45.7	25.5	20
sr	$4 \cdot \eta_{sr}(x)$	0	0	0	400	0	0	400
bf	$4 \cdot \eta_{bf}(x)$	260	233.4	216.5	400	0	253.3	267.4
sd	$4 \cdot \eta_{sd}(x)$	260	369.9	0	177	223	369.9	400
	total	81.8	70.8	61.3	122.5	86.9	73.6	131.3
	rank	4	2	1	6	5	3	7

18.7 Applications and future directions

In this work we are laying the foundations for a new class of adaptive technology that aims to automatically select problem solving representations that are suited both to individuals and the particular class of problems that they wish to solve. The answer to the "why" of this endeavour is that representations are fundamental to human cognition and that good representations are intrinsically hard for humans to pick for themselves without expert instructor assistance. The answer to the "how" of the endeavour is to decompose the problem at multiple levels. The highest level distinguishes informational or epistemic requirements of effective representations and the cognitive requirements concerning how humans use representations. On the next level, we are addressing the informational component in terms of formal properties of representations and have proposed algorithms to assess the sufficiency of competing representations. We address the human user component in terms of cognitive properties, at different spatial and temporal scales. Our pilot work has produced encouraging results, where the suitability values computed by our algorithm rank representations in line with how the human experts would rank them. We are currently conducting a more extensive empirical study with teachers as experts. To take into account competence differences, we are developing profiling tests of individuals' abilities relating to specific cognitive properties of representations, targeting quantity scales first given their fundamental role in representations. It remains future work to test how human users' ability to solve tasks is affected by using representations recommended by our system.

Some representations are more effective for humans and some for machines. Indeed, these could come into conflict, for example, when a machine-centric representation is opaque for humans, or when a human-centric representation is computationally inefficient for machines. Should a representation that our system recommends for a user be the one that the tool uses? Our approach is deliberately focused on benefits for the human user and on developing a system to select representations that are suited to users' individual knowledge and abilities. We envisage that such a system could be used as an interface for any AI tool to enable adaptability and personalisation. If a tool uses a machine-centric representation, then our human-centric recommended representation could serve as a target representation that the interface of the tool could translate its output into. But these are questions for future work.

There are many areas of applications of this work. One area is education, with the potential for AI tutoring systems to be adaptive and personalised for individuals in terms of their level of experience with different representations. Switching representation is an instructional strategy that has received little attention in AI and Education (with one exception (Cox and Brna, 1995)), even though effective representation choice has been acknowledged in the field for decades (Kaput, 1992). Intelligent tutoring systems can achieve learning gains of about 0.5 SDs over conventional instruction using techniques like student modelling to drive tutoring actions (Ma, Adesope, Nesbit and Liu, 2014), but representation switching has far greater potential (e.g., (Cheng, 1999; Cheng, 2002; Cheng, 2011*a*) showed a factor of two in learning gains). For instruction specifically about subject matter content, the system might recommend a familiar representation to the student. For (meta–)instruction about representations themselves, the system could pick representations that are stretching but not beyond the potential capability of the learner. Our tutoring system will host a library of alternative representations and its main intervention will be to recommend representations to the student. Another application area is to tailor the explanations given by knowledge-based or decision-support systems by selecting a representation to meet the user's level of sophistication. For example, suppose the probability problem in Figure 18.1 concerned the interpretation of a test outcome knowing that the base proportion of non flying birds differs for a particular sub-population. We envisage a system would administer a few key representation profiling tests to pick which of the three formats in Figure 18.1, or others, to show to the problem solver, and to trade-off what information to provide that has the best likelihood of being correctly comprehended.

Currently, most mechanised problem solving systems have a single fixed representation available to them, typically logics (Kovács and Voronkov, 2013; Harrison, 2009) and diagrams (Jamnik, Bundy and Green, 1999; Barwise and Etchemendy, 1994). There are a few exceptions like Openbox (Barker-Plummer, Etchemendy, Liu, Murray and Swoboda, 2008) and MixR (Urbas and Jamnik, 2014) that implement multiple representations, but these are either deprecated or not targeted at tutoring but formal reasoning. In machine learning, representation learning has become a rich area of research (Bengio, Courville and Vincent, 2013). However, the gap between the computer processing and the user understanding seems to be increasing.

To what extent has our hypothesis been confirmed? Our framework introduces a language and a number of measures that enable the analysis of diverse formal and informal representations. The user's level of expertise in using and the cognitive load of employing a particular representation are captured by our cognitive properties and the formalisation of cognitive costs. We implemented these concepts in a system that can automatically carry out the suitability analysis. We carried out pilot studies that give us confidence that our system produces results in line with those of human experts. There is much to be further developed including the theoretical characteristics of our framework, further empirical evaluations of the effect of using our system's recommendation on human task solving ability, and the application of this work in an AI tutoring environment.

Why is this work important? AI engines that can choose representations of problems in a similar way to humans will be an essential component of human-like computers. They will give machines a powerful ability to adapt representations so that they are better suited to the particular preferences or abilities of the human user. This is especially important when the machine must give users intelligible explanations about its reasoning. Human choices of representations give us clues of their problem solving approaches. This will aid the construction of the world model that reflects that of a human. Consequently, the machines will be able to not only better adapt to the individual user, but also better interpret instructions or information provided by humans. Ultimately, this will lead to machines collaborating with humans in more intuitive ways, as well as tutoring humans to develop their creative problem-solving skills.

Acknowledgements

We would like to thank our collaborators in the work reported here: Daniel Raggi, Grecia Garcia Garcia, Aaron Stockdill, Gem Stapleton and Holly Sutherland. This work was supported by the EPSRC grants EP/R030650/1, EP/R030642/1, EP/R030642/1, EP/T019034/1 and EP/T019603/1.

References

Anderson, J.R. (2000). *Learning and memory: an integrated approach* (2nd ed. edn). Wiley, New York, N.Y.

Anderson, J.R. (2002). Spanning seven orders of magnitude: A challenge for cognitive modeling. *Cognitive Science*, **26**(1), 85–112.

Anderson, J.R. (2007). *How can the human mind occur in the Physical Universe?* Oxford University Press, Oxford.

Barker-Plummer, D., Etchemendy, J., Liu, A., Murray, M., and Swoboda, N. (2008). Openproof-a flexible framework for heterogeneous reasoning. In *International Conference on Theory and Application of Diagrams*, pp. 347–349. Springer.

Barwise, J. and Etchemendy, J. (1994). *Hyperproof.* CSLI Press, Stanford, CA.

Barwise, J. and Etchemendy, J. (1995). Heterogenous logic. In *Diagrammatic Reasoning: Cognitive and Computational Perspectives* (ed. J. Glasgow, N. Narayanan, and B. Chandrasekaran), pp. 211–234. AAAI Press, Menlo Park, CA.

Bengio, Y., Courville, A., and Vincent, P. (2013). Representation learning: A review and new perspectives. *IEEE transactions on pattern analysis and machine intelligence*, **35**(8), 1798–1828.

Cheng, P.C.H. (1999). Unlocking conceptual learning in mathematics and science with effective representational systems. *Computers in Education*, **33**(2-3), 109–130.

Cheng, P.C.H. (2002). Electrifying diagrams for learning: principles for effective representational systems. *Cognitive Science*, **26**(6), 685–736.

Cheng, P.C.H. (2004). Why diagrams are (sometimes) six times easier than words: benefits beyond locational indexing. In *Diagrams* (ed. A. Blackwell, K. Marriot, and A. Shimojima), LNCS, pp. 242–254. Springer.

Cheng, P.C.H. (2011*a*). Probably good diagrams for learning: Representational epistemic re-codification of probability theory. *Topics in Cognitive Science*, **3**(3), 475–498.

Cheng, P.C.H. (2011*b*). Probably good diagrams for learning: representational epistemic recodification of probability theory. *Topics in Cognitive Science*, **3**(3), 475–498.

Cheng, P.C.H. (2016). What constitutes an effective representation? In *Diagrams* (9781 edn) (ed. M. Jamnik, Y. Uesaka, and S. Schwartz), LNAI, pp. 17–31. Springer.

Cheng, P.C.H. (2020). Truth diagrams versus extant notations for propositional logic. *Journal of Language, Logic and Information*, **29**, 121–161. Online.

Cheng, P.C.H., Lowe, R.K., and Scaife, M. (2001). Cognitive science approaches to diagrammatic representations. *Artificial Intelligence Review*, **15**(1-2), 79–94.

Cheng, P.C.H. and Shipstone, D.M. (2003). Supporting learning and promoting conceptual change with box and avow diagrams. part 2: Their impact on student learning at a-level. *International Journal of Science Education*, **25**(3), 291–305.

Cowan, N. (2001). The magical number 4 in short-term memory: A reconsideration of mental storage capacity. *Behavioral and Brain Science*, **24**(1), 87–114.

Cox, R.J. and Brna, P. (1995). Supporting the use of external representations in problem solving: the need for flexible learning environments. *International Journal of Artificial Intelligence in Education*, **6**, 239–302.

Engelhardt, Y. and Richards, C. (2018). A framework for analyzing and designing diagrams and graphics. In *Diagrams 2018* (ed. C. P., S. G., M. A., P.-K. S., and B. F.), Volume 10871, LNCS, pp. 201–209. Springer.

Finke, R.A. (1989). *Principles of Mental Imagery*. The MIT Press, Cambridge, MA.

Gick, M. L. (1989). Two functions of diagrams in problem solving by analogy. In *Knowledge Acquisition from Text and Pictures* (ed. H. Mandl and J. R. Levin), Advances in Psychology, pp. 215–231. Elsevier (North-Holland), Amsterdam.

Gobet, F., Lane, P.C.R., Croker, S., Cheng, P.C.H., Jones, G., Oliver, I., and Pine, J. M. (2001). Chunking mechanisms in human learning. *Trends in Cognitive Science*, **5**(6), 1236–243.

Gurr, C. (1998). On the isomorphism, or lack of it, of representations. In *Visual language theory*, pp. 293–305. Springer.

Harrison, J. (2009). HOL light: An overview. In *International Conference on Theorem Proving in Higher Order Logics*, pp. 60–66. Springer.

Hegarty, M. (2011). The cognitive science of visual-spatial displays: Implications for design. *Topics in Cognitive Science*, **3**, 446–474.

Jamnik, M., Bundy, A., and Green, I. (1999). On automating diagrammatic proofs of arithmetic arguments. *Journal of logic, language and information*, **8**(3), 297–321.

Kaput, J.J. (1992). Technology and mathematics education. In *Handbook of Research on Mathematics Teaching and Learning* (ed. D. Grouws), pp. 515–556. MacMillan, New York, NY.

Klahr, D., Langley, P., and Neches, R. (1987). *Production System Models of Learning and Development.* MIT Press, Cambridge, Mass.

Koedinger, K.R. and Anderson, J.R. (1990). Abstract planning and perceptual chunks: Elements of expertise in geometry. *Cognitive Science*, **14**, 511–550.

Kotovsky, K., Hayes, J. R., and Simon, H. A. (1985). Why are some problems hard? *Cognitive Psychology*, **17**, 248–294.

Kovács, L. and Voronkov, A. (2013). First-order theorem proving and vampire. In *International Conference on Computer Aided Verification*, pp. 1–35. Springer.

Larkin, J.H. and Simon, H.A. (1987). Why a diagram is (sometimes) worth ten thousand words. *Cognitive science*, **11**(1), 65–100.

Ma, W., Adesope, O.O., Nesbit, J.C., and Liu, Q. (2014, 11). Intelligent tutoring systems and learning outcomes: A meta-analysis. *Journal of educational psychology*, **106**(4), 901–918.

Markman, A.B. (1999). *Knowledge Representation.* Lawrence Erlbaum, Mahwah: NJ.

Miller, G. A. (1956). The magical number seven plus or minus two: Some limits on our capacity for information processing. *Psychological Review*, **63**, 81–97.

Moody, D. (2009). The "physics" of notations: toward a scientific basis for constructing visual notations in software engineering. *IEEE Transactions on Software Engineering*, **35**(6), 756–779.

Newell, Allen (1990). *Unified theories of cognition.* Harvard University Press.

Newell, A. and Simon, H.A. (1972). *Human Problem Solving.* Prentice-Hall.

Peebles, D.J. and Cheng, P.C.H. (2003). Modelling the effect of task and graphical representations on response latencies in a graph-reading task. *Human factors*, **45**(1), 28–45.

Raggi, D., Stapleton, G., Stockdill, A., Jamnik, M., Garcia Garcia, G. and Cheng, P.C.H. (2020). How to (Re)represent it? In *32th IEEE International Conference on Tools with Artificial Intelligence*, pp. 1224–1232. IEEE.

Raggi, D., Stockdill, A., Jamnik, M., Garcia Garcia, G., Sutherland, H.E.A., and Cheng, P.C.H. (2020). Dissecting representations. In *Diagrams* (ed. A. V. Pietarinen, P. Chapman, L. Bosveld-de Smet, V. Giardino, J. Corter and S. Linker), Volume 12169, LNCS, pp. 144–152. Springer.

Schank, R.C. (1982). *Dynamic Memory.* Cambridge University Press, Cambridge.

Shimojima, A. (2015). *Semantic properties of diagrams and their cognitive potentials.* CSLI Press, Stanford, CA.

Stapleton, G., Jamnik, M., and Shimojima, A. (2017). What makes an effective representation of information: A formal account of observational advantages. *Journal of Logic, Language and Information*, **26**(2), 143–177.

Stapleton, G., Shimojima, A., and Jamnik, M. (2018). The observational advantages of Euler diagrams with existential import. In *Diagrams* (ed. P. Chapman, G. Stapleton, A. Moktefi, S. Perez-Kriz, and F. Bellucci), Volume 10871, LNCS, pp. 313–329. Springer.

Stenning, K. and Oberlander, J. (1995). A cognitive theory of graphical and linguistic reasoning: logic and implementation. *Cognitive Science*, **19**(1), 97–140.

Stillings, N.A., Weisler, S.E., Chase, C.H., Feinstein, M.H., Garfield, J.L., and Rissland, E.L. (1995). *Cognitive Science: An Introduction* (2nd edn). MIT press, Cambridge, MA.

Stockdill, A., Raggi, D., Jamnik, M., Garcia Garcia, G., Sutherland, H.E.A., Cheng, P.C.H., and Sarkar, A. (2020). Correspondence-based analogies for choosing problem representations. In *IEEE Symposium on Visual Languages and Human-Centric Computing, VL/HCC 2020* (ed. C. Anslow, F. Hermans, and S. Tanimoto). IEEE. Forthcoming.

Urbas, M. and Jamnik, M. (2014). A framework for heterogeneous reasoning in formal and informal domains. In *Diagrams* (ed. T. Dwyer, H. Purchase, and A. Delaney), Volume 8578, LNCS, pp. 277–292. Springer.

Van Someren, M.W., Reimann, P., Boshuizen, H. et al. (1998). *Learning with Multiple Representations. Advances in Learning and Instruction Series*. ERIC.

Zhang, J. (1997). The nature of external representations in problem solving. *Cognitive science*, **21**(2), 179–217.

Zhang, J. and Norman, D.A. (1995). A representational analysis of numeration systems. *Cognition*, 57(3), 271–295.

19

Human–Machine Collaboration for Democratizing Data Science

Clément Gautrais, Yann Dauxais, Stefano Teso, Samuel Kolb, Gust Verbruggen, and Luc De Raedt

KU Leuven, Department of Computer Science, Leuven, Belgium

19.1 Introduction

Data science is a cornerstone of current business practices. A major obstacle to its adoption is that most data analysis techniques are beyond the reach of typical end-users. Spreadsheets are a prime example of this phenomenon: despite being central in all sorts of data-processing pipelines, the functionality necessary for processing and analysing spreadsheets is hidden behind the high wall of spreadsheet formulas, which most end-users can neither write nor understand (Chambers and Scaffidi, 2010). As a result, spreadsheets are often manipulated and analysed manually. This increases the chance of making mistakes and prevents scaling beyond small datasets.

Lowering the barrier to entry for specifying and solving data science tasks would help in ameliorating these issues. Making data science tools more accessible would lower the cost of designing data processing pipelines and taking data-driven decisions. At the same time, accessible data science tools can prevent non-experts from relying on fragile heuristics and improvised solutions.

The question we ask is then: *is it possible to enable non-technical end-users to specify and solve data science tasks that match their needs*?

We provide an initial positive answer based on two key observations. First, many key data science tasks can be partially specified using *coloured sketches* only. Roughly speaking, a sketch is a collection of entries, rows, or columns appearing in a spreadsheet that are highlighted using one or more colours. A sketch determines some or all of the parameters of a data science task. For instance, while clustering rows, colour highlighting can be used to indicate that some rows belong to the same cluster (by highlighting them with the same colour) or to different clusters (with different colours). This information acts as a partial specification of the data science task. The main feature of sketches is that they require little to no technical knowledge on the user's end, and therefore can be easily designed and manipulated by naïve end-users (Sarkar *et al.*, 2015).

Clement Gautrais, Yann Dauxais, Stefano Teso, Samuel Kolb, Gust Verbruggen, and Luc De Raedt, *Human–Machine Collaboration for Democratizing Data Science* In: *Human Like Machine Intelligence.* Edited by: Stephen Muggleton and Nick Charter, Oxford University Press.
 DOI: 10.1093/oso/9780198862536.003.0019

Second, the data science task determined by a sketch can be solved using automated data science techniques. In other words, since the specification may be missing one or more parameters, the spreadsheet application takes care of figuring these out automatically. The output of this step is a candidate solution, for example a possible clustering of the target rows. The other key feature of sketches is that the result of the data science task can also often be presented through colour highlighting. For instance, row clusters can be captured using colours only.

These two observations enable us to design an interactive framework, VISUALSYNTH, in which the machine and the end-user collaborate in designing and solving a data science task compatible with the user's needs. VISUALSYNTH combines two components: an interaction protocol that allows non-technical people to design partial data science task specifications using coloured highlighting, and a smart framework for automatically solving a partially specified data science task based on inductive models.

In contrast to automation frameworks like AutoML (Thornton *et al.*, 2013; Feurer *et al.*, 2015), VISUALSYNTH does not assume that the data science task is fixed and known a priori.[1] We do not claim that our human–machine interaction strategy is ideal, but we do claim that it is quite minimal and that despite its simplicity, it suffices to guide the system towards producing useful data science results for many central data science tasks, as shown in the remainder of this chapter.

VISUALSYNTH only requires the end-user to check the solution and make sure that it is as expected. This substantially reduces the expertise required of the user: almost everybody can interact using colour highlighting and check whether a solution is compatible with his needs. The bulk of the complexity—namely figuring out the bits that are missing from the user's specification—is handled by the machine itself. The intent of this setup is to combine the respective strengths of end-users, namely their knowledge of the domain at hand, and computers, namely their ability to quickly carry out enormous amounts of computation.

The remainder of this chapter is structured as follows. In section 19.2.2, we motivate our approach using a concrete use case. Section 19.3 discusses sketches for several core data science tasks, including data wrangling, prediction, clustering, and auto-completion, and details how the sketches define interaction. Section 19.3 also describes how tasks partially defined by sketches are solved by the machine. The chapter ends with some concluding remarks.

19.2 Motivation

19.2.1 Spreadsheets

Spreadsheets are used by hundreds of millions of users and are as such one of the most common interfaces that people use to interact with data. The reason for their popularity is

[1] Indeed, VISUALSYNTH supports explorative data science, in which the user is not sure about the task to be performed and tries out different manipulations until it finds one that is interesting or useful. A proper discussion of explorative data science, however, falls outside the scope of this chapter.

their flexibility: (1) spreadsheets are very heterogeneous and can contain arbitrary types of data, including numerical, categorical, and textual values; (2) data can be explicitly organized using tables and operated on using formulas; and (3) the 'data generating process' is almost arbitrary as spreadsheets can be used for anything from accounting to financial analysis to stock management. Since our goal is to enable as many users as possible to perform data science, a natural choice is to bring data science to spreadsheets.

This is very challenging, for two reasons. First and foremost, the vast majority of spreadsheet users have little or no knowledge about how to perform data science. While these naïve users might have heard of data science—at least to some degree—they are likely not technically skilled: most spreadsheet users cannot program even one-line spreadsheet formulas, nor design small data processing pipelines (Chambers and Scaffidi, 2010).

In order to cater to this audience, VISUALSYNTH relies on a visual, concrete, and interactive protocol in which the user and the machine collaborate to explore the data and design a data-processing pipeline. The protocol leverages simple and intuitive forms of interaction that require no or little supervision and almost zero technical knowledge. This is achieved through a combination of interaction and automation.

19.2.2 A motivating example: Ice cream sales

Let us now illustrate interactive data science and VISUALSYNTH with a classic use case of naïve spreadsheet end-users: auto-completion. Tackling this use case requires collaboration between the user and the machine to convey the intentions and the knowledge of the user, as shown below.

Imagine that you are a sales manager at an ice cream factory. You have data about past sales and some information about your shops, as shown in Tables 19.1 and 19.1, respectively.

A first difficulty is that Table 19.1 is not nicely formatted. A first task is therefore to wrangle Table 19.1 into a format such as that listed in Table 19.2 that is more amenable to data analysis. Through interaction, the **data wrangling component** can produce the table presented in Table 19.2.

However, some past sales data are missing. To determine which shops made a profit you need first to obtain an estimate of the missing values. To produce such estimates, you can interact with our system in different ways. First, as the sales manager you know that the profit of a shop depends on the type of ice cream and the characteristics of the city. More precisely, you know that some cities have similar profitability profiles. To convey this knowledge, you can use a colouring scheme to indicate that certain cities belong to the same cluster. This will in turn trigger an **interactive clustering** process which not only allows you to state must-link and cannot-link constraints using colourings but also to correct mistakes that our system might make during the clustering process. Once the clustering is deemed correct, the machine stores this information and displays it as a new column in the spreadsheet. From this enriched data, you can then ask the machine to provide a first **estimate of the missing values**. This can be achieved in different ways.

Table 19.1 *Left: Spreadsheet with ice cream sale numbers. The '?' values are are missing. Right: Spreadsheet containing properties of shops.*

Florence		
Vanilla	June	610
	July	190
	Aug	670
	Total	1470
	Profit	YES
Stracciatella	June	300
	July	250
	Aug	290
	Total	860
	Profit	NO
...	...	...
Milan		
Chocolate	June	430
	July	350
	Aug	?
	Total	?
	Profit	?

City	Touristic	Weather	Country
Florence	High	Hot	IT
Stockholm	High	Cold	SE
Copenhagen	High	Cold	DK
Berlin	Very High	Mild	DE
Aachen	Low	Mild	DE
Brussels	Medium	Mild	BE
Milan	Medium	Hot	IT

First, as a sales manager you could start filling the missing values yourself. After one or two missing values are filled, the machine can infer that the remaining missing values should also be filled. The machine will thus start suggesting values, which you can then either accept them as is or correct them. Corrections will trigger a new auto-completion loop, with additional constraints expressing that the user corrected some values in the previous iteration.

Second, you could trigger the auto-completion by indicating that the machine should fill the missing values. For this, you can use colours to indicate which values should be predicted. Then, human–machine interaction proceeds as described above. Additionally, the machine could provide information about some of the underlying model assumptions. For example, the machine can indicate which columns are used for prediction and you could indicate whether these columns are relevant for predicting profit.

The remainder of this chapter introduces some principles for human–machine collaboration in the context of auto-completion and automated data science. In particular,

Table 19.2 *Spreadsheet with ice cream sale numbers.*

Type	City	June	July	Aug	Total	Profit
Vanilla	Florence	610	190	670	1470	YES
Banana	Stockholm	170	690	520	1380	YES
Chocolate	Copenhagen	560	320	140	1020	YES
Banana	Berlin	610	640	320	1570	NO
Stracciatella	Florence	300	270	290	860	NO
Chocolate	Milan	430	350	?	?	?
Banana	Aachen	250	650	?	?	?
Chocolate	Brussels	210	280	?	?	?

we identify different levels of interaction, discuss how the machine adds user knowledge in its learning mechanisms, and elucidates how different data science tasks fit in our framework.

19.3 Data Science Sketches

We now introduce the interaction strategy of VisualSynth, our framework for interactive data science.

Given a spreadsheet, a *sketch* is simply a set of colours (also known as *colouring*) applied to one or more rows, columns, or cells appearing in the spreadsheet. The key idea is that the colours partially define the parameters (e.g., the type, inputs, and outputs) of a data science task. Hence, taken together, the sketch and the spreadsheet can be mapped onto a very concrete *data science task* (e.g., a clustering task), which can then be solved, and and whose results (e.g., a set of clusters) can be filled into or appended to the original spreadsheet, yielding an extended spreadsheet. This idea is captured in the following schema:

$$\left.\begin{array}{c}\text{spreadsheet}\\ +\\ \text{sketch}\end{array}\right\} \rightarrow \text{data science task} \rightarrow \left.\begin{array}{c}\text{spreadsheet}\\ +\\ \text{model}\end{array}\right\} \rightarrow \text{new spreadsheet}$$

When explaining the different components of VisualSynth we shall adhere to the above scheme, that is our examples and figures will consist of four components: (1) the input sketch and spreadsheet, (2) the data science problem specification, (3) the model, and (4) the resulting spreadsheet.

The above scheme is in line with the closure property of databases and inductive database (Imielinski and Mannila, 1996; De Raedt, 2002). For relational databases, both the inputs and the results of a query are relations, which guarantees that the results of one query can be used as the input for the next. In a similar vein, in our setting, the inputs as well as the result of each operation (or data science task) are tables in a spreadsheet. The closure property guarantees that further analysis is possible after each data science task.

VISUALSYNTH is an example of user-guided interaction that enables the user to convey her intentions by interacting using visual cues. Indeed, the sketches are supplied by and end-user and are gradually refined in an interactive fashion—thus adapting the data science task itself—until the user is satisfied with the result.

Next, we illustrate this interaction protocol using a number of key data science tasks, namely data wrangling, concept learning, prediction, clustering, constraint learning, and auto-completion.

19.3.1 Data wrangling

Wrangling is the task of transforming data in the right format for downstream data science tasks. Colouring cells has already been used to help automated wranglers transform data in a format desired by a user (Verbruggen and De Raedt, 2018). The user has to indicate which cells belong to the same row by colouring them using the same colour. A *wrangling sketch* is therefore a set of coloured cells, where each colour defines a partial example of the expected wrangling result and imposes a constraint on the output, that is that the partial example should be mapped onto a single row into the target spreadsheet.

A commonly used paradigm for data wrangling is programming by example (Lieberman, 2001; Cropper *et al.*, 2015) (PBE), in which a language of transformations $\mathcal{L}$ is defined and the wrangler searches for a program $P \in \mathcal{L}$ that maps the input examples to the corresponding outputs. In the context of VISUALSYNTH, **given** a wrangling sketch and a spreadsheet, the goal is to **find** a program that transforms the spreadsheet in such a way that cells with the same colour end up in the same row, and no row can contain cells with multiple colours.

An example is shown in Figure 19.3a. The data are clearly not in a suitable format for analysis and a novice user might not be able to efficiently transform them. From a small number of coloured cells—the wrangling sketch—the synthesizer is able to learn the program described in Figure 19.3c. This program yields the desired table from Figure 19.3d when applied on the input table.

Finding such a program is a form of *predictive program synthesis*. The desired solution is not known in explicit form, but the wrangling sketch imposes a constraint that the solution should at least satisfy. Additionally, syntactic and semantic properties of the elements in rows and columns are used for heuristically determining the quality of candidate solutions.

In addition to defining constraints on the output, the wrangling sketch can be used to define heuristics for improving the search for a correct program. The relative positions of cells in the same or different colours allow one to impose a strong syntactic bias on the program synthesizer. For example, two consecutive columns with the same number

Table 19.3 *Wrangling tasks take as input a set of coloured cells from a set of tables and produce a transformation program that reformats the data such that cells of different colours end up in different rows and cells of the same colour end up in the same row.*

Florence		
Vanilla	June	610
	July	190
	Aug	670
	Total	1470
	Profit	YES
Stracciatella	June	300
	July	250
	Aug	290
	Total	860
	Profit	NO
...	...	...
Milan		
Chocolate	June	430
	July	350
	Aug	?
	Total	?
	Profit	?

(a) Input data and wrangling sketch where each colour indicates cells that should end up in the same row.

		June	July	Aug	Total	Profit
Vanilla	Florence	610	190	670	1470	YES
Banana	Stockholm	170	690	520	1380	YES
Chocolate	Copenhagen	560	320	140	1020	YES
Banana	Berlin	610	640	320	1570	NO
Stracciatella	Florence	300	270	290	860	NO
Chocolate	Milan	430	350	?	?	?
Banana	Aachen	250	650	?	?	?
Chocolate	Brussels	210	280	?	?	?

(d) Expected output of the wrangling task.

Given the blue and red colourings, a spreadsheet and a language $\mathcal{L}$ in which to express programs,
find a wrangling program $P \in \mathcal{L}$
such that blue cells end up in a single row, and red cells in an other single row.

(b) Wrangling problem statement.

Split(1, empty)	split column 1 into two columns based on whether cells have a non-empty cell to their right
ForwardFill(1)	forward fill column 1
ForwardFill(2)	forward fill column 2
Pivot(3, 4)	pivot columns 3 and 4

(c) High-level description of the transformation program, the model of the data wrangling task. These transformations are detailed in Table 19.4.

of vertically adjacent cells of the same colour are very good candidates for a pivot transformation, as in Table 19.3a, where the two rightmost columns fit this description. A greedy beam search that interleaves heuristically selecting transformations and evaluating the results of these transformations was used in (Verbruggen and De Raedt, 2018) to quickly find spreadsheet transformation programs.

19.3.2 Data selection

Selecting the right data to analyse is one of the essential steps in data science processes (Fayyad *et al.*, 1996). Within VISUALSYNTH we view this as the task to extract a subset of subtables from the original spreadsheet. This is often a necessary step before the machine learning-methods (proposed in the following sections) can be applied.

Table 19.4 *Examples of wrangling functions.* ***Split*** *creates a new column for each value of a given column.* ***Forward fill*** *fills missing values in a column with the value directly above it.* ***Pivot*** *uses unique values of a column as a new set of columns.*

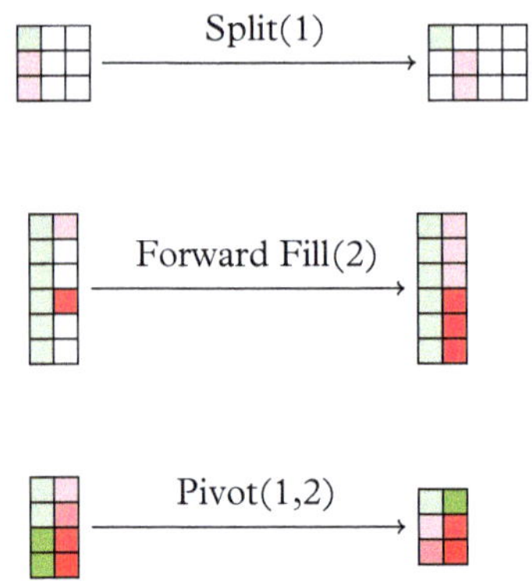

Consider Table 19.5 as a running example. This table can be decomposed as (1) the dataset given as input 19.5a, 2) the problem statement 19.5b, (3) an example of model 19.5c used to represent the selection and 3) the dataset returned as output 19.5d. The dataset is represented by two spreadsheet tables. The *sales* table gathers the log of each ice cream flavour profits in each city and the *provider* table gathers the information about the ice cream providers in each city with a discrete evaluation of the price and quality of their products.

As an example for data selection, if the user wants to predict the missing values for the Chocolate flavour, she could want to predict these using only the known values for Chocolate and Vanilla without considering Banana and Stracciatella based on her knowledge of the ice cream market. However, it would be hard for a non-expert spreadsheet user to perform the selection by hand. Therefore, the set of rows to be used could be induced from a set of examples using a sketch. In a data selection sketch, the user can indicate desirable examples by colouring them in blue, and unwanted or irrelevant ones by colouring them in another colour (say pink). The goal of data selection is then to learn which part of the spreadsheet to retain. The model that is learned will consist of queries that, when performed on the spreadsheet, return the desired selection of the data.

As illustrated on both tables with the columns *Total* and *ProviderID*, if a column or a table does not contain any coloured cell, this column or table will not appear in the final selection. This is an intuitive way to represent the projection operator from relational algebra. It ensures that the user can specify partial examples, that is, examples that do not extend over all coloured columns or tables. These partial examples are then automatically extended over the remaining columns as to consider the full rows in the relevant tables. An example of such a colouring extension is illustrated on the input tables with lighter blue and red for positive and negative examples, respectively.

Table 19.5 *Input, model, and output of the data selection ice cream factory example. The input and output are sets of coloured cells from a set of tables and the output is a set of rules representing the set of coloured cells to be output.*

Type	City	June	July	Aug	Total	Profit
Vanilla	Florence	610	190	670	1470	YES
Banana	Stockholm	170	690	520	1380	YES
Chocolate	Copenhagen	560	320	140	1020	YES
Banana	Berlin	610	640	320	1570	NO
Stracciatella	Florence	300	270	290	860	NO
Chocolate	Milan	430	350	?	?	?
Banana	Aachen	250	650	?	?	?
Chocolate	Brussels	210	280	?	?	?

Type	City	ProviderID	Price	Quality
Vanilla	Florence	1	Cheap	Bad
Vanilla	Florence	2	Regular	Good
Stracciatella	Florence	1	Regular	Great
Chocolate	Copenhagen	3	Cheap	Good
Chocolate	Milan	4	Regular	Good
Chocolate	Milan	5	Expensive	Great
Chocolate	Brussels	6	Regular	Good
Chocolate	Brussels	6	Expensive	Good

(a) Input tables describing ice cream sales and providers and containing coloured examples. The cells coloured in blue and red are the relevant and irrelevant examples, respectively. The cells coloured in lighter gradient are the extension of the partial examples.

Given positive (blue) and negative (pink) tuples in a spread sheet,
the schema of the tables,
find one or more queries that together cover all positives and none of the negatives.

(b) Problem statement

```
?- sales(I0, Type, City, June, July, Aug, 'YES'),
   provider(I1, Type, City, 'Cheap', Quality).

?- sales(I0, Type, City, June, July, Aug, Profit),
   provider(I1, Type, City, 'Regular', 'Good').
```

(c) Queries describing which rows are positive.

Type	City	June	July	Aug	Total	Profit
Vanilla	Florence	610	190	670	1470	YES
Banana	Stockholm	170	690	520	1380	YES
Chocolate	Copenhagen	560	320	140	1020	YES
Banana	Berlin	610	640	320	1570	NO
Stracciatella	Florence	300	270	290	860	NO
Chocolate	Milan	430	350	?	?	?
Banana	Aachen	250	650	?	?	?
Chocolate	Brussels	210	280	?	?	?

Type	City	ProviderID	Price	Quality
Vanilla	Florence	1	Cheap	Bad
Vanilla	Florence	2	Regular	Good
Stracciatella	Florence	1	Regular	Great
Chocolate	Copenhagen	3	Cheap	Good
Chocolate	Milan	4	Regular	Good
Chocolate	Milan	5	Expensive	Great
Chocolate	Brussels	6	Regular	Good
Chocolate	Brussels	6	Expensive	Good

(d) Output tables corresponding to the input tables in which the relevant colours are included.

The data selection sketch can, thus, be decomposed in two steps. First, the colouring of the partial examples is extended to the obtain complete examples. Each example corresponds to a set of rows (or tuples) that can belong to multiple tables. Second, the examples are generalized into queries that should capture the concept underlying the data selection process. Thus, the data selection task can be formalized as an inductive logic programming or logical and relational learning problem (Muggleton and De Raedt, 1994; De Raedt, 2008) such that: **Given** a set of tables in a spreadsheet, a set of partial examples in two colours (representing positive and negative examples), and the schema of the tables in the spreadsheet, **find** one more relational queries whose answers cover all positive examples, and none of the negative tuples. The resulting queries are then run on the tables in the spreadsheet, and all rows that satisfy the query are coloured positively.

It will be assumed that we possess some information about the underlying relational schema, in particular, the foreign key relations need to be known. These can be induced by a learning system such as TACLE (Kolb *et al.*, 2017), which is explained in more detail below.

The use of colours to induce queries was already considered in a database setting (Bonifati *et al.*, 2016). However, the focus was on learning the definition of a single relation, not on performing data selection across multiple tables as we do. Furthermore, partial examples, which provide the user with extra flexibility, were not considered.

Processing the data

The first step is to extend the input colouring of Table 19.5a into a set of examples. This process starts from the template and uses the foreign key relations to indicate the joins. For our running example, the template query is:

```
?-sales(I0, Type, City, June, July, Aug, Profit),
provider(I1, Type, City, Price, Quality).
```

To select the examples, we start by detecting which rows contain at least one colour, and we expand these into the sets of facts we note $Sales^+$ and $Provider^+$, and two other sets matching the irrelevant rows that we note $Sales^-$ and $Provider^-$, respectively. Furthermore, we omit the columns that do not contain any colour, as they are deemed irrelevant.

The next step is then to construct the positive examples by taking every ground atom from one of the positive sets $Sales^+$ and $Provider^+$ and unifying it with the corresponding atom for the same predicate in the template. The set of all answers to the query constitutes an example. For instance, the first tuple in the $Sales^+$ table (having $Type = Vannila$ and $City = Florence$) would yield the example consisting of that tuple and the first two tuples of the $Provider$ table. The negative tables are not expanded, they are only used to prune candidate generalizations.

Relational rule learning

With this setup, we can now define the inductive logic programming problem (De Raedt, 2008). **Given** a set of positive examples (where each example is a set of facts), a set of negative examples (the tuples in the negative set), and the relational structure of the spreadsheet, **find** a set of queries that cover all the positive examples and none of the negative tuples. Such queries can in principle be induced using standard relational learners such as GOLEM (Muggleton *et al.*, 1990) and FOIL (Quinlan, 1990).

What is used in VISUALSYNTH is a simplified GOLEM; VISUALSYNTH uses Plotkin's least general generalization (lgg) operator (Plotkin, 1970; De Raedt, 2008) together with GOLEM's search strategy. The lgg operator takes two examples and produces a generalized set of facts that can serve as the query. More specifically, consider the example related to $Type = Vannila$ and $City = Florence$ and the one related to $Type = Chocolate$ and $City = Copenhagen$. The resulting lgg would be

```
?- sales(I0, Type, City, June, July, Aug, 'YES'),
provider(I1, Type, City, 'Cheap', Quality),
provider(I2, Type, City, Price, 'Good').
```

The strategy followed by GOLEM that we adopt here is to sample positive examples, compute their lgg, verify that the lgg does not cover negative tuples, and if so replace the positive examples (and other positives that are subsumed) by the lgg. This process is continued, until further generalizations yields queries that cover negative tuples and are too general. Applying this strategy to our example yields the two queries shown in Table 19.5.c. Evaluating these queries on the original tables results in Table 19.5.d.

Finally, the result of these rules which represent the rows to colour can be easily matched with the initial template that represents the columns to colour and thus, output the result set of coloured cells.

Implementation choices

We chose for the implementation not to include GOLEM's assumptions in order to extract the complete set of LGGs covering our examples. This implementation is based on the bottom-up search space strategy of GOLEM to extend examples to LGGs until the point where they are too general and, thus, also cover negative examples. Such an approach is not a problem in our context as the number of examples is small. Indeed, this approach is dedicated to extending a set of a few examples to a coherent subset of the whole dataset. Running this approach on an entire dataset would be meaningless. Thereby, the size of the dataset itself, in terms of number of examples, is not a limitation of our approach.

The main limitation, in terms of computation time, would appear while comparing examples that include many relations of the same type. For example, if a lot of providers are available for the pairs ice cream type and city, it would be difficult to compare sets of providers because every combinations of providers would have to be evaluated. Comparing hundreds of providers of chocolate ice cream with hundreds of providers of vanilla ice cream in Florence would then lead to thousands of tuple comparison. In such a case, the assumptions made by GOLEM may be inefficient to constrain the complexity of the algorithm. Using θ-subsumption under object identity (Ferilli *et al.*, 2002) to compute the LGGs would help to constrain the number of generated tuples but may be also inefficient in terms of complexity. Finally, other approaches, like aggregation of tuples, could be used to simplify the dataset itself and, thus, extract some partial information describing such examples. In this case, sets of provider tuples could be aggregated for a given price or a given quality. For example, the term `provider_price('Vanilla', 'Florence', 'Cheap', Count)` can be generated to replace the set of providers selling vanilla in Florence at a cheap price, with $Count$ being the number of aggregated tuples.

19.3.3 Clustering

Clustering is the task of grouping data in different coherent clusters and is a building block of typical data processing pipelines (Xu and Wunsch, 2005). In our use case, we use clustering not only as a way to learn clusters in the data but also as a way to generate new features. Through clustering, a user can express some of her knowledge explicitly and this knowledge can then be used for future data science steps, such as predicting a missing value.

Since clustering is ill-defined, recent developments in this area enable the machine to interactively elicit knowledge from the end-user so as to guide the clustering towards the user's needs (cf. Van Craenendonck *et al.*, 2018). In the simplest case, the machine iteratively presents pairs of (appropriately chosen) examples to the user and asks whether they belong to the same cluster or not. The user's feedback is then translated into pairwise constraints, namely *must-link* and *cannot-link* constraints, which are then used to bias the clustering process according to the elicited knowledge (Wagstaff *et al.*, 2001; Van Craenendonck *et al.*, 2017).

Building on top of such techniques, coloured sketches can be used to implement the interaction: the user colours (a few) objects belonging to the same cluster using the same colour. Hence, items highlighted with the same colour belong to the same cluster. The sketch therefore consists of a set of such colourings, each identifying examples from a given cluster. An example sketch is given in Table 19.6a. In this example, the user coloured a few rows to indicate that the city shops in Milan and Florence (both coloured in green) should belong to the same cluster, while Berlin and Seville belong to a different cluster (coloured in blue). The extra empty column at the end of each table contains the resulting clustering. Although incomplete, this information often suffices to guide the clustering algorithm towards a clustering compatible with the user's requirements (Van Craenendonck *et al.*, 2017).

Problem setting

In section 19.3.1, we presented how data wrangling can map an example to a single row of a table. Hence, we consider that an example in the clustering is a table row. From this observation and the sketch described in the previous paragraph, we get the problem setting for clustering: **Given** a set of set of coloured rows and a set of uncoloured rows, **find** a cluster assignment for all rows such that rows in the same coloured set belong to the same cluster and no rows in different sets belong to the same cluster, or equivalently: **find** a cluster assignment for all rows such that rows in the same coloured set belong to the same cluster and the number of clusters is equal to the number of colours.

Finding a cluster assignment

Current techniques to solve the above problem statement typically start from a partial cluster assignment where all examples in the same colour set are in the same cluster.

Table 19.6 *Input sketch, constraints, and output sketch of the clustering task.*

	City	Touristic	Weather	Nat	
1	Florence	High	Hot	IT	
2	Stockholm	High	Cold	SE	
3	Copenhagen	High	Cold	DK	
4	Berlin	Very High	Mild	DE	
5	Aachen	Low	Mild	DE	
6	Brussels	Medium	Mild	BE	
7	Milan	Medium	Hot	IT	
8	Munich	Medium	Mild	DE	
9	Paris	Very High	Mild	FR	
10	Turin	High	Hot	IT	
11	Seville	High	Hot	ES	
12	Valencia	High	Hot	ES	

(a) Sketch for a clustering task. Rows of the same colour belong to the same cluster

Given the green, pink and blue examples and the constraints in Table 19.6c, **find** a cluster assignment that satisfies the constraints.

(b) Problem setting of the clustering task

mustlink(1,7), mustlink(2,6), mustlink(4,11), cannotlink(1,2), cannotlink(1,6), cannotlink(1,4), cannotlink(1,11), cannotlink(7,2), cannotlink(7,6), cannotlink(7,4), cannotlink(7,11), cannotlink(2,4), cannotlink(2,11), cannotlink(6,4), cannotlink(6,11)

(c) Constraints passed to the clustering algorithm. Arguments are row number, starting from 1.

	City	Touristic	Weather	Nat	Cluster
1	Florence	High	Hot	IT	
2	Stockholm	High	Cold	SE	
3	Copenhagen	High	Cold	DK	
4	Berlin	Very High	Mild	DE	
5	Aachen	Low	Mild	DE	
6	Brussels	Medium	Mild	BE	
7	Milan	Medium	Hot	IT	
8	Munich	Medium	Mild	DE	
9	Paris	Very High	Mild	FR	
10	Turin	High	Hot	IT	
11	Seville	High	Hot	ES	
12	Valencia	High	Hot	ES	

(d) Result of the first clustering task, where each colour represents a cluster. Alight colour means that the cluster assignment has been performed by the clustering algorithm.

This can be achieved by using clustering algorithms using *must-link* and *cannot-link* constraints (Wagstaff *et al.*, 2001; Basu *et al.*, 2004; Van Craenendonck *et al.*, 2017). *Must-link* constraints are enforced between examples of the same colour set, while *cannot-link* constraints are enforced between examples from different colour sets. Then, non-coloured examples have to be assigned according to a learned distance metric (Xing *et al.*, 2003), or generalizations of existing (partial) clusters.

The resulting cluster assignment is mapped back into a set of coloured rows, as depicted in Table 19.6d. The user can then modify the resulting cluster assignment by adding new colours or by putting existing colour on previously colourless rows. Iterative refinements of the cluster assignments and of the sketch are then performed, as the user is unlikely to be able to fix all parameters of the clustering task through a single interaction.

19.3.4 Sketches for inductive models

In this section, we present the use of sketches for learning and using inductive models for auto-completion. In this context, inductive models refer to predictors, constraints or a combination of the two. Learning predictors or constraints typically requires knowing

Table 19.7 *Input sketch, problem setting and output sketch of the inductive model learning task.*

June	July	Aug	Total	Profit
610	190	670	1470	YES
170	690	520	1380	NO
430	350	?	?	?
250	650	?	?	?

June	July	Aug	Total	Profit
610	190	670	B4D	YES
170	690	520	1380	NO
430	350	?	?	?
250	650	?	?	?

June	July	Aug	Total	Profit
610	190	670	1470	YES
170	690	520	1380	NO
430	350	?	?	?
250	650	?	?	?

June	July	Aug	Total	Profit
610	190	670	1470	YES
170	690	520	1380	NO
430	350	?	?	?
250	650	?	?	?

June	July	Aug	Total	Profit
610	190	670	1470	YES
170	690	520	1380	NO
430	350	?	?	?
250	650	?	?	?

(a) Illustration of the inductive models sketch. Top table: simplified ice cream sale numbers. Middle row: excluding a corrupted row from auto-completion using red (left) and selecting of a column as target using blue (right). Bottom row: the machine decided to predict August, in blue, from June and Profit, in green (left); the user improved the system's choice of inputs (right).

Given the green and blue columns
and the red rows,
find a predictive model
that predicts the blue column
from the green ones, while ignoring
the red rows.

(b) Problem setting of the inductive model learning task. The considered sketch is the bottom left from Table 19.7a

For **predictor learning**: Launch an auto ML instance to learn a model predicting August from June and July, without the first row. The loss function in root mean squared error.
For **constraint learning**: Learn constraints using June and July to predict August, using constraint templates **S**
For **auto-completion**: Use inductive models in the system to predict August from June and July, and learn constraints if none are available and predictors if constraints cannot predict missing values of August.

(c) Model learning step solving the problem setting in Table 19.7b

June	July	Aug	Total	Profit
610	190	670	1470	YES
170	690	520	1380	NO
430	350	*460*	?	?
250	650	*540*	?	?

(d) Output sketch, where missing values for August have been filled. Predicted values are in italic formatting to indicate that they come from an inductive model. The learned model (constraints, predictor or a combination of both) is stored in the system and is associated with the spreadsheet.

what data to learn from and what is the target to learn. From this observation, we propose the sketch depicted in Table 19.7a.

First of all, the sketch of Table 19.7a is used to identify target cells and input features. For instance, prior to initiating the learning of inductive models, the user might highlight a target column containing empty cells, as in Table 19.7a (middle right). This prompts the system to ignore other empty regions of the spreadsheet, thus focusing the computation to the user's needs and saving computational resources. After a first round of learning, the system might highlight the columns that the value of $August$ was derived from, as in the Table at bottom left. In the example, the system mistakenly used the Profit information to predict the sales for $August$. Although not technically incorrect, as the two values are correlated, this choice does not help in predicting the missing $August$ sales. The user can improve the choice of inputs by de-selecting irrelevant or deleterious inputs and by adding any relevant columns ignored by the system. A possible result is shown in Table 19.7a (bottom right).

Next, sketches can be used to identify *examples* and *non-examples*. In Table 19.7a (middle left), the $Total$ is corrupted in one row. The user can mark that row (e.g., in red) to ensure that the software does neither use it for inferring predictors and constraints nor for making predictions.

In the next paragraphs, we describe how the sketch of Table 19.7a can be used to define a prediction, a constraint learning task and an auto-completion task.

Prediction

Prediction is one of the most classic tasks in data science. Prediction can be decomposed in two steps. First, a predictor is fit on a dataset to predict targets based on input features. Second, the fit model is used to make predictions on new data using similar input features. A common framework to represent these two steps is *fit-predict*, that is for example used in the scikit-learn library (Buitinck *et al.*, 2013). The fit step typically requires input data (also called training data) and target data. The predict step only requires input data.

The prediction sketch depicted in Table 19.7a indicates the input features, the targets and the excluded examples. From the sketch, the prediction task becomes: **Given** three sets of coloured cells, **find** a predictive model using the columns of the first set of cells to predict the columns of the second set of cells without using rows from the third set of cells.

This prediction task is close to the AutoML task definition (Feurer *et al.*, 2015), with the difference that a loss function usually has to be defined in AutoML. However, we can define default choices for this loss function depending on the type of target feature. Hence, we can use any AutoML system, such as auto sklearn (Feurer *et al.*, 2015), TPOT (Olson *et al.*, 2016) or auto-WEKA (Kotthoff *et al.*, 2017) to perform a prediction task given the sketch presented in Table 19.7a.

If the first set of cells is empty, all columns not in the second set of cells are used as input features. If the second set of cells is empty, all empty cells are automatically added to the second set. The rationale is that we want to predict all empty cells.

Learning constraints and formulas

Formulas and constraints are key elements of spreadsheets. Formulas are used by users to specify how certain cells can be computed from other cells. For example, a formula $C_1 = MAX(C_2, ..., C_n)$ specifies that column C_1 is obtained by, for every row, computing the maximum of columns C_2 to C_n. Constraints can be used to verify whether the data satisfies some invariants and is consistent. Simple constraints are often used by spreadsheet users to perform sanity checks on the data (Hermans, 2013). For example, a constraint could test whether, in a column C_i, the values are ordered in increasing order. However, formulas themselves can also be seen as a type of constraints, specifying that the output values correspond to the values computed by the formula. Therefore, learning constraints and formulas can, in this context, be viewed as simply learning constraints.

In order to assist users in using constraints in their spreadsheets, as well as helping them recover, for example, data exported without formulas from enterprise software packages, existing systems such as TACLE (Kolb *et al.*, 2017) aim at automatically discovering constraints and formulas in spreadsheets across different tables. The authors propose a formalization of spreadsheet content into a hierarchical structure of tables, blocks and single rows or columns. Single rows or columns are denoted as *vectors* to abstract from their orientation and form the minimal level of granularity that constraints can reason about. This means that a constraint such as $C_1 = MAX(C_2, ..., C_n)$ can only span over entire rows or columns. Allowing constraints over subsets of vectors would allow for additional expressiveness at the price of decreased efficiency and a higher risk of finding spurious constraints that are true by accident. The data of every table T is grouped into contiguous blocks of vectors that have the same type and every vector is required to be *type consistent* itself, that is all cells within a vector—and by extension within a block—need to have the same type. In practice, these restrictions prohibit blocks or vectors that contain both textual and numeric cells. Mixed type vectors and blocks will be excluded from the constraint search. Blocks impose a hierarchy on groupings of vectors through the concept of sub-block containment: a block B_1 is a sub-block $B_2 (B_1 \sqsubseteq B_2)$ if B_1 contains a contiguous subsets of the vectors in B_2.

Similar to Inductive Logic Programming (ILP), constraint learning algorithms (De Raedt *et al.*, 2018) construct a hypothesis space of possible constraints. These algorithms then attempt to efficiently search in the hypothesis space for constraints that hold in the example data. TACLE constructs a hypothesis space using a large catalog of *constraint templates*, e.g., $?_1 = MAX(?_2)$. This approach is similar to Modelseeker (Beldiceanu and Simonis, 2012), which uses a catalogue of global constraints. We can now define the tabular constraint problems formally:

Given a set of instantiated blocks $\mathbf{B}$ over tables $\mathbf{T}$ and a set of constraint templates $\mathbf{S}$, **find** all constraints $s(B'_1, ..., B'_n)$ where $s \in \mathbf{S}$, $\forall i : B_i \sqsubseteq B'_i \in \mathbf{B}$ and $(B'_1, ..., B'_n)$ is a satisfied argument assignment of the template s.

We can use the sketch of Table 19.7a to instruct a constraint learning algorithm to learn constraints for the cells of interest. Starting from the given tables $\mathbf{T}$, we can construct a new set of tables $\hat{\mathbf{T}}$ that contains all coloured cells and a minimal number of uncolored cells and no cell coloured in red (the third set of coloured cells). This set of tables is computed by collapsing columns and rows that consist solely of uncoloured cells and removing cells from the third set of coloured cells. The blocks $\hat{\mathbf{B}}$ of these tables could be computed by grouping all neighboring type-consistent vectors. However, to avoid learning constraints over blocks that are not contiguous in the original tables, vectors that are separated in the original tables $\mathbf{T}$ by uncoloured vectors are not grouped within the same block. Additionally, to avoid learning constraints over partial rows or columns, only vectors are considered that are subsets of vectors that were type-consistent in the original set of blocks $\mathbf{B}$. Finally, we can run a tabular constraint learning such as TACLE on blocks $\hat{\mathbf{B}}$ to obtain a set of constraints that hold on these cells and can be mapped back to the original tables $\mathbf{T}$.

We briefly note that, since formulas can also be seen as predictors, and generic constraints—such as those learned by ModelSeeker (Beldiceanu and Simonis, 2012) or INCAL (Kolb *et al.*, 2018)—can also be seen as binary predictors, methods that learn these formulas or constraints can also be used specialized predictors and use the second set of coloured cells as to specify output (predicted) columns or rows.

Auto-completion

In typical spreadsheet applications, whenever the software detects that the user is entering a predictable sequence of values in a row or column (e.g., a constant ID or a sequence of evenly spaced dates), the remaining entries are filled in automatically. This is achieved using propagation rules. This elementary form of *auto-completion*, while useful for automating simple repetitive tasks, is of limited use for data science.

A much more powerful form of auto-completion is *predictive spreadsheet auto-completion under constraints*, or PSA for short (Kolb *et al.*, 2020). PSA can be defined as follows: **Given** a set of tables in a spreadsheet and a set of one or more empty cells, **find** an assignment of values to the cells. The key feature of PSA is that the missing values are inferred using one or more predictive models, often classifiers or regressors (Bishop, 2006), while ensuring that the predictions are compatible with the formulas and the constraints detected in the spreadsheet.

Let us illustrate predictive auto-completion using the sales data in Table 19.7a. Some of the values for $August$ are not yet available, hence $Total$ cannot be computed and no conclusion can be drawn about profitability. Intuitively, PSA auto-completes the table by performing the following steps: (1) find a predictive model for the column $August$ using (some of) the sale numbers for the other months; (2) discover a formula stating that $Total$ is the sum of $June$, $July$, and $August$; (3) find a predictive model for $Profit$ based on both the observed and predicted values; and (4) impute all missing cells.

PSA is significantly more useful for interactive data science than standard auto-completion, because it enables non-experts to make use of automatically extracted formulas and constraints without typing them, and to apply predictive models without specifying them. The assumption is, of course, that an appropriate user interface is available.

From the sketch presented in Table 19.7a, we can derive an auto-completion task, similar to the prediction task described above: **Given** three sets of coloured cells, **find** a predictive model using the columns of the first set of cells to predict the columns of the second set of cells without using rows from the third set of cells.

A general strategy for solving PSA was recently proposed that combines two of the core data science tasks considered above, namely learning predictors and learning constraints (Kolb *et al.*, 2020). At a high level, this strategy consists of two steps. In a first step, a set of predictors and formulas for the target cell(s) as well as a set of constraints holding in the data, are learned from the observed portion of the spreadsheet. Then, the most likely prediction consistent with the extracted constraints is computed. This is achieved by combining the learned predictors and formulas using probabilistic inference under constraints (Koller and Friedman, 2009). Low-performance models are automatically identified and their predictions are ignored.

In order to solve predictive spreadsheet auto-completion, we rely on PSYCHE, the implementation introduced in (Kolb *et al.*, 2020). For ease of exposition, we introduce PSYCHE on the simplest setting, namely auto-completing a single cell. In PSA, auto-completing a cell amounts to determining the most likely value that is consistent with respect to the constraints holding in the spreadsheet. If the machine knew what observed cells determine or influence the missing value (e.g., the $August$ sales) and what formulas and constraints hold in the spreadsheet ($Total$ is the sum of $June$, $July$, and $August$), then the problem would boil down to prediction under constraints. Indeed, one could train a predictive model (e.g., a linear regressor) on the fully observed rows and use it to predict the missing value in the target row. The caveat is that values that violate the constraints (e.g., the prediction for $August$ might be incompatible with the $Total$ revenue) must be avoided. In practice, however, no information is given about the relevant inputs and constraints.

To side-step this issue, PSYCHE extracts a set of candidate predictors and constraints directly from the data. We discuss this process next.

Solving predictive auto-completion under constraints. PSYCHE acquires candidate constraints and formulas from the spreadsheet by invoking TaCLe, a third-party learner specialized for this task (Kolb *et al.*, 2017). As for the predictors, PSYCHE learns a small ensemble of five to ten models, including decision trees, linear regressors, or other models. Since it is unclear which input columns are relevant, each predictor is trained to predict the target value from a random subset of observed columns. The intuition is that, while most input columns are likely irrelevant, some of the predictors will likely look at some of the relevant ones. Of course, some of the predictors may perform poorly. The rest of the pipeline is designed to filter out these bad predictions and retain the good ones. This is achieved with a combination of probabilistic reasoning and robust estimation techniques, as follows.

First, in order to correct for systematic errors, the outputs of all acquired predictors are calibrated on the training data using a robust estimation procedure. For example, in class-unbalanced tasks—like predicting the product ID of a rare ice cream flavour in a sales spreadsheet—predictors tend to favour the majority class. The calibration step is designed so to redistribute probability mass from the over-predicted classes to the under-predicted ones. The calibration is computed using a robust cross-validation procedure (Elisseeff and Pontil, 2003) directly on the data. The resulting estimate is further smoothed to prevent over-fitting. The outcome of this step is a calibrated copy of each base predictor.

In the next step, PSYCHE combines the calibrated predictions to determine the most likely value for the missing cell. The issue is that multiple alternatives are available, one for each predictor. The main goal here is to filter out the bad predictions. In the simplest case, PSYCHE performs the combination using a mixture of experts (Jordan and Jacobs, 1994; Bishop, 2006). At a high level, this means that each calibrated predictor votes one or more values, where the votes are weighted proportionally on the estimated accuracy of the predictors. PSyCHE implements several alternatives which differ in how trust is attributed to the various predictors. This produces a ranking of candidate values for the target cell.

As a final step, the learned constraints are used to eliminate all invalid candidate values and a winner is chosen. This guarantees that the value is both valid and suggested by the majority of high-quality (calibrated) predictors.

Auto-completing multiple cells requires performing the same steps. The only major complication is that, in this case, since the cells being completed may depend on each other (e.g., $August$, $Total$ and $Profit$ are clearly correlated), PSyCHE has to find an appropriate order in which to predict them. Since the rest of the process is intuitively identical to the single-cell case, we do not discuss this further here. The interested reader can find all the technical details in (Kolb *et al.*, 2020).

Integrating the sketches. Let us now consider the effect of coloured sketches. So far, we assumed that no information about the inputs, outputs, and constraints is available to the system. Sketches partially supply this information. In the previous section we discussed two types of sketches: (1) highlighting examples versus non-examples, and (2) identifying and correcting relevant inputs, cf. Table 19.7a. Both can be fit naturally into the design of PSyCHE and greatly simplify the auto-completion process.

In particular, information about invalid examples enables PSyCHE to avoid bad predictive models. The major benefit is that more resources can be allocated to higher-quality models, and that low-quality predictions will be less likely to influence or bias the inference process. Relevant input information has similar consequences.

19.4 Related Work

19.4.1 Visual analytics

Visual analytics refers to technologies that support discovery by combining automated analysis with interactive visual means (Thomas, 2005). VisualSynth is therefore tightly linked with visual analytics, as it combines automated data analysis with visual interaction. Visual analytics is typically used to help a user understand or solve a complex problem. Most approaches are tailored to a specific use case or a particular type of data, see (Kehrer and Hauser, 2012; Amershi *et al.*, 2014; Hohman *et al.*, 2018) for overviews. Some processes of data science have been studied in visual analytics: understanding a machine-learning model (Krause *et al.*, 2016), exploring data visualizations (Wongsuphasawat *et al.*, 2017), or building analysis pipelines (Wang *et al.*, 2016). Because these methods are task specific, a challenge in visual analytics is to design interactions that can handle a range of tasks, through different guidance degrees (Ceneda *et al.*, 2016). VisualSynth provides one way to use simple interaction through colourings across a range of data science related tasks. VisualSynth is therefore a first step towards solving some of the current challenges in visual analytics in the domain of data science.

19.4.2 Interactive machine learning

VisualSynth also has strong ties with the field of Interactive Machine Learning (IML). IML aims at complementing human intelligence by integrating it with computational power (Dudley and Kristensson, 2018). Some of the key challenges of IML are similar to the challenges we are also tackling: inconsistent and uncertain users, intuitive displaying of complex model decisions, and wide range of interesting tasks. To solve some of these challenges, most IML approaches focus on a particular type of data: text (Wallace *et al.*, 2012), images (Fails and Olsen Jr, 2003), or time series (Kabra *et al.*, 2013). In stark contrast, we focus on spreadsheets, which can store arbitrary combinations of numerical and categorical values, text, and time series. Moreover, in our setting the task to be solved (e.g., data wrangling, formula extraction, or clustering) is not given upfront. In explorative tasks, the user herself may not know what she is looking for in the data. Our goal is to help end-users carry out whatever task they have in mind, and which they may have trouble fully articulating.

19.4.3 Machine learning in spreadsheets

Small scale user studies about bringing basic machine learning capabilities for non-expert spreadsheet users have been conducted (Sarkar *et al.*, 2014, 2015). The main conclusion from these studies is that naïve end-users are able to successfully use basic machine learning algorithms to predict missing values or assess the quality of existing values. The user can use one button to indicate the data that can be used for learning (the training examples) and another button to apply the learned model to a specific column (the target variable). Visual feedback, in the form of cell colouring or cell annotations is added to communicate with the user. Colouring is used to indicate cells that should be used for training or whether the values imputed by the model are erroneous. The main difference between these two work (Sarkar *et al.*, 2014, 2015) and VisualSynth is that we present a general framework to perform data science tasks using sketches, while these work focus on user studies for the use of colours in spreadsheet for a specific data science task: prediction using k-Nearest Neighbor.

19.4.4 Auto-completion and missing value Imputation

Spreadsheet applications often implement simple forms of 'auto-completion' via propagation rules (Gulwani, 2011; Harris and Gulwani, 2011; Gulwani *et al.*, 2012). Clearly, even simple predictive auto-completion is beyond the reach of these approaches.

Techniques for missing value imputation focus on completing individual data matrices (Scheuren, 2005; Van Buuren, 2018) using statistics (Van Buuren, 2007) or machine learning (Stekhoven and Bühlmann, 2011). These techniques are not designed for spreadsheet data, which usually involves multiple tables, implicit constraints, and formulas. Several works automate individual elements of the spreadsheet workflow by, for example, extracting and applying string transformations (Gulwani, 2011; Gulwani *et al.*, 2015; Devlin *et al.*, 2017) and acquiring spreadsheet formulas and constraints hidden in the data (Kolb *et al.*, 2017). Psyche (Kolb *et al.*, 2020) combines such tools into a principled predictive auto-completion framework. In order to do so, it leverages probabilistic inference (using a form of 'chaining' (Van Buuren, 2007)) and learned constraints and formulas to fill in the missing values of multiple related tables. Psyche is an integral component of VisualSynth.

19.5 Conclusion

We presented VisualSynth, a framework for interactively modelling and solving data science tasks that combines a simple and minimal interaction protocol based on *coloured sketches* with *inductive models*. The sketches enable naïve end-users to (partially) define data science tasks such as data wrangling, clustering, and prediction. At the same time, the inductive models allow the system to clearly capture and reason with general data transformations. This powerful combination enables even non-experts to solve data science tasks in spreadsheets by *collaborating* with the spreadsheet application. VisualSynth was illustrated through examples on several data science tasks and on concrete use cases.

Building on VisualSynth, an interesting problem is predicting which sketch the user is likely to use given the current state of the spreadsheet. This is the problem of learning to learn, that is learning what knowledge the user would like to learn. To do this, an interesting starting point is to observe how users are using sketches to perform the task they have in mind. Then, learning from these interactions allows us to learn what sketches are typically used in a given state. Finding suitable representations of such a spreadsheet state is a challenging task, but semantic and structural information, as well as available knowledge are likely to play a key role.

Acknowledgements

This work was funded by the European Research Council (ERC) under the European Union's Horizon 2020 research and innovation programme (grant agreement No. 694980) SYNTH: Synthesising Inductive Data Models, by the Research Foundations Flanders and the Special Research Fund (BOF) at KU Leuven through pre- and post-doctoral fellowships for Samuel Kolb and by the Flemish Government (AI Research Program).

References

Amershi, S., Cakmak, M., Knox, W. B. *et al.* (2014). Power to the people: the role of humans in interactive machine learning. *AI Magazine*, 35(4), 105–20.

Basu, S., Banerjee, A., and Mooney, R. J. (2004). Active semi-supervision for pairwise constrained clustering, in *Proceedings of the 2004 SIAM International Conference on Data Mining*. Society for Industrial and Applied Mathematics, 333–44.

Beldiceanu, N. and Simonis, H. (2012). A Model Seeker: Extracting global constraint models from positive examples, in M. Michela, ed., *Principles and Practice of Constraint Programming*. Berlin, Heidelberg: Springer, 141–57.

Bishop, C. M. (2006). *Pattern Recognition and Machine Learning*. Berlin: Springer.

Bonifati, A., Ciucanu, R., and Staworko, S. (2016). Learning join queries from user examples. *ACM Transactions on Database Systems (TODS)*, 40, 24.

Buitinck, L., Louppe, G., Blondel, M. *et al.* (2013). API design for machine learning software: experiences from the scikit-learn project. arXiv preprint arXiv:1309.0238.

Ceneda, D., Gschwandtner, T., May, T. *et al.* (2016). Characterizing guidance in visual analytics. *IEEE Transactions on Visualization and Computer Graphics*, **23**(1), 111–20.

Chambers, C. and Scaffidi, C. (2010). Struggling to Excel: A field study of challenges faced by spreadsheet users, in *2010 IEEE Symposium on Visual Languages and Human-Centric Computing, Leganes*. New York, NY: IEEE, 187–94.

Cropper, A., Tamaddoni-Nezhad, A., and Muggleton, S. H. (2015). Meta-interpretive learning of data transformation programs, in *International Conference on Inductive Logic Programming, Kyoto*. Cham: Springer, 46–59.

De Raedt, L. (2002). A perspective on inductive databases. *ACM SIGKDD Explorations Newsletter*, 4(2):69–77.

De Raedt, L. (2008). *Logical and Relational Learning*. Berlin: Springer.

De Raedt, L., Passerini, A., and Teso, S. (2018). Learning constraints from examples, in *Proceedings of the 32nd AAAI Conference on Artificial Intelligence*, New Orleans. Boston: AAAI Press, 7961–70.

Devlin, J., Uesato, J., Bhupatiraju, S. *et al.* (2017). Robustfill: Neural program learning under noisy i/o, in *Journal of Machine Learning Research*, 70, 990–8.

Dudley, J. J. and Kristensson, P. O. (2018). A review of user interface design for interactive machine learning. *ACM Transactions on Interactive Intelligent Systems (TiiS)*, **8**(2), 8.

Elisseeff, A. and Pontil, M. (2003). Leave-one-out error and stability of learning algorithms with applications. *NATO Science Series Sub Series III Computer and Systems Sciences*, 190, 111–30.

Fails, J. A. and Olsen Jr., D. R. (2003). Interactive machine learning. In *Proceedings of the 8th international conference on Intelligent user interfaces*, Miami. New York, NY: ACM Press, 39–45.

Fayyad, U., Piatetsky-Shapiro, G., and Smyth, P. (1996). The KDD process for extracting useful knowledge from volumes of data. *Communications of the ACM*, **39**(11), 27–34.

Ferilli, S., Fanizzi, N., Di Mauro, N. *et al.* (2002). Efficient θ-subsumption under object identity, in *AI*IA Workshop su Apprendimento Automatico: Metodi e Applicazioni dell'Ottavo Convegno della Associazione Italiana per l'Intelligenza Artificiale*, Siena, Italy, 59–68.

Feurer, M., Klein, A., Eggensperger, K. *et al.* (2015). Efficient and robust automated machine learning, in *Advances in Neural Information Processing Systems*, 2962–70.

Gulwani, S. (2011). Automating string processing in spreadsheets using input-output examples. *ACM Sigplan Notices*, **46**(1), 317–30.

Gulwani, S., Harris, W. R., and Singh, R. (2012). Spreadsheet data manipulation using examples. *Communications of the ACM*, **55**(8), 97–105.

Gulwani, S., Hernández-Orallo, J., Kitzelmann, E. *et al.* (2015). Inductive programming meets the real world. *Communications of the ACM*, **58**(11), 90–9.

Harris, W. R. and Gulwani, S. (2011). Spreadsheet table transformations from examples. *ACM SIGPLAN Notices*, **46**(6), 317–28.

Hermans, F. (2013). Improving spreadsheet test practices, in *Proceedings of the 2013 Conference of the Center for Advanced Studies on Collaborative Research (CASCON'13)*, Toronto. Toronto: IBM Canada, CAS, 56–69.

Hohman, F. M., Kahng, M., Pienta, R. *et al.* (2018). Visual analytics in deep learning: An interrogative survey for the next frontiers. *IEEE Transactions on Visualization and Computer Graphics*, **25**(8), 2674–93.

Imielinski, T. and Mannila, H. (1996). A database perspective on knowledge discovery. *Communications of the ACM*, **39**(11), 58–64.

Jordan, M. I. and Jacobs, R. A. (1994). Hierarchical mixtures of experts and the em algorithm. *Neural Computation*, **6**(2), 181–214.

Kabra, M., Robie, A. A., Rivera-Alba, M. *et al.* (2013). JAABA: Interactive machine learning for automatic annotation of animal behavior. *Nature Methods*, **10**(1), 64.

Kehrer, J. and Hauser, H. (2012). Visualization and visual analysis of multifaceted scientific data: A survey. *IEEE Transactions on Visualization and Computer Graphics*, **19**(3), 495–513.

Kolb, S., Paramonov, S., Guns, T. *et al.* (2017). Learning constraints in spreadsheets and tabular data. *Machine Learning*, **106**(9), 1441–68.

Kolb, S., Teso, S., Dries, A. *et al.* (2020). Predictive spreadsheet autocompletion with constraints. *Machine Learning*, 109, 307–325.

Kolb, S., Teso, S., Passerini, A. *et al.* (2018). Learning SMT(LRA) constraints using SMT solvers, in *Proceedings of the Twenty-Seventh International Joint Conference on Artificial Intelligence*, Stockholm. California: IJCAI Press, 2333–40.

Koller, D. and Friedman, N. (2009). *Probabilistic Graphical Models: Principles and Techniques*. Cambridge, MA: MIT Press.

Kotthoff, L., Thornton, C., Hoos, H. H., Hutter, F. *et al.* (2017). Auto-WEKA 2.0: Automatic model selection and hyperparameter optimization in WEKA. *The Journal of Machine Learning Research*, **18**(1), 826–30.

Krause, J., Perer, A., and Bertini, E. (2016). Using visual analytics to interpret predictive machine learning models. *arXiv preprint arXiv:1606.05685*.

Lieberman, H. (2001). *Your Wish is My Command: Programming by Example*. San Francisco, CA: Morgan Kaufmann.

Muggleton, S. and De Raedt, L. (1994). Inductive logic programming: Theory and methods. *The Journal of Logic Programming*, 19, 629–79.

Muggleton, S., Feng, C. (1990). *Efficient induction of logic programs*. Proceedings of the First International Workshop on Algorithmic Learning Theory, Tokyo. pp. 368–381, Springer/Ohmsha.

Olson, R. S., Bartley, N., Urbanowicz, R. J. *et al.* (2016). Evaluation of a tree-based pipeline optimization tool for automating data science, in *Proceedings of the Genetic and Evolutionary Computation Conference (GECCO)*, Denver. New York, NY: ACM Press, 485–92.

Plotkin, G. D. (1970). A note on inductive generalization. *Machine Intelligence*, 5(1), 153–63.

Quinlan, J. R. (1990). Learning logical definitions from relations. *Machine Learning*, 5(3), 239–66.

Sarkar, A., Blackwell, A. F., Jamnik, M. *et al.* (2014). Teach and try: A simple interaction technique for exploratory data modelling by end users, in *2014 IEEE Symposium on Visual Languages and Human-Centric Computing (VL/HCC)*, Melbourne. New York, NY: IEEE, 53–6.

Sarkar, A., Jamnik, M., Blackwell, A. F. *et al.* (2015). Interactive visual machine learning in spreadsheets, in *2015 IEEE Symposium on Visual Languages and Human-Centric Computing (VL/HCC)*, Atlanta. New York, NY: IEEE, 159–63.

Scheuren, F. (2005). Multiple imputation: How it began and continues. *The American Statistician*, **59**(4), 315–19.

Stekhoven, D. J. and Bühlmann, P. (2011). Missforest—non-parametric missing value imputation for mixed-type data. *Bioinformatics*, **28**(1), 112–18.

Thomas, J. J. (2005). *Illuminating the Path: The Research and Development Agenda for Visual Analytics*. New York, NY: IEEE Computer Society.

Thornton, C., Hutter, F., Hoos, H. H. *et al.* (2013). Auto-WEKA: Combined selection and hyperparameter optimization of classification algorithms, in *Proceedings of the 19th ACM SIGKDD International Conference on Knowledge Discovery and Data Mining*. ACM, 847–55.

Van Buuren, S. (2007). Multiple imputation of discrete and continuous data by fully conditional specification. *Statistical Methods in Medical Research*, **16**(3), 219–42.

Van Buuren, S. (2018). *Flexible Imputation of Missing Data*. Chapman and Hall/CRC, Boca Raton. Florida.

Van Craenendonck, T., Dumančic, S., and Blockeel, H. (2017). COBRA: A fast and simple method for active clustering with pairwise constraints, in *Proceedings of the 26th International Joint Conference on Articial Intelligence, Sydney*. San Francisco, CA: Morgan Kaufmann, 2871–7.

Van Craenendonck, T., Dumančić, S., Van Wolputte, E. *et al.* (2018). COBRAS: Interactive clustering with pairwise queries, in *International Symposium on Intelligent Data Analysis*, 's-Hertogenbosch. Springer, 353–66.

Verbruggen, G. and De Raedt, L. (2018). Automatically wrangling spreadsheets into machine learning data formats, in *International Symposium on Intelligent Data Analysis XVII*, 's-Hertogenbosch. Cham: Springer, 367–79.

Wagstaff, K., Cardie, C., Rogers, S. *et al.* (2001). Constrained k-means clustering with background knowledge, in *Proceedings of the 18th International Conference on Machine Learning*, Williamstown, MA. San Francisco, CA: Morgan Kaufmann, 577–84.

Wallace, B. C., Small, K., Brodley, C. E. *et al.* (2012). Deploying an interactive machine learning system in an evidence-based practice center: ABSTRACKR. In *Proceedings of the 2nd ACM SIGHIT International Health Informatics Symposium*, Miami. New York, NY: ACM Press, 819–24.

Wang, X.-M., Zhang, T.-Y., Ma, Y.-X. *et al.* (2016). A survey of visual analytic pipelines. *Journal of Computer Science and Technology*, 31(4):787–804.

Wongsuphasawat, K., Qu, Z., Moritz, D. *et al.* (2017). Voyager 2: Augmenting visual analysis with partial view specifications, in *Proceedings of the 2017 CHI Conference on Human Factors in Computing Systems*, Denver. New York, NY: ACM Press, 2648–59.

Xing, E. P., Jordan, M. I., Russell, S. J. *et al.* (2003). Distance metric learning with application to clustering with side-information, in *Advances in Neural Information Processing Systems*, Vancouver, British Columbia, 521–8. MIT Press.

Xu, R. and Wunsch, D. (2005). Survey of clustering algorithms. *IEEE Transactions on Neural Networks*, **16**(3), 645–678.

Part 5

Evaluating Human-like Reasoning

20

Automated Common-sense Spatial Reasoning: Still a Huge Challenge

Brandon Bennett and AnthonyG Cohn

University of Leeds

20.1 Introduction

Achieving 'common-sense reasoning' capabilities in a computational system has been one of the goals of Artificial Intelligence since its inception in the 1960s (McCarthy and Hayes, 1969; McCarthy, 1989; Thomason, 1991). However, as Marcus and Davis have recently argued (Marcus and Davis, 2019): 'Common sense is not just the hardest problem for AI; in the long run, it's also the most important problem.' Moreover, it is generally accepted that space (and time) underlie much of what we regard as common-sense reasoning. For example, in the list of common-sense reasoning challenges given at http://www-formal.stanford.edu/leora/commonsense/, most of these rely crucially on spatial information.

From the 1990s onwards, considerable attention has been given to developing theories of spatial information and reasoning, where the vocabulary of the theory was intended to correspond closely with properties and relationships expressed in natural language but the structure of the representation and its inference rules were formulated in terms of computational data and algorithms (Forbus *et al.*, 1991; Egenhofer, 1991; Freksa, 1992; Frank, 1992; Ligozat, 1993; Hernández, 1993; Gahegan, 1995; Zimmermann, 1993; Faltings, 1995; Escrig and Toledo, 1996; Gerevini and Renz, 1998; Moratz *et al.*, 2011; Mossakowski and Moratz, 2012) or in a precise logical language, such as classical first-order logic (Randell *et al.*, 1992; Gotts, 1994; Cohn, 1995; Borgo *et al.*, 1996; Cohn *et al.*, 1997; Galton, 1998; Pratt and Schoop, 1998; Pratt, 1999; Cohn and Hazarika, 2001; Galton, 2004).

However, despite a great number of successes in dealing with particular restricted types of spatial information, the development of a system capable of carrying out automated spatial reasoning involving a variety of spatial properties, of similar diversity to what one finds in ordinary natural language descriptions, seems to be a long way off. The lack of progress in providing general automated common-sense spatial reasoning capabilities suggests that this is a very difficult problem.

Brandon Bennett and AnthonyG Cohn, *Automated Common-sense Spatial Reasoning: Still a Huge Challenge* In: *Human Like Machine Intelligence.* Edited by: Stephen Muggleton and Nick Charter, Oxford University Press.
DOI: 10.1093/oso/9780198862536.003.0020

As with most unsolved problems, there are a variety of opinions about why common-sense spatial reasoning is so difficult to achieve and what might be the best approach to take. A point of particular contention, which will be explored in detail in the current chaper, is what is the role of natural language in relation to common-sense spatial reasoning?

The main purpose of this chapter is to help researchers orient and focus their investigations within the context of a highly complex multifaceted area of research. We believe that research into computational common-sense spatial reasoning is sometimes misdirected for one or both of the following reasons: (1) the goal of the research may incorporate several sub-problems that would be better tackled separately; (2) the methodology of the research may assume that other related problems can be solved much more easily than is actually the case.

The chapter gives a fairly general (though not comprehensive) overview of the goal of automating common-sense reasoning by means of symbolic representations and computational algorithms. Previous work in the area will be surveyed, the nature of the goal will be clarified, and the problem will be analysed into a number of interacting sub-problems. Key difficulties faced in tackling these problems will be highlighted and some possibilities for solving them will be proposed.

The rest of the chapter is structured in terms of the following list of what we consider to be the most important problems that are obstructing the development of automated common-sense spatial reasoning systems:

1. Lack of a precise meaning of 'common-sense reasoning'.
2. Difficulty of establishing a general foundational ontology of spatial entities and relationships.
3. Identification and organization of a suitable vocabulary of formalized spatial properties and relations.
4. How to take account of polysemy, ambiguity, and vagueness of natural language.
5. Difficulty of modelling the role of various forms of implicit knowledge (context, background knowledge, tacit knowledge).
6. Lack of a default reasoning mechanism suited to reasoning with spatial information.
7. Intrinsic complexity of reasoning with spatial information.

Of course we do not claim that there has been no progress in addressing these challenges (and indeed we mention a few examples of works that do below), but it seems to us that these still represent considerable challenges to solve in the general case.

20.2 Common-sense Reasoning

In this section we examine the nature of common-sense reasoning and look at the ways in which research in computational artificial intelligence has sought to model and simulate human common-sense reasoning.

20.2.1 The nature of common-sense reasoning

Although the specific processes by which human reasoning occurs are little understood, the meaning of the word 'reasoning' is relatively clear. It refers to any kind of process by which new implicit information is derived from given or assumed information. However, there are several different forms in which information is manifested and communicated. Figure 20.1 illustrates those types of information and relationships that we consider to be particularly relevant to the understanding of different modes of reasoning.

One might define *common-sense* reasoning as that kind of reasoning that humans use in everyday situations, without explicit use of logical, mathematical, or scientific theories. As such, the ambit of common-sense reasoning corresponds to the lighter-grey region of the diagram, with its primary components being: *mental state*, *perceptual information* and *propositional information*, (expressed in natural language).

Although the idea of an agent's 'mental state' is widely used in explanations of the behaviour of humans and animals, its constitution and function are poorly understood and we will not speculate on these; nor do we have the space here to consider the distinction between short- and long-term representations which are clearly important but not germane to our main argument here. For present purposes, we need only consider

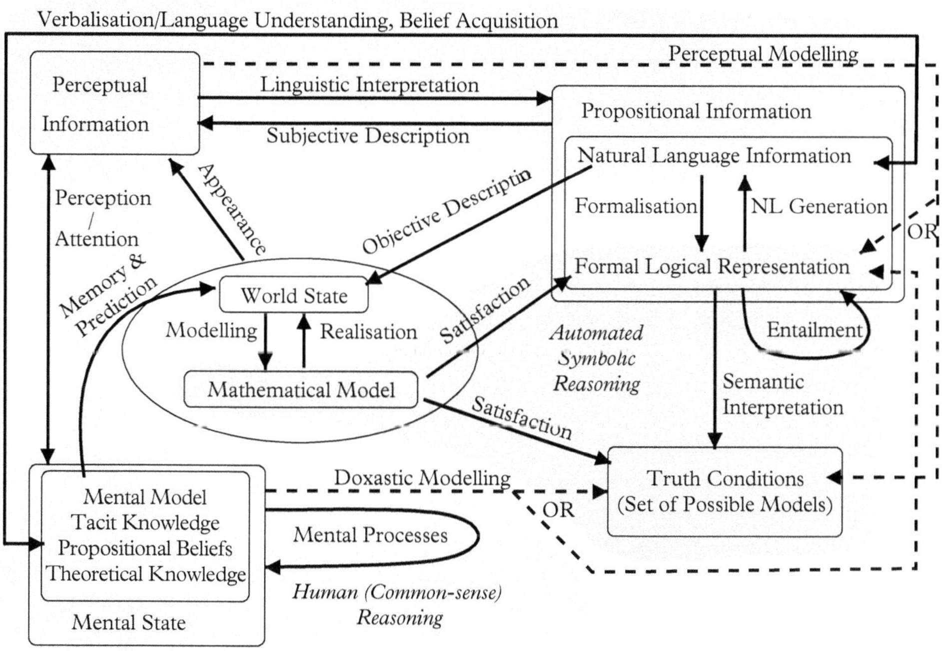

Figure 20.1 *Types of information and the relationships between them. Note that none of the arrows denote causal relationships, perhaps with the exception of the 'mental processes' arrow; rather they denote a wide variety of other kinds of relationships such as epistemological and metalogical relationships.*

what types of information might in some way be stored within an agent's mind. We assume that a mental state includes some kind of *mental model* (Johnson-Laird, 1983), which somehow stores some correlate of received perceptual information in such a way that it can be used to remember or predict useful information about the state of the world. We also assume that the mental state incorporates *tacit knowledge* (Polanyi, 1966; Schacter, 1987; Kimble, 2013), which provides the agent with certain capabilities and skills (either instinctive or learned). Mental models and tacit knowledge are taken to be *non-linguistic* forms of information and hence can be possessed by agents with no linguistic capability. These kinds of information are difficult to articulate in verbal form. Researchers studying tacit knowledge often claim that it is impossible to convert it into an equivalent propositional form. Other researchers (e.g., in symbolic AI) take the view that it is very difficult but not impossible to specify an explicit symbolic correlate of tacit knowledge. We tend towards the latter view.

In the case of beings with linguistic capabilities, their mental state will also somehow store (or be able to generate) propositional information—that is internal correlates of natural language sentences. These correspond to the verbalizable beliefs of the agent. A special case of such beliefs would be theoretical knowledge of logic, mathematics, or science. Although such theoretical knowledge may be part of the mental state of a sophisticated linguistically capable agent and applied in their processes, its use goes beyond what would be considered *common-sense* reasoning. (Hence, in Figure 20.1, 'theoretical knowledge' is not within the light-grey area of the diagram.)

There are several paths that reasoning can take. The most basic is where the appearance of the world generates perceptual information, which is (somehow) absorbed into an agent's mental state. Mental processes then take place that modify the current mental state to produce a new state that may include the results of some kind of inferential process. (We will not speculate on any details of how mental inference might operate.) Finally, the updated mental state may incorporate some prediction about the world state. This prediction is some piece of information that was not directly present in the perceptual information (nor in information derived by low level processing that takes place as part perception) but has been derived by the reasoning process.

The kind of reasoning just described does not necessarily involve any kind of linguistic information. Hence, it could be carried out by languageless animals. However, the diagram also includes a category of *propositional* information and indicates that information expressed in natural language may also play a part in human common-sense reasoning. Perceptual information may be converted into propositional information by linguistic interpretation. This can then be incorporated into an agent's mental state, in the form of propositional beliefs. Mental processes may then make use of this propositional information (often in combination with other types of information in the mental state) in order to draw inferences by some kind of mental argumentation process.

The reader will have noticed that the diagram also indicates a second mode of reasoning, which is ostensibly very different from common sense: the darker-grey area of the diagram demarcates the types of information that are manipulated by automated symbolic reasoning mechanisms. This kind of reasoning is relatively well-understood by mathematicians and computer scientists. However, it is only indirectly linked to the components of the common-sense reasoning system just described. The most overt connection between common-sense reasoning and automated reasoning occurs within the category of *propositional*[1] information. Here we have both natural language propositions (i.e., assertive sentences) and formulae of some logical language. We may map between these by procedures of formalization (natural to formal) and natural language generation (formal to natural). However, establishing appropriate mappings has proved to be extremely difficult, especially in the direction natural-to-formal. There is huge controversy over how this should be done and even over whether it can be done at all in a general and reliable way. Moreover, even the details of a logical language suitable for capturing the meanings of natural language sentences are disputed, both in terms of the non-logical predicates that will be needed and in terms of the logical operators and structures that will be required.

Another linkage between common-sense and automated reasoning processes occurs indirectly *via* reality itself—that is in relation to the *world state* (the physical material and structure of reality). On one side of the connection, the world state interacts with common-sense reasoning in two ways: the world generates perceptual information; and the contents of a mental state somehow enable predictions to be made about the world state. On the other side of the link, the world state is regarded as having some correspondence (albeit usually very coarse grained) with a mathematical model that provides an interpretation (i.e., a *model* in the sense of model theoretic semantics) for the formal logical representation. Here again there is great controversy over what form an appropriate mathematical model should take and even whether it is possible to provide an adequate model at all.

We have also included in the diagram some further linkages indicated by dashed arrows. These are of a more putative nature. One possibility is that one might carry out some kind of 'perceptual modelling', that would map perceptual information either into a formal logic or into some other representation of truth conditions. we also indicate that some kind of 'doxastic modelling' could provide a mapping from mental state either to some formal logical representation or directly to truth conditions (which would then consist of the set of possible worlds that are compatible with the beliefs held within the mental state (Hintikka, 1962)). Our motivation in adding these links is to allow for the possibility that human common-sense reasoning could be simulated by an automated reasoning system without the need to use natural language as a bridging representation. Nevertheless, it is far from obvious how these links could be substantiated by an actual modelling process (although work on mental models has attempted to explicate a linkage to truth conditions and hence to human reasoning processes (Johnson-Laird, 1983)).

[1] By propositional information, we mean information expressed in propositions of any form. We do *not* mean that the logical form of expressions is restricted to only atomic propositions and compounds formed using propositional operators. So any natural language assertive sentence or formula of some logical language would be an example of propositional information.

20.2.2 Computational simulation of commonsense spatial reasoning

A typical approach to developing computational common-sense spatial reasoning within the field of symbolic AI has been to design formal logical representations that are envisaged as being close to natural language forms of spatial description, and therefore similar to the kinds of propositional information used in human common-sense reasoning (Randell *et al.*, 1992; Borgo *et al.*, 1996). Since, reasoning with numerical information is generally considered to be mathematical rather than common-sense reasoning, the formal language is usually restricted to representing qualitative properties and relationships; hence the field is known as Qualitative Spatial Reasoning (QSR) (Cohn and Renz, 2008; Ligozat, 2011) which in turn is part of the larger field of Qualitative Reasoning (Forbus, 2019). For example, the Region Connection Calculus (RCC) has been widely used for a variety of purposes from modelling geographic information to representing activities in video. The two most common variants of RCC are RCC-8 and RCC-5 (8 and 5 referring to the number of jointly exhaustive and pairwise disjoint (JEPD) relationships in the calculus). The RCC-8 relations are depicted in Figure 20.2; Egenhofer (1991) has postulated a similar set of relations from a different mathematical basis. The RCC calculi, along with most other QSR systems, can not only be structured into 'conceptual neighbourhoods' as depicted in the figure, but also one can construct *composition tables* which enable inferences to be made about relationships between spatial relations not already explicit (e.g., from NTPP(a,b) $\wedge$ TPP(b,c) infer NTPP(a,c)).

There are several potential problems with this approach. One is the difficulty of ensuring that the formal language developed is adequate for the kinds of reasoning that can be carried out by human common-sense. Indeed that has certainly not been achieved in a general way. The most that can be claimed is that formal representations have

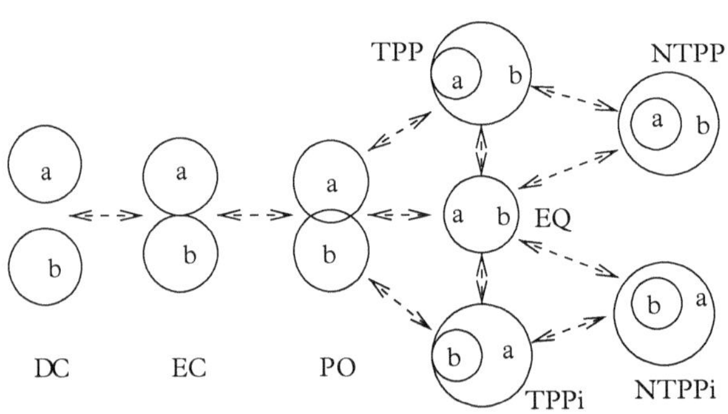

Figure 20.2 *A depiction of the RCC-8 relations. The connecting arcs indicate the 'conceptual neighbourhood', that is those neighbouring relations which one relation can transition to immediately, assuming continuous transitions or deformations. These 'conceptual neighbour' relations can be used to perform qualitative spatial simulations in order to predict possible futures (e.g. see Cui* et al.*, 1992). RCC-5 is formed from RCC-8 by merging* DC *and* EC, TPP *and* NTPP, *and* TPPi *and* NTPPi.

been developed that are capable of simulating some fragment of human common-sense reasoning. Several researchers have conducted psychological experiments to determine the extent to which sets of qualitative relations that have been used as the basis of qualitative spatial representations are comprehensible to human agents and are *cognitively plausible*—that is compatible with human spatial reasoning capabilities e.g., (Knauff *et al.*, 1997; Klippel *et al.*, 2013). The former concluded from their experiments that the more fine-grained RCC-8 relations (rather than RCC-5) 'are actually the most promising starting point for further psychological investigations on human conceptual topological knowledge. However, further evidence will be needed before a detailed modeling of human conceptual knowledge is possible.' Knauff (1999) also investigated the cognitive adequacy of Allen's Interval Calculus (IA), which has 13 JEPD relations between intervals such as 'before', 'meets', 'overlaps', 'during'; the IA has often been used for reasoning about space. Knauff found that some evidence to support the cognitive adequacy of the IA, particular with regard to the associate composition table. However, whereas it had been postulated in the literature that errors in choice of a relation would normally be conceptual neighbours rather than random relations was not upheld in his experiments. Knauff also found that his results agreed with the 'mental model' theory that has been suggested as a human problem-solving paradigm (Knauff *et al.*, 1998).

Another problem is that, as we have elaborated above, propositional information is not the only kind of information involved in human reasoning, and is probably not involved at all in many reasoning tasks, which are effected by manipulation of mental models (Ragni *et al.*, 2005) and/or tacit knowledge. Nevertheless, this need not necessarily be reason not to use formal propositional representations. Such representations have very general expressiveness, so it is plausible that even though the mind of an intelligent agent may be working with non-propositional tacit knowledge and/or neurologically implemented mental models, it might still be possible to encode the relevant information content in a propositional form. This possibility corresponds to the 'doxastic modelling' arrow in Figure 20.1. Of course one cannot directly access the structure and content of a human mind, so the modelling would need to be done indirectly, by a process of hypothesis and testing.

A different approach is to model the perceptual information received by an agent and use this directly within a reasoning system. This is the approach taken within the *situated* approach to AI promulgated by Rodney Brooks, among others (Brooks, 1986).

20.2.3 But natural language is still a promising route to common-sense

Despite the caveats of the last section, we still believe that natural language is likely to provide the most accessible entry point into common-sense spatial reasoning and provide fruitful insight into the semantic distinctions and inference patterns upon which it is based. If we could compute inferences from natural language information (e.g., text) that were judged to be broadly correct by humans, then we would have solved a large part of the problem of automating common-sense reasoning. In so far as non-linguistic information plays a part in common-sense, this would need to be somehow built into the inference generation mechanism. But, given the generality of logical reasoning techniques, there seems to be no obvious reason why this could not be done.

If we do choose to attack the problem of automating common-sense reasoning *via* the analysis of natural language, there are still several different ways in which this can be done. Contrasting views have been put forward by Davis (2013) and Bateman *et al.* (2010); Bateman (2013). Davis' analysis of spatial reasoning required for natural language text understanding begins by analysing the semantics of sentences in terms of the geometrical constraints that they seem to obey (in many cases identifying various constraints corresponding to different interpretations). Bateman's approach is to try to model the semantics of natural language terminology more directly, without attempting to resolve all ambiguities in their geometrical interpretation. The idea is that language-oriented inference rules can be formulated, which generalize over the variety of different ways in which natural language terminology can be employed in spatial descriptions.

Each of these approaches has its own problems.

With the Davis approach, the mapping from natural language to a formal representation is achieved by the expert judgement of a knowledge engineer. This leaves a significant gap in achieving automatic reasoning with natural language. Moreover, questions regarding why a particular interpretation was chosen may be difficult to answer.

The Bateman approach faces the following problem: on the one hand, in many cases it will be difficult (sometimes impossible, we would suggest) to specify the semantics of a natural language term in a way that is sufficiently general to capture all its most common uses. On the other hand, it will be difficult so to do in a way that is not so general that it captures inferences that one would expect to only follow from particular applications of a term, which one would expect to have a more specific semantic interpretation. In other words, the Bateman approach may suffer due to a difficulty in distinguishing generality from ambiguity without recourse to some extra-linguistic semantic view point (such as mathematically specified geometrical constraints). An associated problem is that the Bateman approach seems to fall short of supplying truth conditions for propositions. This means that it is unclear what criteria would be used to judge the validity of an inference.

20.3 Fundamental Ontology of Space

Despite having been an object of enquiry for thousands of years, the constitution and structure of space and the material world is still a subject of much controversy. Although, scientific theories provide detailed accounts in terms of particles, fields, and forces, the mathematical models developed by physicists are far removed from the terminology and informal inference patterns used in everyday description and reasoning about spatial properties and relationships. Over the past couple of decades the need for theories of space and material object that correspond more closely to natural modes of description have been recognised by many researchers in artificial intelligence and information science (Bennett, 2001; Masolo *et al.*, 2003; Grenon and Smith, 2004); nevertheless, some fundamental problems remain.

20.3.1 Defining the spatial extent of material entities

Fundamental to spatial reasoning is the association between material objects and spatial regions. However, determining the spatial extent of a material entity is complicated by the following considerations:

- The conditions for determining whether a particle is a constituent of a particular entity may be vague. For example, the surface of an animal (i.e., its skin) may have an outer layer incorporating dead or damaged cells, which are only loosely attached, and for which it is unclear which of them should be taken as constituents of the animal. Similarly, a rock may be made up of an agglomeration of rock particles, such that it can be unclear which are actually part of the rock and which are separate but ingrained within a cavity of the rock's surface.
- The exact positions of particles are unknown and intrinsically uncertain.
- Matter is made up of particles that are relatively sparsely distributed in space.
- Many materials (e.g., rock) contain tiny voids, such that it is not clear whether the volume of the voids should be considered as part of the material. To complicate matters, the voids may sometimes be filled with other materials such as water (Hahmann and Brodaric, 2012; Hahmann and Brodaric, 2014).

Let us suppose that intrinsic uncertainty and vagueness can be ignored. That is, we *assume* that:

- Although it may be difficult (or even impossible) to determine in practice, each physical entity is associated with a definite set of atoms (often combined into molecules).
- Although the positions of atoms are uncertain and constantly variable, it is possible in principle (though in most cases not in practice) to establish an assignment of a precise spatial location to each atom (e.g., by numerical coordinates), such that the resulting structure of spatially located atoms is sufficient to capture all aspects of the structure of an entity required to characterise its physical properties—except those that depend on sub-molecular scale details.

Even under the assumption that the particles that ultimately constitute matter have definite spatial locations, we still have the problem that these particles are relatively sparsely distributed in space, so that their combined spatial extent would be more like a scattered cloud of almost point-like regions than a continuously filled volume of space. One method of determining the spatial extent of a material entity would be by constructing an α-volume (Edelsbrunner *et al.*, 1983). This gives a well-defined procedure for determining a reasonable containing volume for an arbitrary set of points. The only problem with this is that it depends on the choice of a parameter, α, that determines what size of gap between points gets filled in to form the volume. When considering most physical entities, an α distance that is microscopic but considerably larger than the length of a molecule would be appropriate, since then the α-volumes of cells and larger entities would be continuous and connected, whereas if a smaller α distance were used, their α-volumes would have many gaps and discontinuities arising from the spaces between molecules. But in considering the structure of a molecule or atom, a much smaller α parameter would be needed, otherwise their α-volumes would be too course-grained to exhibit any distinctive spatial structure.

20.4 Establishing a Formal Representation and its Vocabulary

We may divide the analysis of natural language semantics into two parts: the elicitation of logical form (compositional structure) and the specification of content (meanings of terms).

20.4.1 Semantic form

The application of automated symbolic reasoning techniques to natural language sentences requires that they be translated into a form that makes explicit their logical structure. For example, *'The pot contains lead'* would be represented by a formula such as $\exists x[\text{'}Pot\text{'}(x) \wedge \text{'}Contains\text{'}\,(x, \text{'}lead\text{'})]$, which indicates predicative and quantificational structure but retains the vocabulary of the original sentence.[2] Performing this conversion is in general non-trivial. However, even where a sentence conveys spatial information, there is nothing particularly spatial about the logical form of the sentence. Hence, although this conversion is a problem for common-sense reasoning in general (at least if we want to automatically input information expressed in natural language into our common-sense reasoning system) it is not specifically a sub-problem of automating *spatial* common-sense reasoning.

20.4.2 Specifying a suitable vocabulary

By contrast, providing a semantics for the vocabulary of sentences conveying spatial information is a significant part of the common-sense spatial reasoning problem.

Specifying a suitable spatial vocabulary for a common-sense spatial reasoning system is a large and complex task. The concepts and relations used in natural language give a guide to the range of concepts required and the distinctions that one will want to make. However, the expression of these concepts in natural language is often highly ambiguous. Spatial phrases are applied in a wide variety of different situations so that it is not obvious what is the core meaning or whether there are several different interpretations. To make matters worse, there can be considerable overlap between possible interpretations of different phrases. Such differences or overlapping of senses depend very much on the

[2] Note that for simplicity we render the definite article 'The' by the existential quantifier in this example.

specific details of a particular spatial situation: in some situations, two phrases may seem to be equally appropriate, whereas in others one phrase will be much more apt than another.

In order to ensure the consistency and semantic rigour of a formal spatial vocabulary, the classification of meanings will need to be conducted in a systematic way. However, there are several different ways in which such a classification might be organized:

- Taxonomic—concepts are identified by successive differentiation of general concepts into more specific refinements.
- Compositional—a limited set of basic concepts/relations is used to construct a more comprehensive vocabulary. This in turn can be achieved in at least two different ways:
  - Analytic—a set of primitives is identified in order to provide fundamental conceptual units from which more complex concepts can be constructed by definitions that are expressed as structured combinations of the primitives. There is little if any overlap in the meaning of each primitive concept
  - Synthetic—key general concepts are identified from which more specific concepts can be constructed by combination. The key concepts may overlap so that specialisation may be achieved by their conjunction.

20.4.3 The potentially infinite distinctions among spatial relations

A primary reason why a purely taxonomic approach is unlikely to achieve full generality is that ordinary language allows arbitrary elaboration of our descriptions of a spatial situation. An obvious way in which we can express limitless variety in spatial relations is by referring numerically to multiple sub-features of a spatial situation. For example, an entity could have any number of protruding sub-parts and these could be spatially related to some other entity in specific ways. Figure 20.3 illustrates two relatively simple cases of the potentially infinite number of variants of relations between two disconnected regions that can occur when one region has multiple lobes, each of which protrudes into a distinct open cavity of the other region.

Counting sub-features is a rather trivial way in which spatial relationships can be differentiated into more and more sub-types. However, there are other types of relation refinement that give rise to large numbers of distinctive spatial configurations. Figure 20.4(a) illustrates how, in the context of considering the topological relationship

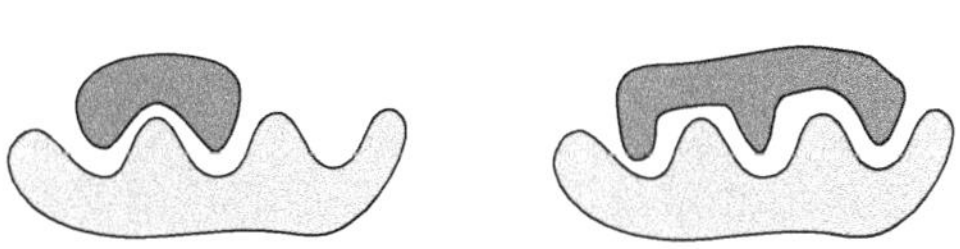

Figure 20.3 *Denumerable variants of disconnectedness with respect to multiple cavities.*

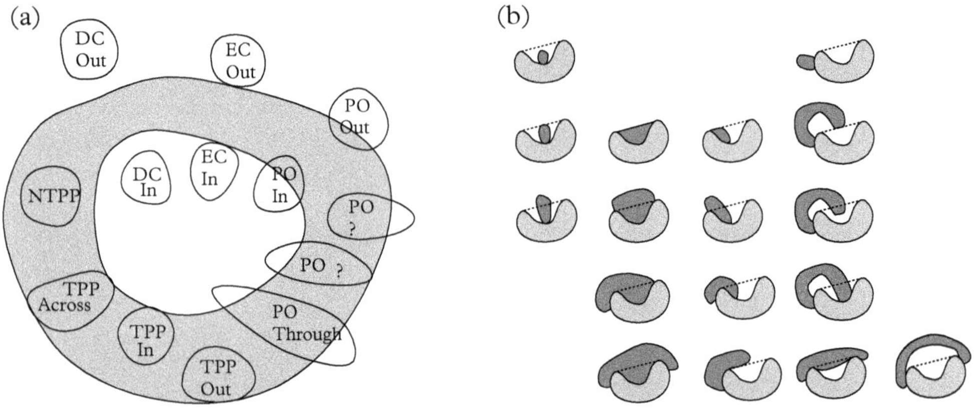

Figure 20.4 *(a) Refining RCC-8 (Randell* et al.*, 1992) with respect to a region with an internal cavity (in 2D). (b) Refining the relation of external connection with respect to a region with an external concavity.*

between an arbitrary self-connected region and an entity with an internal cavity (the shaded 'doughnut'), the RCC-8 relations can be refined into various more specialised relations. Figure 20.4(b) shows a large number of possible refinements of an external connection relation holding with respect to a region with an external concavity (cf. (Cohn *et al.*, 1997)).

These examples suggest to us that the approach used widely in ontology construction, of specifying properties and relations by means of a taxonomy that successively refines concepts from general to specific, may not be the most appropriate for spatial concepts. In Bennett *et al.*, 2013, it is suggested that a structured classification of spatial concepts consisting of an initial shallow hierarchy of topological relations supplemented by an open-ended set of analytic definitions of more specialised relations, formulated by explicitly referring to entities such as surfaces and cavities, may be a better approach.

20.5 Formalizing Ambiguous and Vague Spatial Vocabulary

Many spatial concepts have such a wide variety of uses that it seems to be impossible to give any concise definition that would cover all kinds of application of the concept (e.g., the general concept of 'place' (Bennett and Agarwal, 2007)). Nevertheless, such concepts somehow seem to give the impression of conveying a coherent and unproblematic meaning. As well as often involving concepts that are ambiguous—that is having several distinct (although possibly overlapping) meanings—spatial concepts may also be *vague*—that is subject to gradations of meaning with no clear cut-off point regarding the applicability of the concept (Bennett, 2011). A number of researchers have investigated ways in which typical cut-off points or ranges of cut-off points can be elucidated from human subjects (Mark and Egenhofer, 1994; Mark *et al.*, 1995; Montello *et al.*, 2003).

In this section we shall consider a variety of examples that illustrate the ubiquity of ambiguity and vagueness in natural language spatial vocabulary and thereby indicate the scale of the difficulty that these phenomena pose to the endeavour of formalising common-sense spatial reasoning based on natural language expression of spatial information.

20.5.1 Crossing

Phrases of the form 'x crosses y' are very common in spatial descriptions, and the notion of one entity crossing another seems to convey a basic notion. But such phrases can have a wide range of different interpretations. The nature of the relationship referred to can often be determined (or at least narrowed down) by knowledge of the type of the entities x and y that are involved, but additional background knowledge may also be needed in order to disambiguate the meaning.

Some different interpretations are as follows:

- A flat elongated entity is part of a surface and runs from one edge of the surface to another. For example, a path may cross a park.
- Two elongated entities may intersect (typically, approximately at right angles) at some location that is at a mid-point (or mid-section) of both of them. For example, two roads may cross.
- An entity may cross a barrier by passing through a hole in the barrier.
- An entity may cross a barrier and also be part of that barrier. For example a protein that is part of a cell membrane and protrudes both into the cytoplasm and out to the exterior of the cell.
- A line or linear entity or a three-dimensional entity may cross a surface by having a part on one side of the surface and a part on the other side of the surface.
- An entity may cross another entity by going over it from one side to another.
- There are also many dynamic interpretations of 'cross'—as in 'the runner crossed the finishing line'. The dynamic interpretations vary in similar ways to the static interpretations. (Such interpretations will not be considered further in this chapter.)

20.5.2 Position relative to 'vertical'

The concepts of 'above', 'below', 'over', 'under, 'beneath', depend on having some notion of the directions 'up' and 'down' in relation to the entities being considered. The clearest cases are where 'up' and 'down' are interpreted according to the reference frame of our planet Earth itself. 'Up' normally means away from the centre of the Earth, whereas down is towards the centre of the Earth.

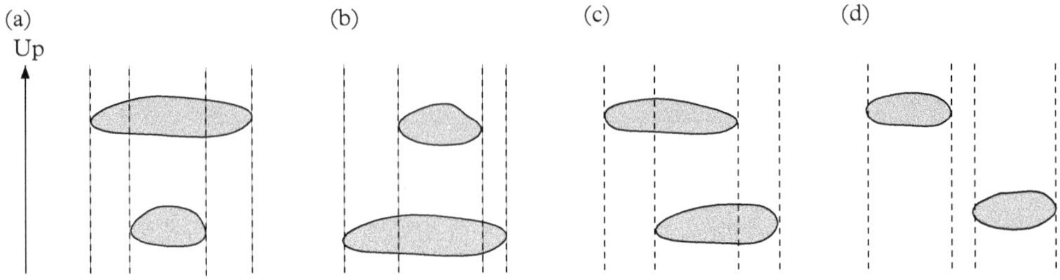

Figure 20.5 *Variants of the above–below relationship.*

However, there is a certain amount of ambiguity in these relations when we consider the variety of situations in which they might be judged to apply. Some possibilities are illustrated in Figure 20.5. The cases shown are as follows:

(Fig. 20.5 a) Here every point of the lower region is directly below some point of the upper region (and no point of the upper region is below any point of the lower region).

(Fig. 20.5 b) Here every point of the upper region is directly above some point of the lower region (and no point of the upper region is below any point of the lower region).

(Fig. 20.5 c) Here some points of the upper region are above some points of the lower region (and no point of the upper region is below any point of the lower region).

(Fig. 20.5 d) Here every point of the upper region is higher than every point of the lower region, even though no point of either region is above (or below) any point of the other region.

It is worth noting that only cases (a) and (b) are transitive. Also, out of all the cases, (a) and (b) seem to be the most typical examples that one would describe by saying one region was above (or below) the other. Hence, is tempting to interpret 'x is above/below y, as holding whenever either of the situations (a) and (b) occurs. However, if we take $\mathsf{above}(x, y)$ as holding when either of the situations (a) and (b) occurs, then this relation is not transitive. By referring explicitly to the relative vertical positions of points (or parts) of the entities concerned, it is possible to define a variety of different relations that describe the relative vertical positions of extended entities. However, these will have different semantics, and hence, different import with respect to logical entailment. Moreover, the mappings between these and relations referred to in natural language will be ill-defined.

20.5.3 Sense resolution

In the works (Bennett *et al.*, 2013; Bennett and Cialone, 2014), the authors investigated (by analysing a text corpus obtained from a large biology text-book (Reece *et al.*,

2011)) the use of the spatial relation terms 'contain', 'enclose', and 'surround'. For each occurrence of any of these words (and cognate forms) the geometrical configuration to which the word was being applied was determined (from the text and with the use of auxiliary reference works). The authors found that around 15 different geometric conditions seemed to cover all usages of the terms. They calculated the frequencies with which each word was used to describe a particular geometrical constraint and found that each word could be applied to a variety of different geometrical constraints and that there was considerable overlap in the use of the terms. However, there were also significant differences in the frequencies that a particular term was applied to a given situation, with each having different typical and atypical uses.

Given the many–many correspondence between natural language spatial terminology and geometrical constraints, common-sense spatial reasoning conducted on the basis of linguistic descriptions must employ some method of ascertaining the intended meaning of a given work use. By *sense resolution* we mean the mechanism by which a word or phrase with multiple possible interpretations is associated with a particular axiomatically defined predicate. The following sentences all use the word 'surround', but in each it refers to a different spatial relation: 'The embryo is surrounded by amniotic fluid'; 'The embryo is surrounded by a shell'; 'The cell is surrounded by its membrane'; 'The garden is surrounded by a wall'; 'The building is surrounded by guards'. To reason on the basis of one of these sentences we need to know what spatial relation is intended. Many factors place constraints on possible interpretations of a lexical predicate. An important consideration is the type(s) of thing to which the predicate is applied. Surrounding by a fluid is different from surrounding by a rigid shell, or a wall or a group of people.

As a further illustration, contrast the meaning of 'contains' in 'This bottle contains wine' and 'the wine contains alcohol'. In the first case, the wine is located within a cavity enclosed by, but not overlapping the bottle, whereas in the second alcohol is an ingredient of the wine. These are very different spatial relations. Establishing a robust automated mechanism for sense resolution would be a significant advance towards achieving automated common-sense reasoning.

20.6 Implicit and Background Knowledge

Suppose we take a convincing chain of reasoning expressed in natural language and translate it into a logical language (e.g., first-order logic): we are unlikely to get a formally valid sequence of inferences. It could be that the formal language that we use does not incorporate the types of logical operations required to articulate the inferences. But even if the logical language is sufficiently expressive in terms of its logical expressivity we are still unlikely to get a valid argument. This is because the reasoning will in typical cases also depend heavily on various kinds of implicit knowledge that is covertly utilised within common-sense reasoning processes.

One source of additional information would be the definitions of spatial (and other) terms and the axioms that specify semantic properties of primitives from which these terms are defined—in other words semantic knowledge. In addition to this there is a huge amount of contingent background information that can be potentially drawn upon to facilitate common-sense reasoning. For instance, knowledge about particular spatial properties and configurations of various kinds of object (tools, buildings, people etc). This kind of knowledge is often called *common-sense knowledge* and is critical to effective common-sense reasoning.[3] Davis (2017) discusses such knowledge at length and categorizes it as, 'roughly, what a typical seven year old knows about the world, including fundamental categories like time and space, and specific domains such as physical objects and substances; plants, animals, and other natural entities; humans, their psychology, and their interactions; and society at large'. Encoding and storing such background information is a major goal of the long-running AI project, CYC (Guha and Lenat, 1990). However, despite several decades of research the original goal of CYC project is still to be achieved.

As well as supplying additional information, background knowledge may also be used to select between different possible interpretations of vague and ambiguous vocabulary terms—that is to facilitate *sense resolution*, as described in the previous section. This seems to be particularly important in the resolution of certain ambiguous spatial relationships.

A kind of tacit knowledge that is particular to spatial reasoning is our ability to transfer information between multiple reference frames without any explicit expression of the relationships between these frames or of the reasoning steps that we must somehow be performing.

Geometry and physics represent space and time by coordinate systems defined relative to some reference frame regarded as fixed. Often a single reference frame will suffice, even for a complex physical situation, and, where multiple frames are used, precise mappings between them are defined. This contrasts sharply with natural language, which typically jumps quickly between multiple reference frames, for example 'the girl hid behind the curtains, but was visible through the window from the front of the house. The policeman in the garden saw the girl but not the lion in the living room.' To model human-like reasoning, we need a representation that can capture such chains of relative location. One object may be used to locate another either directly (using phrases like 'behind', 'through', 'in front of') or indirectly via background knowledge (e.g., the typical relative locations of curtains, windows, houses, gardens and living rooms). Such relative locations have been an important research focus in QSR (Donnelly, 2005) and Scheider's contribution in (Gangemi *et al.*, 2014).

20.7 Default Reasoning

It is widely recognized that common-sense reasoning is often *non-monotonic* in nature. This means that there are cases where is reasonable to infer some conclusion ϕ from some set of information Σ, but if we acquire some additional information α, the conclusion ϕ

[3] Such implicit knowledge is critical in Winograd Schema Challenge problems. An example such problem from the corpus at https://cs.nyu.edu/faculty/davise/papers/WinogradSchemas/WSCollection.html is 'Jim signalled the barman and gestured toward his [**empty glass/bathroom key**]. Whose [empty glass/bathroom key]? Answers: **Jim/the barman**'. Although this is a spatial reasoning problem, background (implicit) knowledge about objects involved is required to perform the appropriate inference.

is no longer warranted. (Symbolically, we may have $\Sigma \mid\!\sim \phi$, but $\Sigma, \alpha \not\mid\!\sim \phi$, where '$\mid\!\sim$' is a common-sense inference relation.) A number of logical calculi have been developed to formalize non-monotonic reasoning, the best known being Reiter's *default logic* (Reiter, 1980) and the *circumscription* theory McCarthy (1986).

Typical examples of default reasoning are inferences such as: If x is a bird, then x can fly (unless you know that x is a penguin or other flightless bird); Gert is German, therefore Gert drinks beer (but not if we know that Gert is three-years old).

It seems that relatively little work had been done on specifically spatial modes of non monotonic reasoning (though see, e.g., (Walega *et al.*, 2017). Yet there are many reasoning examples that suggest that common-sense spatial reasoning is very often supported by simplifying assumptions regarding the spatial properties of objects and configurations. Here are some examples:

- When reasoning with information concerning objects situated in an environment, in many cases we assume that space is empty except for those parts that we know to be occupied by physical objects or matter. (How this affects reasoning about objects moving in space has been considered in detail by Shanahan, 1995.)
- A spatial extent can be assumed to be convex if nothing is known to the contrary. For example, reasoning about objects fitting into containers typically assumes we are dealing with convex objects in containers whose containing space is convex, unless we have explicit information to the contrary.[4]
- If a region is known to be small relative to some other regions then the small region can usually be assumed to behave like a point with regard to inferences involving these regions. For instance, if we know there is a gap between two 'large' objects, we will tend to assume that a 'small' object will fit through it.

There is an interesting relationship between default reasoning and mental models that could be potentially useful in the implementation of common-sense reasoning. A mental model it is often as storing a mental correlate of the most typical way in which a set of beliefs could be realized. Thus construction of a mental model may be seen as the limit of default reasoning, where although one's knowledge does not fully pin down the state of the world, one constructs a prototypical example situation that is compatible with that knowledge and uses that as a basis for reasoning (Knauff *et al.*, 1995). Clearly, such reasoning is not deductively valid; but the inferences drawn could be useful in many cases.

20.8 Computational Complexity

From the point of view of traditional computer science, the most obvious difficulty facing the development of automated common-sense spatial reasoning is computational

[4] This assumption is required in order to reason appropriately about the *Winograd Schema*: 'The trophy would not fit in the case because it was too **big/small**' (Levesque *et al.*, 2012).

complexity. Indeed, many of the intractable or undecidable problems studied in computational complexity theory are spatial in nature. A result of (Grzegorczyk, 1951) proves undecidability of some relatively simple topological theories due to the fact that they can encode arithmetical operators and formulas. It is actually very straightforward to model numbers in terms of multi-piece spatial regions. We simply take the number of components of such a region as representing a number. It is then possible to define equality as a spatial relation and addition and multiplication as spatial constructions.

The extremely high expressive power of spatial concepts is also demonstrated by Tarski's (1956) paper on the definability of concepts, which shows that any concept that can be fully axiomatized within a theory that includes concepts sufficient to axiomatize Euclidean geometry, can actually be defined in terms of the geometrical concepts. This means that the logical properties of any concept can be fully modelled in terms of geometrical properties, with no need for any further axioms, since the standard axioms of geometry together with the geometrical definition of the concept are sufficient (Bennett, 2004).

If we restrict attention to reasoning problems formulated in terms of the limited sets of predicates and operators typically used in QSR, the complexity results are somewhat better, but still rather discouraging. For example, Renz and Nebel (1999) identified maximal tractable subsets of a topological constraint language based on the RCC-8 relation set (Randell *et al.*, 1992). These subsets can be used to carry out useful spatial reasoning tasks, but it is disappointing that it is not possible to reason effectively with more expressive extensions of these languages. Certain tractable extensions have been identified (e.g., by Gerevini and Renz (1998), who devised a reasoning algorithm for a combination of topological and size constraints). However, it seems that as we add expressive power in terms of combining different types of spatial property and relation, we very quickly end up with computationally intractable reasoning problems (see e.g., Davis *et al.*, 1999 unless we strictly limit other aspects of the representation language. Also, introducing natural global constraints to a reasoning problem, such as requiring regions to be self-connected and/or embeddable in the plane tends to raise complexity of spatial reasoning problems and often results in undecidability (Dornheim, 1998).

However, one should bear in mind that the type of problem for which these unpalatable complexity results arise is very different from the circumstances to which one would expect common-sense reasoning to be applied and the types of computation performed (e.g., consistency checking of networks of spatial relations) are quite far removed from everyday reasoning tasks. Whereas existing automated spatial reasoning systems typically carry out exhaustive reasoning with respect to large numbers of spatial constraints expressed a very limited set of spatial relationships, common-sense spatial reasoning typically operates with a small number of spatial facts expressed using a rather wide vocabulary of spatial properties and relations. Thus, although computational complexity is clearly a problem for computational spatial reasoning, it is not necessarily a problem for automated *common-sense* spatial reasoning, since existing spatial reasoning algorithms seem to be doing something very different from what one would expect from common-sense reasoning: humans can clearly make common-sense spatial inferences rather quickly.

20.9 Progress towards Common-sense Spatial Reasoning

The preceding sections have been largely negative in tone: pointing out the challenges in endowing machines with common-sense spatial reasoning. Of course there has been progress towards this goal, and indeed we have already mentioned some of this in the earlier part of this chapter. In this section we briefly mention some of the highlights of such work. Foremost in this direction is the work on qualitative spatial representation and reasoning. There are now a large number of QSR calculi capable of representing spatial information about (mereo)topology, direction, shape, distance, size among other aspects of spatial information. The computational complexity of reasoning with many of these calculi, at least the constraint languages associated with them, has been investigated thoroughly, and tractable subclasses identified (e.g., Renz and Nebel (1999). There are toolkits for reasoning with many of these, such as SparQ (Wolter and Wallgrün, 2013) and for extracting QSRs from video data, for example QSRlib (Gatsoulis *et al.*, 2016). Moreover there are many implemented systems, particularly in the domain of activity understanding which exploit QSR (e.g., Duckworth *et al.*, 2019 or which learn about spatial relations (e.g., Alomari *et al.*, 2017) from real-world data. There is still though a disconnect between much of this work on QSR and the real problems of common-sense reasoning, as noted by Davis and Marcus (2015). Davis has contributed much to the field of common-sense reasoning, and spatial reasoning in particular, for example his work on liquids (Davis, 2008) and containers (Davis *et al.*, 2017).

There has also been work addressing the problem of how to acquire symbolic knowledge from perceptual sensors which are typically noisy and only incompletely observe the world, for example because of occlusion. Approaches in the literature which try to address these issues, include the use of formalisms which explicitly represent spatial vagueness such as Cohn and Gotts (1996), or ways of smoothing noisy detections (e.g., Sridhar *et al.* (2011), building probabilistic models of QSR (e.g., Kunze *et al.*, 2014), or by explicitly reasoning about occlusion (e.g., Bennett *et al.*, 2008).

As is the case for AI in general, the more the task/domain is constrained and well specified, the easier it is to come up with a (spatial) theory that is sufficient for appropriate reasoning and inference. The real challenge is to achieve general common-sense (spatial) reasoning.

20.10 Conclusions

In this chapter we have decomposed the problem of achieving automated commonsense spatial reasoning into a number of sub-problems (seven to be precise), which we consider to be key to solving the general problem, and are sufficiently independent from each other as to be addressed separately. Possibly, we have missed out further important problems, or conflated issues that would be best treated separately. For example one issue that we have little discussed is how a common-sense knowledge could be acquired by an automated reasoning system, and in particular spatially related knowledge. One approach, adopted by the CYC system already mentioned above is to manually specify such knowledge; the challenge here is the enormity of the knowledge and it is clear that despite several decades of research and development this remains an unfinished enterprise. The alternative is to try to acquire such knowledge via a process of learning. The NELL project (Mitchell *et al.*, 2018) aims to learn such knowledge by learning from text. An alternative is to learn from multimodal data, which has the advantage in simultaneously learning a semantic grounding in the perceptual world. For example, Alomari *et al.* (2017) show how the meaning of object properties, spatial relations, and actions, as well as a grammar, can be learned from paired video-text clips, while Richard-Bollans *et al.* (2020) demonstrate how the different senses of spatial prepositions such as in, above, against, and under can be acquired from human annotations in a virtual reality setting.

Another issue we have hardly discussed is how *embodiment* affects perception and spatial awareness. Tversky, among others, has discussed at length how embodiment affects the human reasoning: 'Spatial thinking comes from and is shaped by perceiving the world and acting in it, be it through learning or through evolution' (Tversky, 2009). There is work in AI which takes an embodied approach to spatial cognition and spatial common-sense (e.g., (Spranger *et al.*, 2014; Alomari *et al.*, 2017)) but more research on this is certainly needed.

Most of the problems we have discussed actually apply to common-sense reasoning in general, rather than exclusively to spatial reasoning, and yet in the examples we have considered, it is primarily in the spatial aspects of semantics and reasoning where the difficulties lie. This is because the spatial domain is extremely rich and manifests huge variety and complexity. Issues relating to ambiguity vagueness are particularly apparent for spatial relationships because, although we have well-developed mathematical theories within which geometrical constraints can be precisely defined, there is no direct mapping from natural language terms to these precise constraints. And, even if these interpretative problems are circumvented, reasoning about space involves many highly intractable computations (though perhaps these go beyond the realm of common sense).

Our analysis was not intended to be prescriptive of a particular research direction or methodology.[5] As well as exposing a large number of problems, we have indicated a variety of different approaches that might lead to their solution. Our aim was primarily to provide an overview that would help researchers progress effectively by focusing their attention on some particular aspect of the highly complex problem of achieving automated common-sense spatial reasoning.

Acknowledgements

This research was partly inspired by prior collaboration by the first author with Vinay K. Chaudhri and Nikhil Dinesh at SRI International, while working on Project HALO. The

[5] Davis and Marcus (2015) suggests some research directions including the development of benchmarks and evaluation metrics, integration of different AI methodologies which have complementary strengths (e.g., facts gathered from web mining with mechanisms for formal reasoning), and a better understanding of human common-sense reasoning (as the second author has been attempting in robotic manipulation (Hasan *et al.*, 2020)

second author was partially supported by the EPSRC under grant EP/R031193/1, and the EU under H2020 project 824619, and by a Turing Fellowship from the Alan Turing Institute. We also thank two anonymous referees for their helpful comments.

References

Alomari, M., Duckworth, P., Hogg, D. C, and C. *et al.* (2017). Natural language acquisition and grounding for embodied robotic systems, in *Proceedings of the 31st AAAI Conference on Artificial Intelligence*, San Francisco. New York, NY: AAAI Press, 4349-56.

Bateman, J. A. (2013). Space, language and ontology: a response to Davis. *Spatial Cognition & Computation*, **13**(4), 295–314.

Bateman, J. A., Hois, J., Ross, R. J. *et al.* (2010). A linguistic ontology of space for natural language processing. *Artificial Intelligence*, **174**(14), 1027–71.

Bennett, B. (2001). Space, time, matter and things, in *Proceedings of the 2nd International conference on Formal Ontology in Information Systems*, FOIS'01, Cape Town. New York, NY: ACM, 105–16.

Bennett, Brandon (2004*b*). Relative definability in formal ontologies. In *Proceedings of the 3rd International Conference on Formal Ontology in Information Systems (FOIS-04)* (ed. A. Varzi and L. Vieu), pp. 107–118. IOS Press, Amsterdam.

Bennett, B. (2011). Spatial vagueness. In *Methods for Handling Imperfect Spatial Information*. Springer-Verlag.

Bennett, Brandon and Agarwal, Pragya (2007). Semantic categories underlying the meaning of 'place'. In *Spatial Information Theory: proceedings of the 8th international conference (COSIT-07)*, Volume 4736, Lecture Notes in Computer Science, pp. 78–95. Springer, Berlin, Heidelberg.

Bennett, B., Chaudhri, V., and Dinesh, N. (2013). A vocabulary of topological and containment relations for a practical biological ontology, in J. Stell, T. Tenbrink, and Z. Wood, eds, *International Conference on Spatial Information Theory, COSIT 2017*, L'Aquila, Italy. Cham: Springer, 418–37.

Bennett, B. and Cialone, C. (2014). Corpus guided sense cluster analysis: a methodology for ontology development (with examples from the spatial domain). In *Proc. FOIS-14* (ed. P. Garbacz and O. Kutz). IOS Press.

Bennett, B., Magee, D. R., Cohn, A. G. *et al.* (2008). Enhanced tracking and recognition of moving objects by reasoning about spatio-temporal continuity. *Image and Vision Computing*, **26**(1), 67–81.

Borgo, S., Guarino, N., and Masolo, C. (1996). A pointless theory of space based on strong congruence and connection, in *Proceedings of the Fifth International Conference on Principles of Knowledge Representation and Reasoning, KR'96*, Cambridge, MA. San Francisco, CA: Morgan Kaufmann, 220–9.

Brooks, R. A. (1986). A robust layered control system for a mobile robot. *IEEE Journal on Robotics and Automation*, 2(1), 14–23.

Cohn, A. G. (1995). A hierarchical representation of qualitative shape based on connection and convexity, in *Proceedings International Conference on Spatial Information Theory, COSIT'95, Semmering*, Austria. Berlin, Heidelberg: Springer, 311–26.

Cohn, A. G., Bennett, B., Gooday, J. *et al.* (1997). RCC: a calculus for region based qualitative spatial reasoning. *GeoInformatica*, **1**, 275–316.

Cohn, A. G. and Gotts, N. M. (1996). The 'egg-yolk' representation of regions with indeterminate boundaries, in P. Burrough and A. M. Frank, eds, *Proceedings GISDATA Specialist Meeting on Geographical Objects with Undetermined Boundaries*. Francis Taylor, 171–87.

Cohn, A. G. and Hazarika, S. M. (2001). Qualitative spatial representation and reasoning: an overview. *Fundamenta Informaticae*, **46**(1-2), 1–29.

Cohn, A. G. and Renz, J. (2008). Qualitative spatial representation and reasoning, in F. van Harmelen, V. Lifschitz, and B. Porter, eds, *Handbook of Knowledge Representation*. Amsterdam, Netherlands: Elsevier, 513–96.

Cui, Z., Cohn, A. G., and Randell, D. A. (1992). Qualitative simulation based on a logical formalism of space and time, in *AAAI Proceedings Tenth National Conference on Artificial Intelligence, AAAI-92*, San Jose, CA. Boston, MA: AAAI Press, 679–84.

Davis, Ernest (2008). Pouring liquids: A study in commonsense physical reasoning. *Artificial Intelligence*, **172**(12), 1540–78.

Davis, E. (2013). Qualitative spatial reasoning in interpreting text and narrative. *Spatial Cognition & Computation*, **13**(4), 264–94.

Davis, E. (2017). Logical formalizations of commonsense reasoning: a survey. *Journal of Artificial Intelligence Research*, **59**, 651–723.

Davis, E., Gotts, N., and Cohn, A. G. (1999). Constraint networks of topological relations and convexity. *Constraints*, **4**(3), 241–80.

Davis, E. and Marcus, G. (2015). Commonsense reasoning and commonsense knowledge in artificial intelligence. *Communications of the ACM*, **58**(9), 92–103.

Davis, E., Marcus, G., and Frazier-Logue, N. (2017). Commonsense reasoning about containers using radically incomplete information. *Artificial Intelligence*, **248**, 46–84.

Donnelly, M. (2005). Relative places. *Applied Ontology*, **1**(1), 55–75.

Dornheim, C. (1998). Undecidability of plane polygonal mereotopology, in *Proceedings of the Sixth International Conference on Principles of Knowledge Representation and Reasoning, KR'98*, Toronto, Canada. San Francisco, CA: Morgan Kaufmann, 342–53.

Duckworth, P., Hogg, D. C., and Cohn, A. G. (2019). Unsupervised human activity analysis for intelligent mobile robots. *Artificial Intelligence*, **270**, 67–92.

Edelsbrunner, H., Kirkpatrick, D. G., and Seidel, R. (1983). On the shape of a set of points in the plane. *IEEE Transactions on Information Theory*, **29**(4), 551–9.

Egenhofer, M. (1991). Reasoning about binary topological relations, in *Proceedings Second Symposium on Spatial Databases*, Zurich. Berlin: Springer, 141–60.

Escrig, M. T. and Toledo, F. (1996). Qualitative spatial orientation with constraint handling rules, in *Proceedings of the 12th European Conference on Artificial Intelligence, ECAI'96*. London: Wiley, 486–90.

Faltings, B. (1995). Qualitative spatial reaoning using algebraic topology, , in A. U. Frank and W. Kuhn, eds, *International Conference on Spatial Information Theory, COSIT'95*, Semmering, Austria. Berlin, Heidelberg: Springer, 17–30.

Forbus, K. D. (2019). *Qualitative Representions*. Cambridge, MA: MIT Press.

Forbus, K. D., Nielsen, P., and Faltings, B. (1991). Qualitative spatial reasoning: The clock project. *Artificial Intelligence*, **51**, 417–71.

Frank, A. (1992). Qualitative spatial reasoning with cardinal directions. *Journal of Visual Languages and Computing*, **3**, 343–71.

Freksa, C. (1992). Using orientation information for qualitative spatial reasoning, in *Proceedings of the International Conference on Theories and Methods of Spatio-Temporal Reasoning in Geographic Space*, Pisa. Berlin: Springer, 162–78.

Gahegan, M. (1995). Proximity operators for qualitative spatial reasoning, in *Proc. COSIT*. Berlin: Springer-Verlag, 31–44.

Galton, A. P. (1998). Modes of overlap. *Journal of Visual Languages and Computing*, **9**, 61–79.

Galton, A. (2004). Multidimensional mereotopology, in *Proceedings of the Ninth International Conference Knowledge Representation and Reasoning, KR2004, Whistler BC*. Boston, MA: AAAI Press, 45–54.

Gangemi, A. Hafner, V. V., Kuhn, W. (2014). Spatial reference in the Semantic Web and in Robotics (Dagstuhl Seminar 14142). In *Dagstuhl Reports* (Vol. 4, No. 3). Schloss Dagstuhl-Leibniz-Zentrum fuer Informatik.

Gatsoulis, U. Y., Alomari, M., Burbridge, C, Dondrup, C. *et al.* (2016). QSRlib: a software library for online acquisition of qualitative spatial relations from video, in *Proceedings 29th International Workshop on Qualitative Reasoning*, New York. Netherlands: University of Amsterdam, 36–41.

Gerevini, A. and Renz, J. (1998). Combining topological and qualitative size constraints for spatial reasoning, in *Proceedings International Conference on Principles and Practice of Constraint Programming*, Pisa. Berlin, Heidelberg: Springer, 220–34.

Gotts, N. M. (1994). How far can we C? defining a 'doughnut' using connection alone, in *Proceedings Principles of Knowledge Representation and Reasoning*, Bonn. San Francisco, CA: Morgan Kaufmann, 246–57.

Grenon, P. and Smith, B. (2004). Snap and span: towards dynamic spatial ontology. *Spatial cognition and computation*, **4**(1), 69–104.

Grzegorczyk, A. (1951). Undecidability of some topological theories. *Fundamenta Mathematicae*, **38**, 137–152.

Guha, R. V. and Lenat, D. B. (1990). CYC: a mid-term report. *AI Magazine*, **11**(3), 32–59.

Hahmann, T. and Brodaric, B. (2012). The void in hydro ontology, in *Proceedings of the Seventh International Conference, FOIS 2012, Graz*, Austria. Frontiers in Artificial Intelligence and Applications 239. Amsterdam: IOS Press.

Hahmann, Torsten and Brodaric, B. (2014). Voids and material constitution across physical granularities, in *Formal Ontology in Information Systems: Proceedings of the Eighth International Conference*, FOIS 2014, Rio de Janeiro, Brazil. Frontiers in Artificial Intelligence and Applications 267. Amsterdam: IOS Press, 51–64.

Hasan, M., Warburton, M., Agboh, W. C. *et al.* (2020). Human-like planning for reaching in cluttered environments, in *IEEE Proceedings International Conference on Robotics and Automation, ICRA'20*, Paris. New York, NY: IEEE, 7784–90.

Hernández, D (1993). Reasoning E. S. J Doyle and P. Torasso). with qualitative representations: Exploiting the structure of space, in *Proceedings of the Third International Workshop on Qualitative Reasoning and Decision Technology. QUARDET '93*. Barcelona: International Centre for Numerical Methods in Engineering (CIMNE) Press, Technical University of Catalonia, 493–502.

Hintikka, J. (1962). *Knowledge and Belief: An Introduction to the Logic of the Two Notions*. Ithaca, NY: Cornell University Press.

Johnson-Laird, P. (1983). *MentalModels: Toward a Cognitive Sience of Language, Inference and Consciousness*. Cambridge, MA: Harvard University Press.

Kimble, C. (2013). Knowledge management, codification and tacit knowledge. *Information Research*, **18**(2), paper 577.

Klippel, A., Wallgrün, J. O., Yang, J. *et al.* (2013). Fundamental cognitive concepts of space (and time): Using cross-linguistic, crowdsourced data to cognitively calibrate modes of overlap, in *11th International Conference on Spatial Information Theory, COSIT 2013*, Scarborough, UK. Cham: Springer, 377–96.

Knauff, M. (1999). The cognitive adequacy of Allen's interval calculus for qualitative spatial representation and reasoning. *Spatial Cognition and Computation*, **1**(3), 261–90.

Knauff, M., Rauh, R., and Renz, J. (1997). A cognitive assessment of topological spatial relations: results from an empirical investigation, in *Proceedings International Conference on Spatial Information Theory, COSIT'97*, Laurel Highlands, Pennsylvania. Berlin, Heidelberg: Springer, 193–206.

Knauff, M., Rauh, R., and Schlieder, C. (1995). Preferred mental models in qualitative spatial reasoning: a cognitive assessment of Allen's calculus, in *Proceedings of the Seventeenth Annual Conference of the Cognitive Science Society*, Pittsburgh, USA. Mahwah, NJ: Lawrence Erlbaum Associates, 200–5.

Knauff, M., Rauh, R., Schlieder, C. *et al.* (1998). Mental models in spatial reasoning, in *Proceedings of International Symposium on Spatial Cognition*, Trier, Germany. Lecture Notes in Computer Science 1404. Berlin, Heidelberg: Springer, 267–91.

Kunze, L., Burbridge, C., and Hawes, N. (2014). Bootstrapping probabilistic models of qualitative spatial relations for active visual object search, in *Proceedings 2014 AAAI Spring Symposium*, San Francisco, CA. Boston, MA: AAAI Press, 81–88.

Levesque, H. J., Davis, E., and Morgenstern, L. (2012). The Winograd Schema challenge, in *Proceedings Thirteenth International Conference on the Principles of Knowledge Representation and Reasoning*, Rome, Italy. Boston, MA: AAAI Press, 552–61.

Ligozat, G. F. (1993). Qualitative triangulation for spatial reasoning, in A. U. Frank and I. Campari, eds, *Proceedings European Conference on Spatial Information Theory*, LNCS 716, Elba, Italy. Berlin, Heidelberg: Springer, 54–68.

Ligozat, G. F. (2011). *Qualitative Spatial and Temporal Reasoning*. London, UK: Wiley-ISTE.

Marcus, G. and Davis, E. (2019). *Rebooting AI*. Cambridge, MA: MIT Press.

Mark, D., Comas, D., Egenhofer, M. *et al.* (1995). Evaluating and refining computational models of spatial relations through cross-linguistic human-subjects testing, in *Proceedings International Conference on Spatial Information Theory*, Semmering, Austria. Berlin, Heidelberg: Springer, 553–68.

Mark, D. and Egenhofer, M. (1994). Calibrating the meanings of spatial predicates form natural language: line-region relations, in *Proceedings Sixth International Conference on Spatial Data Handling*, Vol. 1, Edinburgh, UK. Oxford, UK: Taylor Francis, 538–53.

Masolo, C., Borgo, S., Gangemi, A. *et al.* (2003). WonderWeb Deliverable D18: DOLCE Ontology Library. Technical report, Laboratory For Applied Ontology - ISTC-CNR. http://wonderweb.semanticweb.org/deliverables/documents/D18.pdf.

McCarthy, J (1986). Applications of circumscription to formalizing common-sense knowledge. *Artificial Intelligence*, **28**, 89–116.

McCarthy, J (1989). Artificial intelligence, logic and formalizing common sense, in R. Tomason, ed., *Philosophical Logic and Artificial Intelligence* Dordrecht: Springer, 161–90.

McCarthy, J. and Hayes, P. J. (1969). Some philosophical problems from the standpoint of artificial intelligence, in B, Meltzer and D. Mitchie, eds, *Machine Intelligence*, Vol. 4. Edinburgh: Edinburgh University Press, 463–502.

Mitchell, T., Cohen, W., Hruschka, E. *et al.* (2018, April). Never-ending learning. *Communication of the Association of Computing Machinery*, **61**(5), 103–15.

Montello, D. R., Goodchild, M. F., Gottsegen, J. *et al.* (2003). Where's downtown?: Behavioral methods for determining referents of vague spatial queries. *Spatial Cognition & Computation*, **3**(2–3), 185–204.

Moratz, R., Lücke, D., and Mossakowski, T. (2011). A condensed semantics for qualitative spatial reasoning about oriented straight line segments. *Artificial Intelligence*, **175**(16–17), 2099–127.

Mossakowski, T. and Moratz, R. (2012). Qualitative reasoning about relative direction of oriented points. *Artificial Intelligence*, **180–181**, 34–45.

Polanyi, M. (1966). *The Tacit Dimension.* Abingdon, UK: Routledge and Kegan Paul.

Pratt, I. (1999). First-order qualitative spatial representation languages with convexity. *Spatial Cognition and Computation*, **1**, 181–204.

Pratt, I. and Schoop, D. (1998). A complete axiom system for polygonal mereotopology of the real plane. *Journal of Philosophical Logic*, **27**, 621–58.

Ragni, M., Knauff, M., and Nebel, B. (2005). A computational model for spatial reasoning with mental models, in *Proceedings of the* 27th *Annual Conference of the Cognitive Science Society*, Vol. 27, Stresa, Italy. Mahwah, NJ: Lawrence Erlbaum Associates, 1064–70.

Randell, D. A., Cui, Z., and Cohn, A. G. (1992). A spatial logic based on regions and connection, in *Proceedings of 3rd International Conference on Knowledge Representation and Reasoning*, Cambridge, MA. San Francisco, CA: Morgan Kaufmann, 165–76.

Reece, Jane B., Urry, Lisa A., Cain, Michael L., Wasserman, Steven A., Minorsky, Peter V., and Jackson, Robert B. (2011). *Campbell Biology* (9th Edition edn). Pearson, New York.

Reiter, R. (1980). A logic for default reasoning. *Artificial Intelligence*, **13**(1,2), 81–132.

Renz, J. and Nebel, B. (1999). On the complexity of qualitative spatial reasoning: a maximal tractable fragment of the Region Connection Calculus. *Artificial Intelligence*, **108**(1–2), 69–123.

Richard-Bollans, A., Bennett, B., and Cohn, A. G. (2020). Automatic generation of typicality measures for spatial language in grounded settings, in *Proceedings 24th European Conference on Artificial Intelligence*, Santiago de Compostela. Amsterdam, Netherands: IOS Press, 2164–71.

Schacter, D. L. (1987). Implicit memory: history and current status. *Journal of Experimental Psychology*, **13**(3), 501–18.

Shanahan, M. (1995). Default reasoning about spatial occupancy. *Artificial Intelligence*, **74**(1), 147–63.

Spranger, M., Suchan, J., Bhatt, M. *et al.* (2014). Grounding dynamic spatial relations for embodied (robot) interaction, in *Proceedings 14th Pacific Rim International Conference on Artificial Intelligence*, Gold Coast, Australia. Cham: Springer, 958–71.

Sridhar, M., Cohn, A. G., and Hogg, D. C. (2011). From video to RCC8: exploiting a distance based semantics to stabilise the interpretation of mereotopological relations, in *Proceedings 10th International Conference on Spatial Information Theory*, Belfast, ME. Berlin, Heidelberg: Springer, 110–25.

Tarski, A. (1956). Some methodological investigations on the definability of concepts, in *Logic, Semantics, Metamathematics*, Vol. 52. Transl. J. H. Woodger. Oxford Clarendon Press.

Thomason, R. H. (1991). Logicism, AI, and common sense: John McCarthy's program in philosophical perspective, in V. Lifschitz, ed., *Artificial Intelligence and Mathematical Theory of Computation: Essays in Honor of John McCarthy*. London: Academic Press, 449–66.

Tversky, B. (2009). Chapter 12: Spatial cognition: embodied and situated, in *Cambridge Handbook of Situated Cognition.* Cambridge, UK: Cambridge University Press, 201–16.

Walega, P. A., Schultz, C., and Bhatt, M. (2017). Non-monotonic spatial reasoning with answer set programming modulo theories. *Theory and Practice of Logic Programming*, **17**(**2**), 205–25.

Wolter, D. and Wallgrün, J. O. (2013). Qualitative spatial reasoning for applications: New challenges and the sparq toolbox, in J. Rodrigues, ed., *Geographic Information Systems: Concepts, Methodologies, Tools, and Applications*. Hershey, PA: IGI Global, 1639–64.

Zimmermann, K. (1993). Enhancing qualitative spatial reasoning – combining orientation and distance, in *Proceedings European Conference on Spatial Information Theory*, Elba, Italy. Berlin, Heidelberg: Springer, 69–76.

21

Sampling as the Human Approximation to Probabilistic Inference

Adam Sanborn,[1] Jian-Qiao Zhu,[1] Jake Spicer,[1] Joakim Sundh,[1] Pablo León-Villagrá,[1] and Nick Chater[2]

[1] *University of Warwick and* [2] *Warwick Business School*

We live in an uncertain world, which makes it difficult to know what we should believe. In the absence of certainty, the Bayesian approach provides a formal framework that results in assigning each possible state of the world a probability, and using the laws of probability to calculate what to believe and what to do. This theoretical framework was developed in the 1940s and 1950s to provide prescriptions for human behaviour, and has been particularly influential in developing theories of how people behave in economic situations (von Neumann and Morgenstern, 1947; Savage, 1954; Edwards, 1961; Peterson and Beach, 1967).

Advances in methodology and computational power in the past two decades has seen researchers begin to compare human performance to Bayesian models in complex domains, such as vision, motor control, language, categorisation or common-sense reasoning. In these domains, people's performance has been found to be similar to that of highly complex probabilistic models, if these models assume the same sensory limitations that people have (Anderson, 1991; Chater and Manning, 2006; Yuille and Kersten, 2006; Griffiths *et al.*, 2007; Oaksford and Chater, 2007; Wolpert, 2007; Goodman *et al.*, 2008; Sanborn *et al.*, 2010*a*; Griffiths and Tenenbaum, 2011; Houlsby *et al.*, 2013; Pantelis *et al.*, 2014; Petzschner *et al.*, 2015). Some of the most compelling demonstrations have been in the domain of intuitive physics, in which participants are asked to make judgements about various physical quantities, such as blocks in motion or liquids. Despite the complexities of these predictions, complex probabilistic models often explain people's judgements better than competing frameworks like heuristics (Battaglia *et al.*, 2013; Sanborn *et al.*, 2013).

These results are surprising as they run counter to an extensive literature showing that people make systematic errors when reasoning about probabilities (Tversky and Kahneman, 1974; Gigerenzer and Gaissmaier, 2011; Kahneman, 2011). First, there are many demonstrations of how asking a question in different ways will alter the probability

Adam Sanborn, Jian-Qiao Zhu, Jake Spicer, Joakim Sundh, Pablo León-Villagrá, and Nick Chater, *Sampling as the Human Approximation to Probabilistic Inference* In: *Human Like Machine Intelligence.* Edited by: Stephen Muggleton and Nick Charter, Oxford University Press.
 DOI: 10.1093/oso/9780198862536.003.0021

judgement that a person makes. For example, *unpacking effects* show that people's estimate of the probability that someone 'can buy a gun in a hardware store' is more than the probability that they 'can buy an antique gun or some other type of gun in a hardware store', but less than the probability that they 'can buy a staple gun or some other type of gun in a hardware store', despite all three questions asking about the same set of events and therefore having the same probabilities (Sloman *et al.*, 2004; Dasgupta *et al.*, 2017). Perhaps more damning, however, is the observation that people's probability estimates are not consistent with one another—that combinations of estimates do not follow the rules of probability theory as they should (e.g., Costello and Watts, 2014). One salient demonstration of this is the conjunction fallacy made famous by Tversky and Kahneman (1983): people will more often than not judge the probability that a highly educated, liberal-seeming person is a bank teller to be lower than the probability that this person is both a feminist and a bank teller, despite the fact that the group of bank tellers includes all feminists who are also bank tellers. As probability theory is at the heart of complex probabilistic models, it appears a paradox that people's judgements in complex tasks match those of probabilistic models, yet their probability judgements disagree with probability theory.

How can we explain this apparent paradox? First, we note that the idealized way of implementing complex probabilistic models, representing all possible probabilities and making exact calculations with these probabilities, is implausible for any physical system, including brains (Aragones *et al.*, 2005; Sanborn and Chater, 2016). An example of why this is the case is to consider the problem of categorizing objects in the world into different natural kinds, and then making a decision in the light of that categorization. A common Bayesian approach to this problem is to represent all possible ways of dividing the observed objects into different categories, and then summing over all these possible partitions to make a decision (Anderson, 1991; Sanborn *et al.*, 2010*a*). This calculation becomes intractable long before we reach a number that could realistically correspond to a lifetime of experience: even for just 100 objects there are over 4.7×10^{115} ways to divide them into categories, which is far greater than the number of atoms in the observable universe. Therefore, explicitly using probabilities in categorization or other complex domains where Bayesian models have been successful such as vision, intuitive physics, and language, is thus clearly impossible.

But how can a Bayesian model of categorization, vision, intuitive physics, or language possibly work without explicitly representing probabilities? A key insight is that it is not necessary to represent probabilities explicitly in order to implement complex probabilistic models. Instead, these models can be approximated, and a straightforward way in which to do so is to draw samples from the probability distribution rather than representing it explicitly. Using sampling as an approximation to complex probabilistic models has a long history beginning in the 1940s and 1950s (Metropolis *et al.*, 1953); and as the computational resources available to researchers have increased, it has become a common way in which to approximate these models in both cognitive science and artificial intelligence (Griffiths *et al.*, 2007; Susskind *et al.*, 2008). The major attraction of sampling is that it comes with a theoretical guarantee: an infinite number of samples will provide the same answer as exactly calculating with explicit probabilities. Additionally, using a finite and achievable number of samples provides useful approximations, though this can lead to erroneous answers in some situations. Cognitive science researchers have recently been intrigued by the possibility that the mind implements a sampling algorithm, both for their positives and negatives: the positives could explain how people with finite brains could approximate complex probabilistic models, while the negatives could explain the systematic errors people show when reasoning about probabilities.

One particularly revealing bias is *probability matching*. For example, on a multiple-choice test if a student believes that Option A has a 90% chance of being the right answer, instead of always choosing Option A he or she will still choose an alternative 10% of the time (Mosteller and Nogee, 1951; Vulkan, 2000). This behaviour contradicts a key supposition of rationality—that people always choose the option they consider best—and instead shows that human decision making is stochastic. This puzzling bias can though be explained, at least as a first approximation, by sampling: drawing a single independent sample from the probability distribution of which option is correct will result in Option A being sampled 90% of the time (Vul *et al.*, 2014). A host of other reasoning fallacies have also begun to be explained by sampling, including the unpacking effect described above (Sanborn and Chater, 2016; Dasgupta *et al.*, 2017; Lieder *et al.*, 2018*a*). However, research in this area is only beginning, and the current state of the art is that different algorithms are used to explain different effects, as the tasks investigated thus far make it very difficult to distinguish between sampling algorithms. Ignorance of the sampling process makes it then difficult to arrive at a coherent explanation of how sampling can produce biases, and also prevents precise quantitative predictions from being made.

21.1 A Sense of Location in the Human Sampling Algorithm

The best-known and often most efficient method for drawing samples is to draw them independently from the probability distribution of interest—we term this *direct sampling*. In statistics, there are a variety of methods for drawing samples independently. Computer algorithms have been developed to generate samples from simple distributions such as Gaussian or uniform distributions, and with more complex distributions there are other methods that can generate independent samples, such as rejection or importance sampling. However, to take advantage of these efficient sampling methods requires knowing a fair amount about the distribution of interest—either characterizing it exactly or, as is the case for rejection or importance sampling, knowing it well enough to be able to identify another distribution that is very similar (Bishop, 2006).

However, it does not seem likely the mind or the brain directly samples from probability distributions. To develop an intuition for why this is the case, consider the task of unscrambling a jumbled-up string of letters to make a word, knowing that each string can only be unscrambled so as to make a single word. In this example, the three strings of letters are "CIBRPAMOLET", "NNLNRIEITOAAT", and "AABRMSTENMESR". We can think of this problem as implying a probability distribution where the different hypotheses are each of the possible orderings of a letter string. This means that there are 11 factorial or 39,916,800 orderings for "CIBRPAMOLET" and 13 factorial or

6,227,020,800 orderings each for "NNLNRIEITOAAT" and "AABRMSTENMESR". The probability of each ordering is then simply the probability that that ordering of the letters is a word. As there is only one way that each string can be unscrambled to be a word, this probability distribution must be concentrated on the single correct ordering, with the small remainder divided amongst the huge number of non-word orderings.

If it was possible to sample from the probability distribution over hypotheses directly and efficiently, it would be easy to unscramble each of the letter strings—as samples are generated according to their probabilities, almost all of the generated samples would be of the correct ordering. However, as will be obvious, we cannot immediately generate the correct answers. This might be because samples are just generated very slowly in this task, so another observation is useful: changing the task to be to unscramble the mildly scrambled strings "PROBELMATIC", "INTERNATOINAL", and "EMABRRASS-MENT" makes it a lot easier, even though these are the same sets of letters. Therefore, taken from the perspective of sampling, sampling the correct answer is much easier when starting from a mildly scrambled string, but this cannot arise through direct sampling, which is independent of the starting point.

While direct sampling does not seem tenable as a result of these observations, there are sampling algorithms for which new samples do depend on previous samples. One very well-known algorithm that has this property is Markov Chain Monte Carlo (MCMC; Metropolis *et al.*, 1953). MCMC works by constructing a Markov chain that is characterized by a set of transition probabilities between potential states of the chain. During any one iteration, the chain is in a specific state, and a nearby state is blindly proposed as the next potential state. The ratio of the probability of the new state to the probability of the current state is then calculated, and this ratio is used to decide stochastically whether the chain stays put or transitions to the proposed state.

An MCMC chain generates a series of states through many iterations of this procedure, and under mild assumptions this series of states can be treated as samples from the probability distribution. Of course, the order in which samples arise from this algorithm is not independent, as in direct sampling. This is because the proposed state is selected from those states that are nearby the current state, and so the states of the Markov chain change more slowly than do the states in direct sampling. The greater chance MCMC has of transitioning to nearby hypotheses, that is, a 'sense of location', helps explain our observations in the example above, where it is much easier to unscramble the letter strings "PROBELMATIC", "INTERNATOINAL", and "EMABRRASSMENT" than it is to unscramble the strings we initially presented.

MCMC's sense of location has led to this algorithm being used to explain a variety of cognitive biases, such as the anchoring effect. In anchoring experiments, participants are first asked to make a decision about whether a quantity is higher or lower than an irrelevant number. For example, participants are asked to add 400 to the last three digits of their phone number, to think of the resulting number as a date, and to decide whether Atilla the Hun was defeated before or after that date. Finally, participants were asked to provide the specific year in which Atilla the Hun was defeated. Despite the fact that the numbers generated from the participants' phone numbers were transparently irrelevant, participants' estimates were pulled toward these values (Russo and Schoemaker, 1989).

MCMC has been used to explain these results by assuming that the decision about whether a quantity is higher or lower than an irrelevant number sets the initial state of the Markov chain. Once this initial state is set, the algorithm samples from the probability distribution (e.g., of possible dates when Atilla the Hun was defeated), and the last sample generated is taken as the estimated date of defeat. If the number of iterations is great enough, then distribution of estimates will be unbiased. However, a limited number of iterations will result in an estimate distribution that is biased by the algorithm's starting point, producing an anchoring effect. MCMC can also explain how various manipulations affect the strength of the anchoring effect, including whether the anchor is provided or self-generated, the level of participant expertise, cognitive load, and financial incentives (Lieder *et al.*, 2012; Lieder *et al.*, 2018*a*; Lieder *et al.*, 2018*b*).

The final example of explaining cognitive biases with MCMC we will discuss here is the unpacking effect. In experiments by Dasgupta *et al.* (2017), participants were told that their friend sees a table in a visual scene that they themselves cannot see, and in the first condition were asked to judge the probability that 'any object starting with a C' is also in the scene. In the second condition, participants were asked to judge the probability that a 'chair, computer, curtain, or any other object starting with a C' shares the scene with the table. Finally, in the third condition, participants were asked to judge the probability that a 'cannon, cow, canoe, or any other object starting with a C' shares the scene with the table. These three questions are formally identical: the two unpacked versions of the questions just list kinds of objects that are implicit in the packed question 'any object starting with a C'. Despite this, average estimates are highest when the question is unpacked as 'chair, computer, curtain, or any other object starting with a C', intermediate for the simple question 'any object starting with a C', and lowest for unpacking 'cannon, cow, canoe, or any other object starting with a C'.

As with anchoring, MCMC has been used to explain this effect as the result of its starting point. First, it is assumed that object names are arranged in a semantic space and that asking participants which objects share a scene with a table induces a probability distribution over objects. Then the question that is asked helps position the starting point of the sampler: towards a region in which objects that begin with C are likely as in 'chair, computer, curtain, or any other object starting with a C', or towards a region in which objects that begin with C are unlikely as in 'cannon, cow, canoe, or any other object starting with a C'. This starting point bias can thus explain how this unpacking effect depends on the probability of the unpacked hypotheses (Dasgupta *et al.*, 2017).

The list above is illustrative but certainly not complete. A variety of other biases have also been explained by MCMC, including the base-rate fallacy, conjunction fallacy, the weak evidence effect, the dud alternative effect, the self-generation effect, and wisdom of the crowd effects (Sanborn and Chater, 2016; Dasgupta *et al.*, 2017). Even perceptual effects, such as switching times in bistable perception, have also been explained by MCMC (Gershman *et al.*, 2012).

21.2 Key Properties of Cognitive Time Series

While MCMC has been used to explain a variety of judgement biases, it is certainly not the only sampling algorithm with a sense of location. MCMC is often slow to converge, particularly for multimodal probability distributions, properties that have been exploited for explaining cognitive biases. These weaknesses have resulted in computer scientists and statisticians developing a variety of algorithms based on simple MCMC that mitigate these problems. Various proposals include methods that learn to adapt state proposals to the problem at hand, methods that involve running multiple chains, and methods that use the gradient of the probability distribution (Robert *et al.*, 2018). This list only considers elaborations of MCMC algorithms, and additionally there are alternative algorithms such as particle filtering that allow for changing posterior distributions (Doucet *et al.*, 2001).

For the most part, cognitive scientists comparing sampling algorithms to human data have evaluated the qualitative properties of these algorithms, though a few researchers have quantitatively fit individual sampling algorithms to human data (Abbott and Griffiths, 2011; Lieder *et al.*, 2018*a*). What has been lacking are quantitative comparisons between sampling algorithms to determine which algorithm best matches human behaviour amongst the many possible candidates.

Part of the problem stems from the fact that the types of data that researchers have been using sampling algorithms to explain are not very diagnostic. First, the qualitative finding that the starting point has an influence (e.g., the algorithm has a sense of location) can be produced by a number of algorithms. Second, cognitive biases are generally biases in decision-making, and decisions are commonly understood to be the result of the aggregation of a number of samples (Bogacz *et al.*, 2006). As sampling algorithms all converge to the correct distribution in the limit, the sample aggregates produced by various sampling algorithms will often be similar, and this problem is compounded by the fact that sampling algorithms can closely mimic one another, given suitable choices of parameters (Lieder *et al.*, 2018*a*).

Intuitively, there should be more power to discriminate between sampling algorithms if individual samples are observed, rather than only observing a decision based on an aggregation of samples. In particular, different sampling algorithms will generate different proposals, and will show different dependencies on previously generated hypotheses, often for a wide range of the settings of their parameters. Characterizing the properties of the time series of candidate sampling algorithms and comparing them against the properties of "cognitive time series" is thus a promising avenue for distinguishing between algorithms.

Fortunately, there exists a body of work by psychologists investigating such cognitive time series which we can repurpose to compare and contrast sampling algorithms. Classic work by Bousfield and Sedgewick (1944) asked participants to generate responses in a task in which the number of potential responses was large but limited, for example asking participants to generate the names of quadruped mammals. While the focus of this work was quantifying the rate at which quadruped animal names and other responses were produced, it was noted in passing that responses tended to be clustered. For example, participants would first produce a set of animals that could be found on a farm, and then produce a set of animals that could be found on a safari.

More recent work in this paradigm by Rhodes and Turvey (2007) looked more closely at the time intervals between successive recalls of animal names. While retrieval intervals lengthened as the pool of unreported animal names shrank, there were also bursts of short retrieval intervals interleaved with long waits, which were perhaps due to participants slowly searching for a new cluster of animal names to report and then quickly reporting the names in that cluster. Qualitatively, there were many more short retrieval times than long retrieval times. Quantitatively, the retrieval intervals were examined in the raw data and in data that were de-trended to remove the effect of slowing retrieval intervals. In both datasets, the retrieval intervals l were well characterized by Lévy probability density distributions

$$P(l) \sim l^{-u} \tag{21.1}$$

showing a power–law relationship between the length of retrieval times and their probabilities. In particular, the best-fitting values of the exponent were $u \approx 2$ for most individual participants.

Lévy distributions with $u \approx 2$ suggest an interesting correspondence with the animal foraging literature. This same distribution (or a truncated version of it) has been used to characterize the mobility patterns of a wide array of species, including Albatrosses, marine predators, monkeys, and people (Viswanathan *et al.*, 1996; Ramos-Fernández *et al.*, 2004; González *et al.*, 2008; Sims *et al.*, 2008). The theoretical justification for Lévy distributions of mobility patterns is that when resources are patchy (i.e., clustered), steps that follow this distribution are more likely to result in successful foraging than Gaussian-distributed steps are. In particular, in environments with patchy resources, $u = 2$ has been analytically shown to be the exponent that produces the most effective foraging (Viswanathan *et al.*, 1999). As a result of this correspondence, Rhodes and Turvey (2007) suggested that human memory retrieval is essentially a foraging task within a mental representation, with response times equated with the distances between samples.

Aside from the distances between successive responses, researchers have also found long-range dependencies in cognitive time series. These dependencies have been investigated in a separate line of work from that on step sizes, at least in the literature on sampling from internal representations, though the two have been found to co-occur in investigations of eye movements, which is a process of sampling information from the external world (Rhodes *et al.*, 2011). Gilden *et al.* (1995) first gave participants one minute's worth of training with a metronome that was set to produce a target temporal interval, such as one second. Following this short period of training, participants were asked to repeatedly press the spacebar on a computer keyboard every time they believed the target interval had elapsed. Participants then continued to 'drum' the keyboard at the target interval 1,000 times in a row, which generated enough responses to characterize how a new response depended on previous responses.

There are various ways in which responses can depend on one another. Most cognitive models assume that responses are independent of one another. For example, standard drift-diffusion models of response times assume that people make independent responses on each trial (Ratcliff, 1978).[1] Standard models of categorization assume that responses are independent given what has been learned (Nosofsky, 1986). Alternatively, the next response may depend solely on the most recent response, as would result from a model that produces a random walk over the space of possibilities (e.g., Abbott *et al.*, 2015).

However, the temporal production task of Gilden *et al.* (1995) showed neither independence nor short-range dependencies, but instead showed long-range dependencies, termed $1/f$ noise. This name comes from the process of quantifying long-range dependencies: performing a Fourier transform and examining how the spectral power $S(f)$ depends on frequency f. For independent responses, $S(f) = 0$, for random walk responses, $S(f) = 1/f^2$, and for long-range dependencies $S(f) = 1/f$. These long-range $1/f$ dependencies are much more difficult to generate than independent responses or the dependencies found in random walks. As such they are often considered the hallmarks of complex processes, and have been found in the dynamics of leaky faucets, heart rates, turbulence, and stock markets (Bak, 1996).

$1/f$ noise is also not unique to temporal production tasks, and it has been found in a variety of similar cognitive tasks, including reproducing complex drumming patterns (Hennig *et al.*, 2011). It has also been found in the estimation of spatial intervals, the time taken for mentally rotating objects, the time taken for lexical decision, the time required for either serial or parallel visual search, and in measures of implicit bias (Gilden *et al.*, 1995; Gilden, 1997; Correll, 2008). Interestingly, these long-range dependencies disappear if a different task is interleaved with the task of interest (Gilden, 2001), or if the task is both very simple and unpredictable (Gilden *et al.*, 1995).

21.3 Sampling Algorithms to Explain Cognitive Time Series

These two properties of human samples, Lévy distributed distances between samples, and $1/f$ noise, are diagnostic for any theory of human inference via sampling. In probabilistic terms, a patchy representation is one that is multimodal: it has regions of high probability separated by troughs of low probability. These types of distributions are a difficult challenge for sampling algorithms with a sense of location, and distances between samples that follow a Lévy probability density distribution are a sign that the sampling algorithm used is successfully navigating this challenge.

However for a sampling algorithm, $1/f$ noise is not at all desirable. Direct sampling, as described above, would be the most efficient in terms of sample size: N independent samples contain more information than the same number of dependent samples. For a set of dependent samples, we can estimate the number of independent samples that they would be equivalent to in terms of the information they contain, which is termed the effective sample size (ESS)

[1] However, these models can be augmented to produce long-range dependencies (Wagenmakers *et al.*, 2004).

$$ESS = \frac{N}{1 + 2\sum_k^\infty C(k)} \tag{21.2}$$

where $C(k)$ is the degree of autocorrelation in the sample sequence at lag k. Thus, for an equivalent level of autocorrelation at the first lag, $1/f$ noise in the samples is also less efficient in terms of sample size than a random walk.

What algorithm could produce both of these properties, and why? As we have noted, direct sampling is the most efficient in uncovering the underlying distribution. Directly drawing independent samples from the underlying distribution will result in a posterior distribution resembling the true distribution, as long as sufficiently many samples are used. Thus, direct sampling will explore even far-apart modes of the true distribution. However, because these samples are drawn independently, direct sampling will not produce characteristic human autocorrelation patterns. Furthermore, while direct sampling allows exploration of far-apart modes, the rate at which distant and close regions of the landscape are visited does not resemble Lévy distributions, but instead will resemble uncorrelated white noise (Zhu *et al.*, 2018). For an example of direct sampling, see Figure 21.1.

In contrast to direct sampling, MCMC does not require knowledge of the true underlying distribution and samples are not drawn independently. Instead, MCMC samplers are initialized at some random location and sequentially explore the probability landscape by performing a random walk, moving to nearby locations proportionally to the probability of the underlying space. For an illustration of the sampling behaviour

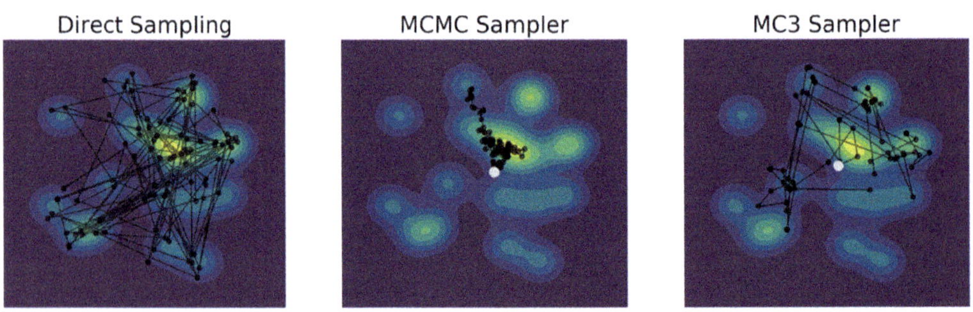

Figure 21.1 *The behaviour of the three sampling procedures in a patchy environment. The underlying distribution that the samplers are exploring corresponds to a mixture of 20 Gaussian distributions. Direct sampling does not require a starting position and subsequent samples are drawn independently from the underlying distribution. As a result, successive samples cover the whole distribution. In contrast, MCMC requires a starting state (solid-grey point). Successive samples are proposed by performing a random walk, biased towards regions of high probability of the underlying distribution. However, these proposals do not allow the rapid exploration of multiple modes. Instead, the sampler will slowly traverse the space, and in many cases, never explore far-away modes. Finally, Metropolis-coupled Markov chain Monte Carlo (MC^3, explained below) also rests on a starting state and iteratively explores the underlying distribution. However, since it does so in multiple parallel chains at higher temperatures, it will occasionally jump into far-away modes.*

of MCMC in multi modal environments, see Figure 21.1. However, while these samples are correlated, the correlations do not exhibit long-range dependencies. Instead, MCMC samplers produce random-walk (Brownian) noise. Furthermore, distances between subsequent MCMC states do not resemble Lévy distributions. Importantly, this cannot be alleviated by replacing the common Gaussian proposal with a heavy-tailed distribution, as the resulting far-ranging proposals are very unlikely to ever be accepted, since these proposals tend to correspond to regions of very low probability.[2]

Other algorithms can produce both long-range autocorrelations and Lévy distributed distances. We have previously explored Metropolis-coupled Markov chain Monte Carlo (MC^3), a type of MCMC sampler.[3] As in MCMC, MC^3 starts at a random location and sequentially traverses the underlying probability landscape, producing a chain of locations, that, given enough samples, will be proportional to the true distribution. However, to allow the sampler to explore far-away areas of the distribution, MC^3 maintains several of these chains, each chain corresponding to an MCMC random walk.

The key idea underlying MC^3 is that of annealing (Kirkpatrick *et al.*, 1983)—to allow the sampler to explore far-away modes each of its parallel chains explores an increasingly flatter version of the underlying distribution by applying different temperatures and thereby 'melting down' modes of the underlying space. MC^3 generates chains in parallel at increasing temperatures and occasionally swaps the states of these chains, therefore allowing the sampler to jump to far-off modes of the distributions. For an example of MC^3 samples, see Figure 21.1. The resulting posterior distribution is then commonly obtained from the first chain, for which no temperature is applied. As we have shown previously, this kind of sampler will sometimes produce long-range jumps, but commonly stay close to the previous location, thus producing Lévy distributed distances. Furthermore, samples obtained by MC^3 will produce slowly decaying autocorrelations resembling those of human data. Essentially, MC^3 pays the price of $1/f$ noise in order to generate the Lévy distributed distances that signify successful jumps between modes.

As MC^3 is a type of MCMC procedure, it can account for the cognitive biases outlined above by manipulating its starting point. Furthermore, inferences based on MC^3 will be strongly biased when the number of samples is reduced, for example due to the temporal constraints or cognitive load.

21.3.1 Going beyond individuals to markets

Interestingly, the properties arising in the structure of human behaviour, including Lévy distributed distances and $1/f$ noise, also can arise in complex real-world tasks. In particular, many (although not all) financial time series, such as asset prices and currency exchange rates, show these properties. It is therefore interesting to see if MC^3 can explain some of the excess variability seen in these prices. To do so, though, requires finding a bridge between internal samples (which might be an individual trader's estimate of the

[2] Heavy-tailed proposal distributions in a uniform space do however produce Lévy-distributed distances, as *every* proposal is equally likely to be accepted.

[3] This algorithm is also sometimes called parallel-tempering or replica-exchange MCMC.

price at the next time step) and the asset prices. It turns out that, using a classic model from behavioural finance (De Long *et al.*, 1990), it is possible to map samples (from traders) to prices in a straightforward way, such that it turns out that prices have the same statistical properties as the samples themselves (Sanborn *et al.*, 2019).[4]

Before doing so, though, we consider the surprising empirical parallels between cognitive and financial time series. For example, the log price changes of cotton and stocks traded in the New York Stock Exchange has been modelled as a stochastic process with Lévy stable non-Gaussian increments (Fama, 1965; Mandelbrot, 1963). This indicates that large price changes in speculative markets happen far more frequently than a simple random-walk market would predict. That is, a person trading in a hypothetical random-walk market would expect a financial crisis of magnitudes greater than four standard deviations to occur merely once every 126 years. The random-walk assumption cannot, though, be correct, as that same person trading with a portfolio of the largest 100 UK companies listed in the London Stock Exchange would have experienced such losses 11 times just between 22 October 1987 and 21 January 2008, even excluding the 2008 financial crisis (Frain, 2009).

Another well-studied property of financial markets is volatility clustering (Granger and Ding, 1995; Mandelbrot, 1963). Qualitatively, this describes how large changes are more likely to be followed by large changes of both positive and negative changes, and similarly for small changes (Mandelbrot, 1963). That is, markets do not allocate volatile time periods randomly across economic periods but the volatility of price changes is serially correlated. The long-range correlations in volatility has also been examined by the power spectrum analysis, and the absolute value of price changes of the Standard & Poor 500 Index measured in one-hour intervals can be characterized as $1/f$ noise with the power-law exponent estimated equal to 0.7 (Mantegna and Stanley, 1997; Liu *et al.*, 1999).

The heavy-tailed distributions of price changes and long-range dependence in the magnitudes of price changes resemble the Lévy distributed distances and $1/f$ noise that psychologists have observed in time estimation and animal naming tasks, where people's change in hypothesis space is measured (Bousfield and Sedgewick, 1944; Gilden, 1997; Gilden, 2001). We have begun to undertake a more careful parallel analysis of price dynamics and cognitive time series in order to establish appropriate correspondences and differences between price changes in the market and hypothesis changes in the mind (Sanborn *et al.*, 2019). Here, we envision that a large part of variability in price changes can be attributed to the variability in opinion changes among market participants. As searching the mental space of hypotheses can be understood through sampling in a mental space, the price dynamics could reflect the stochastic behaviour of a sampler searching in a mental space regarding the future prospects of a commodity, a stock, or a financial portfolio.

[4] The mapping between sampled expected future prices and actual future prices is linear, at least in the simplest case.

21.4 Making the Sampling Algorithm more Bayesian

While sampling algorithms are commonly employed to approximate the answer that a complex probabilistic model would produce under uncertainty, they themselves are not Bayesian: the algorithms have no sense of the uncertainty in the answers that they produce. The algorithms can however be augmented to have probabilistic models over their outputs, therefore giving the sampling algorithms a way to incorporate uncertainty. In statistics and machine learning, this has been called Bayesian Monte Carlo (Rasmussen and Ghahramani, 2003), but for consistency with the above work we term it the *Bayesian sampler*.

Although the aforementioned sampling processes can explain a range of biases in human judgement as the consequence of dependent samples drawn from a large and unevenly distributed hypothesis space, this does not explain why biases also arise when the hypothesis space is small and easy to explore, such as outcomes of six-sided dice (Wedell and Moro, 2008). Human probability judgements in particular tend to exhibit a conservatism bias, in the sense that people's probability estimates tend to be less extreme than one would expect (Peterson and Beach, 1967; Fiedler, 1991; Erev *et al.*, 1994; Hilbert, 2012; Costello and Watts, 2014, 2017). This effect cannot be explained by sampling in itself, but it can be shown that such conservatism is a natural consequence of reasoning with samples of limited size.

Imagine an urn with an unknown proportion of red and/or blue balls. If we draw one ball that turns out to be blue, then presumably we would not on that basis alone conclude that the urn contained only blue balls. Assuming that we lack any prior information regarding the proportion of red and blue balls (i.e., assuming a uniform prior distribution), the optimal Bayesian estimate is that the urn has a proportion of 0.67 blue balls, that is, that the probability of drawing a blue ball is 0.67. More generally, for a prior defined by the beta distribution Beta(α, β) the optimal Bayesian probability estimate $\hat{P}$ based on S outcomes in a sample of size N is

$$\hat{P} = \frac{S+\alpha}{N+\alpha+\beta} \tag{21.3}$$

From this equation, it is easy to see that for any prior distribution where $\alpha = \beta > 0$ (i.e., any prior that is both symmetric and continuous) a Bayesian estimate must necessarily be moderated towards the middle of the distribution; even if we observe ten blue balls in a row, and assuming a uniform prior, the optimal Bayesian estimate of the probability of drawing a blue ball is approximately 0.92 rather than 1. Thus, if we presume that people generally make judgements based on a relatively small number of samples, and there is evidence suggesting this is the case (Goodman *et al.*, 2008; Mozer *et al.*, 2008; Vul *et al.*, 2014), then, in order to minimize average error, conservatism is not a bias but a necessity.

This adjustment to the sampled proportions will sometimes result in incoherence in the sense that estimates for mutually exclusive events will not necessarily sum to one (De Finetti, 1937), and it has been shown that such adjustments will produce the same quantitative conservatism biases as have been observed empirically (Zhu *et al.*, 2020). Indeed, if one assumes that conjunctions require more effort and time to sample than singular events, in turn resulting in relatively fewer samples, then conjunctions will quite naturally be subject to greater Bayesian adjustment, producing conjunction fallacies. For example, sampling the proportions of liberal-seeming persons who are bank tellers is arguably more straightforward than sampling the proportion of liberal-seeming persons who are both feminists and bank tellers and, as a consequence, the latter proportion is likely to be based on a smaller sample and therefore subject to greater adjustment.

21.4.1 Efficient accumulation of samples explains perceptual biases

This Bayesian sampler can also be extended to provide potential explanations for previously observed perceptual biases. When making estimates of perceptual features such as stimulus motion or numerosity, an initial decision regarding this feature can bias subsequent estimates: for example, deciding whether the linear direction of motion of a set of dots is clockwise or counter-clockwise of some boundary line pushes direct estimates of the direction of that motion further from the considered boundary compared with estimates made without such preceding decisions (Jazayeri and Movshon, 2007; Zamboni *et al.*, 2016; Luu and Stocker, 2018). These results then contrast with the anchoring effects described in the previous sections: both tasks observe an impact of a decision on subsequent estimates, but in the cognitive domain, estimates move towards the queried boundary, while in the perceptual domain, estimates move away from the queried boundary.

While other explanations have been offered for these effects, we suggest that one explanation could be the reuse of samples between decisions and estimates, known as amortization (Gershman and Goodman, 2014): learners may draw samples from a sensory representation to make their initial decision regarding the boundary, then reuse those samples in their estimates rather than expending further cognitive resources on additional sampling. This then creates a consistency between the two responses as both the decision and estimate are based on the same set of observations, and so will both reflect any pattern contained in the sample. Simple amortized sampling with a fixed number of samples will not however produce such repulsion effects as there is no bias in this sample: the samples taken for a decision and reused for an estimate would be roughly equivalent to those used for an estimate alone, leading to no systematic difference between these two cases. There is however the possibility that the number of samples is not fixed but adapts to the strength of evidence collected to that point: if a set of samples provides compelling evidence towards a particular decision, the cost of further sampling may outweigh any potential gain in information, encouraging the early termination of sampling. This would then produce a bias in the sample, as sampling is more likely to stop where a high number of samples favour one decision: in the direction of motion example, if several successive samples are clockwise of the decision boundary, we may conclude that the true direction of motion is on this side, and stop sampling.

This then raises the question of how the threshold for the termination of sampling is decided. In keeping with the above sections, this could use a Bayesian updating process in which prior beliefs are updated with each sample to provide a posterior probability for each potential decision. This then allows for the comparison of the cost of terminating sampling with the cost of continuing sampling. Thus, the cost of termination is the expected probability of making an error based on the currently collected evidence, given by the posterior at that point. The cost of continuation, meanwhile, is the sum of the inferred costs of the outcomes of future samples, plus a fixed cost for the generation of the sample itself. As with the probability estimates described above, we assume a Beta prior across the two potential sampling outcomes, here being the two sides of the decision boundary. The sampler therefore begins in a position of ambiguity, and updates this belief with each piece of evidence until the value of further information is outweighed by the cost of its generation. We term this system the Bayesian Amortized Sequential Sampler, or BASS. In comparisons with empirical data, we find BASS provides a better match to behaviour than previously offered candidate models: while other methods are able to predict the decision bias, BASS also explains the strong consistency between decisions and estimates shown by real learners, and more closely matches belief distributions collected from participants regarding their estimates (Zhu *et al.*, 2019).

A question remaining to be answered regarding this process however is how these samples are drawn, as discussed previously in this chapter; amortization describes the reuse of samples in decision-making, but makes no assumptions about the mechanism by which these samples are originally generated. As noted previously, one possibility is direct sampling from the sensory representation; indeed, the results described above were based on direct sampling, and show that such a mechanism is able to predict previously observed perceptual biases. If however the BASS system were to use a sampler with a sense of location such as the MCMC algorithm noted in the previous sections, the resulting estimation system could then capture both these perceptual biases as well as the more traditional anchoring effects found in the cognitive domain (e.g., Russo and Schoemaker, 1989). Specifically, as noted above, MCMC can account for anchoring effects under the suggestion that the anchor provides a starting point for the sampler which, under a limited number of samples, the chain is unable to move far from (Lieder *et al.*, 2012; Lieder *et al.*, 2018*a*; Lieder *et al.*, 2018*b*). Combining an MCMC sampling algorithm with an adaptive stopping rule such as BASS could then provide a single estimation system able to produce both the attraction and repulsion effects observed in existing research. Future work may then wish to examine whether both effects can appear in the same task as a test of such a system, including the potential cross-over in these effects between the cognitive and perceptual domains.

21.5 Conclusions

In this chapter, we have explored the idea that the brain carries out approximate probabilistic reasoning through local sampling rather than through intractable Bayesian calculations. This approach has many of the virtues of a Bayesian analysis of cognition, because it explains why the cognitive system will reason successfully, if the number of samples is sufficiently large. In practice, though, the probability distributions that the brain must deal with will be enormously complex and cannot possibly be sampled in their entirety. If the Bayesian sampling perspective is correct, we might hope that the biases observed in human cognition, including those directly involved the probabilistic estimation, might be those that would be expected from limited Bayesian sampling, where there will be excessive influence of the starting point (as in the anchoring effect in probability judgement). Moreover, a concrete sampling account requires choosing a specific sampling algorithm. We have argued that the characteristic statistics of successive samples (e.g., $1/f$ autocorrelation between durations in rhythmic tapping; and a Lévy distribution on the sizes of jumps between successive durations) provide powerful empirical constraints on the sampling process. We suggest that a specific sampling algorithm, Metropolis-coupled Markov chain Monte Carlo (MC^3), designed to deal with complex multimodal distributions, may be a good candidate sampling mechanism, able to capture patterns in both human judgements and financial time series, which presumably arise from the aggregation of many judgements. We note that the brain should not simply read off the relative frequencies from any sample that it generates. Instead, the correction of such a sample based on prior knowledge is likely to be appropriate, leading to what appears to be conservatism in some cognitive tasks. Moreover, given that sampling is likely to be cognitively slow and costly, an intelligent sampler will actively continue or terminate the sampling process, depending on how results are accumulating. As we have seen, this can lead to estimation biases that push away from a decision boundary, in some ways yielding the opposite pattern to that observed in anchoring.

From the perspective creating human-like computation, we suggest that sampling algorithms provide an attractive research direction. Such algorithms provide a mechanism for approximating complex calculations required to deal with a rich and highly uncertain world, a challenge as relevant for artificial intelligence as for the human brain. Even for designers of machine intelligence that only aspire to effectively interact with people rather than imitate them, samples can be a common framework for collaboration (e.g., Sanborn *et al.*, 2010*b*). Finally, an interesting commonality in our work is that people seem to utilise only a handful of samples—far fewer than what is considered the minimum in statistical applications (e.g., Gelman and Rubin, 1992)—but use them effectively. As people of course operate effectively in the world despite theses restrictions, this may offer a broad lesson for designers of machine intelligence that needs to operate in real time: careful use of a few samples can provide a rough but effective characterization of uncertainty.

Acknowledgements

Sanborn, Zhu, Spicer, and Chater were partially supported by a grant from the ESRC Rebuilding Macroeconomics program. Sanborn, Zhu, Spicer, Sundh, and León-Villagrá were supported by a European Research Council consolidator grant [817492-SAMPLING]. Chater was partially supported by the ESRC Network for Integrated Behavioural Science [ES/P008976/1].

References

Abbott, J. T., Austerweil, J. L., and Griffiths, T. L. (2015). Random walks on semantic networks can resemble optimal foraging. *Psychological Review*, **122**(3), 558–69.

Abbott, J. T. and Griffiths, T. L. (2011). Exploring the influence of particle filter parameters on order effects in causal learning, L. Carlson, ed., *Proceedings of the 33rd Annual Conference of the Cognitive Science Society*, 20–23 July 2011, Boston, MA. Red Hook, NY: Curran Associates, 2950–5.

Anderson, J. R. (1991). The adaptive nature of human categorization. *Psychological Review*, **98**(3), 409–429.

Aragones, E., Gilboa, I., Postlewaite, A. *et al.* (2005). Fact-free learning. *American Economic Review*, **95**, 1355–68.

Bak, P. (1996). *How Nature Works: The Science of Self-organised Criticality*. New York, NY: Springer-Verlag.

Battaglia, P. W., Hamrick, J. B., and Tenenbaum, J. B. (2013). Simulation as an engine of physical scene understanding. *Proceedings of the National Academy of Sciences*, **110**(45), 18327–32.

Bishop, C. M. (2006). *Pattern Recognition and Machine Learning*. New York, NY: Springer.

Bogacz, R., Brown, E., Moehlis, J. *et al.* (2006). The physics of optimal decision making: a formal analysis of models of performance in two-alternative forced-choice tasks. *Psychological Review*, **113**(4), 700–65.

Bousfield, W. A. and Sedgewick, C. H. W. (1944). An analysis of sequences of restricted associative responses. *The Journal of General Psychology*, **30**(2), 149–65.

Chater, N. and Manning, C. D. (2006). Probabilistic models of language processing and acquisition. *Trends in Cognitive Sciences*, **10**(7), 335–44.

Correll, J. (2008). 1/f noise and effort on implicit measures of bias. *Journal of Personality and Social Psychology*, **94**(1), 48–59.

Costello, F. and Watts, P. (2014). Surprisingly rational: probability theory plus noise explains biases in judgment. *Psychological Review*, **121**(3), 463–80.

Costello, F. and Watts, P. (2017). Explaining high conjunction fallacy rates: the probability theory plus noise account. *Journal of Behavioral Decision Making*, **30**(2), 304–21.

Dasgupta, I., Schulz, E., and Gershman, S. J. (2017). Where do hypotheses come from? *Cognitive Psychology*, **96**, 1–25.

De Finetti, B. (1937). La prévision: ses lois logiques, ses sources subjectives. *Annales de l'institut Henri Poincaré*, 7(1), 1–68.

De Long, J. B., Shleifer, A., Summers, L. H. *et al.* (1990). Noise trader risk in financial markets. *Journal of Political Economy*, **98**(4), 703–38.

Doucet, A., de Freitas, N., and Gordon, N. (2001). *Sequential Monte Carlo Methods in Practice*. New York, NY: Springer.

Edwards, W. (1961). Behavioral decision theory. *Annual Review of Psychology*, **12**(1), 473–98.

Erev, I., Wallsten, T. S., and Budescu, D. V. (1994). Simultaneous over-and underconfidence: the role of error in judgment processes. *Psychological Review*, **101**(3), 519–27.

Fama, E. F. (1965). The behavior of stock-market prices. *The Journal of Business*, **38**(1), 34–105.

Fiedler, K. (1991). Heuristics and biases in theory formation: on the cognitive processes of those concerned with cognitive processes. *Theory & Psychology*, **1**(4), 407–30.

Frain, J. C. (2009). *Studies on the Application of the Alpha-stable Distribution in Economics*. PhD Thesis, Department of Economics, Trinity College Dublin.

Gelman, A. and Rubin, D. B. (1992). Inference from iterative simulation using multiple sequences. *Statistical Science*, 7(4), 457–72.

Gershman, S. J. and Goodman, N. D (2014). Amortized inference in probabilistic reasoning, in P. Bello, M. Guarini, M. McShane, and B. Scassellati, eds, *Proceedings of the 36th Annual Conference of the Cognitive Science Society*, 23–26 July 2014, Quebec City, Canada. Red Hook, NY: Curran Associates, 517–22.

Gershman, S. J., Vul, E., and Tenenbaum, J. B. (2012). Multistability and perceptual inference. *Neural Computation*, **24**, 1–24.

Gigerenzer, G. and Gaissmaier, W. (2011). Heuristic decision making. *Annual Review of Psychology*, **62**, 451–82.

Gilden, D. L. (1997). Fluctuations in the time required for elementary decisions. *Psychological Science*, **8**(4), 296–301.

Gilden, D. L. (2001). Cognitive emissions of 1/f noise. *Psychological Review*, **108**(1), 33–56.

Gilden, D. L., Thornton, T., and Mallon, M. W. (1995). 1/f noise in human cognition. *Science*, **267**, 1837–9.

González, M. C., Hidalgo, C. A., and Barabási, A.-L. (2008). Understanding individual human mobility patterns. *Nature*, **453**(7196), 779–82.

Goodman, N. D., Tenenbaum, J. B., Feldman, J. *et al.* (2008). A rational analysis of rule-based concept learning. *Cognitive Science*, **32**, 108–54.

Granger, C. W. J. and Ding, Z. (1995). Some properties of absolute return: an alternative measure of risk. *Annales d'Economie et de Statistique*, 67–91.

Griffiths, T. L., Steyvers, M., and Tenenbaum, J. B. (2007). Topics in semantic representation. *Psychological Review*, **114**(2), 211–44.

Griffiths, T. L. and Tenenbaum, J. B. (2011). Predicting the future as Bayesian inference: people combine prior knowledge with observations when estimating duration and extent. *Journal of Experimental Psychology: General*, **140**(4), 725–43.

Hennig, H., Fleischmann, R., Fredebohm, A. *et al.* (2011). The nature and perception of fluctuations in human musical rhythms. *PLoS ONE*, **6**(10), e26457.

Hilbert, M. (2012). Toward a synthesis of cognitive biases: How noisy information processing can bias human decision making. *Psychological Bulletin*, **138**(2), 211.

Houlsby, N. M. T., Huszár, F., Ghassemi, M. M. *et al.* (2013). Cognitive tomography reveals complex, task-independent mental representations. *Current Biology*, **23**(21), 2169–75.

Jazayeri, M. and Movshon, J. A. (2007). A new perceptual illusion reveals mechanisms of sensory decoding. *Nature*, **446**(7138), 912–15.

Kahneman, D. (2011). *Thinking, Fast and Slow*. New York, NY: Macmillan.

Kirkpatrick, S., Gelatt Jr., C. D., and Vecchi, M. P. (1983). Optimization by simulated annealing. *Science*, **220**, 671–80.

Lieder, F., Griffiths, T. L., and Goodman, N. D. (2012). Burn-in, bias, and the rationality of anchoring, in F. Pereira, C. J. C. Burges, L. Bottou, and K. Q. Weinberger, eds, *Proceedings of Advances in Neural Information Processing Systems*, 3–8 December 2012, Lake Tahoe, CA. Burlington, MA: Morgan Kaufmann, 2690–798.

Lieder, F., Griffiths, T. L., Huys, Q. J. *et al.* (2018*a*, Feb). The anchoring bias reflects rational use of cognitive resources. *Psychonomic Bulletin & Review*, **25**(1), 322–49.

Lieder, F., Griffiths, T. L., Huys, Q. J. *et al.* (2018*b*). Empirical evidence for resource-rational anchoring and adjustment. *Psychonomic Bulletin & Review*, **25**(2), 775–84.

Liu, Y., Gopikrishnan, P., Cizeau, P. *et al.* (1999). Statistical properties of the volatility of price fluctuations. *Physical Review E*, **60**(2), 1390–400.

Luu, L. and Stocker, A. A. (2018). Post-decision biases reveal a self-consistency principle in perceptual inference. *eLife*, 7, e33334.

Mandelbrot, B. B. (1963). The variation of certain speculative prices. *The Journal of Business*, **36**(4), 394–419.

Mantegna, R. N. and Stanley, H. E. (1997). Physics investigation of financial markets, in F. Mallamace and H. E. Stanley, eds, *Proceedings of the International School of Physics Enrico Fermi, Course CXXXIV*, 9–19 July 1996, Varenna, Italy. Amsterdam: IOS Press.

Metropolis, A. W., Rosenbluth, A. W., Rosenbluth, M. N. *et al.* (1953). Equations of state calculations by fast computing machines. *Journal of Chemical Physics*, **21**, 1087–92.

Mosteller, F. and Nogee, P. (1951). An experimental measurement of utility. *Journal of Political Economy*, **59**(5), 371–404.

Mozer, M. C., Pashler, H., and Homaei, H. (2008). Optimal predictions in everyday cognition: The wisdom of individuals or crowds? *Cognitive Science*, **32**(7), 1133–47.

Nosofsky, R. M. (1986). Attention, similarity, and the identification-categorization relationship. *Journal of Experimental Psychology: General*, **115**, 39–57.

Oaksford, M. and Chater, N. (2007). *Bayesian Rationality: The Probabilistic Approach to Human Reasoning*. Oxford: Oxford University Press.

Pantelis, P. C., Baker, C. L., Cholewiak, S. A. *et al.* (2014). Inferring the intentional states of autonomous virtual agents. *Cognition*, **130**(3), 360–79.

Peterson, C. R. and Beach, L. R. (1967). Man as an intuitive statistician. *Psychological Bulletin*, **68**, 29–46.

Petzschner, F. H., Glasauer, S., and Stephan, K. E. (2015). A Bayesian perspective on magnitude estimation. *Trends in Cognitive Sciences*, **19**(5), 285–93.

Ramos-Fernández, G., Mateos, J. L., Miramontes, O. *et al.* (2004). Lévy walk patterns in the foraging movements of spider monkeys (ateles geoffroyi). *Behavioral Ecology and Sociobiology*, **55**(3), 223–30.

Rasmussen, C. E. and Ghahramani, Z. (2003). Bayesian Monte Carlo, in S. Becker, S. Thrun, and K. Obermayer, eds, *Advances in Neural Information Processing Systems 15*. Cambridge, MA: 505–12.

Ratcliff, R. (1978). A theory of memory retrieval. *Psychological Review*, **85**, 59–108.

Rhodes, T., Kello, C., and Kerster, B. (2011). Distributional and temporal properties of eye movement trajectories in scene perception, in L. Carlson, ed., *Proceedings of the 33rd Annual Conference of the Cognitive Science Society*, Boston, MA, 20–23 July 2011. Red Hook, NY: Curran Associates, 178–84.

Rhodes, T. and Turvey, M. T. (2007). Human memory retrieval as Lévy foraging. *Physica A: Statistical Mechanics and its Applications*, **385**(1), 255–60.

Robert, C. P., Elvira, V., Tawn, N. *et al.* (2018). Accelerating MCMC algorithms. *Wiley Interdisciplinary Reviews: Computational Statistics*, **10**(5), e1435.

Russo, J. E. and Schoemaker, P. J. H. (1989). *Decision Traps: Ten Barriers to Brilliant Decision-making and How to Overcome Them*. New York, NY: Simon and Schuster.

Sanborn, A. N. and Chater, N. (2016). Bayesian brains without probabilities. *Trends in Cognitive Sciences*, **20**(12), 883–93.

Sanborn, A. N., Chater, N., Zhu, J.-Q. *et al.* (2019). Macroeconomics implications of the sampling brain. Technical report, National Institute of Economics and Social Research.

Sanborn, A. N., Griffiths, T. L., and Navarro, D. J. (2010*a*). Rational approximations to the rational model of categorization. *Psychological Review*, **117**, 1144–67.

Sanborn, A. N., Griffiths, T. L., and Shiffrin, R. M. (2010*b*). Uncovering mental representations with Markov chain Monte Carlo. *Cognitive Psychology*, **60**, 63–106.

Sanborn, A. N., Mansinghka, V., and Griffiths, T. L. (2013). Reconciling intuitive physics and Newtonian mechanics for colliding objects. *Psychological Review*, **120**, 411–37.

Savage, L. J. (1954). *Foundations of Statistics*. John Wiley & Sons.

Sims, D. W., Southall, E. J., Humphries, N. E. *et al.* (2008). Scaling laws of marine predator search behaviour. *Nature*, **451**(7182), 1098.

Sloman, S., Rottenstreich, Y., Wisniewski, E. *et al.* (2004). Typical versus atypical unpacking and superadditive probability judgment. *Journal of Experimental Psychology: Learning, Memory, and Cognition*, **30**(3), 573–82.

Susskind, J. M., Hinton, G. E., Movellan, J. R. *et al.* (2008). Generating facial expressions with deep belief nets, in V. Kordic, ed., *Affective Computing, Focus on Emotion Expression, Synthesis and Recognition*. London, UK: IntechOpen, 421–40.

Tversky, A. and Kahneman, D. (1974). Judgment under uncertainty: heuristics and biases. *Science*, **185**, 1124–31.

Tversky, A. and Kahneman, D. (1983). Extensional vs. intuitive reasoning: The conjunction fallacy in probability judgment. *Psychological Review*, **90**, 293–315.

Viswanathan, G. M., Afanasyev, V., Buldyrev, S. V. *et al.* (1996). Lévy flight search patterns of wandering albatrosses. *Nature*, **381**(6581), 413.

Viswanathan, G. M., Buldyrev, S. V., Havlin, S. *et al.* (1999). Optimizing the success of random searches. *Nature*, **401**(6756), 911–14.

von Neumann, L. J. and Morgenstern, O. (1947). *Theory of Games and Economic Behavior*, Vol. 60. Princeton, NJ: Princeton University Press.

Vul, E., Goodman, N., Griffiths, T. L. *et al.* (2014). One and done? Optimal decisions from very few samples. *Cognitive Science*, **38**, 599–637.

Vulkan, N. (2000). An economist's perspective on probability matching. *Journal of Economic Surveys*, **14**, 101–18.

Wagenmakers, E.-J., Farrell, S., and Ratcliff, R. (2004). Estimation and interpretation of $1/f^{\alpha}$ noise in human cognition. *Psychonomic Bulletin & Review*, **11**(4), 579–615.

Wedell, D. H. and Moro, R. (2008). Testing boundary conditions for the conjunction fallacy: Effects of response mode, conceptual focus, and problem type. *Cognition*, **107**(1), 105–36.

Wolpert, D. M. (2007). Probabilistic models in human sensorimotor control. *Human Movement Science*, **26**(4), 511–24.

Yuille, A. and Kersten, D. (2006). Vision as Bayesian inference: analysis by synthesis? *Trends in Cognitive Sciences*, **10**, 301–8.

Zamboni, E., Ledgeway, T., McGraw, P. V. *et al.* (2016). Do perceptual biases emerge early or late in visual processing? Decision-biases in motion perception. *Proceedings of the Royal Society B*, **283**, 20160263.

Zhu, J.-Q., Sanborn, A. N., and Chater, N. (2018). Mental sampling in multimodal representations, in S. Bengio, H. Wallach, H. Larochelle, *et al.*, eds, *Proceedings of Advances in Neural Information Processing Systems*, 2–8 December 2018, Montreal, Canada. Burlington, MA: Morgan Kaufmann, 5748–59.

Zhu, J.-Q., Sanborn, A. N., and Chater, N. (2019). Why decisions bias perception: an amortised sequential sampling account, in A. Goel, C. Seifert, and C. Freksa, eds, *Proceedings of the 41st Annual Conference of the Cognitive Science Society*, 24–27 July 2019, Montreal, Canada. Red Hook, NY: Curran Associates, 3220–6.

Zhu, J.-Q., Sanborn, A. N., and Chater, N. (2020). The Bayesian sampler: Generic Bayesian inference causes incoherence in human probability judgments. *Psychological Review*, **127**(5), 719–48.

22

What Can the Conjunction Fallacy Tell Us about Human Reasoning?

Katya Tentori
CIMeC, University of Trento

In what follows, I will briefly summarize and discuss the main results obtained from more than three decades of studies on the conjunction fallacy (hereafter CF) and will argue that this striking and widely debated reasoning error is a robust phenomenon that can systematically affect the probabilistic inferences of both laypeople and experts, with potentially relevant real-life consequences. I will then introduce what is, in my view, the best explanation for the CF and indicate how it allows the reconciliation of some classic probabilistic reasoning errors with the outstanding reasoning performances that humans have been shown capable of. Finally, I will tackle the open issue of the greater accuracy and reliability of evidential impact assessments over those of posterior probability and outline how further research on this topic might also contribute to the development of effective human-like computing.

22.1 The Conjunction Fallacy

When presented with the following scenarios (Tversky and Kahneman, 1983), the great majority (80–90%) of participants ranked the conjunctions ('Linda is a bank teller and is active in the feminist movement' and 'Mr. F. has had one or more heart attacks and he is over 55 years old') as more probable than their less-representative constituents ('Linda is a bank teller' and 'Mr. F. has had one or more heart attacks'):

Linda scenario
Linda is 31 years old, single, outspoken, and very bright. She majored in philosophy. As a student, she was deeply concerned with issues of discrimination and social justice, and also participated in anti-nuclear demonstrations. [Evidence *e*]
Please rank the following statements by their probability:

- Linda is a teacher in elementary school.
- Linda works in a bookstore and takes yoga classes.

Katya Tentori, *What Can the Conjunction Fallacy Tell Us about Human Reasoning?* In: *Human Like Machine Intelligence.* Edited by: Stephen Muggleton and Nick Charter, Oxford University Press. © Oxford University Press 2021.
DOI: 10.1093/oso/9780198862536.003.0022

- Linda is active in the feminist movement. [Single hypothesis h_2]
- Linda is a psychiatric social worker.
- Linda is a member of the League of Women Voters.
- Linda is a bank teller. [Single hypothesis h_1]
- Linda is an insurance salesperson.
- Linda is a bank teller and is active in the feminist movement. [Conjunction $h_1 \wedge h_2$]

(For this and all following scenarios: the order of the response options was randomized; headings and square brackets are for reference only, and they were not included in the original experimental material.)

Mr. F scenario
A health survey was conducted in a representative sample of adult males in British Columbia of all ages and occupations. Mr. F. was included in the sample. He was selected by chance from the list of participants.
Which of the following statements is more probable? (check one)

- Mr. F. has had one or more heart attacks. [h_1]
- Mr. F. has had one or more heart attacks and he is over 55 years old. [$h_1 \wedge h_2$]

Similar results have been documented not only with a variety of hypothetical scenarios, but also in many real-life domains, both for laypeople and experts who had been asked for probability judgements in their own fields of specialization (e.g., (Tversky and Kahneman, 1983; Frederick and Libby, 1986; Ho and Keller, 1994; Adam and Reyna, 2005; Garb, 2006; Crupi *et al.*, 2018)).

22.2 Fallacy or No Fallacy?

From the very first description, judgements such as those described above have been considered violations of 'the simplest and the most basic qualitative law of probability' (i.e., the *conjunction rule*, Tversky and Kahneman, 1983, 293; but already mentioned in Wyer, 1976; Goldsmith, 1978; Beyth-Marom, 1981; Tversky and Kahneman, 1982). This emphasis is easy to understand, since the ordinal comparison between the probability of a conjunction and the probability of one if its conjuncts does not pose a great challenge to cognitive resources and can rest on elementary class-inclusion relationships, without requiring the mastery of formal logic or probability theory. The CF became then a key topic in the fervent debate on human rationality, and a remarkable body of empirical studies inspired by the pragmatics of communication tried to control for whether participants' responses were manifestations of a genuine reasoning error or were, rather, generated by interpretations of the CF stimuli that deprived them of their normative relevance. The evidence provided during this long and heated debate is too extensive to be fully discussed here, however, to give the reader the flavor of it, I will describe the three main candidate misunderstandings and how they have been controlled for.

The first potential misunderstanding has to do with participants' interpretation of the verbal descriptions concerning the *isolated conjunct* h_1. Various researchers (e.g., Adler, 1984; Dulany and Hilton, 1991) have pointed out that the comparison between the relative probability of a set with its superset is anomalous, and widely shared principles of cooperative communication (Grice, 1975) might lead participants to interpret h_1 as $h_1 \wedge \neg h_2$. In such a case, participants' selection of $h_1 \wedge\ h_2$ could be justified, since in a typical CF scenario $P(h_1 \wedge h_2) > P(h_1 \wedge \neg h_2)$. Several experimental techniques have been put forward to block such a conversational implicature. Among these are rephrasing of the single conjunct (e.g.,'Linda is a bank teller *whether or not she is active in the feminist movement*', emphasis added), controlling for the interpretation of h_1 after the CF task, and above all, changing the set of options offered to participants by explicitly including among the response options the conjunction $h_1 \wedge \neg h_2$ (along with h_1 and conjunction $h_1 \wedge h_2$). The idea, in this case, is that it does not make sense to interpret the conjunct h_1 as $h_1 \wedge \neg h_2$ if this latter option is already available among the choice options, since, from a conversational point of view, it would be uncooperative to repeat one of the options in a different form. When this technique was applied (as in Tentori *et al.*, 2004; Wedell and Moro, 2008), the CF occurred at a lower rate than first reported in the original CF scenarios but remained prevalent. Such a pattern makes clear that the misunderstanding of the single conjunct should indeed be avoided in order to distinguish proper and improper fallacy answers, but also that it cannot be considered the primary reason for the occurrence of the CF.

According to a second line of thought (e.g., Fiedler, 1988; Gigerenzer, 1996), the linguistic misunderstanding between the experimenter and participants concerns the term *probable*. The conjunction rule is not violated, of course, if participants interpret this word not in its technical sense as assigned by modern probability theory, but rather as *plausible, believable, or imaginable*—all legitimate meanings, according to well-respected dictionaries. The vagueness of the term *probable* in everyday language can be overcome by asking participants to rate the hypotheses according to their 'willingness to bet' on them (e.g., Tversky and Kahneman, 1983) or, even more directly, by asking participants to bet real money on hypotheses that concern future events (e.g., Sides *et al.*, 2002; Bonini *et al.*, 2004). The implicit rationale is that participants would be aiming to maximize their winnings, and, in order to do so, they should bet on the most probable hypothesis in the intended mathematical meaning of the word. When this technique has been applied, a drop in the CF with respect the original scenarios has been observed, but still, most participants committed it.

A final and major objection to the existence of a CF involves the interpretation of the connective *and*. This objection stems from an uncontroversial fact: the conjunction rule concerns the logical connective $\wedge$ while its experimental tests typically rely on a natural language sentential connective *and*, which, as opposed to the former, can reflect various set-theoretical operators and convey a wide range of *temporal* or *causal* relationships between the conjuncts. For example, the word *and* in the sentence 'Tom invited friends and colleagues to his party' suggests a *union* rather than an *intersection* of sets, while the *and* in a sentence like 'Sara will go to the party, and Mike will be extremely happy' clearly expresses more than the mere co-occurrence of two events. Moving from this premise, a number of authors have argued that responses commonly taken as manifestations of CF might in fact emerge from 'reasonable pragmatic and semantic inferences' induced by the ambiguity of the *and* conjunction (Hertwig *et al.*, 2008, but see also Gigerenzer, 1996). The point is relevant because if the *and* were interpreted as suggesting a union rather than as an intersection operator, or if it were interpreted as indicating a conditional probability instead of the corresponding conjunctive probability (i.e., the probability that h_2 happened given that h_1 did), then there would of course be no fallacy (as already observed by Tversky and Kahneman themselves, 1983). Fortunately, there are various ways to prevent (ex-ante) a misinterpretation of the conjunction or to check (ex-post) whether such a misinterpretation did actually take place. With regard to the former, for example, (Bonini *et al.*, 2004) overtly point out the conjunctive meaning of *and* by including a reminder in the description of the bets that they offered to their participants (it read 'both events must happen for you to win the money placed on this bet'). The control for the interpretation of *and* after the CF task can been accomplished using Venn diagrams (e.g., Tentori and Crupi, 2012*b*) or questions that check whether participants hold that the *and* statement at issue implied the truth of both the conjuncts, and therefore, the corresponding $\wedge$ statement (e.g., Tentori *et al.*, 2004). Yet again, when these various techniques have been applied, the CF remained prevalent and affected a great number of judgements. The CF has also been proven to be resistant to linguistic training aimed at improving participants' accuracy in distinguishing proper conjunctions from other meanings that may be conveyed by the word *and* (Crandall and Greenfield, 1986).

In summary, although the CF rates reported in the original scenarios (like Linda or Mr. F above) were somewhat inflated, none of the techniques that has been developed to prevent or control for the various sources of misinterpretations mentioned in the literature proved able ultimately to dissipate the effect (for a more comprehensive review on this topic and a similar conclusion, see Moro, 2009). Therefore, the CF is a real cognitive bias that can, through careful phrasing of stimuli and with a suitable scenario, easily affect more than 50% of judgements (for a clarification of what makes a good CF scenario, see the following section). The first major point of this chapter is that reasoning errors do not always originate from computational difficulties, inexperience or carelessness. Probabilistic reasoning can exhibit systematic departures from relevant standards of rationality when very simple tasks are at issue and logically correct answers are rewarded, and even in statistically sophisticated individuals.

22.3 Explaining the Fallacy

Once the CF is established as a real fallacy, it becomes interesting to explain why people are prone to such an elementary error in reasoning about chance. A good starting point is to observe that only a limited number of comparisons between the probability of a

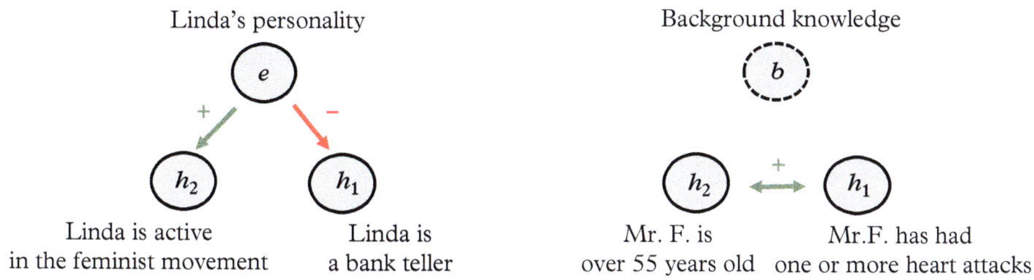

Figure 22.1 *Diagrams representing the Linda (left) and Mr. F (right) scenarios. According to Tversky and Kahneman (1983), in the former case there exists some psychologically salient connection between evidence e and the added conjunct h_2, while in the latter case what is crucial is the relation between the two conjuncts h_1 and h_2.*

conjunction and that of one of its conjuncts results in a CF. A survey of the literature reveals that the added conjuncts typically employed in successful CF scenarios are both *highly probable* and *positively supported*, as specified by Bayesian theories of confirmation (Carnap, 1962; Earman, 1992; Crupi and Tentori, 2016).[1] Let's consider, for example, the Linda scenario introduced above: the hypothesis that 'Linda is active in the feminist movement' (h_2) is probable in the light of Linda's description (e), but also positively supported (or inductively confirmed) by this evidence; similarly, in the Mr. F scenario, the hypothesis that 'Mr. F is over 55 years old' (h_2) is probable in the light of the other hypothesis 'Mr. F. has had one or more heart attacks' (h_1) but also positively supported by it. (See Figure 22.1.) The association between posterior probability and confirmation in CF scenarios is not surprising, given that these two variables are also often positively correlated in real life, that is high [low] probability hypotheses are typically confirmed [disconfirmed] by available evidence. However, the posterior probability of a hypothesis and the support for it can be dissociated, so one may wonder which of these two variables is the one crucial for the CF to occur. Tentori *et al.* (2013) designed four experiments to disentangle the perceived probability and confirmation of the added conjuncts in order to contrast them in a CF task, as illustrated by the following scenarios:

Violinist scenario
O. has a degree in violin performance. [e]
Which of the following hypotheses do you think is the most probable?

- O. is an expert mountaineer. [h_1]
- O. is an expert mountaineer and gives music lessons. [$h_1 \wedge h_2$]
- O. is an expert mountaineer and owns an umbrella. [$h_1 \wedge h_3$]

[1] A common way to formalize the notion of evidential impact (or confirmation) is to devise a function $C(h, e)$ assuming a positive value *iff* $P(h|e) > P(h)$, value zero *iff* $P(h|e) = P(h)$, and a negative value *iff* $P(h|e) < P(h)$. A variety of such functions have been proposed, for example $\log P(e|h)/P(e|\neg h)$ (Good, 1984; but see also Fitelson, 1999; Crupi *et al.*, 2007; Tentori *et al.*, 2007).

Swiss person scenario
Which of the following hypotheses do you think is the most probable?

- M. is Swiss. $[h_1]$
- M. is Swiss and can ski. $[h_1 \wedge h_2]$
- M. is Swiss and has a driving licence. $[h_1 \wedge h_3]$

Assuming h_1 = 'O. is an expert mountaineer' as a (largely irrelevant) piece of background evidence, the majority of participants judged e = 'O. has a degree in violin performance' as supporting h_2 = 'O. gives music lessons' more than h_3 = 'O. owns an umbrella', that is, they judged $C(h_2, e|h_1) > C(h_3, e|h_1)$. However, they also judged h_2 to be less probable than h_3 (in the light of e and h_1), that is, $P(h_2|e \wedge h_1) < P(h_3|e \wedge h_1)$, since almost everybody (even expert mountaineers who have a degree in violin performance) owns an umbrella. A similar dissociation was obtained without providing explicit evidence e. For example, the majority of participants judged h_1 = 'M. is Swiss' as supporting h_2 = 'M. can ski' more than h_3 = 'M. has a driving licence', that is, they judged $C(h_2, h_1) > C(h_3, h_1)$. At the same time, they judged the overall probability of h_2 given h_1 to be lower than that of h_3, that is, $P(h_2|h_1) < P(h_3|h_1)$. Once the perceived probability of the added conjunct (higher for h_3) and the perceived support for it (stronger for h_2) were dissociated, participants were presented with a CF task in which they had to select the most likely among $h_1, h_1 \wedge h_2$, and $h_1 \wedge h_3$. As a result, Tentori *et al.* (2013) found that a large majority of the fallacious responses targeted $h_1 \wedge h_2$ rather than $h_1 \wedge h_3$ (83% vs 17% and 79% vs 21%, respectively for the two above scenarios), a pattern that supported the role of inductive confirmation for the added conjunct rather than its probability as a major determinant of the CF (see also Crupi *et al.*, 2008; Tentori and Crupi, 2012*a*). This outcome is incompatible with most of the alternative explanations of the CF, from those that ascribe it to *non-normative combination rules* for calculating the conjunctive probability from the probabilities of the two conjuncts (e.g., *weighted average*, Fantino *et al.*, 1997; *configural weighted average*, Nilsson *et al.*, 2009; *signed summation*, Yates and Carlson, 1986), to various *models of rationality rescue*, which consider the CF a consequence of participants' normative patterns of reasoning (e.g., *random error variation*, Costello, 2009; *source reliability*, Bovens and Hartmann, 2003). Indeed, although very different from each other, all these proposals predict that CF rates would have risen as the perceived probability of the added conjunct increased.

Tversky and Kahneman's (1983) original explanation of the CF by means of the *representativeness* heuristic deserves separate discussion. According to this account, a conjunction (e.g., 'Linda is a bank teller and is active in the feminist movement') can appear more probable than one of its constituents ('Linda is a bank teller') because it is more *representative* of the evidence provided (Linda's description) than the latter. Such a reading of the CF is flexible enough to accommodate a number of findings. Critics (e.g., Gigerenzer, 1996) have countered, however, that the notion of representativeness is too vague and imprecisely characterized to serve as a full explanation. It falls short in accounting for the underlying cognitive processes (what drives the representativeness assessment?) and the antecedent conditions that could elicit or suppress it (when should a CF be expected to occur and to what extent?). In reply to these critiques, (Tenenbaum and Griffiths, 2001) provided a formal account of representativeness in the context of Bayesian inference by quantifying how much evidence e is representative of hypothesis h in terms of $\log P(e|h)/P(e|\neg h)$. Such a proposal is completely in line with the confirmation-theoretic account of the CF introduced above, since it focuses on the Bayesian support for the hypotheses at issue. Indeed, the logarithm of the likelihood ratio is not only one of the most popular confirmation measures (Fitelson, 1999) but has also been shown to be one of the two measures that best captures people's intuitive judgements of impact (Crupi *et al.*, 2007; Tentori *et al.*, 2007). In this sense, the confirmation explanation of the CF can be seen as a formalization and generalization of the original representativeness account and, as such, could be extended to other phenomena that have been traced to this heuristic, from other reasoning errors (e.g., the base rate fallacy, Kahneman and Tversky, 1973), to information retrieval (Gennaioli and Shleifer, 2009), and even to stereotype formation (Bordalo *et al.*, 2016).

22.4 The Pre-eminence of Impact Assessment over Probability Judgements

The explanation of the CF provided in the previous section suggests that common probability errors can be determined by a pre-eminence of evidential reasoning over probabilistic reasoning. In this regard, it is worth noting that people's judgements of evidential impact have been reported to be accurate, both when applied to the evaluation of abstract arguments concerning, for example, urns and balls of different colours (Tentori *et al.*, 2007), and in everyday tasks that require participants to quantify the impact of uncertain evidence (Mastropasqua *et al.*, 2010) or the value of evidence with regard to competing hypotheses (Crupi *et al.*, 2009; Rusconi *et al.*, 2014). These results are consistent with those from the category-based induction literature, according to which adults, and even children as young as five, when evaluating argument strength, follow popular principles of evidential impact, such as the *similarity* between premises and conclusion and the *diversity* of premises (Osherson *et al.*, 1990; Lopez *et al.*, 1992; Heit and Hahn, 2001; Lo *et al.*, 2002; Zhong *et al.*, 2014). The spontaneous, and often implicit, appreciation of evidential impact has been shown, under various names, to play a fundamental role in a variety of other higher- and lower-level cognitive processes, including causal induction (Cheng and Novick, 1990; Cheng, 1997), conditional reasoning (Douven and Verbrugge, 2012; Krzyzanowska *et al.*, 2017), learning (Danks, 2003), language processing (Bullinaria and Levy, 2007; Paperno *et al.*, 2014; Bhatia, 2017; Nadalini *et al.*, 2018), and even perception (Mangiarulo *et al.*, 2021).

In the light of the aforementioned results, one may wonder whether the updating of the probability of the hypothesis on new evidence and the estimation of the impact of the new evidence on the credibility of the hypothesis are equally reliable cognitive assessments. Tentori *et al.* (2016) tried to answer this question by directly comparing impact and probability judgements on the very same arguments. More specifically, they asked 200 undergraduates (100 females and 100 males) drawn from various UCL departments to fill in a survey with dozens of personal questions, such as the following: *Do you have a driving licence? Do you own (at least) one videogame console? Can you ski? Do you support any football team? Do you like cigars? Do you like shopping? Do you have freckles?* Response frequencies were used to derive objective conditional probabilities (e.g., the probability that a UCL student has a driving licence given that s/he is female/male) and corresponding impact values (e.g., the impact of the evidence that a UCL student is female/male on the hypothesis that s/he has a driving licence). Fifty-six arguments were than generated by combining two complementary pieces of evidence ('X is a male / female student') with 28 different hypotheses (e.g., 'X has a driving licence', 'X likes cigars', etc.). The hypotheses were selected so as to have (together with the two pieces of evidence) all possible combinations of high/low posterior probability and positive/neutral/negative impact, that is, an identical number of arguments with high (> .5) and low (< .5) posterior probability of the hypotheses, and, for each of these two classes, the same number of arguments with high, neutral, and low impact. A new sample of participants belonging to the same population (i.e., UCL undergraduates) came to the laboratory twice, with an interval of 7–10 days. The two sessions were identical, and, on both occasions, participants were presented with the 56 arguments generated and were asked to judge, for each of them, the probability of the hypothesis in the light of the evidence provided and the impact of the evidence provided on the credibility of the hypothesis. The results showed that, compared to probability judgements, impact judgements were more consistent over time and more accurate. Impact judgements also predicted the direction of errors in probability judgements.

The conclusions of the studies above converge in suggesting that human inductive reasoning relies more on the estimation of evidential impact than of posterior probability. They also offer a novel approach to bridge the so-called *reality-laboratory gap*, that is the alleged clash between the body of experimental work in the heuristics and biases tradition (Tversky and Kahneman, 1974), that has deeply challenged the assumption of people's rationality, and the claims of various evolutionary psychologists, who have argued that it is implausible that humans would have evolved with no 'instinct for probability' (Pinker, 1997). According to the latter view, reasoning experiments have been designed to 'trick our probability calculators', and when people are given 'information in a format that meshes with the way they naturally think about probability, they can be remarkably accurate' (Pinker, 1997). Tentori *et al.* (2016) propose a third perspective that reconciles these two views in that, in dealing with everyday uncertainty, people may appear more rational than in experimental psychology laboratories because they can derive posterior probability from impact. In most situations, indeed, these two kinds of assessments often yield similar results, that is when evidence has a strong positive [negative] impact on a hypothesis, then the probability of the latter in the light of the former is rather high [low]. For example, think of a physician who has to diagnose a patient's disease. Usually, when the available evidence (e.g., symptoms, clinical signs, results of laboratory tests) strongly supports [opposes] the diagnosis of a certain disease, then the probability that the patient has that disease is high [low]. Because this association is so common in real life, one may use impact as a proxy for posterior probability without making critical errors. As shown in the previous section, however, posterior probability and evidential impact can be dissociated, a typical occurrence in classical experimental reasoning tasks, not only those in which the CF has been observed, but also those that showed other well-known fallacies, such as base-rate neglect (in which scenarios the target hypothesis retains a low probability because of its extremely low prior, even in the light of supporting evidence). When such a dissociation between probability and confirmation takes place, people cannot derive correct posterior probability judgements from impact and appear to be particularly exposed to biased probability reasoning, whose direction and magnitude seems to depend precisely on perceived impact.

22.5 Implications for Effective Human-like Computing

The greater consistency over time of judgements of evidential impact over those of posterior probability and, above all, their greater accuracy is of course an empirical finding of interest *per se*, but how can it inform human-like computing?

One of main targets of human-like computing is to improve data mining and machine-learning explainability, that is, the process by which intelligent systems explain their outputs to humans, so to generate a shared understanding and, ultimately, increase trust. However, the concept of explanation, despite being a traditional topic in the philosophy of science and a central notion in human reasoning, lacks a unique definition and consensus on the features that would make an explanation 'good' or at least 'satisfactory'. Classical Bayesian confirmation theory makes no explicit reference to explanation, however, the strength of an explanation (i.e., the degree of *explanatory power* of a candidate explanans h relative to its explanandum $e, E(e, h)$) can be expressed in a like manner to the quantification of confirmation, that is, as a function of probability values involving evidence e (e.g., a certain symptom) and hypothesis h (e.g., a disease that can cause the observed symptom). The connection between explanation and confirmation is not new, and the general idea is that, *ceteris paribus*, the greater the statistical relevance between evidence e and hypothesis h, the greater the strength with which h can explain e (a condition named *positive relevance* by Schupbach and Sprenger, 2011; for some well-known probabilistic measures of explanatory power, see Good, 1960, and Schupbach and Sprenger, 2011).

To appreciate the relationship between confirmation and explanation, it might be of help to refer to the following CF scenario, which has been recently presented to 82 experienced internists (Crupi *et al.*, 2018).

Anaemia scenario

A 50-year-old man from northern Italy has chronic anaemia. Currently, the only additional information available comes from a blood exam: haemoglobin 10 g/dL and normal values of leukocytes and platelets. Mean corpuscular volume (MCV) is also in normal range. (Such values are essentially unchanged from a previous test two months earlier.) [*e*]

Please consider the following clinical conditions and rank them from the most to the least probable (ties are allowed).

- thalassaemia trait $[h_1]$
- no thalassaemia trait and alcoholism $[\neg h_1 \wedge h_2]$
- thalassaemia trait and alcoholism $[h_1 \wedge h_2]$
- thalassaemia trait and no alcoholism $h_1 \wedge \neg h_2]$
- alcoholism $[h_2]$

The conjunction $h_1 \wedge h_2$ ('thalassaemia and alcoholism') was evaluated by 68% of internists as more likely than h_1 ('thalassaemia'), by 60% of internists as more likely than h_2 ('alcoholism'), and by 49% of internists as more likely than h_1 as well as more likely than h_2 (i.e, around half of participants committed a *double* CF). These results show once again that experts can make defective probability judgements, which nevertheless rely on a sound intuitive assessment of relations of evidential impact (between diagnostic conditions and clinical signs). Indeed, in the Anaemia scenario, each of the two conjuncts is disconfirmed by the available evidence: thalassaemia (h_1) because it typically produces low MCV, while alcoholism (h_2) because it typically produces high MCV. However, the conjunction $h_1 \wedge h_2$ is supported by the clinical evidence e (that is $P(h_1 \wedge h_2|e) > P(h_1 \wedge h_2)$) because thalassaemia and alcoholism together can *explain* MCV being at normal levels overall. (See Figure 22.2.)

The implications of these results for the debate over explainability are, at least, twofold. First, they offer a new perspective on the factors driving the explanation quality (the so-called explanatory virtues). In particular, they challenge the mainstream view (e.g., Miller, 2019; Lombrozo, 2007) that simpler explanations are judged better and more likely to be true. Indeed, although a conjunction of causes undoubtedly represents a more complex explanation, from both a syntactic and a semantic point of view, than each of the two causes mentioned in the individual conjuncts, most participants ranked the

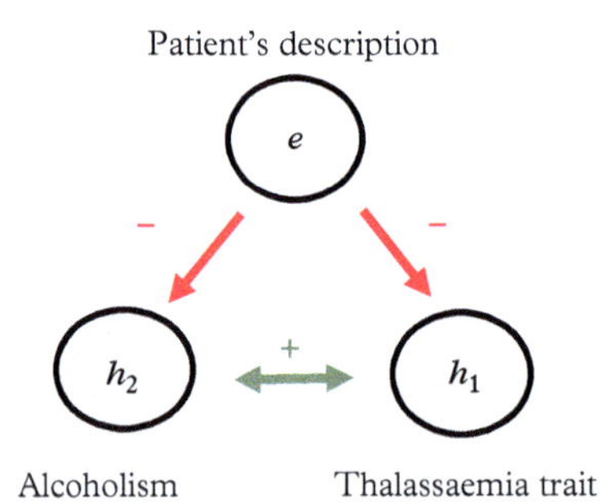

Figure 22.2 *Diagram representing the Anaemia scenario. The patient's description disconfirms each of the two conjuncts h_1 and h_2 occurring alone, however, it confirms the conjunction $h_1 \wedge h_2$ (and the two conjuncts confirm each other in the light of the evidence provided).*

former explanation over the latter two. Therefore, the common assumption in the data mining and machine-learning literature that users always find simpler models easier to understand and more convincing to believe is not necessarily well founded. For a similar conclusion that questions the predominant 'simplicity bias' when the plausibility of rule-based models is at issue (intended as the likeliness that a user will accept the model as an explanation for a prediction), see also Fürnkranz *et al.* (2019).

Second, the results of the Anaemia scenario suggest that the perceived plausibility of a list of causal explanations does not depend on their probability given the available evidence as much as on their being supported by the available evidence. This is in line with Miller's (2019) claim that the probability of the explanation being true is not that important for having a good explanation. Moreover, it embraces one of the main tenets of the inference to the best explanation model (see Lipton, 2014), which is the idea that inferences are often guided by explanatory considerations, yet it also suggests an opposing tendency: causal explanations need to rest on strong confirmation relations to be perceived as convincing.[2]

In this regard, it is worth noting that the constructs of confirmation and explanation share a number of interesting properties. To mention one, they both typically involve asymmetric relations: apart from some rare exceptions, if h explains e, the 'backward' inference from e to h does not appear equally explanatory; similarly most popular Bayesian confirmation models do not classify inversely symmetric confirmatory arguments as equally strong (i.e., $C(h,e) \neq C(e,h)$, for more details on this, see Eells and Fitelson, 2002; Crupi *et al.*, 2007). However, the strength of the explanans-explanandum relation between h and e should not be equated to the strength of the impact of e on h. A formal argument supporting this statement is provided by Crupi (2012); for the purposes of this chapter, it is enough to observe that, when e confirms h, the conjunction of e with a piece of evidence x that is probabilistically independent from both e and h, as well as from $e \wedge h$, leaves the degree of confirmation on h unaffected while weakening the explanatory power of h, that is, $C(h,e) = C(h, e \wedge x)$ but $E(e,h) > E(e \wedge x, h)$.

Crupi and Tentori (2012) presented a treatment of the relation between confirmation and explanation, according to which for any e_1, e_2, h_1, and h_2 such that e_1 confirms h_1 (i.e., for which $P(h_1|e_1) > P(h_1)$) and e_2 confirms h_2 (i.e., $P(h_2|e_2) > P(h_2)$) then $C(h_1, e_1) \gtreqless C(h_2, e_2)$ *iff* $E(e_1, \neg h_1) \lesseqgtr E(e_2, \neg h_2)$. Such a principle postulates an inverse (ordinal) correlation between the degree of positive confirmation that a successful explanatory hypothesis h receives from the occurrence of explanandum e and the degree to which e fails to be explained by $\neg h$. In other words, an explanatory hypothesis h is confirmed by evidence e to the extent that the latter appears inexplicable (i.e., a sort of 'miracle') assuming the falsity of the former.

2 Note that this conclusion concerns only the *necessity* of relevant impact relation(s) for causal explanations to be perceived as convincing, while it does not mean that confirmation should be intended as a *sufficient* condition for explanations. In fact, various statistical relations (for example the association between shapes and colours in a set of figures) allow inductive inferences that do not have an explanatory nature (and for those it seems weird even to talk about a 'cause' in the strict sense).

Future empirical studies might quantify more precisely the role of evidential reasoning in the understandability and acceptability of explanations by examining to what extent equally probable explanations that are supported to various degrees by relevant evidence are perceived to be more or less convincing. Such an experimental manipulation might be of interest also with respect to the purpose of exploring how explanations are generated in both cognitive science and artificial intelligence. In particular, one of the strongest claims in Miller's (2019) review on explanation in artificial intelligence is that people are 'cognitively wired to process contrastive explanations', in the sense that they do not explain the causes for an event *per se* but only relative to some other counterfactual event that did not occur (i.e., an explanation is always of the form 'why event e_1 rather than e_2?'). Contrastive explananda are, without a doubt, important for defining the specific *nature* of what has to be explained. However, Crupi and Tentori's (2012) proposal argues that convincing causal explanations make use of an additional type of contrast, namely among candidate explanantia. According to this, the value of an explanation would be determined not only by how well it accounts for evidence but also by its 'advantage' in doing so over available alternatives. Note that such an idea has a straightforward prediction to offer: when an event equally admits multiple competing explanations, despite the fact that they each account for the event, none would gain any particular credit.

22.6 Conclusion

Cognitive scientists have long been interested in people's systematic violations of basic principles of probability theory, not only because these occurrences detail specific limitations of human reasoning but also because, by elucidating the underlying cognitive processes through which people make judgements, they may offer insight into how to improve the quality of thinking (see on this point the 'negative' and 'positive' agendas of the heuristics and biases program). In continuity with this tradition, the current chapter reviewed some of the main findings generated in more than 30 years of studies on the CF and to discuss how they can be extended beyond cognitive science by informing human-like computing. To sum up, the results of CF experiments make clear that, when it comes to human reasoning, difficulties do not depend on computational overload or poor statistical numeracy alone. Moreover, the understanding of why laypeople and experts alike commit such an elementary error provides useful information on how inductive inferences are made and shows that the very same cognitive processes that are responsible for errors in one instance allow for gains and accurate performances in another. Finally, what we know about the weaknesses and strengths of human probabilistic and evidential reasoning could offer some inspiration in designing machine-learning algorithms that provide understandable and convincing explanations. Future empirical and modelling studies might delve more deeply into these possibilities and reveal if they can be implemented in enhancing the reasoning capacity of machines and in allowing them to communicate with humans in a better way.

References

Adam, M. B. and Reyna, V. F. (2005). Coherence and correspondence criteria for rationality: Experts' estimation of risks of sexually transmitted infections. *Journal of Behavioral Decision Making*, **18**(3), 169–86.

Adler, J. E. (1984). Abstraction is uncooperative. *Journal for the Theory of Social Behaviour*, **14**(2), 165–81.

Beyth-Marom, R. (1981). *The Subjective Probability of Conjunctions*, Decision Research Report No. 81–12. Eugene, Oregon: Decision Research Inc.

Bhatia, S. (2017). Associative judgment and vector space semantics. *Psychological Review*, **124**(1), 1–20.

Bonini, N., Tentori, K., and Osherson, D. (2004). A different conjunction fallacy. *Mind and Language*, **19**(2), 199–210.

Bordalo, P., Coffman, K., Gennaioli, N. *et al.* (2016). Stereotypes. *The Quarterly Journal of Economics*, **131**(4), 1753–94.

Bovens, L. and Hartmann, S. (2003). *Bayesian Epistemology*. Oxford: Oxford University Press.

Bullinaria, J. A. and Levy, J. P. (2007). Extracting semantic representations from word co-occurrence statistics: A computational study. *Behavior Research Methods*, **39**(3), 510–26.

Carnap, R. (1950/1962). *Logical Foundations of Probability*. Chicago, IL: University of Chicago Press.

Cheng, P. W. (1997). From covariation to causation: A causal power theory. *Psychological Review*, **104**(2), 367–405.

Cheng, P. W. and Novick, L. R. (1990). A probabilistic contrast model of causal induction. *Journal of Personality and Social Psychology*, **58**(4), 545–67.

Costello, F. J. (2009). How probability theory explains the conjunction fallacy. *Journal of Behavioral Decision Making*, **22**(3), 213–34.

Crandall, C. and Greenfield, B. (1986). Understanding the conjunction fallacy: A conjunction of effect. *Social Cognition*, **4**(4), 408–19.

Crupi, V. (2012). An argument for not equating confirmation and explanatory power. *The Reasoner*, **6**(3), 39–40. Erratum: The Reasoner, 6, 68.

Crupi, V., Elia, F., Aprà, F. *et al.* (2018, Jun). Double conjunction fallacies in physicians' probability judgment. *Medical Decision Making*, **38**(6), 756–60.

Crupi, V., Fitelson, B., and Tentori, K. (2008). Probability, confirmation, and the conjunction fallacy. *Thinking & Reasoning*, **14**(2), 182–99.

Crupi, V. and Tentori, K. (2012). A second look at the logic of explanatory power (with two novel representation theorems). *Philosophy of Science*, **79**(3), 365–85.

Crupi, V. and Tentori, K. (2016). Confirmation theory, in A. Hajek and C. Hitchcock, eds, *Oxford Handbook of Philosophy and Probability*, Oxford: Oxford University Press, 650–65.

Crupi, V., Tentori, K., and Gonzalez, M. (2007). On bayesian measures of evidential support: Theoretical and empirical issues. *Philosophy of Science*, **74**(2), 229–52.

Crupi, V., Tentori, K., and Lombardi, L. (2009). Pseudodiagnosticity revisited. *Psychological Review*, **116**(4), 971–85.

Danks, D. (2003). Equilibria of the rescorla–wagner model. *Journal of Mathematical Psychology*, **47**(2), 109–21.

Douven, I. and Verbrugge, S. (2012). Indicatives, concessives, and evidential support. *Thinking & Reasoning*, **18**(4), 480–99.

Dulany, D. E. and Hilton, D. J. (1991). Conversational implicature, conscious representation, and the conjunction fallacy. *Social Cognition*, **9**(1), 85–110.

Earman, J. (1992). *Bayes or Bust?* Cambridge, MA: MIT Press.

Eells, E. and Fitelson, B. (2002). Symmetries and asymmetries in evidential support. *Philosophical Studies*, **107**(2), 129–42.

Fantino, E., Kulik, J., Stolarz-Fantino, S. *et al.* (1997). The conjunction fallacy: a test of averaging hypotheses. *Psychonomic Bulletin & Review*, **4**(1), 96–101.

Fiedler, K. (1988). The dependence of the conjunction fallacy on subtle linguistic factors. *Psychological Research*, **50**(2), 123–9.

Fitelson, B. (1999). The plurality of bayesian measures of confirmation and the problem of measure sensitivity. *Philosophy of Science*, **66**, S362–S378.

Frederick, D. M. and Libby, R. (1986). Expertise and auditors judgments of conjunctive events. *Journal of Accounting Research*, **24**(2), 270–90.

Fürnkranz, J., Kliegr, T., and Paulheim, H. (2019). On cognitive preferences and the plausibility of rule-based models. *Machine Learning*, 1–46.

Garb, H. N. (2006). The conjunction effect and clinical judgment. *Journal of Social and Clinical Psychology*, **25**(9), 1048–56.

Gennaioli, N. and Shleifer, A. (2009). What comes to mind. *The Quarterly Journal of Economics*, **125**(4), 1399–433.

Gigerenzer, G. (1996). On narrow norms and vague heuristics: A reply to Kahneman and Tversky. *Psychological Review*, **103**(3), 592–6.

Goldsmith, R. W. (1978). Assessing probabilities of compound events in a judicial context. *Scandinavian Journal of Psychology*, **19**, 103–10.

Good, I. J. (1960). Weight of evidence, corroboration, explanatory power, information and the utility of experiments. *Journal of the Royal Statistical Society, Series B (Methodological)*, **22**, 319–31.

Good, I. J. (1984). C197. the best explicatum for weight of evidence. *Journal of Statistical Computation and Simulation*, **19**(4), 294–299.

Grice, H. P. (1975, Dec). Logic and conversation, in P. Cole and J. Morgan, eds, *Syntax and Semantics (Volume 3)*. Cambridge, MA: Academic Press.

Heit, E. and Hahn, U. (2001). Diversity-based reasoning in children. *Cognitive Psychology*, **43**(4), 243–73.

Hertwig, R., Benz, B., and Krauss, S. (2008). The conjunction fallacy and the many meanings of and. *Cognition*, **108**(3), 740–53.

Ho, J. L. and Keller, L. R. (1994). The effect of inference order and experience-related knowledge on diagnostic conjunction probabilities. *Organizational Behavior and Human Decision Processes*, **59**(1), 51–74.

Kahneman, D. and Tversky, A. (1973). On the psychology of prediction. *Psychological Review*, **80**(4), 237–51.

Krzyzanowska, K., Collins, P. J., and Hahn, U. (2017). Between a conditional's antecedent and its consequent: Discourse coherence vs. probabilistic relevance. *Cognition*, **164**, 199–205.

Lipton, P. (2014). *Inference to the Best Explanation*, 2nd edn. London: Routledge.

Lo, Y., Sides, A., Rozelle, J. *et al.* (2002). Evidential diversity and premise probability in young childrens inductive judgment. *Cognitive Science*, **26**(2), 181–206.

Lombrozo, T. (2007). Simplicity and probability in causal explanation. *Cognitive Psychology*, **55**(3), 232–57.

Lopez, A., Gelman, S. A., Gutheil, G. *et al.* (1992). The development of category-based induction. *Child Development*, **63**(5), 1070.

Mangiarulo, M., Pighin, S., Polonio, L. *et al.* (2021). The effect of evidential impact on perceptual probabilistic judgments. *Cognitive Science*, **45**, e12919.

Mastropasqua, T., Crupi, V., and Tentori, K. (2010). Broadening the study of inductive reasoning: Confirmation judgments with uncertain evidence. *Memory & Cognition*, **38**(7), 941–50.

Miller, T. (2019). Explanation in artificial intelligence: Insights from the social sciences. *Artificial Intelligence*, **267**, 1–38.

Nadalini, A., Marelli, M., Bottini, R. *et al.* (2018). Local associations and semantic ties in overt and masked semantic priming, in E. Cabrio, A. Mazzei, and F. Tamburini, eds, *Proceedings of the Fifth Italian Conference on Computational Linguistics*, 10–12 December 2018, Torino, Italy. Torina, Italy: Accademia Press, 283–7.

Nilsson, H., Winman, A., Juslin, P. *et al.* (2009). Linda is not a bearded lady: Configural weighting and adding as the cause of extension errors. *Journal of Experimental Psychology: General*, **138**(4), 517–34.

Osherson, D. N., Smith, E. E., Wilkie, O. *et al.* (1990). Category-based induction. *Psychological Review*, **97**, 185–200.

Paperno, D., Marelli, M., Tentori, K. *et al.* (2014). Corpus-based estimates of word association predict biases in judgment of word co-occurrence likelihood. *Cognitive Psychology*, **74**, 66–83.

Pinker, S. (1997). *How the Mind Works*. New York, NY: Norton.

Rusconi, P., Marelli, M., D'Addario, M. *et al.* (2014). Evidence evaluation: Measure z corresponds to human utility judgments better than measure l and optimal-experimental-design models. *Journal of Experimental Psychology: Learning, Memory, and Cognition*, **40**(3), 703–23.

Schupbach, J. N. and Sprenger, J. (2011). The logic of explanatory power. *Philosophy of Science*, **78**(1), 105–27.

Sides, A., Osherson, D., Bonini, N. *et al.* (2002). On the reality of the conjunction fallacy. *Memory & Cognition*, **30**(2), 191–8.

Tenenbaum, J. B. and Griffiths, T. L. (2001). The rational basis of representativeness, in J. Moore and K. Stenning, eds, *Proceedings of 23rd Annual Conference of the Cognitive Science Society*, 1–4 August 2001, Edinburgh, Scotland. Red Hook, NY: Curran Associates, 1036–41.

Tentori, K., Bonini, N., and Osherson, D. (2004). The conjunction fallacy: a misunderstanding about conjunction? *Cognitive Science*, **28**(3), 467–77.

Tentori, K., Chater, N., and Crupi, V. (2016). Judging the probability of hypotheses versus the impact of evidence: which form of inductive inference is more accurate and time-consistent? *Cognitive Science*, **40**(3), 758–78.

Tentori, K. and Crupi, V. (2012*a*). How the conjunction fallacy is tied to probabilistic confirmation: Some remarks on schupbach (2009). *Synthese*, **184**(1), 3–12.

Tentori, K. and Crupi, V. (2012*b*). On the conjunction fallacy and the meaning of and, yet again: A reply to Hertwig, Benz, and Krauss (2008). *Cognition*, **122**(2), 123–34.

Tentori, K., Crupi, V., Bonini, N. *et al.* (2007). Comparison of confirmation measures. *Cognition*, **103**(1), 107–19.

Tentori, K., Crupi, V., and Russo, S. (2013). On the determinants of the conjunction fallacy: probability versus inductive confirmation. *Journal of Experimental Psychology: General*, **142**(1), 235–55.

Tversky, A. and Kahneman, D. (1974). Judgment under uncertainty: heuristics and biases. *Science*, **185**(4157), 1124–31.

Tversky, A. and Kahneman, D. (1982). Judgments of and by representativeness, in D. Kahneman, P. Slovic, and A. Tversky, eds, *Judgment under Uncertainty: Heuristics and Biases*. Cambridge: Cambridge University Press, 84–98.

Tversky, A. and Kahneman, D. (1983). Extensional versus intuitive reasoning: The conjunction fallacy in probability judgment. *Psychological Review*, **90**(4), 293–315.

Wedell, D. H. and Moro, R. (2008). Testing boundary conditions for the conjunction fallacy: Effects of response mode, conceptual focus, and problem type. *Cognition*, **107**(1), 105–36.

Wyer, R. S. (1976). An investigation of the relations among probability estimates. *Organizational Behavior and Human Performance*, **15**, 1–18.

Yates, J. F. and Carlson, B. W. (1986). Conjunction errors: Evidence for multiple judgment procedures, including 'signed summation'. *Organizational Behavior and Human Decision Processes*, **37**(2), 230–53.

Zhong, L., Lee, M. S., Huang, Y. *et al.* (2014). Diversity effect in category-based inductive reasoning of young children: Evidence from two methods. *Psychological Reports*, **114**(1), 198–215.

23

Logic-based Robotics

Claude Sammut[1], Reza Farid[2], Handy Wicaksono[3], and Timothy Wiley[4]

[1]*University of New South Wales,* [2]*WiseTech Global,* [3]*Petra Christian University, and* [4]*RMIT University, Australia*

23.1 Introduction

Robot software architectures are often characterized as hierarchical systems where the lower layers handle motor control and feature extraction from sensors, and the higher layers deal with problem solving and planning. Because the lower layers interact directly with the environment, data are usually continuous and noisy, and control decisions operate on short time scales; whereas the upper layers work on longer time scales and usually assume the world is more discrete and deterministic.

Figure 23.1 is adapted from Nilsson's (2001) triple tower architecture. Sensors, such as cameras, LIDAR, ultrasonics, microphones, and tactile sensors deliver their raw data into a working memory. The sensor data are interpreted by programs in the 'perception tower', transforming them into successively more abstract representations. For example, a raw camera image may become a collection of edge points, that then become sets of lines, then corners, then while objects. Relations between these objects may then be added to the working memory, which also represents the robot's world model. The 'action tower' contains planners and plan libraries that produce actions to be sent as motor commands to the robots actuators. Like the perception tower, the action tower is also hierarchical. At its highest level, a task planner generates high-level actions, such as 'pick up the cup', which must be translated into trajectories for the mobile robot platform and arms. This requires motion planning in a continuous domain, and monitoring to perform error correction and, if necessary, instigate replanning if the original plan fails.

Early attempts at building integrated robot systems tended to focus on the higher levels of the architecture, using logic-based representations, but were often slow and not able to handle the uncertainty inherent in the physical world (Brooks, 1986). Recent progress in robotics owes much to the development of probabilistic and behaviour-based methods that overcome some of the shortcomings those early approaches. Deep learning systems (Bengio *et al.*, 2015), to some degree, automatically create hierarchical representations, at the cost of explainability. However, high-level symbolic reasoning and

Claude Sammut, Reza Farid, Handy Wicaksono, and Timothy Wiley, *Logic-based Robotics* In: *Human Like Machine Intelligence.* Edited by: Stephen Muggleton and Nick Charter, Oxford University Press.
DOI: 10.1093/oso/9780198862536.003.0023

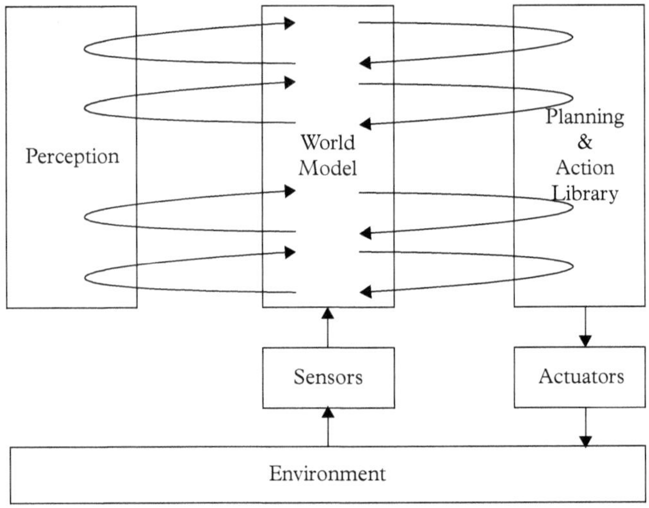

Figure 23.1 *Nilsson's triple tower as an example of a hierarchical robot software architecture.*

learning still have important roles to play. In addition to being more readable, they are capable of more powerful generalizations.

In the following chapter, we give several examples from the work of Farid (2104), Wicaksono (2020), and Wiley (2017) of how symbolic and sub-symbolic systems can be combined to take advantage of the strengths of each approach. In perception, Inductive Logic Programming (ILP) (Muggleton, 1991) can be used to learn descriptions of classes of objects and to find relations between objects. We also discuss examples of learning plans and behaviours for robots. Relational learning is used to acquire an abstract model of robot actions that is then used to constrain sub-symbolic learning for low-level control. Models can be variously expressed in the classical STRIPS representation, or as qualitative models. We give examples of each in the context of RoboCup Rescue and @Home competitions.

23.2 Relational Learning in Robot Vision

Farid (2014) describes an ILP system that learns an object classifier for an autonomous robot in an urban search and rescue operation. The primitive input to the classifier is a range image, which is transformed into a set of three-dimensional (3D) coordinates for each pixel, producing a point cloud. From that point cloud, we can extract features used in learning. Figure 23.2 shows a point cloud of a staircase. Using a plane detector, the point cloud is segmented into planes that are identified by unique colours. Planes are useful features in built environments, including indoor urban search and rescue for identifying floors, walls, staircases, ramps, and other terrain that the robot is likely to encounter. An ILP system is used to discover the properties and relationships between the planes that form an object, representing them as Prolog clauses.

Figure 23.2 *Point cloud of a staircase.*

This method is most closely related to Shanahan (2002) who used a logic program as a relational representation for 3D objects in 2D line drawings, and abduction is used in object recognition. We have extended this representation, replacing the 2D lines with 3D planes. After extracting the planes, they are labelled according to the class to which they belong. We have use two ILP systems, ALEPH (Srinivasan, 2002) and Metagol (Muggleton *et al.*, 2014) to build classifiers of objects such as staircases, ramps and other objects that occur in the rescue arena (Figure 23.3).

A plane is represented by the spherical coordinates of its normal vector (θ and ϕ) and other attributes are derived from the convex hull of the plane. These are the diameter and width of the convex hull and the ratio between these values. The plane's bounding cube is used to calculate the ratios between the three axes, two by two. The final plane feature is the axis along which the plane is most distributed.

After planes are found and their individual attributes are calculated, we then construct relations between each pair of planes. The first relation is the angle between the normal vectors of each pair of adjacent planes. The second is a directional relationship that describes how two planes are located with respect to each other in the point cloud view. For example, a plane may be located to the east of another, or one plane may cover another or be connected to another. Since planes exist in 3D space, we project the 3D view onto two 2D views and find spatial-directional relationships in each 2D view.

Figure 23.4 shows the point cloud segmentation with each region's convex hull and normal vector and the corresponding colour image. The red region, region 1, is the wall, while the yellow region, region 4, is a desktop. For this example, we define a positive example for the class 'box' that includes regions 2, 3, and 5, creating the predicate *box([pl00_2, pl00_3, pl00_5])*. That is, predicates of the form *box(+plane-set)* represent a set of planes in an image that form an object.

A set of predicates describes the individual attributes for each plane. All ratios and angles are discretized.

Figure 23.3 *Elements of a robot rescue arena.*

Figure 23.4 *Planes and plane features. The legend at the top shows the numerical label associated with each coloured plane.*

distributed_along(*plane*, *axis*) This is the long axis of the plane.

ratio_yz(*plane*, *ratio*) The aspect ratio of the planes bounding box in the yz plane.

ratio_xz(*plane*, *ratio*) The aspect ratio of the planes bounding box in the xz plane.

ratio_xy(*plane*, *ratio*) The aspect ratio of the planes bounding box in the xy plane.

normal_spherical_theta(*plane*, *angle*) Spherical coordinate of the planes normal vector.

normal_spherical_phi(*plane*, *angle*) Spherical coordinate of the planes normal vector.

Relations between pairs of planes are created for the angle between the normal vectors of two planes and the directional relationship for two adjacent planes from XY and XZ views. The relations can be one of: *north*, *south*, *east*, *west*, *connected*, *overs*.

angle(*plane1*, *plane2*, *angle*) The angle between the normal vectors of two planes.

dr_xy(*plane1*, *plane2*, *reln*) The directional relationship of two adjacent planes in the xy plane.

dr_xz(*plane1*, *plane2*, *reln*) The directional relationship of two adjacent planes in the xz plane.

The number of planes that form an object may differ. For example, staircases may differ in the number of steps:

```
staircase([pl02_06, pl02_08, pl02_10, pl02_11]).
staircase([pl02_01, pl02_03, pl02_04, pl02_05, pl02_06, pl02_08]).
staircase([pl02_04, pl02_05, pl02_06, pl02_08, pl02_10, pl02_11]).
```

An example of a learned classifier for the concept of 'staircase', is shown below. It was constructed from 237 positive examples and 656 negative examples.

```
staircase(B):
    member(C, B), member(D, B), member(E, B),
    angle(E, C, '0~15'),
    dr_xz(D, C, east), dr_xy(E, D, south).
staircase(B) :-
    member(C, B), member(D, B), member(E, B),
    angle(D, C, '0~15'), angle(E, D, '90~15'),
    angle(E, C, '90~15'), distributed_along(E, axisX).
staircase(B) :-
    member(C, B), member(D, B), member(E, B),
    angle(E, D, '0~15'), angle(E, C, '0~15').
    dr_xy(D, C, south).
```

In these rules, the set of planes that constitute an object is denoted by variable B. Thus, *member(X, B)* means X is a plane from plane set B. The '$\sim$' represents $\pm$, e.g. 0 ± 15.

Rule 1 recognizes plane set B as a *staircase* if it has two planes C and D and that D is to the east of C in the XZ-view. It also contains plane E, which is approximately parallel to plane C. Also, the spatial-directional relationship between planes E and D in the XY-view is south. This rule covers 186 positive examples (above 78.48% of all positive examples) and no negative examples.

Rule 2 recognizes B as a *staircase* if B has planes C, D parallel to each other and E is distributed mostly along X-axis and perpendicular to C and D. This rule covers 213 positive examples (above 89.87% of total positives) and no negative examples.

Rule 3 represents plane sets having at least three planes C, D, and E where E is parallel to C and D, while D is to the south of C in the XY-view. This rule covers more than 53.58% of the positive examples.

A limitation of ALEPH is that it was not able to discover a recursive relation to represent a staircase. However, Metagol (Muggleton and Lin, 2013) and MIL, are capable of predicate invention and learning recursive definitions. The application of Metagol resulted in the following description of a staircase:

```
staircase(B) :-
    p_a(B).
staircase([X, Y, Z|B]) :-
    p_a([X, Y, Z]), staircase([Z|B]).

p_a(B) :-
    member(X, B), member(Y, B), member(Z, B),
    angle(X, Y, '90~15'), angle(X, Z, '0~15').
```

Here, the invented predicate can be interpreted as 'step'. That is, a staircase is a step or a step followed by another staircase. A step consists of three planes, two of which are roughly parallel, and the third orthogonal to the parallel planes.

Here we see one of the most important reasons for using a symbolic description of a visual concept. The recursive description of staircase is sufficiently general that it can recognize quite different kinds of staircase without any further training examples, including circular stairs. This generality is difficult to achieve with any propositional or network-based learner. It is true that we have had to do a substantial amount of feature engineering, however, once the primitive features are in place, predicate invention can introduce changes of representation, analogous to those created in a layered network, with the difference that the invented predicates can be interpreted by a human being.

In Nilsson's triple tower architecture, both the perception and action towers contain a hierarchy. In the perception tower, raw input data are presented at the bottom and programs in the tower progressively interpret the data in more abstract ways to obtain a world model that can be understood and reasoned about. For example, once the step and staircase concepts have been learned, a new scene may trigger the step rule, in a forward chaining manner. That, in turn, may trigger the staircase rules, and so on. Perception need not be entirely bottom-up as the higher level concepts may also create expectations of what can be seen, and so call lower-level routines to look for expected objects. In complex scenes, this is more efficient than trying to derive all possible properties and relations in the world. The lower-levels may consist of pre-built feature detectors, such as the plane detector, or the features may be learned a deep learning technique. A promising field of study is the combination of symbolic and sub-symbolic methods, taking advantage of the strengths of both approaches (Mao *et al.*, 2019).

23.3 Learning to Act

Like the perception tower, the action tower is also layered. The top-level generally plans how to act at an abstract level, while the lowest level deals with direct motor commands. Just as the perception tower is not strictly bottom-up, the action tower is not strictly top-down. The process may begin by a robot being given a goal to achieve. A planner, in the upper, deliberative layers, may generate a sequence of actions to perform. For example, if a person asks the robot to fetch a glass of water, it must first go to the kitchen, pick up a glass, place it under the tap, turn the tap, close the tap, return to the person, hand over the cup, etc. Each of these discrete actions is handed down to a set of motion planners to navigate the robot to the correct position, to control the arm and gripper, etc. Below that, each motor command must be monitored by a feedback control system to correct errors that inevitably occur. Errors may be so bad that the robot cannot complete its task and replanning is required.

This is the 'classical' approach to robot action planning, but as Brooks (1986) pointed out, often unexpected things happen at short time scales and going through a complete planning sequence is too slow. For example, if a cat suddenly jumps out in front of the robot, by time the robot goes through its planning sequence, the cat, or the robot, may not be in very good shape. This is the reason that control decisions are made in layers. If a rapid response is required, like a reflex action, there may be a very short circuit between perception action, bypassing the upper layers of the architecture.

Long-term planning must assume that the world is reasonably predictable. In games like chess, the outcome of each move is completely predictable and the state of the world after a move can be fully observed. Therefore, it is possible to look ahead and plan moves many steps in advance, subject to the opponent's response. In other domains, such as robot soccer, the world is highly unpredictable. The robot can easily slip or fall over. A kick may not go in the intended direction or the path of the ball may curve due to irregularities in the field. Uneven lighting may confuse the vision system. When robots are close to each other, it is difficult for the vision system to distinguish them. All of these things make planning almost impossible. The only recourse is to program behaviours as situation–action action rules, decision trees, or state machines that respond to the immediate state of the world without any, or very limited, foresight. In such unpredictable domains, reinforcement learning or similar stochastic methods are most commonly used in an attempt to avoid laborious handcraft of behaviours. Unfortunately, this kind of learning usually requires many trials before an adequate set of behaviours is produced. This means that we need a high-fidelity simulator because performing many trials on a real robot is very slow and the robot will almost certainly break down before the trials are complete. Alternatively, we can try to use, or learn, knowledge about the domain to significantly reduce the number of trials needed. This is the approach we take here.

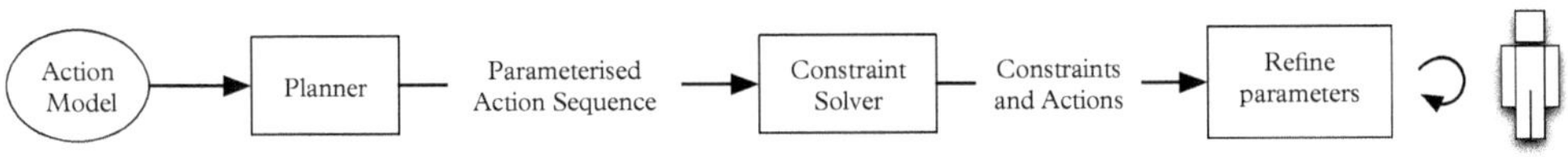

Figure 23.5 *A hierarchical architecture for action selection for a robot.*

Figure 23.5 shows the performance element of our architecture. As an example, consider a robot with bipedal locomotion. We can create a rough plan of how the robot should walk, such as: transfer weight to the left leg; lift the right leg; swing the right leg forward. transfer weight to the right leg, and so on. This high-level sequence of actions must be translated into actual motor commands. We derive a set of constraints from the pre and post-conditions in the planning steps and pass them to a constraint solver. In this example, the constraints may limit the range of angles that the leg joints. Some kind of parameter refinement through trial-and-error learning may still be required, but the constraints reduce the parameter search space to a very large degree. By this method, (Yik, 2007; Sammut and Yik, 2010), a bipedal robot was about the learn a stable walk in an average of 40 trials. Considering that the robot had 23 degrees of freedom, this is a massive reduction in effort.

In this scheme, learning may occur in two different stages. The action models used by the planner may be programmed or learned, and parameter refinement may be achieved by reinforcement learning, genetic algorithms, or hill climbing. In the following subsections we described several variants. At the planning stage, we can treat the world as being mostly discrete and deterministic and therefore use a classical planner and attempt to learn a STRIPS-like action model (Fikes and Nilsson, 1971). Alternatively, we can treat the world as continuous and either build a numerical model, or in our case, a qualitative model (Kuipers, 1986). Section 23.3.1 describes how classical action models can be learned and Section 23.3.3 describes learning qualitative models and how the can be refined by reinforcement learning.

23.3.1 Learning action models

Inductive Logic Programming can be used to learn STRIPS-like action models. We will illustrate this with a system that learns to use and create tools for a robot (Brown and Sammut, 2013; Wicaksono and Sammut, 2018). For these experiments, we use a Baxter robot, shown in Figure 23.6. To learn to use a tool, the robot observes a tool use action by another agent and proceeds to learn by imitation. That is, it tries to reproduce the other agent's action to achieve a similar goal, but since the starting conditions may be different, the robot must generalize its action model to accommodate changes. We wiil use a simple example of a the robot using a hook to pull a box from a tube. The concepts the robot must learn include the shapes and dimensions appropriate for the tool.

An extension of tool use learning is tool creation. The tools we use are simple hooks, wedges, and levers. With a 3D printer at its disposal, the robot is able to create new tools, if no suitable tool is available. For example, if a new tube is longer than previously encountered, through its generalization, the robot can instruct the 3D printer to create a new hook with a longer handle. Further generalizations may lead to the construction of

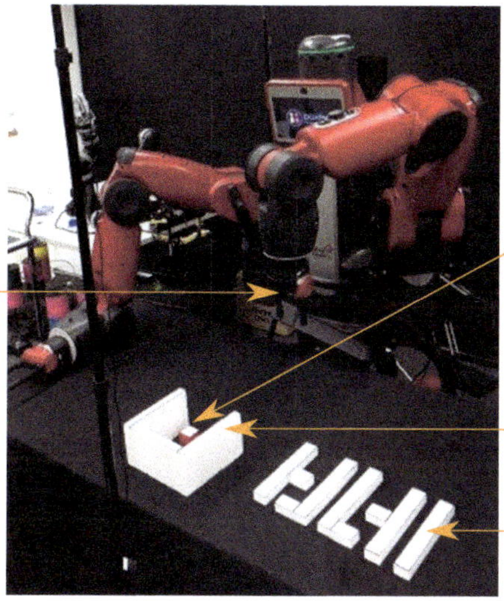

Figure 23.6 *Experimental setup with Baxter robot.*

tools with different shapes. These inventions are tested in simulation before printing the physical tool, reducing the cost of experimentation in the real world.

An action model can be learned by observing another agent performing some task and then employing a form of explanation-based learning, using ILP. An 'explanation' consists of matching an observed action to an existing action model. If an action cannot be explained, in this fashion a new action model is created. A new tool can be created by a form of generalization similar to that employed in Marvin (Sammut, 1981; Sammut and Banerji, 1986).

The state of a system is represented by a set of predicates, which may describe primitive features obtained from the vision system, including shape and pose of the objects. The state representation may also contain derived features, such as the alignment of a tool being along the same axis as the target object (e.g., if the tool is a hook used to pull an object).

As in STRIPS an action model includes the pre and post-conditions of an action. For example, a 'position_tool' action may be described as.

```
position_tool(Tool, Box)
    PRE:       in_gripper(Tool)
    EFFECTS:   tool_pose(Tool,Box)

tool_pose(Handle,Hook,Box,State) :-
    attached_end(Handle,Hook,back),
    attached_side(Handle,Hook,Side),
    in_tube_side(Box,Tube,Side,State),
    in_tube_side(Hook,Tube,Side,State),
    touching(Hook,Box,back,State).
```

This describes a hook that is positioned inside a tube and behind an object that is to be pulled out of the tube.

An action model can be learned by observing the behaviour of another agent (Brown, 2009). The process is as follows:

1. Record a demonstration by a trainer.
2. Segment the continuous time series into discrete states and extract their features and relations.
3. Match the segments with existing action models.
4. If no match is found, construct a new action model using the unmatched segments.

Trace recording

Camera images are captured and objects are recognized using methods described elsewhere (Wicaksono, 2020). Measurements consist of a stream of values of quantitative variables, such as position and orientation.

Segmentation of states

The stream of continuous variable values must be divided into qualitatively meaningful segments. For example, we may be interested in the start and stop positions of a robot, but not all the positions in between. If we regard position as a qualitative variable (Kuipers, 1986), then the qualitative value of that variable is the interval between the start and stop positions.

We define a *qualitative state*, denoted by Q_t, as the set of predicates that hold over some period. A qualitative state corresponds to a segment in a behavioural trace. Each time a state variable undergoes a qualitative change (e.g., change in motion or contact with another object), a segment boundary is created (Figure 23.7). The logical conditions that hold during a segment are conjoined to create a clause that describes the qualitative state.

Matching the segments with existing action models

To perform matching, we use a form of Explanation-Based Learning (EBL), a learning mechanism that finds a hypothesis that follows deductively from the background theory and also covers the training data (Mitchell *et al.*, 1986). In our case, an 'explanation' consists of finding an action that can transform one qualitative state into another, that is

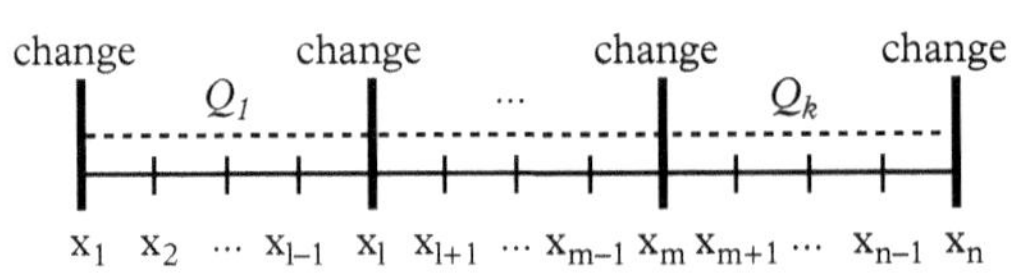

Figure 23.7 *A segment boundary is created when a qualitative variable, Q, changes, i.e., a condition that is currently true no longer holds, or a new condition becomes true. For example, when the underlying quantitative variable, x, undergoes a non-montonic change, e.g., it changes from increasing to steady or decreasing, the corresponding qualitative variable changes.*

the action explains the transition from one segment to the next. A candidate action is one whose preconditions match the initial qualitative state and whose postconditions match the successor state, that is, if the preconditions of an action A, denoted by $Pre(A)$, are theta-subsumed by Q_{t-1}, and effects the action, denoted by $Eff(A)$, are theta-subsumed by Q_t.

$$Pre(A) \preceq_\theta Q_{t-1} \wedge Eff(A) \preceq_\theta Q_i \tag{23.1}$$

Theta-subsumption is possible if and only if there exists a variable substitution θ such that $Pre(A)\theta \subseteq Q_{t-1}$.

If the system has a complete set of action models, this process should result in a sequence:

$$Q_1 \xrightarrow{A_1} Q_2 \xrightarrow{A_2} Q_3 \xrightarrow{A_3} \dots \xrightarrow{A_n} Q_{n+1} \tag{23.2}$$

Building a new action model

If the system cannot find an action that connects a pair of qualitative states, the system must build a new action model that can complete the explanation. The precondition of the new action model is constructed so that it is satisfied by the first qualitative state in the pair. The postconditions, or effects, must be satisfied by the succeeding qualitative state.

Where the system is trying to learning how to use a tool, we assume that tool actions have two parts: an action that positions the tool, denoted by A_{pos}, and another action that applies the tool, denoted by A_{app}. Each part has its own action model, where its preconditions and effects are acquired from corresponding qualitative states.

$$\dots \rightarrow Q_m \xrightarrow{A_{pos}} Q_{m+1} \xrightarrow{A_{app}} Q_{m+2} \rightarrow \dots \tag{23.3}$$

The *tool_pose* predicate, introduced earlier, appears in the effects of the positioning action and the precondition of the application action. For example, the action corresponding to the hook application is:

```
apply_tool(Tool, Box):
    PRE: tool_pose(Tool)
    EFF: not(in_tube(Object))
```

The initial definition of *tool_pose* is derived from the observation of the tool action by another agent. This will be too specific as it is derived from a single example. The next step is to make it the initial hypothesis for the action model, then generalize and test it by the robot performing experiments.

Learning by experimentation

The robot tries to find the correct *tool_pose* hypothesis by experimenting with various objects and poses, enabling the system to generalize *tool_pose* so that the action model can be applied under a broader range of conditions. An experiment consists of constructing an instance of the current hypothesis and using the resulting action model in a new task. To be able to conduct a tool-use experiment, we start by running a planner to generate a sequence of actions that includes tool use.

The actions produced by a classical planner are qualitative, for example 'pull the hook out of the tube'. However, to be executed by the robot, these must be turned into quantitative motor commands. To do this, we use a constraint solver to convert logical constraints, such as 'behind' or 'on the right' into numerical parameters that can be given to an inverse kinematic solver as a target position. Thus, 'behind' can be interpreted as a range between an object and the back of the scene. We use the CLP(R) library in SWI Prolog (Triska, 2012) to find the ranges of variable values for the tool's structure and pose. Since any value within the range is valid, we can arbitrarily chose the mid-point. Once the plan has been converted into a set motor commands, it is passed to an 'executor', which monitors the states and deciding which action should be activated at a particular time.

The system learns incrementally, as only one example can be acquired in an experiment. A positive example is saturated with respect to the background knowledge and the hypothesis is generalized by finding the least general generalization (*LGG*) of itself and the saturated positive example. This generalization is repeated whenever a positive example is acquired. We also use negative-based reduction to eliminate redundant literals (Muggleton and Feng, 1992). This tries to eliminate one literal at a time from the clause body, checking if the result covers any negative examples. This process is repeated until every literal has been tested.

Experimentation in simulation and real world

We conduct learning experiments in simulation and in the real world. A physics simulator is used to minimize learning in the real world, saving time and reducing the risk of damage to a physical robot. However, the experiments in the real world are still needed to validate the final result.

The algorithm for performing experiments in both worlds is simple. We conduct the learning experiments in a simulation where each trial is repeated five times and the result (success or failure) that occurred more often is assumed to be the label for the example. After learning has finished in the simulator, the hypothesis and action models are used in real-world experiments. If the robot can perform tool use successfully, then the previous results are assumed to be valid and learning is complete.

It is not practical to perform exhaustive testing, therefore it is possible to learn an incorrect action model. If this happens, a theory repair mechanism is needed, either specializing an over-general theory, or continuing to generalize a theory that is too specific. This can be implemented in much the same way as in Marvin (Sammut, 1981).

The system has been tested in a variety of domains using different types of tools. For example, the system has learned how to use a lever to lift an object. To do this, it must conduct experiments to determine that the lever requires a fulcrum and it must determine the correct position for the fulcrum. The experiments include the following generated configurations:

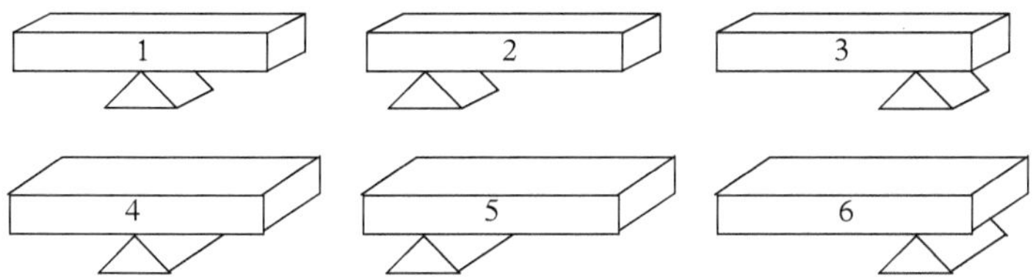

Figure 23.8 *Experiments to learn the positioning of a fulcrum.*

23.3.2 Tool creation

Tool use assumes that a suitable tool is available for a given task. What happens if no such tool exists? Fortunately, we now have 3D printers that give a robot the means to create a new tool. This is done as follows:

1. An existing tool model is generalized.
2. The new qualitative model is converted to numerical model need to send to a 3D printer.
3. The manufactured tool is used in an experiment, providing another example for learning.

Tool generalizer

To aid innovation in tool design, it is useful to have an ontology of tools. In our case, these are simple tools like hooks, levers, and wedges. The main function of this ontology is to limit the search space when a generalization is performed. The ontology can be used to suggest generalizations by climbing the hierarchy. For example, if the current model requires a hook to be on the right-hand side of the tool, a generalization may be that the hook can be on either side. In this case, the new hypothesis is tested by generating a new instance of the hypothesis that does not match previous instances seen or constructed. The generalization mechanism is based on Marvin (Sammut, 1981):

1. Find a tool that is closest to the requirement for the given task, using the scoring mechanism in section 23.3.1.
2. Generate a new instance of the hypothesis which is different from any other positive example.
3. Test if this new object belongs to the target concept by performing an experiment
   - If the new object belongs, the generalization is accepted.
   - If it does not, perform specialization by conjoining two generalizations of a trial concept, creating a new instance and carrying out another experiment until it is successful.
4. Store the learned concept.
5. Return to the first step until there are no more possible generalizations.

Generalization is done by a replacement operation, similar to absorption (Muggleton and Buntine, 1988). We use background knowledge to deduce additional relations from the example predicates. The deduced literals are added to the example predicates, while the example literals that implied the deduced literal are removed, this creating a more general concept. For example, suppose the background knowledge contains the following predicates:

```
length(short).
length(long).

attached_side(Object1,Object2,left).
attached_side(Object1,Object2,right).
```

If the description of the handle of a hook contains a literal like $length(short)$, it is replaced by $length(X)$, allowing the system to test a long handle. Similarly, if it contains $attached\_side(handle, hook, left)$, a generalization will allow the system to test attachedment of the hook on the right hand side.

As with tool use learning, testing is first done in simulation, but then examples are manufactured by 3D printing and tested with the Baxter robot (Wicaksono, 2020).

23.3.3 Learning to plan with qualitative models

In the previous section, we described a learning system for a robot in which we assumes the world is reasonably deterministic and high-level actions are discrete. In scenarios such as tool use or service robots in the home or office, these assumptions are justifiable. However, that is not the case in other domains. Consider a robot designed for urban search and rescue. These are generally tracked vehicles with reconfigurable sub-tracks or flippers. Driving such a vehicle over rough terrain requires considerable judgement because the operator must make decisions about the configuration of the flippers, as well as steering and speed. Thus, sub-tracks give the robot greater terrain traversal capabilities at the expense of greater control complexity. Since the interactions of the robot with the terrain in a disaster site are extremely difficulty to predict, we use a learning system to build a model of how control actions, including changing flipper angles, affect the robot's state. Once we have the model, the driving system can plan a sequence of actions to achieve the desired goal state.

An important requirement is that the learning system must be able to acquire the model in a small number of trials. A naïve application of reinforcement learning (Sutton and Barto, 1998), which is commonly used for such tasks, may need thousands of trials, which would be very slow and eventually break the robot. Therefore, a more economical approach is required. As in learning action models, we use a symbolic learning system to acquire a high-level model of actions and use this to constrain a reinforcement learning system.

To handle a continuous domain, we switch from a STRIPS-like representation of actions to a qualitative model, based on Kuipers (1986) QSIM. The qualitative model describes each action at an abstract level but does not specify exact numerical values for any parameters. This two-stage learning process, acquiring an approximate abstract model, followed by reinforcement learning to refine parameters, greatly reduces the overall search space, and therefore reduces the number of trials required to learn a new skill. However, it is possible that the learned qualitative model is incorrect, in the sense that it does not provide the constraints needed to learn an operational behaviour. In this case, system acquires more training data to refine the qualitative model. Thus, it is a *closed-loop* learning system that can continually improves its behaviour.

The experimental platform we use is an iRobot Negotiator, shown in Figure 23.9 climbing a step. In this case, the step is too high for the robot to climb with the flippers forward, since they are not long enough lift to robot over the step. Instead, the robot reverses up to the step and uses the flippers to raise the body, which is long enough to reach over the edge of the step. The robot's planning system should be able to reason about the geometry of the vehicle and make appropriate decisions about what sequence of actions will achieve it's goal. To do so, the planner must have a model of how actions affect the robot and its environment. The model does not need to be highly accurate for the planner to get the right sequence. An approximate qualitative model will suffice.

The qualitative model extends QSIM so that it can be used for planning. QSIM represents the dynamics of a system by a set of *qualitative differential equations* (QDE). An example of a model for this domain is shown in Figure 23.10. The graph shows the relationship of the angles of the robot's body to the floor, θ_b, and the flipper angle, relative to the body, θ_f. The relation, $M^+(\theta_f,\theta_b)$, indicates that if one of the arguments increases (θ_f), the other also increases (θ_b). The relation, $M^-(\theta_f,\theta_b)$, says that when one variable increases, the other decreases; that is, they change in opposite directions. The $const(\theta_b, 0)$ relation states that θ_b remains steady at 0.

Each segment in the graph represents an *operating region*; that is, a region of the state space in which one model holds. As the flippers rotate clockwise through 360°, the body is raised and lowered, while the flippers are angled below the body, but have no effect when they are above the body. If the flippers are rotated anticlockwise, they can raise the body. To accommodate different operating regions, we extend the QSIM notation so that each QDE has an associate *guard*, which is the condition under which that QDE holds.

(a) (b) (c)

Figure 23.9 *The Negotiator robot reversing up a high step.*

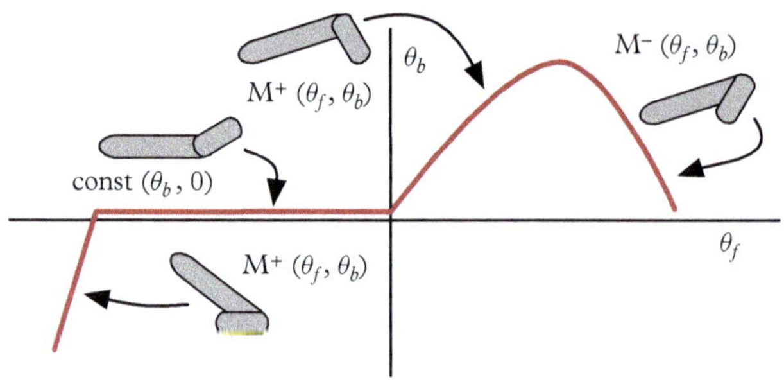

Figure 23.10 *Qualitative models for the Negotiator Robot, describing the relationship between the angle of the flippers, θ_f, and base, θ_b, through four operating regions.*

$$Guard \rightarrow Constraint \tag{23.4}$$

Since a QDE does not specify a numerical relationship between variables, we regard a QDE as a *constraint* on the possible values of the variables. For example, in a particular region, if the flipper angle decreases, the body angle increases. QSIM was originally intended to perform qualitative simulation; that is, given an initial condition, QSIM estimates how the system's state evolves over time. A state in QSIM is represented by the qualitative values of the variables. However, a qualitative variable does not take on a numerical value. Instead its value is a pair: *magnitude/direction*, where the magnitude may be a fixed landmark value or an interval, and the direction is one of increasing, decreasing or steady. For example, the robot's position may be given by $x = 0..x_{step}/inc$, which states that the x position of the robot is between its initial position and the position of the step, and its value is increasing. A set of transition rules specifies how one state evolves into the next. For example, when the robot reaches the step, the position becomes $x = x_{step}/std$. A detailed explanation is given by Wiley (2017).

Planning with qualitative models

Kuiper's QSIM action, which is needed for planning. We extend the QSIM representation by distinguishing certain variables as *control variables*, whose values can be changed by the planner. A change in a control variable corresponds to an action. For example, changing the flipper angle, θ_f, corresponds to the motor action that moves the flipper. Like classical planning, given an initial state and a goal state, the qualitative planner searches for a sequence of actions that leads to the goal state. We briefly describe the planner below but details of the implementation are given elsewhere (Wiley *et al.* 2014; Wiley *et al.* 2016).

Qualitative planning differs from classical planning and numerical simulation because the variables can specify a range of values. Therefore, as in learning action models, a qualitative state can be thought of as defining constraints on regions of valid quantitative states, that is, where the variables take on specific values. For example, a qualitative state may be $x = 0..x_{step}/inc, \theta_b = 0/std, v = 0...max/std, \theta_f = 0...90/std$, which describes the robot driving up to the step. To find a sequence of actions the qualitative planner must propagate the constraints from the initial to the final state. Therefore, planning can be seen as a constraint satisfaction problem. We take advantage of this, translating the planning problem into an Answer Set Programming (ASP) problem (Gebser *et al.*, 2013) and using the Clingo-4 solver (Gebser *et al.*, 2011) to generate the plan, described in Wiley *et al.* (2014).

The search space for the qualitative planner is considerably smaller than the search space for the corresponding continuous domain. This can be seen from the previous observation that one qualitative state covers a region of quantitative states. Thus, qualitative planning is reasonably efficient in finding a sequence of actions. However, these actions are only approximate. In the case of the robot, as it approaches an obstacle, the plan may say that the flippers should be raised, but not by how much. We will see in section 23.3.3, that 'how much' can be found by reinforcement learning, but for this to be efficient, that is require only a small number of trials, the planner must pass on its constraints to the reinforcement learning system.

In addition to the sequence of actions, the planner generates the state transitions caused by those actions, giving us a sequence very similar to that shown in Equation 23.2. For each action, we have the preceding and succeeding states. As these are qualitative states, they effectively specify the preconditions and postconditions of the action. Thus, when the reinforcement learning system searches for the angle to set for the flippers, it need only search within the constraints of the pre- and postconditions. In the following sections we explain how the qualitative model is learned and then how reinforcement learning is used to find operational parameters for the actions generated by the planner.

Learning a qualitative model

To learn a qualitative model of the robot, the system must acquire samples of the robot's interaction with its environment. In the experiments described in section 23.3.3, a human operator drives the robot, performing random actions. This could equally be done by the robot 'playing' by itself. Each time an action is performed, the before and after states are recorded so that the changes effected by the action can be determined. An example of flipper actions is shown in Figure 23.11. The figure also shows qualitative relations induced by Padé (Žabkar *et al.* 2011). This systems uses a form of regression, called *tubed regression*, to find regions of a graph where neighbouring points have the same qualitative relation. In Figure 23.11, Padé has identified regions where the body angle increases with the flipper angle, decreases, or remains steady. In this case, the binary relation between the angles has been rewritten in functional form so that the body angle is dependent on the flipper angle. Note that this plot corresponds to the graph in Figure 23.10.

The data show that there are several operating regions where these relations apply. Recall that we express the qualitative model as a set of rules, whose left hand side specifies the operating region and the right hand side give the qualitative constraints (Equation 23.1). These rules can be automatically generated from the graph by applying a symbolic rule-learning system or decision tree learner. In this case, we use Quinlan's C4.5 (Quinlan, 1993). We make a kind of closed-world assumption where the space

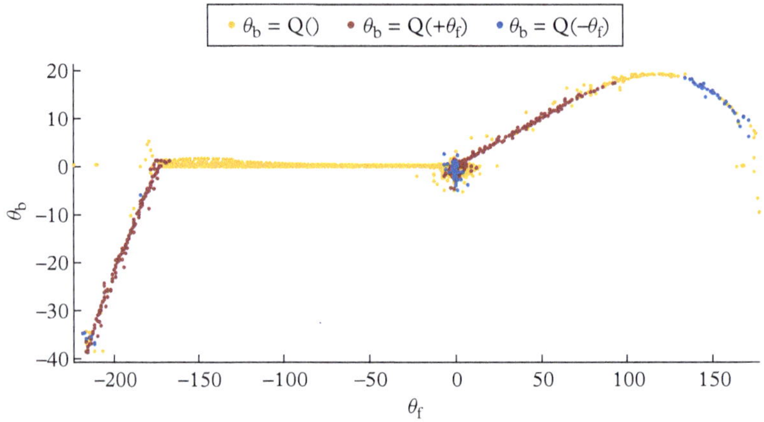

Figure 23.11 *Body angle versus flipper angle from actions.*

```
theta_b<=-5.072
|       theta_f<=-178.415: Q(+theta_f)
|       theta_f>-178.415: Q(null)
theta_b>-5.072
|       theta_b<=1.628
|       |       theta_f>17.279: Q(null)
|       |       theta_f<=17.279
|       |       |       theta_b<=-2.772: Q(+theta_f)
|       |       |       theta_b>-2.772: Q()
|       theta_b>1.628
|       |       theta_f<=133.524
|       |       |       theta_b>18.704: Q()
|       |       |       theta_b<=18.704
|       |       |       |       theta_f<=-10.273: Q(null)
|       |       |       |       theta_f>-10.273
|       |       |       |       |       theta_f<=98.500: Q(+theta_f)
|       |       |       |       |       theta_f>98.500: Q(null)
|       |       theta_f>133.524
|       |       |       theta_f<=154.558: Q(-theta_f)
|       |       |       theta_f>154.558: Q()
```

Figure 23.12 *The C4.5 decision tree.*

outside the sampled area is assumed to contain only negative examples. Figure 23.12 shows the decision tree induced from the flipper data. Since the training data are noisy, the decision tree is not as clean as a model that a human might write.

With the qualitative model represented by the decision tree, the planner can determine the qualitative state of the system. QSIM's transition rules tell the planner what possible next states are reachable depending on which action is applied in the current state. With this information, the planner can search for a sequence of actions that will achieve its goal.

Refining actions by reinforcement learning

Like learning action models in section 23.3.1, the actions generated by the planner are qualitative. For example, to climb a low step, the plan may indicate that the velocity should be forward, non-zero and the flipper angle should be between 0 and 90°. To find values of velocity and flipper angle that actually work, the robot performs trial-and-error learning. Figure 23.13 illustrates this process. For each action generated by the plan, the robot has many options for executing that action (e.g., selecting an angle between 0 and 90°). Through trial-and-error learning, it must discover the parameter settings that will result in the robot achieving its goal.

Selecting actions for the robot is a Semi-Markov Decision Process (SMDP) over Options (Sutton *et al.*, 1999). In an SMDP, time is continuous and actions have a duration, which may be of variable length. The robot tries to select a set of options, that is, numerical control value settings, that will allow the robot to succeed in its task. The unique part of this process is how the SMDP is constructed from a plan. Since a qualitative variable specifies an interval, numerical values are sampled, constrained by the interval ranges. For each action, the qualitative states satisfying the precondition and postcondition are required. Options are formed from every combination of sampled numerical states that within the bounds of the qualitative states. QSIM constrains a variable's rate-of-change, as well as its magnitude. If the qualitative state satisfying the postcondition has an increasing rate-of-change, its quantitative value must increase during the option, likewise for decreasing or steady rates of change. Only options that satisfy the rates-of-change constraints are added to the SMDP.

Once the system has found the available options, it can perform its trial and error learning to turn the planning actions into operational motor commands. However, the quality of the plan depends on the quality of the model that was induced from the original training data, which was randomly sampled in the beginning. If the sampling does not yield sufficient training data in the regions of the state space relevant to the task, then

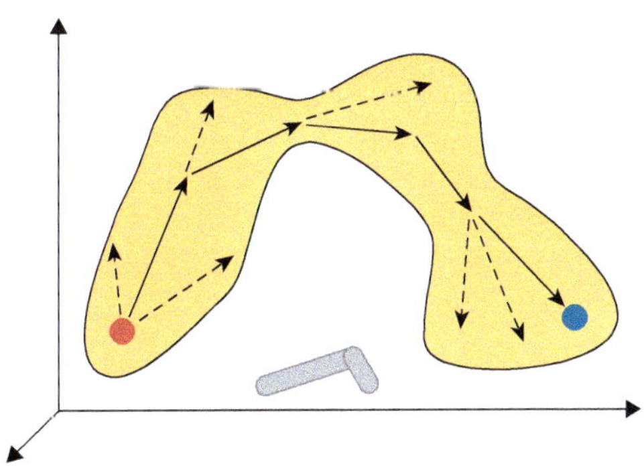

Figure 23.13 *The planner gives an approximation of actions, (yellow region) that must be refined by trial and error learning into precise motor actions (arrows).*

the model may be poor, resulting in inefficient plans. This problem can be reduced by closed-loop learning.

Closed-loop learning and experiments

The learning process begins with the collection of training examples for learning the qualitative model. In these experiments, the data were generated by a human operator instructing the robot to perform mostly random actions. Experiments were performed for several tasks using the Negotiator robot (Wiley *et al.*, 2016; Wiley, 2017). These included driving over different height steps and climbing a staircase. Of these, climbing a high step, which requires the reversing manoeuvre, was the most difficult.

In closed-loop learning, the system can repeat the entire learning process to improve its performance. Over 3 repetitions, the average number of trials needed to learn to climb the step was 104 and the average amount of time needed to complete the task is 32.7 seconds. The actions and their effects are recorded so that they can be added to the training examples for the model learner. After the data from the first pass of learning to climb a high step are added, over 5 repetitions, the average number of trials needed to learn to climb the step was reduced to 31 and the average amount of time needed to complete the task is 25.1 seconds, indicating that a better plan was produced. The systems performs better because the training data for closed-loop learning contains more training examples in areas that the human operator failed to properly sample.

23.4 Conclusion

The work described here demonstrates the usefulness of combining logical representations, reasoning and learning with numerical methods. While the latter have proved essential in enabling a robot to perform in the real world, symbolic methods are better suited to generalizing over large classes of objects and behaviours. They provide more powerful planning and problem-solving mechanisms, and they are more easily explained than sub-symbolic methods.

Acknowledgements

We have benefited greatly from collaborations with our colleagues: Ivan Bratko, Keith Clark, Dianhuan Lin, Stephen Muggleton, Max Ostrowski, Torsten Schaub, and Jure Žabkar. This work was supported by the Australian Research Council.

References

Apt, K. R., and Wallace, M. (2006). *Constraint Logic Programming using ECLiPSe*. Cambridge, UK: Cambridge University Press.

Bengio, Y., LeCun, Y., and Hinton, G. (2015). Deep learning. *Nature*, **521**(7553), 436–44.

Brooks, R. A. (1986). A robust layered control system for a mobile robot. *IEEE Journal of Robotics and Automation*, **RA-2**(1), 14–23.

Brown, S. (2009). *A Relational Approach to Tool-use Learning in Robots*. Ph.D. thesis, School of Computer Science and Engineering, The University of New South Wales.

Brown, S. and Sammut, C. (2013). A Relational Approach to Tool-use Learning in Robots,, in F. Riguzzi and F. Zelezny), eds, *Inductive Logic Programming*. Berlin: Springer, 1–15.

Dechter, R. (1986). Learning while searching in constraint-satisfaction problems, in *Proceedings of the Fifth National Conference on Artificial Intelligence*, Philadelphia, PA. New York, NY: AAAI Press, 178–83.

Farid, R. (2104). *Generic 3D Object Recognition Using Multi-view Range Data*. Ph.D. thesis, School of Computer Science and Engineering, The University of New South Wales.

Farid, R. and Sammut, C. (2014, Jan). Plane-based object categorisation using relational learning. *Machine Learning*, **94**(1), 3–23.

Fikes, R. and Nilsson, N. (1971). Strips: a new approach to the application of theorem proving to problem solving. *Artificial Intelligence*, **2**, 189–208.

Gebser, M., Kaminski, R., Kaufmann, B. *et al.* (2013). *Answer Set Solving in Practice*. Synthesis Lectures of Artificial Intelligence and Machine Learning. San Rafael, CA: Morgan & Claypool Publishers.

Gebser, M., Kaufmann, B., Kaminski, R. *et al.* (2011). Potassco: The Potsdam Answer Set Solving Collection. *AI Communications*, **24**(2), 107–24.

Kuipers, Benjamin (1986). Qualitative simulation. *Artificial Intelligence*, **29**, 289–338.

Mao, J., Gana, C., Kohli, P. *et al.* (2019). The neuro-symbolic concept learner: Interpreting scenes, words, and sentences from natural supervision, in *Proceedings of the 7th International Conference on Learning Representations*, New Orleans. OpenReview.net, University of Massachusetts Amherst.

Mitchell, T. M. (1977). Version spaces: A candidate elimination approach to rule learning, in *Proceedings of the 5th International Joint Conference on Artificial intelligence*, Vol. 1, Massachusetts Institute of Technology, MA. San Francisco, CA: Morgan Kaufmann, 305–10.

Mitchell, T. M., Keller, R. M., and Kedar-Cabelli, S. T. (1986). Explanation-based generalisation: a unifying view. *Machine Learning*, **1**, 47–80.

Muggleton, S. (1991). Inductive logic programming. *New Generation Computing*, **8**, 295–318.

Muggleton, S. and Buntine, W. (1988). Machine invention of first-order predicates by inverting resolution, in *Proceedings of the Fifth International Conference on Machine Learning*, University Michigan, Ann Arbor. San Francisco, CA: Morgan Kaufmann, 339–52.

Muggleton, S. and Feng, C. (1992). Efficient induction in logic programs, in S. Muggleton, ed., *Inductive Logic Programming*. London: Academic Press, Harcourt Brace Jovanovich, 281–98.

Muggleton, S. H. and Lin, D. (2013). Meta-interpretive learning of higher-order dyadic datalog: Predicate invention revisited. In F. Rossi, ed, *Proceedings of the 23rd International Joint Conference Artificial Intelligence (IJCAI 2013)*, pp. 1551–1557, Beijing, 3–9 August 2013. AAAI Press: Menlo Park, Caliifornia.

Muggleton, S. H., Lin, D., Pahlavi, N. *et al.* (2014). Meta-interpretive learning: application to grammatical inference. *Machine Learning*, **94**(1), 25–49.

Nilsson, N. J. (2001). Teleo-reactive programs and the triple-tower architecture. *Electronic Transactions on Artificial Intelligence*, 5(B), 99–110.

Plotkin, G. D. (1971). A further note on inductive generalization, in B. Meltzer and D. Michie, eds, *Machine Intelligence 6*. Edinburgh UK: Edinburgh University Press, 101–24.

Plotkin, G. D. (1970). A note on inductive generalization, in B. Meltzer and D. Michie, eds, *Machine Intelligence 5*. Edinburgh: Edinburgh University Press, 153–63.

Quinlan, J. R. (1993). *C4.5: Programs for Machine Learning*. San Manteo, CA: Morgan Kaufmann.

Sammut, C. and Yik, T. F. (2010). Multistrategy learning for robot behaviours, in J. Koronacki, Z. W. Ras, S. T. Wierzchon, and J. Kacprzyk, eds, *Advances in Machine Learning I*, Vol. 262, Studies in Computational Intelligence. Berlin: Springer, 457–76.

Sammut, C. A. (1981). *Learning Concepts by Performing Experiments*. Ph.D. thesis, Department of Computer Science, University of New South Wales.

Sammut, C. A. and Banerji, R. B. (1986). Learning concepts by asking questions, in R. S. Michalski, J. Carbonell, and T. Mitchell, eds, *Machine Learning: An Artificial Intelligence Approach, Vol 2*. Los Altos, CA: Morgan Kaufmann, 167–92.

Shanahan, M. (2002). A logical account of perception incorporating feedback and expectation, in D. Fensel, F. Giunchiglia, D. McGuinness, and M.-A. Williams, eds, *Proceedings of 8th International Conference on Principles of Knowledge Representation and Reasoning*, Toulouse, France. San Francisco, CA: Morgan Kaufmann, 3–13.

Srinivasan, A. (2002). The ALEPH Manual (Version 4 and above). Technical report, University of Oxford.

Sutton, R. S. and Barto, A. G. (1998). *Reinforcement Learning: An Introduction* Cambridge, MA: MIT Press.

Sutton, R. S., Precup, D., and Singh, S. P. (1999, December). Between MDPs and semi-MDPs: a framework for temporal abstraction in reinforcement learning. *Artificial Intelligence*, **112**(1–2), 181–211.

Triska, M. (2012). The finite domain constraint solver of SWI-Prolog, in G. Goos, J. Harmanis, and J. van Leeuwen, eds, *Proceedings of the 11th International Symposium on Functional and Logic Programming*, Kobe, Japan. Berlin, Heidelberg: Springer, 307–16.

Wicaksono, H. (2020). *A Relational Approach to Tool Creation by a Robot*. Ph.D thesis, School of Computer Science and Engineering, The University of New South Wales.

Wicaksono, H. and Sammut, C. (2018). Tool use learning for a real robot. *International Journal of Electrical and Computer Engineering (IJECE)*, **8**(2), 1230–7.

Wiley, T. (2017). *A Planning and Learning Hierarchy for the Online Acquisition of Robot Behaviours*. Ph.D. thesis, School of Computer Science and Engineering, University of New South Wales.

Wiley, T., Sammut, C., and Bratko, I. (2014, August). Qualitative simulation with answer set programming, T. Schaub, G. Friedrich, and B. O'Sullivan, eds, *Proceedings of the Twenty-First European Conference on Artificial Intelligence*, Prague, Czech Republic. Amsterdam, Netherlands: IOS Press, 915–20.

Wiley, T., Sammut, C., Hengst, B. *et al.* (2016). A planning and learning hierarchy using qualitative reasoning for the on-line acquisition of robotic behaviors. *Advances in Cognitive Systems*, **4**, 93–112.

Yik, T. F. (2007). *Locomotion of Bipedal Humanoid Robots: Planning and Learning to Walk*. Ph.D. thesis, School of Computer Science and Engineering, The University of New South Wales.

Žabkar, J., Mozina, M., Bratko, I. *et al.* (2011). Learning qualitative models from numerical data. *Artificial Intelligence*, **175**(9-10), 1604–19.

24

Predicting Problem Difficulty in Chess

Ivan Bratko[1], Dayana Hristova[2], and Matej Guid[1]

[1] *University of Ljubljana,* [2] *University of Vienna*

24.1 Introduction

A question relevant to explainable AI and human-like computing is: How can we automatically predict the difficulty of a given problem for humans? The practical motivation for predicting task difficulty arises for example in intelligent tutoring systems and computer games. In both cases, the difficulty of problems has to be adjusted to the user. In general, understanding the difficulty for humans of problems that AI tries to solve is a relevant question for human-like computing. If AI systems find problems easy while humans find them hard, and vice versa, then this is evidence that the AI systems are solving the problems in a different way from humans. Also, for computation to be "human-like", it should be easy to understand by humans. Ideally, the system should be able to recognise when the problem or computation gets difficult for humans.

The difficulty of a problem for a human depends on the human's expertise in the domain of the problem, and consequently on how the human would go about solving the problem. The automatic prediction of difficulty could therefore involve a kind of simulation of human problem-solving, which would make prediction of difficulty particularly hard.

In this chapter we discuss an approach to the automatic prediction of difficulty for humans, of problems that are typically solved through informed search. Our experimental domain is the game of chess. Chess has often proved to be an excellent environment for research in human problem-solving. One reason for this, which is important for the present study, is the existence of the FIDE chess federation's rating system for registered players worldwide, and the Chess Tempo website with a large number of chess problems with measured difficulty ratings.

In this chapter we analyse experimental data of human chess players who attempted to solve tactical chess problems and also assess the difficulty of these problems. We carry out an experiment with an approach to predicting the difficulty of problems for humans automatically in this domain.

Ivan Bratko, Dayana Hristova, and Matej Guid, *Predicting Problem Difficulty in Chess* In: *Human Like Machine Intelligence.*
Edited by: Stephen Muggleton and Nick Charter, Oxford University Press.
DOI: 10.1093/oso/9780198862536.003.0024

Solving tactical problems in chess requires search among available alternative moves. The size of the search space is typically much too large for humans to search exhaustively. Good chess players therefore use pattern-based knowledge to guide their search extremely effectively. Problem-solving thus consists of detecting chess patterns—motifs, and the calculation of concrete chess variations trying to exploit these motifs to the player's advantage. What could be such motifs and how motifs are used in chess problem solving is explained in Section 3, where concrete examples of motifs and corresponding problem-solving are given. In our analysis we take into account players' comments on how they tackled individual problems.

Automated estimation of difficulty for humans in chess is hard because it requires the understanding of how humans solve chess problems. Strong chess players use large amount of pattern-based knowledge acquired through experience. To duplicate this vast amount of largely tacit knowledge in the computer is a formidable task that has never been accomplished. Therefore we are interested in alternative ways: estimating difficulty for humans without the use of chess-specific knowledge. In an experiment with such an approach, described in the second part of this chapter, we reduce this need for human players' pattern knowledge to a speculated equivalent: properties of game-tree search deemed to be carried out by strong players. We believe that this approach is applicable to estimating the problem difficulty in other domains where problems are solved through expert knowledge and search.

Related research into the issue of estimating problem difficulty of specific types of puzzles includes the following: Tower of Hanoi (Kotovsky *et al.*, 1985), Chinese rings (Kotovsky and Simon, 1990), 15-puzzle (Pizlo and Li, 2005), Traveling Salesperson Problem (Dry *et al.*, 2006), Sokoban puzzle (Jarušek and Pelánek, 2010), Sudoku (Pelánek, 2011), puzzle games played on grids (Van Kreveld *et al.*, 2015), mathematical puzzles (Sekiya *et al.*, 2019). Kegel and Haarh (2019) review techniques for procedural contents generation for games, paying attention to the difficulty of generated problems.

An early attempt at automated estimation of the difficulty of chess problems was made in Guid and Bratko, 2006. In that paper the authors analysed the quality of chess games played at world championship level. The positions in the analysed games were submitted to a strong chess-playing program, and the best moves (according to the program) were computed. For each player, the average difference per move between the value of the move suggested by the chess program and the value of the move actually played by the player (average loss per move) was computed. It would now be inappropriate simply to rank the players according to their average loss per move because the players' playing styles were different. Some players naturally tended towards quiet, simple positions, and others towards complex positions. In simple positions it is much easier to achieve a small loss than in complex positions. In order to allow a fair comparison, the difficulty of the positions had to be taken into account. An approach to automatic difficulty estimation of a position was therefore designed, essentially based on the amount of search required by the chess program to find the best move in the position. This made it possible to compute average loss per move for each player if all the players were faced with positions of equal difficulty. This approach to difficulty estimation was analysed in detail in Guid and Bratko, 2013. However, it was found that this approach does not produce realistic estimates of the difficulty for humans in tactical chess problems. Therefore, in Stoiljkovikj *et al.* (2015) a more suitable approach for estimating the difficulty of tactical problems was developed, which will also be used in the experiment in the present chapter.

24.2 Experimental Data

In this study we used the data obtained in an experiment in which 12 chess players of various chess strengths were asked to solve 12 tactical chess problems (Hristova *et al.*, 2014*a*). A chess position is said to be *tactical* if finding the best move in the position requires the calculation of variations, and the solution typically leads to an obvious win after a relatively short sequence of moves.

The chess strength of our players, measured by the FIDE chess ratings, was in the range between 1845 and 2279 rating points. The strength of the registered chess players is officially computed by (World Chess Federation) using the Elo rating system. This rating system was designed by Arpad Elo (1978). This rating is calculated for each player and updated regularly according to the tournament results of the players. The rating range of our players, between 1845 and 2279, means that there were big differences in chess strength between the players. The lowest end of this range corresponds to club players, and the highest end to chess masters (to obtain the FIDE master title, the player must reach at least 2300 points at some point in his career). Among our participants there were actually two chess masters, one of whom also had the title of a female grandmaster. The expected result in a match between the top ranked player in our experiment and our lowest ranked player would be about 92% against 8% (the stronger player winning 92% of all possible points). According to the definition of the Elo rating system, the expected outcome between two players is determined only by the *difference* between their ratings, and not by the ratings themselves. For example, consider two players with ratings 2200 and 2000. The difference is 200 rating points, which determines that the expected success rate of the higher rated player playing against the lower rated player is 76% and the expected success rate of the lower rated player is 24%. The same success rates could be expected if the players' ratings were, say, 2350 and 2150.

In addition to the differences in chess strength expressed by chess ratings, other differences between players could also be taken into account. One such factor might be the chess school where a player was taught, or the particular instructor who trained the player. However, in this chapter we did not explore the effects of such additional factors.

The 12 chess problems were selected from the Chess Tempo website,[1] which is intended for tactical chess training. At Chess Tempo, the problems are rated according to their difficulty. Chess problems are rated in a similar way as the players, except that the evidence does not come from chess games played, but from attempts by chess

[1] The website Chess Tempo is at www.chesstempo.com.

players to solve problems at Chess Tempo. Thus at Chess Tempo a problem's rating is determined by the success of the players in solving the problem. The principle is as follows: if a weak player has solved a problem, this is considered a strong indication that the problem is easy. So the problem's rating goes down. If a stronger player has solved the problem, the rating of the problem still decreases, but not as much as with a weak player. On the contrary, if a strong player failed to solve the problem, this is considered a strong evidence that the problem is hard, and the rating of a problem increases. More specifically, a problem's rating in Chess Tempo is determined by the Glicko rating system (Glickman, 1999), which is similar to the Elo system. Unlike Elo, the Glicko system takes into account the time a player has been inactive. In cases of prolonged inactivity, the player's rating becomes uncertain. It should be noted that the ratings of players—Chess Tempo users—are determined by the evidence of their success in solving problems, and not by their chess-playing results. Otherwise, the meaning of ratings in Chess Tempo is similar to the FIDE ratings of players. So a player with rating 2000 has a 50% chance of correctly solving a problem with rating 2000. The same player has a 76% chance to solve a problem rated 1800, and a 24% chance to solve a problem rated 2200.

In our selection of 12 chess problems we ensured a mixture of problems that largely differ in their difficulty. The problems were randomly selected from Chess Tempo according to their difficulty ratings. Based on their Chess Tempo ratings, our problems can be divided into three classes of difficulty: 'easy' (2 problems; their average Chess Tempo rating was 1493.9), 'medium' (4 problems; average rating 1878.8), and 'hard' (6 problems; average rating 2243.5). While the problems within the same difficulty class have very similar difficulty rating, each of the three classes is separated from their adjacent classes by at least 350 Chess Tempo rating points. Some problems have more than one correct solution. To ensure correctness, all the solutions were verified by a chess-playing program.

The experimental set-up was as follows. Chess problems, that is chess positions, were displayed to a participating player one after the other as chess diagrams on a monitor. For each problem, the player was asked to find a winning move, and the player's solution moves were recorded. The problem-solving time per position was limited to three minutes.

While the player was solving the problem, the player's eye movements were tracked with an eye-tracking device, EyeLink 1000, and recorded in a database. The processing of recorded eye movements roughly reveals on which squares of the chessboard the participant was focussing at any time during the problem-solving process. Observing eye movements has often been used in chess decision-making (Sheridan and Reingold, 2017).

After the player had finished with the 12 problems, a retrospection interview was conducted in which the player described how he or she approached the problem. From these retrospections, one could see which motifs were considered by the player, and roughly how the calculation of variations driven by the motifs was carried out. Finally, the players were asked to sort the 12 problems according to the difficulty of the problems perceived by the players. Further details of the experiment are described in (Hristova *et al.*, 2014*a*, *b*).

The relevant experimental data include the following. For every player and every position we have: (1) the correctness of the solution proposed by the player, (2) the motifs considered by the player compared to the motifs required to solve the problem, and (3) the correctness of the calculation of variations. The motifs considered were found through the players' retrospections, and to some extent verified by the eye movement data, although this verification cannot be done completely reliably.

The data concerning the correctness of the recognition of motifs and the calculation of variations were mostly constructed manually from the submitted solutions of the players and from their retrospections. To decide whether the player detected a complete set of motifs needed to carry out correct calculation, we defined for each position and each possible solution of the position, the 'standard' set of motifs necessary and sufficient to find the solution. In defining the standard sets of relevant motifs, we took into account all the motifs mentioned by all the players. In very rare cases when needed, we had to add motifs that fully enabled correct calculation for each possible solution. In doing so, we used our own chess expertise (two of us have chess ratings over 2300 and 2100 respectively). We verified all the solutions and corresponding chess variations by a chess program, and we believe that it would be hard to come up with reasonable alternative standard sets of motifs.

24.3 Analysis

24.3.1 Relations between player rating, problem rating, and success

We first consider some correlations between success in solving a problem, a player's chess rating, and problem's Chess Tempo rating. We represent the success in solving a problem by 1, and the failure to solve by 0. The total number of data points of the form (Rating, Success) in our experimental data was 142. For 12 problems and 12 players, there are altogether 12*12 = 144 such pairs, however due to misunderstandings during the experiment in two cases invalid results were obtained, so that they were excluded, which finally results in 142 data points.

Sample correlation coefficient between problem's Chess Tempo rating and success was:

```
r(ProblemRating, Success) = -0.345 (P = 0.000027)
```

This is basically as expected: higher problem rating means lower chances of success. Sample correlation between player's chess rating and success was:

```
r(PlayerRating, Success) = 0.077 (P = 0.36)
```

This result is not statistically significant. According to this, there is almost no correlation between player's Elo rating and success in solving a problem, which appears rather surprising. Success depends much more on the Chess Tempo difficulty of the problem than on the players' rating. One attempt at explaining this difference can be that the differences in Chess Tempo ratings were higher than differences between players' ratings. The ranges were:

```
players' ratings: 2279 - 1845 = 434;
problems' ratings: 2243 - 1492 = 751.
```

Another, more plausible explanation is based on the observation of an important difference between the players' FIDE rating and Chess Tempo problem ratings. Chess Tempo problem ratings and players' FIDE ratings measure different things. The first measures success of individual moves, and the second measures success over long sequences of moves. The players' FIDE ratings are based on the results of complete games (won, or drawn, or lost). Winning a game is the outcome of a sequence of moves (usually about 40 moves). The success in winning a game depends on the sum of the correctness of all moves in the game, and not on the correctness of a single move in the game. A 40 moves game is typically decided by one or two moves where decisive mistakes are made, while the rest of the moves by the two players are of very similar quality. On the other hand, to solve a (single) problem successfully in our experiment, just a correct single first move of a tactical combination was required. This is similar to scoring a correct solution in Chess Tempo, although, to be precise, not exactly the same. The Chess Tempo system, to accept an answer as correct, requires from a player a correct first move, possibly followed by one or more moves in the main variation of the combination. The point of requiring additional moves is to verify that the player actually saw the whole variation and indeed played the first move for the right reasons (and was not just lucky). So, what counted as success in Chess Tempo was not exactly the same as what counted as success in our experiment. Nevertheless, both notions of success refer to solving a single position, which is considerably different from success in winning a game.

24.3.2 Relations between player's rating and estimation of difficulty

The next question of interest is how good chess players are at estimating the difficulty of problems. We take the Chess Tempo (CT for short) difficulty ratings as the gold standard, because they are based on observing large numbers of players' attempts at solving these problems, and the ratings are computed from these observations using an accepted method. So we compare the players' rankings of problems with the CT rankings.

In the experiment, the players did not directly estimate the difficulty ratings of the problems, but each player was asked to rank the 12 problems according to his or her perceived difficulty of the problems. We used Kendall's Tau rank correlation coefficient as a statistical measure of agreement between rankings. Given two rankings, Kendall's Tau is defined as:

$$\tau = (n_c - n_d)/(n_c + n_d) \tag{24.1}$$

Here n_c and n_d are the number of concordant pairs and discordant pairs, respectively. A pair of chess positions is concordant if their relative rankings are the same in both rankings; that is, given two problems, the same problem precedes the other one in both rankings. Otherwise the pair is discordant. In our data, subsets of the positions were, according to Chess Tempo, of very similar difficulty. Such positions belong to the same difficulty class, either 'easy' (CT-rating between 1492 and 1495) or 'medium' (between 1875 and 1883) or 'hard' (between 2231 and 2275). Within the three difficulty classes, we consider any ordering by the players to be acceptable. To account for this, we used a variation of the Tau formula above. When determining n_c and n_d, we only counted the pairs of positions that belong to different Chess Tempo classes. In view of the distribution of problems over the three classes (easy: 2, medium: 4, hard: 6), we have therefore only considered 2*4 + 2*6 + 4*6 = 44 problem pairs.

In (Hristova *et al.*, 2014*a*), we computed Kandall's Tau in this way for each of the 12 players. Then we computed sample correlation between the players' Tau and the players' FIDE ratings. There was a moderate positive relationship (not statistically significant) between Kendall's Tau and the FIDE ratings. We can strengthen this result by considering separately all pairs of positions of different difficulty class, and correctness of the relative rankings of these pairs in the 12 players' difficulty rankings. We represented correctly ordered pairs by 1, and incorrectly by 0. This way we have 44 pairs of problems and 12 players, which gives 12 * 44 = 528 data points. Each data point is of the form (PlayerRating, OrderCorrect), where OrderCorrect is 1 or 0 as stated above. Sample correlation coefficient for this data set is r = −0.196, which is significant (P = 0.000095). This result is as one would expect. It indicates that stronger players are indeed better able to assess the difficulty of problems than weaker players, although the correlation is quite weak, indicating that this relation is rather noisy. Overall, in all the difficulty rankings by all the players, 72.5% of relevant pairs are concordant (a 'relevant pair' is a pair of problems of different Chess Tempo difficulty class).

Now let us consider values of Tau for individual players. Tau is between −1 and 1. Tau = 1 indicates a perfect ranking, and Tau = −1 indicates a ranking which is "as wrong as possible'. Kendall's Tau coefficients of the players were in the large interval between −0.18 and 0.95 (two payers actually ordered more of the relevant pairs of positions incorrectly than correctly). It is interesting to consider the 'average ranking' by all 12 players. This can be obtained by a kind of players' voting, considering the average rank of each problem over all the players' rankings. The obtained ranking order of positions was: 2, 3, 1, 6, 10, 7, 4, 5, 9, 8, 12, 11; that is, overall, position 2 was perceived as the easiest, followed by position 3, etc., with position 11 perceived as the hardest. According to Chess Tempo ratings, the sets of positions belonging to the three difficulty classes are as follows:

```
easy: {1, 2}
medium: {3, 4, 5, 6}
hard: {7, 8, 9, 10, 11, 12}
```

Kendall's Tau for the joint ranking by the players is 0.77. This can be compared with the individual players' Tau coefficients. The highest player's Tau was 0.95, and the second highest 0.68. Average Tau over the 12 players was 0.45. This is also in agreement with the result that overall, the players correctly ordered 72.5% of pairs of problems that were taken into account.

The results regarding the players' difficulty estimation require careful interpretation. The task of the players in the experiment was stated simply as follows: rank the 12 given problems according to their difficulty, from the easiest to the most difficult. The players were not told that there were essentially three difficulty classes, and that there were many pairs of positions of practically equal difficulty. Given this circumstance, the following cases were possible regarding players' rankings. Consider a pair of positions A and B. If position A was easier than B according to Chess Tempo then: if the player ranked A before B then this counted as a concordant pair, otherwise this counted as a discordant pair (incorrect order). If A and B belonged to the same Chess Tempo difficulty class then this pair was not included in the calculation of Tau, so it did not matter whether the player ordered the problems A before B or B before A. Both cases were treated as acceptable, and did not affect the player's evaluated ranking performance. This is reasonable because in this case there is no evidence of ranking error. However, in such cases we do not actually know. Suppose that the player ranked A before B. In this case, there are two possibilities: (1) the player actually considered both problems to be equally difficult, and arbitrarily ordered A before B (just because a total ordering was required, and he had to order them one way or the other); or (2) the player actually believed that A was easier than B; in this case he was wrong, but there is no way to detect this from experimental data. Our modified Tau measure can therefore be interpreted as a potentially optimistic assessment of the player's ranking accuracy.

24.3.3 Experiment in automated prediction of difficulty

In this section we carry out an experiment, using our 12 experimental positions, with a program for automatically estimating the difficulty of tactical chess positions. We used the approach to estimating difficulty proposed in (Stoiljkovikj *et al.*, 2015), which will be referred to as the SBG method. This method is based on machine learning about difficulty for humans, using features of search trees that are searched by good human players when solving a tactical chess problem.

The size of the combinatorial search space involved in solving the problem is the most obvious source of difficulty. For an uninformed problem-solver without problem-specific knowledge, the size of the search space would indeed be a useful indicator of difficulty. For experienced chess players the situation is quite different. Such players employ their knowledge to search this space very selectively so that only a small fraction of the entire space is actually searched.

When solving tactical chess problems, human players use a repertoire of common motifs that allow such a highly selective and effective search. Figure 10.1 illustrates this. In this example, the solution consists in spotting the well-known motif of a pinned piece. Brute force search, realized as, say, iterative deepening to depth 5 (which would be an adequate depth in this example) would in this implementation require searching millions of positions. Using the chess-specific motif of a pin, this is reduced to the order of

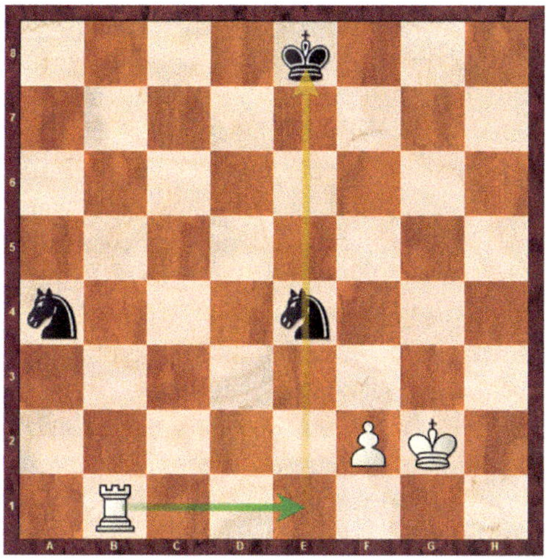

Figure 24.1 *White to move and win. An experienced player will immediately notice the motif that Black king and Black knight on e4 are on the same file. This gives rise to the motif of pinning Black knight with the move 1.Re1 (green arrow). Knight on e4 is now attacked and cannot escape due to the pin. Black may try to defend knight e4 with moving the other knight: 1...Na4-c5. Now a common mechanism of exploiting a pin is used by White: attack the pinned piece with yet another piece. In this case this can be accomplished by White pawn move to f3. On the next move, Black knight on e4 will be captured, giving White a decisive advantage.*

10 or 20 positions. It is this latter number that is indeed relevant for the difficulty of the position for good players.

We will be referring to such a reduced search space as 'meaningful search tree'. It should be noted that all the players in our experiment easily had enough knowledge to solve the position of Figure 10.1 quickly, exploring a small meaningful tree, as explained in the caption of Figure 10.1. In this position, there is another common motif for White: double attack with White rook move to b4, simultaneously attacking both Black knights. However, a trivial search shows that in this position Black knights can defend each other with the move Ne4 c5, so double attack motif does not work in this case.

To estimate the difficulty for an expert human player, the estimation program would ideally simulate the search actually performed by the player and predict the difficulty based on this simulated search. To simulate such a search, the program would have to possess similar chess knowledge as the player. However, this knowledge consists of a very large library of chess motifs, or patterns, of the kind illustrated in Figure 10.1. Some of this knowledge is acquired by players through explicit instruction, and that part can be found in chess books. The larger part of that pattern-based knowledge is however tacit knowledge that a player has acquired through experience, but does not exist in formalized and documented form. The difficulty in predicting the difficulty for experts lies in the question: how to take into account such tacit knowledge?

The main idea of the SBG method is the concept of a 'meaningful search tree' (defined later). This is based on the assumption that the search of the human chess expert can be simulated by a standard chess-playing program without knowing the extensive pattern knowledge. Hopefully, for a given chess position, the meaningful tree approximates the tree actually searched by a chess expert when searching for the best move in a position. Accordingly, a meaningful tree is formally defined with this aim in mind. For a given position P, the meaningful tree is a subtree of the game tree rooted in P. Suppose that a player is given position P and is asked to find a winning move in P. The player will try to solve the problem by economical search, so he or she will only investigate moves that come into consideration and discard other moves. The player's pattern knowledge and detected motifs will help the player to identify promising moves. The idea is to use a standard chess engine like Stockfish to carry out a relatively shallow search (e.g., 10 ply) and evaluate the positions in the corresponding game tree by backing-up heuristic values of the positions in the leaves of this search tree. These backed-up heuristic evaluations are hopefully indicative of what an expert player can (approximately) evaluate without search, just by using his or her pattern knowledge. The meaningful tree is the game tree up to a chosen depth limit (in our experiments set to 5 ply), with 'unpromising' moves removed from the tree (unpromising from the player's point of view, or from the opponent's point of view, depending on whose move it is). Formally, for the task of winning in P, the meaningful tree consists of the root position P, and all the player-to-move positions whose backed-up heuristic value exceeds w ('winning threshold'), and the opponent-to-move positions whose value differs from the value of the best sibling (from the opponent's point of view) by no more than m ('margin'). These parameters were set to $w = 200$ centipawns, $m = 50$ centipawns in our experiment.

This design can be debated in the light of the question: how well do so defined meaningful trees approximate trees that are actually searched by chess players? Another question can be: the SBG approach is mainly concerned with the 'meaningful complexity', and ignores some other sources of difficulty discussed in the next section, such as 'invisible moves' (Neiman and Afek, 2011). Another contentious issue could be: is it appropriate to assume that all the players (at least of the chess strength comparable to our group of players) search more or less the same search tree? Or does this depend on certain players and their chess knowledge, especially on their specific repertoire of chess motifs? The classical study by De Groot (1965) on human problem-solving in chess suggests that players solve chess problems in a similar way over large ranges of chess rating (such as 400 rating points, as in the case of our 12 players). Experiments in a related study (Gobet, 1998) also generally confirm this. The following is a relevant result concerning this latter question. A quantitative model of chess problem-solving of tactical problems as a Bayesian network was proposed in (Bratko *et al.*, 2016). The network is structured according to the classical chess problem-solving model in (De Groot, 1965). Standard sets of chess motifs required to solve the 12 experimental positions were defined and were needed to solve each problem. In most positions, more motifs than one are relevant. Also, relevant chess moves to be searched by players that corresponded to the positions' motifs were defined. It was possible to observe success of the players at detecting relevant motifs, and also at carrying out the calculations. The players successfully detected relevant motifs in 88% of all the cases (Bratko *et al.*, 2016). Here we add how this percentage depends on the players' ratings. This percentage was somewhat higher for the top-half ranked players (92%, average rating 2198), and somewhat lower for the bottom-half ranked players (84%, average rating 1980). In spite of this difference, this indicates that a large majority of our players were able to detect relevant motifs correctly. The fact that the large majority of players detected relevant motifs (standard sets for the experimental positions) supports the assumption that, at least roughly, the players searched similar trees.

Some properties of a meaningful tree are naturally indicative of difficulty. For example, the total number of nodes in a meaningful tree or the branching factors at different levels of the tree. A more sophisticated indication of difficulty, is the attribute of a tree denoted by *NarrowSolution(L)*. This is defined as the number of opponent's moves at level *L* in the tree for which the winning player has only one good reply. A high value of *NarrowSolution* indicates situations where the opponent has many promising moves, and each of them requires to be met by the player with a unique reply.

In Stoiljkovikj *et al.* (2015), 10 attributes of a meaningful tree of this kind were defined. Another 10, chess-specific attributes of a position were defined, such as the number of chess pieces in the position or the existence of 'long moves' in the meaningful tree. Long moves are moves in which a chess piece moves by a long distance on the board, and sometimes such moves are suspected of being harder to notice by chess players, so they are one kind of 'invisible moves'. They contribute to the difficulty. Definitions of all the attributes can be found in (Stoiljkovikj *et al.*, 2015).

These 20 attributes of a position define a space for machine learning, and the problem of learning to predict the difficulty of chess positions can be formulated as follows. The learning data consists of a set of chess positions together with their difficulty class, where each position is described by the 20 attributes.

An experiment with learning to predict problem difficulty using this setting was carried out by (Stoiljkovikj *et al.*, 2015). Nine-hundred chess problems from Chess Tempo were randomly selected for learning. The difficulty class (easy, medium, or hard) was determined according to the Chess Tempo ratings of the problems, resulting in a balanced learning set with 300 examples of each class. In that experiment, the average Chess Tempo ratings of problems in the three learning subsets belonging to the three classes were as follows: easy: 1254.6, medium: 1669.3, hard: 2088.8. The reported classification results were very high (up to 83%, depending on the learning method used). However, these results cannot be trusted due to a suspected methodological slippage, which became apparent later when these experimental results could not be completely reproduced. In this chapter we repeat the learning experiment with the same set of learning problems (not including our 12 experimental positions), and the same positions' attribute values. However, to make the trained classifiers applicable to the 12 experimental problems of this chapter, we redefined the difficulty classes in the learning data, so that the new classes are appropriate for the three difficulty classes in our 12 positions. To this end we moved the thresholds for class separation to the midpoints between the Chess Tempo ratings of the three classes in the present experimental set, as given in Table 10.1.

After this redefinition of the thresholds between classes, the class distribution became imbalanced (which is less favourable for learning), as follows. Class easy: 479 examples,

Table 24.1 *The difficulty classes were determined according to the Chess Tempo ratings.*

1	rating < 1685	easy
2	$1685 \leq$ rating < 2055	medium
3	$2055 \leq$ rating	hard

medium: 231 examples, hard: 190 examples. We used several learning methods implemented in the scikit-learn machine learning library.

The best classification accuracy was obtained with Gradient Boosting Trees learning method (60%, measured by 10-fold cross-validation). We will refer to this predictor of difficulty as SBG2020. We applied this classifier to our 12 experimental problems. The results are given in Table 10.1. For each position, the table also gives the position's ranks according to average players' rankings, and the number of players that successfully solved the position.

Here are some quick observations from the table. The actual success rates by our players do not correlate very well with CT classes. Success rates of 100% (solved by all 12 players, see the column Success in Table 10.1) for problems 3 and 6, both medium difficulty by Chess Tempo, are surprising. A closer look at position 3 gives a likely explanation for what happened with this position. There are several winning moves in position 3 which all counted as success in our study, while for an unclear reason Chess Tempo only accepted as correct one of these alternative solutions. A similar explanation is possible for position 6. A closer look at position 6 suggests that this position is in fact relatively easy. According to this, the predicted class 'easy' by the SBG2020 classifier seems to be more appropriate. There are other discrepancies: problem 4, and some problems in CT class hard. But for these we could not find any simple explanation other than chance.

The sample correlation coefficients between the variables in Table 'classes' were computed by representing the three classes easy, medium and hard with 1, 2, and 3 respectively. The correlations are as follows:

```
r( CT-class, Success) = -0.60 (P = 0.0383)
r( SBG2020-class, Success) = -0.79 (P = 0.0023)
r( PlayersRanking, Success) = -0.78 (P = 0.0029)
r( CT-class, SBG2020-class) =  0.79 (P = 0.0024)
```

Also of interest are relations between the perceived difficulty of the positions by the players, represented by the joint players' rankings (average ranking of the positions), and the measured difficulty (CT-class) and automatically estimated difficulty (SBG2020-class):

```
r( Rank-by-players, CT-class) = 0.74
r( Rank-by-players,  SBG2020-class) = 0.94
```

Table 24.2 *Basic description of the Chess Tempo problem set.*

Position	*CT class*	*SBG2020*	*Rank*	*Success*
1	easy	easy	3	11
2	easy	easy	1	12
3	medium	easy	2	12
4	medium	medium	7	4
5	medium	medium	8	5
6	medium	easy	4	12
7	hard	medium	6	3
8	hard	hard	10	7
9	hard	medium	9	8
10	hard	medium	5	9
11	hard	hard	12	4
12	hard	hard	11	4

This is surprising as it suggests that the difficulty, as perceived by the human players, in fact better correlates with the automatically predicted difficulty by the SBG2020 approach, than with the actually measured Chess Tempo difficulty. This can be however at least partially explained by the problems mentioned above with positions 3 and 6, whose solutions seem to have been treated too harshly in Chess Tempo. The average ranking of the 12 positions by the 12 chess players is, interestingly, completely consistent with the SBG2020 classification.

Finally, we can try to compare the appropriateness of SBG2020 classification with respect to Chess Tempo classification by using Kendall's Tau coefficient. This is useful for comparison of the individual players' rankings (earlier assessed by Kendall's Tau) with SBG2020 rankings. There is a difficulty in that players' rankings are complete orderings, whereas SBG2020 classes only define a partial ordering. There are many total orderings consistent with the SBG2020 partial ordering. Now imagine a human player whose perceived position difficulties were exactly as by SBG2020. When asked to produce a total ordering, as in our experiment, this player could answer with any of the total rankings consistent with SBG2020. Assuming that all these rankings are equally likely, the expected value Tau over all these rankings is 0.78. Over all consistent rankings, Tau is between 0.56 and 1, with standard deviation 0.106. This is practically equal to Kendall's Tau of the average players' ranking. Even if the SBG approach is based on a very crude approximation to human players' game-tree search, it does seem to capture well the difficulty of problems as perceived by humans.

24.4 More Subtle Sources of Difficulty

There are some other sources of difficulty in chess problems, in addition to the size and other properties of meaningful trees, which were used in the SBG method in the previous section. In this section we point out other sources of difficulty, illustrated by examples from our experimental positions. These sources of difficulty were not considered in the SBG method.

24.4.1 Invisible moves

Some moves are hard to see by good chess players. Neiman and Afek (2011) investigated the properties of chess moves that are difficult to find and anticipate. It is precisely good chess player's knowledge which is so successfully used to make search more selective, that prevents the player from seeing such moves and is occasionally the cause of bad mistakes. For example, novice players are taught from the beginning that chess pieces should be developed as quickly as possible, therefore they have to move forward and preferably towards the centre where they are generally the most powerful. This cliché makes the players more likely to consider forward moves and sometimes automatically disregard moves away from the centre. Thus some moves become more difficult to see simply for geometrical reasons. Bent Larsen even points out that backward moves on diagonals are particularly difficult to detect 'except on the long diagonal' (Larsen, 2014).

Of all possible backward moves, those of the knight are the most difficult to find (Neiman and Afek, 2011). There is also a technical reason for this: As a short-range piece, the knight in particular has to be centralized. It takes too long to bring it back into the critical areas once it is out of play. There was an example of this kind of invisible move in our experimental position no. 4 (Figure 10.2), which was only solved by four players. Although the players who failed to solve it were in fact considering the right idea (described in players' retrospections), they simply could not see the winning move by a knight into the corner of the board.

24.4.2 Seemingly good moves and the 'Einstellung' effect

Clearly, difficulty should not be confused with complexity. Sometimes a problem may seem easy because the position does not seem complex at all. There may be an attractive move that seems to lead to victory, but in reality it does not. It is the presence of such a 'seemingly good' move (Stoiljkovikj *et al.*, 2015) that diverts the players attention from a truly good move and thus makes the problem difficult. In our experimental position no. 7 (Figure 10.3), there is a seemingly good move: 1 ... Qd8-b6. But the real solution requires the insertion of the move 1 ... Bf8-h6 before pinning the knight. It would be much easier to spot the correct move sequence if the above mentioned move with the queen did not look so attractive. In fact, only three players correctly solved this problem.

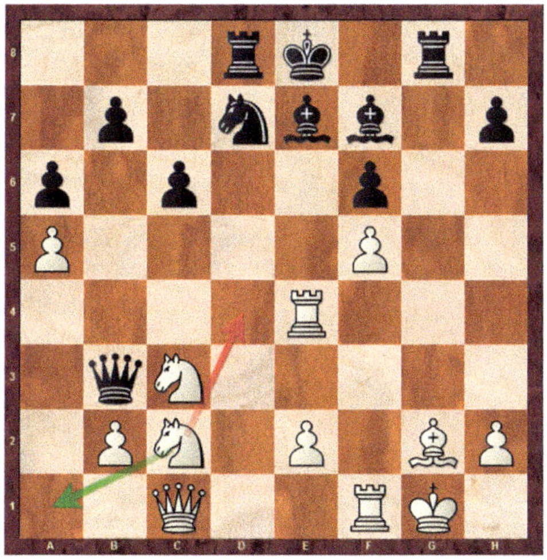

Figure 24.2 *There are two main motifs for White here. The first is to attack Black king via open e file, and part of this idea is the pinned Black bishop at e7. An obvious move to exploit that is by move 1.Qe3, increasing the pressure on Be7. However, this does not work. Black can successfully defend with 1...Ne5, which can be determined by relatively complex calculation. Another, completely different motif is triggered by a complex pattern: Black queen is surrounded by many White pieces and does not have any safe square to move to. This gives rise to the idea of trapping Black queen. To this end, queen has to be attacked, and White knight on c2 can do that, in two ways. One way is to move to d4 (red arrow). This move however disables the control of square c4 by White rook's on e4. So Black queen can now escape to c4. Now White has another familiar powerful pattern at disposal: discovered attack on Black queen with move 2.Ne6, also attacking Black rook d8. All that looks very strong for White, but as it turns out not sufficient for a clear win. This was calculated by many players who did play Nd4 and eventually failed to solve this problem. Much more straightforward and effective is the invisible move 1.Na1 (green arrow), immediately winning Black queen, but not seen by many players.*

When faced with a decision in chess, people are sometimes misled by familiar patterns and motifs, so that they miss better solutions. When we solve problems, our prior knowledge usually helps us by efficiently leading us to solutions that have worked for us in the past. However, if a problem requires a new solution, it can sometimes be surprisingly difficult to find the new solution because of our prior knowledge. This problem-solving effect was discovered by a psychologist Abraham Luchins (1942). He called this effect the 'Einstellung' effect. Bilalic *et al.*, (2008) experimentally confirmed that the Einstellung effect also exists in chess. A familiar pattern in a chess position drew the attention of the players to find a familiar solution (which did not work) and prevented them from finding a real solution that could be linked to a completely different pattern.

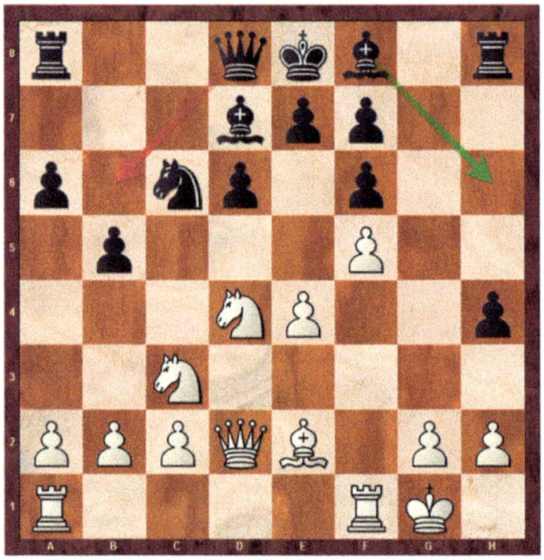

Figure 24.3 *Black to move wins. It is trivial for a good player to immediately notice the possibility of pinning White knight on d4 against White king with Qb6 (red arrow). The seemingly straightforward variation is thus 1... Qb6 2.Rfd1 Bh6 (another common method: attack the piece defending the pinned knight on d4) 3.Qd3 Nxd4 4.Qxd4 Be3+ winning White queen. This looks excellent for Black, but fails to notice that instead of 3.Qd3 White can unexpectedly strike back with 3.Nd5. After that it is no longer clear whether Black can win. The clear winning line for Black is 1...Bh6 (green arrow) 2.Qd3 Qb6 and now this indeed wins.*

24.5 Conclusions

This is a summary of the results of the analysis of our experimental data in expert problem-solving in chess. The results apply to solving tactical chess problems with Chess Tempo ratings roughly between 1500 and 2300, and players with FIDE ratings between 1800 and 2300:

1. A negative correlation was found between player's success in solving problems and the Chess Tempo rating of the problems, which is as expected.
2. There was no evidence of a correlation between the success of the players and the FIDE rating of the players. This is surprising. A plausible explanation is that success refers to finding a winning move in one position, whereas the FIDE rating measures success over entire games; that is, over a sequence of positions. Winning a game often means making a better decision than your opponent in only one or two positions in the whole game.
3. There is a statistically significant positive correlation between the players' ratings and the correctness of the ranking by the players of the 'relevant' position pairs according to their difficulty, although this relationship is quite weak.

We carried out an experiment in which the difficulty of the 12 experimental positions was automatically estimated using the SBG method. The main idea of the SBG is to use the properties of a 'meaningful' search tree as attributes for learning to estimate the difficulty of example positions, which are divided into difficulty classes. A meaningful search tree is defined as an attempt to automatically construct approximations to trees searched by human experts without knowing human expertise. The learned classifier was applied to our experimental positions, and the resulting classification of the positions compared well with the Chess Tempo difficulty classes, and also to the average perceived difficulty by the players. A question for future work is to explore why the SBG has done surprisingly well, even though it is based on a rather crude approximation to problem-solving by human experts.

Acknowledgements

This work was in part supported by the Research Agency of Republic of Slovenia (ARRS), research program Artificial Intelligence and Intelligent Systems. The authors would like to thank Peter Cheng for pointing out relevant research, and anonymous reviewers for their comments and suggestions.

References

Bilalić, M., McLeod, P., and Gobet, F. (2008). Why good thoughts block better ones: The mechanism of the pernicious einstellung (set) effect. *Cognition*, **108**(3), 152–61.

Bratko, I., Hristova, D., and Guid, M. (2016). Search versus knowledge in human problem solving: a case study in chess, in *Model-Based Reasoning in Science and Technology*. Berlin: Springer, 569–83.

De Groot, A. D. (1965). *Thought and Choice in Chess*. The Hague: Mouton.

De Kegel, B. and Haahr, M. (2019). Procedural puzzle generation: A survey. *IEEE Transactions on Games*, **12**(1), 21–40.

Dry, M., Lee, M. D., Vickers, D. (2006). Human performance on visually presented traveling salesperson problems with varying numbers of nodes. *Journal of Problem Solving*, **1**(1), 20–32.

Elo, A. E. (1978). *The Rating of Chessplayers, Past and Present*. London: Arco Publications.

Glickman, M. E. (1999). Parameter estimation in large dynamic paired comparison experiments. *Applied Statistics*, **48**, 377–94.

Gobet, F. (1998). Chess players' thinking revisited. *Swiss Journal of Psychology*, **57**, 18–32.

Guid, M. and Bratko, I. (2006). Computer analysis of world chess champions. *ICGA Journal*, **29**(2), 65–73.

Guid, M. and Bratko, I. (2013). Search-based estimation of problem difficulty for humans, in H. Lane, K. Yacef, J. Mostow, *et al.*, eds, *Artificial Intelligence in Education*, Vol. 7926, Lecture Notes in Computer Science. Berlin: Springer, 860–3.

Hristova, D., Guid, M., and Bratko, I. (2014*a*). Assessing the difficulty of chess tactical problems. *International Journal on Advances in Intelligent Systems*, 7(3&4), 728–38.

Hristova, D., Guid, M., and Bratko, I. (2014*b*). Toward modeling task difficulty: the case of chess, in *Proceedings of the Sixth International Conference on Advanced Cognitive Technologies*

and Applications, Venice, Italy. Wilmington: International Academy Research and Industry Association (IARIA), 211–4.
Jarušek, P. and Pelánek, R. (2010). Difficulty rating of sokoban puzzle, in T. Agotnes, ed., *Proceedings of the Fifth Starting AI Researchers' Symposium (STAIRS 2010)*, Lisbon, Portugal. Amsterdam, Netherlands: IOS Press, 140–50.
Kotovsky, K., Hayes, J. R., and Simon, H. A. (1985). Why are some problems hard? Evidence from tower of Hanoi. *Cognitive Psychology*, **17**(2), 248–94.
Kotovsky, K., and Simon, H. A. (1990). What makes some problems really hard: Explorations in the problem space of difficulty. *Cognitive Psychology*, **22**(2), 143–83.
Larsen, Bent (2014). *Bent Larsen's Best Games: Fighting Chess with the Great Dane*. New in Chess.
Luchins, A. S. (1942). Mechanization in problem solving: the effect of Einstellung. *Psychological Monographs*, **54**(6), i.
Neiman, E. and Afek, Y. (2011). *Invisible Chess Moves: Discover Your Blind Spots and Stop Overlooking Simple Wins*. Amsterdam, Netherlands: New in Chess.
Pelánek, R. (2011). Difficulty rating of sudoku puzzles by a computational model, in R. Murray and P. McCarthy, eds, *Proceedings of Florida Artificial Intelligence Research Society Conference*, Palm Beach. New York, NY: AAAI Press, 434–9.
Pizlo, Z. and Li, Z. (2005). Solving combinatorial problems: the 15-puzzle. *Memory and Cognition*, **33**(6), 1069–84.
Sekiya, R., Oyama, S. and Kurihara, M. (2019). User-adaptive preparation of mathematical puzzles using item response theory and deep learning, in F. Wotawa, G. Friedrich, I. Pill, R. Koitz-Hristov, and M. Ali, eds, *Proceedings International Conference on Industrial, Engineering and Other Applications of Applied Intelligent Systems*, Graz, Austria. Cham: Springer, 530–7.
Sheridan, Heather and Reingold, Eyal M. (2017). Chess players' eye movements reveal rapid recognition of complex visual patterns: Evidence from a chess-related visual search task. *Journal of vision*, **17**(3), 4–4.
Stoiljkovikj, S., Bratko, I., and Guid, M. (2015). A computational model for estimating the difficulty of chess problems, in A. Goel and M. Riedl, eds, *Proceedings of the Third Annual Conference on Advances in Cognitive Systems*, Atlanta, Georgia. Auckland: Cognitive Systems Foundation, 7.
Van Kreveld, M., Löffler, M., and Mutser, P. (2015). Automated puzzle difficulty estimation, in C. Lee, I. Wu, M. Wang, eds, *2015 IEEE Conference on Computational Intelligence and Games (CIG 2015)*, Tainan, Taiwan. New York, NY: IEEE Press, 415–22.

Index

Note: Tables and figures are indicated by an italic *t* and *f* following the page number.